T0348736

ENGINEERED BIOMIMICRY

ENGINEERED BIOMIMICRY

Edited by

AKHLESH LAKHTAKIA
Department of Engineering Science and Mechanics,
Pennsylvania State University, USA

RAÚL J. MARTÍN-PALMA
Departamento de Física Aplicada,
Universidad Autónoma de Madrid, Spain

AMSTERDAM • BOSTON • HEIDELBERG • LONDON • NEW YORK • OXFORD • PARIS
SAN DIEGO • SAN FRANCISCO • SINGAPORE • SYDNEY • TOKYO

ELSEVIER

Elsevier
225 Wyman Street, Waltham, MA 02451, USA
The Boulevard, Langford Lane, Kidlington, Oxford, OX5 1GB, UK
Radarweg 29, PO Box 211, 1000 AE Amsterdam, The Netherlands

Notice

No responsibility is assumed by the publisher for any injury and/or damage to persons or property as a matter of product liability, negligence or otherwise, or from any use or operation of any methods, products, instructions or ideas contained in the material herein. Because of rapid advances in the medical sciences, in particular, independent verification of diagnoses and drug dosages should be made.

Library of Congress Cataloging-in-Publication Data
Engineered biomimicry / edited by Akhlesh Lakhtakia and Raúl Jose Martín-Palma.
　　　pages cm
　ISBN 978-0-12-415995-2
　1. Biomimicry.　I. Lakhtakia, A. (Akhlesh), 1957– editor of compilation.
II. Martín-Palma, R. J. (Raúl J.) editor of compilation.
　T173.8.E53　2013
　660.6—dc23

2013007422

British Library Cataloguing in Publication Data
A catalogue record for this book is available from the British Library.

ISBN: 978-0-12-415995-2

For information on all **Elsevier** publications
visit our web site at store.elsevier.com

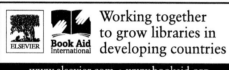

ELSEVIER　**Book Aid** International

Working together
to grow libraries in
developing countries

www.elsevier.com • www.bookaid.org

Dedication

To all creatures, great and small (except politicians and CEOs of airlines)

Contents

13. Biomimetic Self-Organization and Self-Healing

TORBEN LENAU AND THOMAS HESSELBERG

14. Solution-Based Techniques for Biomimetics and Bioreplication

ADITI S. RISBUD AND MICHAEL H. BARTL

15. Vapor-Deposition Techniques

RAÚL J. MARTÍN-PALMA AND AKHLESH LAKHTAKIA

16. Atomic Layer Deposition for Biomimicry

LIANBING ZHANG AND MATO KNEZ

17. Evolutionary Computation and Genetic Programming

WOLFGANG BANZHAF

Contributors

Michael H. Bartl   Department of Chemistry, University of Utah, Salt Lake City, UT 84112, USA

Steven F. Barrett   Department of Electrical and Computer Engineering, University of Wyoming, Laramie, WY 82071, USA

Francois Barthelat   Department of Mechanical Engineering, McGill University, Montreal, QC H3A 2K6, Canada

Wolfgang Banzhaf   Department of Computer Science, Memorial University of Newfoundland, St. John's, NL A1B 3X5, Canada

Narayan Bhattarai   Department of Chemical and Bioengineering, North Carolina A&T State University, Greensboro, NC 27411, USA

Princeton Carter   Department of Chemical and Bioengineering, North Carolina A&T State University, Greensboro, NC 27411, USA

Javaan Chahl   School of Engineering, University of South Australia, Adelaide, SA 5001, Australia

Shantanu Chakrabartty   Department of Electrical and Computer Engineering, Michigan State University, East Lansing, MI 48824, USA

Natalia Dushkina   Department of Physics, Millersville University, Millersville, PA 17551, USA

Michael S. Ellison   School of Materials Science and Engineering, Clemson University, Clemson, SC 29634, USA

Stanislav N. Gorb   Zoological Institute, University of Kiel, 24118 Kiel, Germany

Thomas Hesselberg   Department of Zoology, University of Oxford, Oxford OX1 3PS, United Kingdom

Thamira Hindo   Department of Electrical and Computer Engineering, Michigan State University, East Lansing, MI 48824, USA

Peng Jiang   Department of Chemical Engineering, University of Florida, Gainesville, FL 32611, USA

Mato Knez   CIC nanoGUNE Consolider, Tolosa Hiribidea 76, 20018 Donostia-San Sebastian, Spain

Akhlesh Lakhtakia   Department of Engineering Science and Mechanics, Pennsylvania State University, University Park, PA 16802, USA

Torben Lenau   Department of Mechanical Engineering, Technical University of Denmark, DK2800 Lyngby, Denmark

Raúl J. Martín-Palma   Department of Materials Science and Engineering, Pennsylvania State University, University Park, PA 16802, USA

Mohammad Mirkhalaf   Department of Mechanical Engineering, McGill University, Montreal, QC H3A 2K6, Canada

Akiko Mizutani   Odonatrix Pty. Ltd., One Tree Hill, SA 5114, Australia

Blayne M. Phillips   Department of Chemical Engineering, University of Florida, Gainesville, FL 32611, USA

Aditi S. Risbud   Lawrence Berkeley National Laboratory, MS 67R3110, Berkeley, CA 94720, USA

Mohsen Shahinpoor   Mechanical Engineering Department, University of Maine, Orono, ME 04469, USA

Jayant Sirohi   Department of Aerospace Engineering and Engineering Mechanics, University of Texas at Austin, Austin, TX 78712, USA

Ranjan Vepa   School of Engineering and Materials Science, Queen Mary, University of London, London E1 4NS, United Kingdom

Erwin A. Vogler   Department of Materials Science and Engineering, Pennsylvania State University, University Park, PA 16802, USA

H. Donald Wolpert Bio-Optics, 1933 Comstock Avenue, Los Angeles, CA 90025, USA

Cameron H.G. Wright Department of Electrical and Computer Engineering, University of Wyoming, Laramie, WY 82071, USA

Lianbing Zhang CIC nanoGUNE Consolider, Tolosa Hiribidea 76, 20018 Donostia-San Sebastian, Spain

Deju Zhu Department of Mechanical Engineering, McGill University, Montreal, QC H3A 2K6, Canada

Foreword

Biology inspires - Diversity matters

Never before has biology been such an inspiration for innovations. Today, the vast majority of engineers and materials scientists are aware of this source of ideas for technological developments. However, since the functionality of biological systems is often based on their extreme complexity, there are not very many examples of successfully implemented ideas in modern technology. For this reason, Bioinspiration, Biomimetics, and Bioreplication require not only intense experimental research in the field of biology, but also an abstraction strategy for extracting essential features responsible for a particular function of a biological system. Therefore, before transferring biological principles into technology, we must recognize and distinguish these principles from among a tremendous biological complexity. There are several approaches to deal with this problem.

The first approach is a classical biological one, where one particular functional system is comparatively studied in various organisms. In this case, we can potentially recognize the same functional solutions having evolved several times, independently, in the evolution of different groups of organisms. Additionally, we obtain information about the diversity of solutions. Such a comparative approach is time consuming, but may be very effective for Bioinspiration later leading to Biomimetics and Bioreplication.

The second approach deals with the concept of the model organism. Here, instead of the diversity of organisms studied by one or two

methods, different methods can be applied to one particular organism/system. This approach aids in revealing broad and detailed information about structure-function relationships in one system.

Unfortunately, not just any kind of experiments can be performed with biological systems. A third approach relies on theoretical and numerical modeling. By doing virtual experiments, we can gradually improve our knowledge about a biological system and explore it in a much wider range of experimental conditions, as would be possible with the real biological system. Additionally, sometimes there is the possibility of mimicking the biological system into an artificial but partially real model, keeping some essential features of the biological original and then performing experiments with this artificial imitation. This is the fourth approach, which usually leads to a generation of the first laboratory prototypes that can later be used for further industrial developments. The latter approach has a strong link to Biomimetics.

The present book reports on a broad diversity of biological systems and their biomimetic systems studied using various combinations of the approaches mentioned above. The chapters have been written by prominent specialists in materials science, engineering, optics, surface science, computation, etc. Their correlation stems from the idea of solving old problems by applying new ideas taken from biology. Readers will enjoy the great creativity of the authors in making links between biological observations and technological implementations. This book can not only inspire engineers with countless ideas reported therein, but also biologists to further explore biology. A wonderful example of modern science without boundaries!

Stanislav N. Gorb,
*Functional Morphology and Biomechanics
Group, Zoological Institute,
University of Kiel, Germany*

Preface

"Look, Ma! I am flying!" Flapping arms stretched sideways and weaving a zigzag while running down the sidewalk, many a child has imagined soaring in air like a bird. Not only have most of us pretended as children that we could fly, from times immemorial adults have looked up at flying birds with envy. If humans could fly, they could swoop down on enemies and wooly mammoths alike. And how free would they be, unshackled from the ground.

Greek mythology provides numerous examples of our eagerness to fly. Krios Khrysomallos was a fabulous, flying, golden-fleeced ram. He was sent by the nymph Nemphale to rescue her children Phrixos and Helle when they were about to be sacrificed to the gods. The rescuer went on to become the constellation Aries. The Drakones of Medea were a pair of winged serpents harnessed to her flying chariot. Pegasus, the thundering winged horse of Zeus, was the offspring of Poseidon and the gorgon Medusa. When Pegasus died, Zeus transformed him into a constellation. But the classical example of a flying human is that of Icarus, who escaped from a Cretan prison using wings of feather and wax. Exhilarated with freedom, he flew too close to the sun—and perished because his wings melted, inspiring poets and engineers alike.

Leonardo Da Vinci (1452–1519) was probably the first historic individual who attempted an engineering approach to flying. A student of avian flight, he conjured up several mechanical contraptions, some practical, others not. As professors, neither of us can ignore the legend that he attached wings to the arms of one of his (graduate?) students, who took off from Mt. Ceceri, but crashed and broke a leg.

Three centuries later, mechanical flight was demonstrated by Sir George Cayley (1773–1857). He made a glider that actually flew—without a pilot. Orville and Wilbur Wright are credited as the first people to successfully fly an aeroplane with a person onboard, on December 17, 1903. Today flying has progressed far beyond dreams and myths into the quotidian, so much so that with perfunctory apologies incompetently run airlines routinely deprive numerous passengers of their own beds.

The development of powered flying machines that was inspired by birds in self-powered flight is an excellent example of bioinspiration. But there are significant differences: aeroplanes do not flap their wings, and the tails of birds do not have vertical stabilizers. Although very close to the dreams of Leonardo da Vinci, hang gliders too have fixed wings. Helicopters, also anticipated by the Renaissance genius, are rotorcraft completely unlike birds.

The goal in bioinspiration is to reproduce a biological function but not necessarily the biological structure. Our history is marked by numerous approaches to the solution of engineering problems based on solutions from nature. All of these approaches are progressions along the same line of thought: Engineered Biomimicry, which encompasses bioinspiration, biomimetics, and bioreplication.

Biomimetics is the replication of the functionality of a biological structure by approximately reproducing an essential feature of that structure A terrific example is the hook-and-loop structure of Velcro coming from the hooked barbs on a burdock seed. When an animal brushes against the seed, the hooks attach into the fur of the animal and the seed is carried along until it is either pulled off or drops out of the fur. Velcro often replaces traditional fasteners in apparel and footwear.

Bioreplication is the direct replication of a structure found in natural organisms, and thereby aims at copying one or more functionalities. To date, there are no commercial bioreplicated devices, but engineers have been able to replicate structures such as the compound eyes of insects, the wings of butterflies, and the elytrons of beetles. Having emerged only within the last decade with the spread of nanofabrication techniques, bioreplication is in its infancy.

Engineered systems are rapidly gaining complexity, which makes it difficult to design, fabricate, test, reliably operate, repair, reconfigure, and recycle them. But elegant, simple, and optimal solutions may often exist in nature. Although not ignored in the past, solutions from nature—especially from the realm of biology—are being increasingly taught, emulated, and enhanced. "Biology is the future of engineering" is a refrain commonplace in engineering colleges today.

The ongoing rise of engineered biomimicry in research communities has encouraged a few specialist conferences, new journals, and special issues of existing journals. Very few technoscientific books have been published, in part because of the multi-disciplinarity innate in engineered biomimicry. Following three specialist conferences organized by both of us under the aegis of SPIE, we decided to edit a technoscientific book that would expose the richness of this approach. Colleagues handsomely responded to our requests to write representative chapters that would at once be didactic and expose the state of the art. The result is the book entitled *Engineered Biomimicry*.

The reader may expect this book to be divided into three parts of engineered biomimicry—namely, bioinspiration, biomimetics, and bioreplication. But the boundaries are not always evident at research frontiers to permit a neat division, and the progression from bioinspiration to biomimetics to bioreplication has been followed loosely by us.

The book begins with an introductory article entitled "The world's top Olympians" written by H. Donald Wolpert (Bio-Optics, Inc.). The overview of the amazing capabilities of insects, birds, and other animals by Wolpert is bound to inspire researchers to emulate natural mechanisms and functionalities in industrial contexts.

Six chapters are more or less devoted to bioinspiration. Thamira Hindo and Shantanu Chakrabartty (Michigan State University) have entitled their chapter "Noise exploitation and adaptation in neuromorphic sensors". They describe several important principles of noise exploitation and adaptation observed in neurobiology, and show that these principles can be systematically used for designing neuromorphic sensors. In the chapter "Biomimetic hard materials", Mohammad Mirkhalaf, Deju Zhu, and Francois Barthelat (McGill University) state and exemplify that a very attractive combination of stiffness, strength, and toughness can be achieved by using several staggered structures. The properties and characteristics of ionic-biopolymer/metal nano-composites for exploitation as biomimetic multi-functional distributed nanoactuators, nanosensors, nanotransducers, and artificial muscles are presented in "Muscular biopolymers" by Mohsen Shahinpoor (University of Maine). Princeton Carter and Narayan Bhattarai (North Carolina A&T State University) discuss scaffolding in tissue engineering and regenerative medicine in "Bioscaffolds: fabrication and performance". Biomimicry within the context of the core mechanisms of the biological response to materials *in vivo* is discussed in "Surface modification for biocompatibility" by Erwin A. Vogler (Pennsylvania State University). In a departure from materials science to computer science, the chapter "Evolutionary computation and genetic programming" by Wolfgang Banzhaf (Memorial University of Newfoundland) is focused on evolutionary computation—in particular, genetic programming—which draws inspiration from the discipline of evolutionary biology.

Eight chapters form a group on biomimetics. Steven F. Barrett and Cameron H. G. Wright (University of Wyoming) discuss the strengths and weaknesses of vision sensors based on the vision systems of both mammals and insects, and present guidelines for designing such sensors in their chapter entitled "Biomimetic vision sensors". The distinguishing features of biomimetic robotics and facilitating technologies are discussed by Ranjan Vepa (Queen Mary College, University of London) in "Biomimetic robotics". Man-made microflyers are described in the chapter "Bioinspired and biomimetic microflyers" of Jayant Sirohi (University of Texas at Austin). Also related to mechanical flight, the chapter "Flight control using biomimetic optical sensors" by Javaan S. Chahl (University of South Australia) and Akiko Mizutani (Odonatrix Pty Ltd.) reports on flight trials of insect-inspired maneuvers by unmanned aerial vehicles. Bioinspiration has resulted in improved fibrous materials, as discussed by Michael S. Ellison (Clemson University) in "Biomimetic textiles". Ellison has also penned his thoughts on the prospects of continued progress in this direction. "Structural colors" by Natalia Dushkina (Millersville University) and Akhlesh Lakhtakia (Pennsylvania State University) is a comprehensive but succinct account of the origin and use of structural colors. Blayne M. Phillips and Peng Jiang (University of Florida) discuss the fabrication, characterization, and modeling of moth-eye antireflection coatings grown on both transparent substrates and semiconductor wafers in "Biomimetic antireflection surfaces". Finally in this group of chapters, "Biomimetic self-organization and self-healing" has been written by Torben A. Lenau (Technical University of Denmark) and Thomas Hesselberg (University of Oxford) on eight different self-organizing and self-healing approaches present in

nature. The authors also take a look at realized and potential applications.

The last group of chapters is a compilation of three different fabrication methodologies for bioreplication. The chapter "Solution-based techniques for biomimetics and bioreplication" by Aditi S. Risbud and Michael H. Bartl (University of Utah) illustrates how structural engineering in biology can be replicated using sol-gel chemistry, resulting in optical materials with entirely new functionalities. Physical vapor deposition, chemical vapor deposition, atomic layer deposition, and molecular beam epitaxy are succinctly described in the context of engineered biomimicry by Raúl J. Martín-Palma and Akhlesh Lakhtakia (Pennsylvania State University) in "Vapor-deposition techniques". Lianbing Zhang and Mato Knez (CIC nanoGUNE Consolider) provide a comprehensive description of the fundamentals of atomic layer deposition and its applications to biomimicry in the chapter entitled "Atomic layer deposition for biomimicry".

We thank all authors for timely delivery of their chapters as well as during the subsequent splendid production of this volume. Not only did they write their chapters, several of them also contributed by reviewing other chapters. We are also grateful to the following colleagues for reviewing a chapter each (in alphabetical order): Stephen F. Badylak (University of Pittsburgh), Satish T.S. Bukkapatnam (Oklahoma State University), Francesco Chiadini (Università degli Studi di Salerno), Hyungjun Kim (Yonsei University), Roger J. Narayan (North Carolina State University), Michael O'Neill (University College Dublin), Oskar Paris (Montanuniversität Leoben), Maurizio Porfiri (Polytechnic Institute of New York University), Akira Saito (Osaka University), Kazuhiro Shimonomura (Ritsumeikan University), Thomas Stegmaier (Zentrum der bionischen Innovationen für die Industrie), and Douglas E. Wolfe (Pennsylvania State University). Stanislav N. Gorb (University of Kiel) is thanked for writing an informative foreword that provides a biologist's perspective on engineered biomimicry.

Louisa Hutchins, Kathryn Morrissey, Paula Callaghan, Patricia Osborn, Donna de Weerd-Wilson, Danielle Miller, and Poulouse Joseph efficiently shepherded *Engineered Biomimicry* through different stages at Elsevier. Our families graciously overlooked the time we did not spend with them. Our universities were indifferent, but they did foot the additional bills for electricity in our offices. Skype provided free communication.

We do hope that the insects we caught for our bioreplication research forgave us for translating them from the miseries of life to the serenity of death. Some of them were immortalized on Youtube. Who could ask for anything more!

Akhlesh Lakhtakia
Pennsylvania State University
Raúl José Martín-Palma
Universidad Autónoma de Madrid
February 2013

The World's Top Olympians

H. Donald Wolpert

Bio-Optics, 1933 Comstock Avenue, Los Angeles, CA 90025, USA

Prospectus

Animals, insects, and birds are capable of some amazing feats of speed, jumping, weight carrying, and endurance capabilities. As Olympic contestants, the records of these competitors challenge and, in many cases, exceed the best of human exploits and inspire us to emulate natural mechanisms and functionalities.

Keywords

Animal, Animal Olympians, Bioinspiration, Biomimicry, Insect and bird record holders

1 INTRODUCTION

Some of the world's top Olympians are not who you might imagine. They are the animals, insects, and birds that inhabit the Earth. The feats they achieve are truly worthy of Olympic medals. In this prolog to *Engineered Biomimicry*, the exploits of insects, animals, and birds in the sprint, middle-distance, and long-distance events; their training at high altitudes; records in the long-jump and high-jump categories; records in swimming and diving events; and record holders in free-weight and clean-and-jerk contests are discussed.

2 SPRINTS, MIDDLE-DISTANCE, AND LONG-DISTANCE EVENTS

Like the hare and tortoise, there are Olympic athletes that are sprinters, capable of reaching high speeds in a short distance, whereas others are long-distance experts, in it for the long haul.

Cheetahs (Figure 1), the sprint-champion species of the animal kingdom, have been clocked at 70–75 mph. Their stride can reach 10 yards when running at full tilt. It is said they can reach an impressive 62 mph from a standing start in 3 s [1].

The peregrine falcon is often cited as the fastest bird, cruising at 175 mph and diving in attacks at 217 mph. But in level horizontal flight,

FIGURE 1 Cheetah. (Image Courtesy of the U.S. Fish and Wildlife Service, Gary M. Stolz)

FIGURE 2 Pronghorn antelope. (Image Courtesy of the U.S. Fish and Wildlife Service, Leupold James)

FIGURE 3 Monarch butterfly. (Image Courtesy of the U.S. Fish and Wildlife Service)

the white-throated swift, topping out at 217 mph, is the all-around winner [2].

In the marathon you would most likely see the pronghorn antelope (Figure 2) on the award stand. The pronghorn weighs about as much as a grown human but can pump three times as much blood, which is rich in hemoglobin. It has extra-large lungs and a large heart, which provides much-needed oxygen to its muscles [3].

Aerobic performance is often evaluated on the basis of the maximal rate of oxygen uptake during exercise in units of milliliters of oxygen per kilogram of mass per minute. An elite human male runner might measure in the 60's or low 70's, whereas a cross-country skier may be in the low 90's. But the pronghorn antelope tops out at about 300 ml/kg/min [3, 4].

The wandering albatross, in a different marathon class, would leave the competition in the dust. Satellite imagery has revealed that these birds, with a wing span of 12 ft, travel between 2,237 and 9,321 miles in a single feeding trip, often sleeping *on the wing*.

If the race were handicapped for size and weight, the ruby-throated hummingbird and monarch butterfly would rank in the top tier. The ruby-throated hummingbird, being faster, flies 1,000 miles between seasonal feeding grounds, 500 of those miles over the featureless Gulf of Mexico. On average the male hummingbird has a mass of 3.4 g. The monarch butterfly (Figure 3),

with a mass of a mere 2–6 g, migrates 2,000 miles, flying up to 80 miles per day during its migration between Mexico and North America.

There are two types of human ultra-marathoners: those that cover a specific distance (the most common are 50 km and 100 km) and those who participate in events that take place over a specific interval of time, mainly 24 h or multiday events. These events are sanctioned by the International Association of Athletics Federation.

Although they are not sanctioned as Olympic contenders, there are some contenders in the animal kingdom that are in line for first place in the ultra-marathon. The Arctic tern (Figure 4) flies from its Arctic breeding grounds in Alaska to Tierra del Fuego in the Antarctic and back

FIGURE 4 Arctic tern. (Image Courtesy of Estormiz)

again each year, a 19,000-km (12,000-mile) journey each way.

The longest nonstop bird migration was recorded in 2007. A bar-tailed godwit flew 7,145 miles in nine days from its breeding grounds in Alaska to New Zealand. Without stopping for food or drink, the bird lost more than 50% of its body mass on its epic journey [5].

3 HIGH-ALTITUDE TRAINING

In the autumn, the bar-headed goose migrates from its winter feeding grounds in the lowlands of India to its nesting grounds in Tibet. Like Olympic long-distance runners that train at high altitudes, the bar-headed goose develops mitochondria that provide oxygen to supply energy to its cells. This journey takes the bar-headed goose over Mount Everest, and the bird has been known to reach altitudes of 30,000 ft to clear the mountain at 29,028 ft. At this altitude, there is only about a quarter of the oxygen available that exists at sea level and temperatures that would freeze exposed flesh [6].

Other high-altitude trainers are whooper swans, which have been observed by pilots at 27,000 ft over the Atlantic Ocean. The highest flying bird ever observed was a Ruppell's griffon that was sucked into the engines of a jet flying at 37,900 ft above Ivory Coast [6].

4 LONG JUMP AND HIGH JUMP

There are two basic body designs that enable animals to facilitate their jumping capabilities. The long legs of some animals give them a leveraging power that enables them to use less force to jump the same distance as shorter-legged animals of the same mass. Shorter-legged animals, on the other hand, must rely on the release of stored energy to propel themselves. And then there are those animals that combine the features of both approaches.

FIGURE 5 American bullfrog. (Image Courtesy of U.S. Fish and Wildlife Service, Gary M. Stolz)

The red kangaroo, with a capacity to jump 42 ft, and the Alpine chamois that can clear crevasses 20 ft wide and obstacles 13 ft high, certainly have impressive jumping capabilities. But when you handicap animals, you discover that bullfrogs, fleas, and froghoppers vie for the title of best jumper.

One long-jump specialist is the American bullfrog (Figure 5). Trained for the Calaveras Jumping Frog Jubilee held annually in Angeles Camp, California (USA), Rosie the Ribeter won the event in 1986 with a recorded jump of 21 ft 5¾ in. Muscles alone cannot produce jumps that good. The key to the frog's jumping ability lies in its tendons. Before the frog jumps, the leg muscle shortens, thereby loading energy into the tendon to propel the frog. Its long legs and energy-storing capabilities are key to the jumping capabilities of Rosie the Ribeter [2, 5].

Although not a record holder, the impala or African antelope (Figure 6) is a real crowd pleaser. This animal, with its long, slender legs and muscular thighs, is often seen jumping around just to amuse itself, but when frightened it can bound up to 33 ft and soar 9 ft in the air [1].

The leg muscles of the flea are used to bend the femur up against the coxa or thigh, which contains resilin. Resilin is one of the best materials known for storing and releasing energy

FIGURE 6 Impala. (Image Courtesy of U.S. Fish and Wildlife Service, Mimi Westervelt)

efficiently. Cocked and ready, a trigger device in the leg keeps it bent until the flea is ready to jump. Its jumping capability is equal to 80 times its own body length, equivalent to a 6-ft-tall person jumping 480 ft! Once thought to be the champion of its class, the flea has lost its ranking as top jumper to the froghopper [7].

The froghopper or spittle bug jumps from plant to plant while foraging. To prepare to jump, the insect raises the front of its body by its front and middle legs. Thrust is provided by simultaneous and rapid extension of the hind legs. The froghopper exceeds the height obtained by the flea relative to its body length (0.2 in., or 5 mm) despite its greater weight. Its highest jumps reach 28 in. A human with this capability would be able to clear a 690-ft building [2, 8].

5 SWIMMING AND DIVING

Birds are not the only long-distance competitors. A great white shark pushed the envelope for a long-distance swimming event by swimming a 12,400-mile circuit from Africa to Australia in a journey that took nine months. This trip also

included the fastest return migration of any known marine animal [9].

The Shinkansen bullet train runs from Osaka to Hakata, Japan, through a series of tunnels. On entering a tunnel, air pressure builds up in front of the train; on exiting, the pressure wave rapidly expands, causing an explosive sound. To reduce the impact of the expanding shock wave and to reduce air resistance, design engineers found that the ideal shape for the Shinkansen is almost identical to a kingfisher's beak. Like any good Olympic diver, the kingfisher streamlines its body and enters the water vertically, thereby minimizing its splash and leading to a perfect score of 10. Taking inspiration from nature, the Shinkansen engineers designed the train's front end to be almost identical in shape to the kingfisher's beak, providing a carefully matched pressure/impedance match between air and water [10] (Figure 7).

Without a dive platform, Cuvier's and Blainville's beaked whales can execute foraging dives that are deeper and longer than those reported for any other air-breathing species. Cuvier's beaked whales dive to maximum depths of nearly 6,230 ft with a maximum duration of 85 min; the Blainville's beaked whale dives to a

FIGURE 7 Kingfisher. (Image Courtesy of Robbie A)

maximum depth of 4,100 ft. Other Olympic dive contestants are sperm whales and elephant seals. The sperm whale can dive for more than 1 h to depths greater than about 4,000 ft, and it typically dives for 45 min. The elephant seal, another well-known deep diver, can spend up to 2 h in depths over 5,000 ft, but these seals typically dive for only 25–30 min to depths of about 1640 ft [11].

6 PUMPING IRON

Olympic weightlifting is one of the few events that separates competitors into weight classes. In the +231 lb class, a competitor might lift weights approximately 2.2 times his body weight. The average bee, on the other hand, can carry something like 24 times its own body weight, and the tiny ant is capable of carrying 10–20 times its body weight, with some species able to carry 50 times their body weight [2].

Ounce for ounce, the world's strongest insect is probably the rhinoceros beetle (Figure 8). When a rhinoceros beetle gets its game face on, it can carry up to 850 times its own body weight on its back [12].

FIGURE 8 European rhinoceros beetle. (Image Courtesy of George Chernilevsky)

7 CONCLUDING REMARKS

When you consider some of the running, jumping, flying, diving, and weightlifting capabilities of some animals, insects, and birds, you have to be awed. Many of Earth's creatures are certainly worthy of world-class status and could certainly vie for Olympic gold medals. How exactly do these animals and insects achieve their fabulous performances? The answer to this question is not necessarily clear, but through multidisciplinary research we are beginning to comprehend these Olympic achievements. Although the ability to swim or fly long distances is an achievement in itself, what is more intriguing is how some animals navigate day and night, in bad weather or clear and over large distances. How elapsed time, distance traveled, and the sun's position are used in this navigation process is important to understand. Visual clues such as star patterns and the sun's position, along with the time of day, may be used solely or used in conjunction with other aids in navigation. For some creatures, the Earth's magnetic field or sky polarization is as important as any navigational aid. Some or all of these tools may be used to cross-calibrate one navigation tool to another in order to more precisely locate an animal's or insect's position and determine its heading. The more we study natural approaches to problems, the more we will discover clever solutions to vexing problems.

References

[1] Cheetah, http://en.wikipedia.org/wiki/Cheetah (accessed 27 January 2013).

[2] B. Sleeper, Animal Olympians. *Animals*, July/August 1992.

[3] M. Zeigler, The world's top endurance athletes ply the US plains, *San Diego Union Tribune*, 2 July 2000.

[4] National Geographic, Geographia, May 1992.

[5] Frogs' amazing leaps due to springy tendons, http://news.brown.edu/pressreleases/2011/11/frogs (accessed 27 January 2013).

[6] G.R. Scott, S. Egginton, J.G. Richards, and W.K. Milsom, Evolution of muscle phenotype for extreme high altitude flight in the bar-headed goose, *Proc R Soc Lond B* **276** (2009), 3645–3654.

[7] The flea, the catapult and the bow, http://www.ftexploring.com/lifetech/flsbws1.html (accessed 27 January 2013).

[8] M. Burrows, Biomechanics: Froghopper Insects Leap to new heights, *Nature* **424** (2003), 509.

[9] Animal record breakers, http://animals.nationalgeographic.com/animals/photos/animal-records-gallery/ (accessed 27 January 2013).

[10] The Shinkansen bullet train has a streamlined forefront and structural adaptations to significantly reduce noise resulting from aerodynamics in high-speed trains, http://www.asknature.org/product/6273d963ef015b98f641fc2b67992a5e (accessed 27 January 2013).

[11] Beaked whales perform extreme dives to hunt deepwater prey, Woods Hole News Release, October 19, 2006, http://www.whoi.edu/page.do?pid=9779&tid=3622&cid=16726 (accessed 27 January 2013).

[12] Geek Wise, What is the strongest animal, http://www.wisegeek.com/what-is-the-strongest-animal.htm (accessed 27 January 2013).

ABOUT THE AUTHOR

H. Donald Wolpert obtained a BS degree in mechanical engineering from Ohio University in 1959 and then began an industrial career. He worked for E.H. Plesset Associates on electro-optic devices; Xerox Electro-Optical Systems on laser scanners; and TRW and TRW-Northrop Grumman on three-dimensional imaging and the design and development of electro-optical space payloads. Early on, he became interested in bio-optics, on which subject he continues to publish many articles and deliver many lectures and seminars.

Biomimetic Vision Sensors

Cameron H.G. Wright and Steven F. Barrett

Department of Electrical and Computer Engineering, University of Wyoming,
Laramie, WY 82071, USA

Prospectus

This chapter is focused on vision sensors based on both mammalian and insect vision systems. Typically, the former uses a single large-aperture lens system and a large, high-resolution focal plane array; the latter uses many small-aperture lenses, each coupled to a small group of photodetectors. The strengths and weaknesses of each type of design are discussed, along with some guidelines for designing such sensors. A brief review of basic optical engineering, including simple diffraction theory and mathematical tools such as Fourier optics, is followed by a demonstration of how to match an optical system to some collection of photodetectors. Modeling and simulations performed with tools such as Zemax and MATLAB® are described for better understanding of both optical and neural aspects of biological vision systems and how they may be adapted to an artificial vision sensor. A biomimetic vision system based on the common housefly, *Musca domestica*, is discussed.

Keywords

Apposition, Biomimetic, Camera eye, Compound eye, Fly eye, Hyperacuity, Lateral inhibition, Light adaptation, Mammal eye, Motion detection, Multiaperture, Multiple aperture, Neural superposition, Optical flow, Optical superposition, Photoreceptor, Retina, Single aperture, Vision sensor

1.1 INTRODUCTION

Biomimetic vision sensors are usually defined as imaging sensors that make practical use of what we have learned about animal vision systems. This approach should encompass more than just the study of animal eyes, because, along with the early neural layers, neural interconnects, and certain parts of the animal brain itself, eyes form a closely integrated vision system [1–3]. Thus, it is inadvisable to concentrate only on the eyes in trying to design a good biomimetic vision sensor; a systems approach is recommended [4].

This chapter concentrates on the two most frequently mimicked types of animal vision systems: ones that are based on a mammalian *camera eye* and ones that are based on an insect *compound eye*. The camera eye typically uses a single large-aperture lens or lens system with a relatively large, high-resolution focal plane array of photodetectors. This is similar to the eye of humans and other mammals and has long been mimicked for the basic design of both still and video cameras [1, 3, 5]. The compound eye

http://dx.doi.org/10.1016/B978-0-12-415995-2.00001-5

instead uses many small-aperture lenses, each coupled to a small group of photodetectors. This is the type of eye found in insects in nature and has only recently been mimicked for use as alternative vision sensors [1, 3, 5]. However, knowledge of the optics and sensing in a camera eye is very helpful in understanding many aspects of the compound eye.

Using just two categories—camera eyes and compound eyes—can be somewhat oversimplified. Land and Nilsson describe at least 10 different ways in which animal eyes form a spatial image [1]. Different animals ended up with different eyes due to variations in the evolutionary pressures they faced, and it is believed that eyes independently evolved more than once [1]. Despite this history, the animal eyes we observe today have many similar characteristics. For example, a single facet of an apposition compound eye in an insect is quite similar to a very small version of the overall optical layout of the camera eye in a mammal.

Mammals evolved to have eyes that permit a high degree of spatial acuity in a compact organ, along with sufficient brain power to process all that spatial information. While mammals with foveated vision have a relatively narrow field of view for the highest degree of spatial acuity, they evolved ocular muscles to allow them to scan their surroundings, thereby expanding their effective field of view; however, this required additional complexity and brain function [1]. Insects evolved to have simple, modular eyes that could remain very small yet have a wide field of view and be able to detect even the tiniest movement in that field of view [1]. The insect brain is modest and cannot process large amounts of spatial information, but much pre-processing to extract features such as motion is achieved in the early neural layers before the visual signals reach the brain [1].

In general, the static spatial acuity of compound eyes found in nature is less than most camera eyes. Kirschfeld famously showed that a typical insect compound eye with spatial acuity equal to that of a human camera eye would need to be approximately 1 m in diameter, far too large for any insect [6]. Each type of eye has specific advantages and disadvantages. As previously mentioned, the camera eye and the compound eye are the two most common types of eye that designers have turned to when drawing upon nature to create useful vision sensors.

Before getting into the specifics of these two types of vision systems, we first need to discuss image formation and imaging parameters in general, using standard mathematical techniques to quantify how optics and photodetectors interact, and then show how that translates into a biomimetic design approach. Separate discussions of biomimetic adaptations of mammalian vision systems and insect vision systems are provided, along with strengths and weaknesses of each. The design, fabrication, and performance of a biomimetic vision system based on the common housefly, *M. domestica*, are presented.

1.2 IMAGING, VISION SENSORS, AND EYES

We have found that one of the most common problems encountered in designing a biomimetic vision sensor is a misunderstanding of fundamental optics and image-sampling concepts. We therefore provide a brief overview here. This chapter is by no means an exhaustive reference for image formation, optical engineering, or animal eyes. In just a few pages, we cover information that spans many books. We include only enough detail here that we feel is important to most vision sensor designers and to provide context for the specific biomimetic vision sensor discussion that follows. For more detail, see [1–3, 5, 7–17]. We assume incoherent light in this discussion; coherent sources such as lasers require a slightly different treatment. Nontraditional imaging modalities such as light-field cameras are not discussed here.

1.2.1 Basic Optics and Sensors

1.2.1.1 Object and Image Distances

An image can be formed when light, reflected from an object or scene (at the object plane), is brought to focus on a surface (at the image plane). In a camera, the film or sensor array is located at the image plane to obtain the sharpest image. One way to create such an image is with a converging lens or system of lenses. A simplified diagram of this is shown in Figure 1.1, which identifies parameters that are helpful for making some basic calculations. One such basic calculation utilizes the Gaussian lens equation

$$\frac{1}{s_o} + \frac{1}{s_i} = \frac{1}{f}, \tag{1.1}$$

which assumes the object is in focus at the image plane. Equation (1.1) is based on the simple optical arrangement depicted in Figure 1.1 containing a single thin lens of focal length f but can be used within reason for compound lens systems (set to the same focal length) where the optical center (i.e., nodal point) of the lens system takes the place of the center of the single thin lens [7]. Note that focal length and most other optical parameters are dependent on the wavelength λ.

The focal length is usually known, and given one of the two axial distances (s_o or s_i) in Figure 1.1, the other axial distance is easily calculated. When the distance to the object plane s_o is at infinity, the distance to the image plane s_i is equal to the focal length f. The term *optical infinity* is used to describe an object distance that results in an image plane distance very close to the focal length; for example, some designers use $s_o \geqslant 100f$ as optical infinity, since in this case s_i is within 1% of f. On the other hand, for visual acuity exams of the human eye, optometrists generally use $s_o \approx 338f$ as optical infinity.

Equation (1.1) is also useful for calculating distances perpendicular to the optical axis (i.e., transverse distances). Similar triangles provide the relationship

$$\frac{x_o}{s_o} = -\frac{x_i}{s_i}, \tag{1.2}$$

which allows calculation of x_o or x_i when the other three values are known. The minus sign accounts for the image inversion in Figure 1.1. Modern cameras and vision sensors based on the mammalian camera eye typically place a focal plane array (FPA) of photodetectors (e.g., an array of either charge-coupled devices (CCD) or CMOS sensors) at the image plane. This array introduces spatial sampling of the image, where the center-to-center distance between sensor locations (i.e., the spatial sampling interval) equals the reciprocal of the spatial sampling frequency. Spatial sampling, just like temporal sampling, is limited by the well-known sampling theorem: Only spatial frequencies in the image up to

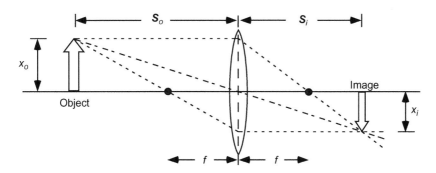

FIGURE 1.1 Optical distances for object (s_o) and image (s_i) with a single lens of focal length f.

one-half the spatial sampling frequency can be sampled and reconstructed without aliasing [18].

Aliasing is evident when the reconstructed image shows incorrect spatial frequencies that are lower than the true image. If the spatial sampling frequency in a given direction in the image is F_s, and the true spatial frequency in the same direction of some pattern in an image is f_o, then if $f_o > F_s/2$, and the aliased frequency will be

$$F_s/2 - (f_o - F_s/2) = F_s - f_o. \qquad (1.3)$$

Aliasing in an image is most noticeable to humans with regard to periodic patterns, such as the stripes of a person's tie or shirt, which when aliased tend to look broader and distorted [18]. Note that most real-world images are not strictly band-limited, so some amount of aliasing is usually inevitable.

Fourier theory tells us that even a complex image can be modeled as an infinite weighted sum of spatially sinusoidal frequencies [18, 19]. Knowledge of how these spatial frequencies are sampled can help predict how well a vision sensor may perform. Equation (1.2) allows us to map transverse distances between the object plane and the image plane and understand how the spatial sampling interval compares to the various transverse distances in the image.

Example Problem: As a real-world example that highlights the use of these relationships, suppose you need to remotely monitor the condition of a small experimental snow fence at a certain point along a northern-route interstate highway, using a digital camera. The individual slats of the snow fence are made of a new environmentally green recycled material that may or may not hold up under the wind pressures expected during the winter; that is the purpose of the monitoring. The imaging only needs to detect if a slat breaks and thus appears to be missing. The camera will be an inexpensive webcam that will periodically capture images of the snow fence and transmit the images back to a monitoring station. The webcam uses a 1.3-megapixel CCD rectangular sensor array (1280 H × 720 V pixels), and the physical size of the CCD array inside the camera is 19.2 mm horizontally by 10.8 mm vertically.[1] The aspect ratio of each image frame taken with this camera is 16:9, and the aspect ratio of each individual photodetector (pixel) in the CCD array is 1:1 (i.e., square), such that the center-to-center pixel spacing is the same in the x direction and the y direction. The webcam uses a built-in 22.5 mm focal length lens that is permanently fixed at a distance from the CCD array such that it will always focus on objects that are relatively far away (i.e., optical infinity). The snow fence is made up of very dark-colored slats that are 2.44 m (about 8 f) high and 200 mm wide, with 200 mm of open space between adjacent slats. In the expected snowy conditions, the contrast of the dark slats against the light-colored background should allow a good high-contrast daytime image of the snow fence, within the limits of spatial sampling requirements. No night-time images are needed.

See Figure 1.2 for a simple illustration of what the snow fence might look like, not necessarily drawn to scale nor at the actual viewing distance, with snow at the base obscuring some unknown part of the slat height. Assume the fence extends to the right and left of the figure a considerable distance beyond what the simple figure shows. You would like to view as wide a section of the experimental snow fence as possible, so you want to place the camera as far away from the fence as possible. Thus, you need to calculate the maximum distance you can place the webcam from the fence and yet still be able to easily make out individual slats in the image, assuming the limiting factor is the spatial sampling frequency of the image. Assume the

---

[1] It is traditional for manufacturers of cameras, monitors, televisions, etc. to provide size specifications as (horizontal, vertical) and aspect ratios as H:V, so this is how the information is provided here. However, this is the *opposite* of most image-processing and linear algebra books, where dimensions are usually specified as (row, column).

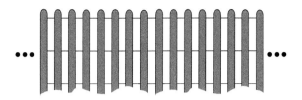

FIGURE 1.2 Illustration of the snow fence to be imaged by the digital camera.

optical axis of the camera is perpendicular to the fence, so you can neglect any possible angular distortions.

Solution: The periodic nature of the slats is not a sinusoidal pattern (it is actually closer to a square wave), but the spatial period of the slats is equal to the fundamental frequency of a Fourier sum that would model the image of the fence, and the individual slats will be visible with acceptable fidelity (for this specific application) if this fundamental frequency is sampled properly [18]. The sampling theorem requires a minimum of two samples per cycle; one complete cycle at the fundamental frequency is a single slat/opening pair. Thus, the 200 mm slat plus a 200 mm opening at the object plane must span two (or more) pixels at the image plane for adequate sampling to occur. In other words, the pixel spacing, mapped to the object plane, must be 200 mm or less in the horizontal direction (the vertical direction is not as important for this image). At the image plane, the pixel spacing is $19.2 \, \text{mm}/1280 = 15 \, \mu\text{m}$. Referring back to Figure 1.1, we know that $s_i = 22.5$ mm since the object plane is at optical infinity. Using similar triangles, we get $(22.5 \, \text{mm}/15 \, \mu\text{m}) = (s_0/200 \, \text{mm})$, thus $s_0 = 300$ m, which is the maximum distance allowed from the camera to the snow fence. If the camera is placed farther away than 300 m, the fundamental spatial frequency of the snow fence will alias as described by Eq. (1.3), and the image would likely be unacceptable.

How is this pertinent to someone developing biomimetic vision sensors? For any type of vision sensor (biomimetic or traditional), the basic trade-offs of the optics and the spatial sampling remain the same, so knowledge of these concepts is needed to intelligently guide sensor development.

1.2.1.2 Effect of Aperture Size

Another basic concept that is often important to sensor development is diffraction. No real-world lens can focus light to an infinitesimally tiny point; there will be some minimum *blur spot*. Figure 1.3 depicts the blur spots due to two simple optical setups with circular apertures, where the top setup has a larger aperture than the bottom one. In the figure, if the object plane is at optical infinity, then $d = f$. The diameter of the aperture is shown as D; this could be due to the physical diameter of the lens in a very simple optical setup, or to the (sometimes variable) aperture diaphragm of a more complex lens system.[2] As light travels from the lens to the image plane, differences in path length are inevitable. Where the difference in path length equals some integer multiple of $\lambda/2$, a lower-intensity (dark) region appears; where the difference in path length equals some integer multiple of λ, a higher-intensity (bright) region appears. With a circular aperture, the blur spot will take the shape of what is often called an *Airy disk*. The angular separation between the center peak and the first minimum of an Airy disk, as shown in Figure 1.3, is $\theta = 1.22\lambda/D$, which confirms the inversely proportional relationship between the blur spot diameter and the aperture diameter. The value of θ is often referred to as the *angular resolution*, assuming the use of what is known as the *Rayleigh criterion* [7].

A cross-section of an Airy disk is shown in Figure 1.4. Though the angular measure from the peak to the first minimum of the blur spot

---

[2] A variable circular aperture is often called an *iris diaphragm*, since it acts much in the same way as the iris of an eye.

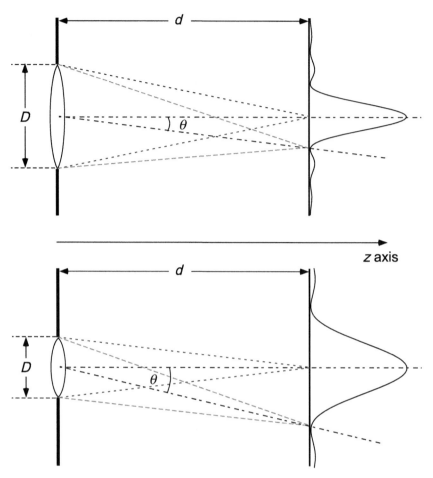

FIGURE 1.3 Minimum blur spot due to diffraction of light when $d = f$. Notice how a larger aperture D results in a smaller blur spot.

is $1.22\lambda/D$, the diameter of the blur spot at the half-power point, shown in Figure 1.4 to be λ/D, is often of interest. The size of the blur spot is what determines what is often called the *diffraction limit* of an optical system; however, keep in mind that the blur spot size may be dominated by lens aberrations, discussed later. The diffraction limit assumes the use of some resolution criterion, such as the ones named after Rayleigh and Sparrow [7]. Note that certain highly specialized techniques can result in spatial resolution somewhat better than the

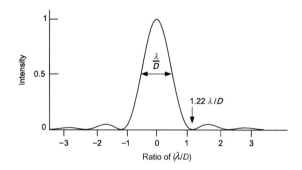

FIGURE 1.4 Cross-section of a normalized Airy disk.

diffraction limit, but that is beyond the scope of this discussion [20].

Any discussion of aperture should mention that the various subfields of optics (astronomy, microscopy, fiber optics, photography, etc.) use different terms to describe the aspects related to the effective aperture of the system [21]. In astronomy, the actual aperture size as discussed before is typically used. In microscopy, it is common to use *numerical aperture* (NA), defined as $NA = n \sin \phi$, where n is the index of refraction of the medium through which the light travels, and ϕ is the half-angle of the maximum cone of light that can enter the lens. The angular resolution of a standard microscope is often specified as $\lambda/2NA$. For multimode fiber optics, numerical aperture is typically defined as $NA = n \sin \phi \approx \sqrt{n_1^2 - n_2^2}$, where n_1 is the index of refraction of the core and n_2 is the index of refraction of the cladding. This can provide an approximation for the largest acceptance angle ϕ for the cone of light that can enter the fiber such that it will propagate along the core of the fiber. Light arriving at the fiber from an angle greater than ϕ would not continue very far down the fiber. In photography, the more common measure is called *f-number* (written by various authors as f# or F), defined as $F = f/D$, where f is the focal length and D is the effective aperture. A larger F admits less light; an increase in F by a factor of $\sqrt{2} \approx 1.414$ is called an increase of one f-stop and will reduce the admitted light by one-half. Note that to obtain the same image exposure, an increase of one f-stop must be matched by twice the integration time (called the *shutter speed* in photography) of the photosensor. Typical lenses for still and video cameras have values of F that range from 1.4 to 22. Whether the designer uses D, NA, F, or some other measure is dependent on the application.

How is this pertinent to someone developing biomimetic vision sensors? We sometimes desire to somewhat match the optics to the photosensors. For example, if the optics design results in a blur spot that is significantly smaller than the photosensitive area of an individual photodetector (e.g., the size of a single pixel in a CCD array), then one could say that the optics have been *overdesigned*. A blur spot nearly the same size as the photosensitive area of an individual photodetector results when the optics have been tuned to match the sensors (ignoring for the moment the unavoidable spatial sampling that a photodetector array will impose on the image). There are many instances, sometimes due to considerations such as cost, or weight, or size of the optical system and sometimes due to other reasons as described in the case study of the fly-eye sensor, in which the optical system is purposely designed to result in a blur spot larger than the photosensitive area of an individual photodetector.

Example Problem: For the webcam problem described earlier, what aperture size would be needed to approximately match a diffraction-limited blur spot to the pixel size?

Solution: The angular blur spot size is approximately (λ/D), so the linear blur spot size at the image plane is $(\lambda/D)s_i$. The pixel size was previously found to be 15 μm. If we assume a wavelength near the midband of visible light, 550 nm, then the requirement is for $D = 825$ μm. Since the focal length of the lens was given as 22.5 mm, this would require a lens with an f-number of $f/D = 27.27$, which is an achievable aperture for the lens system. However, the likelihood of a low quality lens in the webcam would mean that aberrations (discussed later) would probably dominate the size of the blur spot, not diffraction. Aberrations always make the blur spot larger, so if aberrations are significant then a larger aperture would be needed to get the blur spot back down to the desired size.

1.2.1.3 Depth of Field

The size of the effective aperture of the optics not only helps determine the size of the blur spot, but also helps determine the depth of field (DOF) of the image. While Figure 1.1 implies there is only a single distance s_o for which an

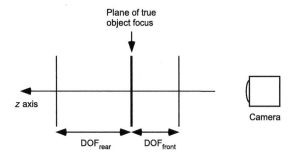

FIGURE 1.5 Front and rear DOF.

object can be brought to focus at distance s_i, DOF describes the practical reality that there is an axial distance over which objects are imaged with acceptable sharpness. Thus, an object within the range of $(s_0 - \text{DOF}_{\text{front}})$ to $(s_0 + \text{DOF}_{\text{rear}})$ would be imaged with acceptable sharpness (see Figure 1.5). With reference to Figure 1.3, note that, for a given focal length, a larger aperture results in a larger angle of convergence of light from the aperture plane to the image plane. This larger angle means that any change in d will have a greater blurring effect than it would for a smaller aperture. Thus, the DOF is smaller for larger apertures. Combining what is shown in both Figures 1.1 and 1.3, along with Eq. (1.1), we can show that focal length f also affects DOF; longer focal-length lenses have a smaller DOF for a given aperture size. There is no specific equation for DOF, since it is based on what is considered acceptable sharpness, and that is very much application dependent.

Photographers often manipulate DOF for artistic purposes, but sensor system designers are usually more interested in maximizing the DOF. In general, the $\text{DOF}_{\text{front}}$ is always less than DOF_{rear}. At relatively small values of s_0, the ratio of $\text{DOF}_{\text{front}}/\text{DOF}_{\text{rear}}$ is close to unity. As s_0 is increased, the ratio of $\text{DOF}_{\text{front}}/\text{DOF}_{\text{rear}}$

decreases, and there is a certain point at which DOF_{rear} extends to optical infinity; this value of s_0 is called the *hyperfocal distance*. At this setting, all objects from $(s_0 - \text{DOF}_{\text{front}})$ to ∞ would be imaged with acceptable sharpness. When the sensor system is set to the hyperfocal distance, the DOF extends from half the hyperfocal distance to infinity, and the DOF is the largest possible for a given focal length and aperture size. Therefore, the hyperfocal distance is often of interest to the sensor system designer.

1.2.1.4 Field of View

The field of view (FOV) for a sensor system is the span over which a given scene is imaged. Although it may seem at first that the aperture size might determine FOV, in typical imaging situations it does not.[3] The approximate FOV is determined only by the geometry of Figure 1.1, where x_i would be one-half the size (in that dimension) of the imaging sensor array or of the film, and x_0 would be one-half the spatial FOV at distance s_0. Since *angular* FOV is independent of object distance, it is the more frequently used form of FOV. For an imaging sensor (or film) of size a in a given direction, the angular FOV in that direction is $2\arctan(a/2s_i)$. When the system is set to focus at optical infinity, this takes on the familiar form of $2\arctan(a/2f)$. The shape of the FOV matches the shape of the sensor array or film that is used to capture the image, not the shape of the aperture. Although optics are typically transversely circular, sensor arrays and film are more often rectangular, so the FOV would then also be rectangular.

Example Problem: For the webcam problem described earlier, assume that the diameter of the lens aperture is approximately 8 mm. (a) What is the F for this camera? (b) What is the angular FOV of the camera? (c) How much of the snow fence will be imaged at the maximum distance of the camera from the fence?

---

[3] If too small of an aperture is used at the wrong point in an optical system, it can restrict the FOV to less than the full sensor dimensions. This is almost always unintentional.

Solution: (a) The focal length of the lens was given as 22.5 mm, yielding $F = f/D \approx 2.8$. (b) The FOV is determined from the sensor array dimensions, not the lens aperture. The sensor size was given as 19.2 mm horizontally by 10.8 mm vertically. The horizontal angular FOV is found by calculating $2\arctan(19.2\text{mm}/2(22.5 \text{ mm})) = 0.807$ rad $= 46.2°$. The vertical angular FOV is found by calculating $2\arctan(10.8 \text{ mm}/2(22.5 \text{ mm})) = 0.471$ rad $= 27.0°$. (c) The maximum distance for s_o was found previously to be 300 m. From Figure 1.2, it is obvious that the horizontal FOV is what determines how much of the fence will be imaged. At a range of $s_o = 300$ m and an angular FOV of 0.807 rad, the horizontal distance of the fence that will be imaged is $(2)(300)\tan(0.807/2) \approx 256$ m. This same answer could also be found using Eq. (1.2).

1.2.1.5 Aberrations

No optical system is perfect, and the imperfections result in what are called *aberrations* [7]. Most aberrations are due to imperfections in lenses and are usually categorized as either *monochromatic* or *chromatic*. Up to this point in Section 1.2, the discussion has centered on aspects of an optical system that must be considered one wavelength at a time.[4] This is how monochromatic aberrations must be treated. A chromatic aberration, on the other hand, is a function of multiple wavelengths. See additional references such as Smith [8] for more information on aberrations.

Common forms of monochromatic aberrations include spherical, coma, astigmatism, field curvature, defocus, barrel distortion, and pincushion distortion. Monochromatic aberrations are primarily due to either an unintended imperfection (which is usually caught and eliminated in the lens manufacturing stage) or an intentional mismatch between the actual geometry of

the lens and the geometry of the lens that would be required to take into account the exact nature of the propagation of light. This intentional mismatch is due to the much higher cost of producing a geometrically perfect lens. For example, a common form of monochromatic aberration is spherical aberration, which occurs when the lens is manufactured with a radius of curvature that matches a sphere; it is much cheaper to fabricate this type of lens than what is called an *aspheric* lens, which more closely matches the physics related to the propagation of light. Spherical aberration causes incorrect focus, but the effect is negligible near the center of the lens. A typical spherical lens made from crown glass exhibits spherical aberration such that only 43% of the center lens area (i.e., 67% of the lens diameter) can be used if objectional misfocus due to spherical aberration is to be avoided.

Chromatic aberration is primarily due to the unavoidable fact that the refractive index of any material, including lens glass, is wavelength dependent. Therefore, a single lens will exhibit slightly different focal lengths for different wavelengths of light. The fact that monochromatic aberrations are also wavelength dependent means that even differences in monochromatic aberrations due to differences in wavelength can be considered contributors to chromatic aberration. Chromatic aberration appears in a color image as fringes of inappropriate color along edges that separate bright and dark regions of the image. Even monochrome images can suffer from degradation due to chromatic aberration, since a typical monochrome image using incoherent light is formed from light intensity that spans many wavelengths. A compound lens made of two materials (e.g., crown glass and flint glass), called an *achromatic lens*, can correct for a considerable amount of chromatic aberration over a certain range of wavelengths. Better

---

[4] Depending on the needs of the application, a designer can make individual calculations at many wavelengths, two calculations at the longest and shortest wavelengths, or one calculation at the approximate midpoint of the range of wavelengths.

correction can be achieved with low-dispersion glass (typically containing fluorite), but these lenses are quite expensive.

In multiple-lens optical systems, aberrations of all types can be mitigated using special combinations of convex and concave lenses and specific types of glass; this is usually described as *corrected optics*. High-quality corrected optics, although expensive, can achieve images very close to the theoretical ideal. Spherical aberration, by itself, can be minimized even in a single-lens system by using a relatively expensive aspheric lens.

1.2.1.6 Reflection and Refraction

When light enters an optical system, it encounters a boundary between two different indices of refraction n. For example, when light propagates through air $(n_1 = 1)$ and enters a lens made of crown glass $(n_2 = 1.5)$, it is both refracted and reflected[5] at this boundary (Figure 1.6). For a multiple-lens system using different types of glass, there may be many such boundaries. The familiar Snell's law $(\sin\theta_1 / \sin\theta_2 = n_2/n_1)$ predicts the angle of refraction; the angle of reflection is equal to the angle of incidence.[6] The reflectance R is the fraction of incident light intensity (i.e., power) that is reflected, and the transmittance T is the fraction of incident light intensity that is refracted. Obviously, $R + T = 1$. At this point, we need to take into account the polarization state of the incident light to determine how much of the incident light will be reflected.

Assume the page on which the plot of Figure 1.6 appears is the plane of incidence for the

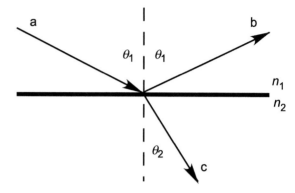

FIGURE 1.6 Reflection (ray b) and refraction (ray c) of incident light (ray a) encountering a boundary.

incoming light. If the incident light is polarized such that the electric field is perpendicular to the plane of incidence, then

$$R_s = \left| \frac{n_1 \cos\theta_1 - n_2 \cos\theta_2}{n_1 \cos\theta_1 + n_2 \cos\theta_2} \right|^2, \quad T_s = 1 - R_s.$$
(1.4)

If the incident light is polarized such that the electric field is parallel to the plane of incidence, then

$$R_p = \left| \frac{n_1 \cos\theta_2 - n_2 \cos\theta_1}{n_1 \cos\theta_2 + n_2 \cos\theta_1} \right|^2, \quad T_p = 1 - R_p.$$
(1.5)

If the light is unpolarized (i.e., randomly polarized), then a common estimate is $R = (R_s + R_p)/2$.

One ramification of reflection for a vision sensor designer is that the fraction of light intensity that is reflected at the boundary never makes it to the photodetectors. For example, for incident

---

[5] We consider here specular reflection, where any irregularities in the boundary surface are small compared to the wavelength (i.e., an optically smooth surface). If this is not true, diffuse reflection (i.e., scattering) occurs. We also assume $n_2 > n_1$. If the converse is true, then there exists a critical angle for which total internal reflection will occur [7].

[6] *Snell's law* is named after Willebrord Snellius (born Willebrord Snel van Royen; 1580–1626). Note that the spelling of Snell's name has been Anglicized; the more correct Dutch spelling is Snel (or Snellius), but Snell is overwhelmingly found in the literature [22]. It is appropriate here to mention that Snell's law and many other significant discoveries in optics were made by Middle Eastern scientists such as Ibn Sahl (c. 940–1000) and Ibn al-Haytham (c. 965–1039) many hundreds of years before Snel, Descartes, or Newton were born [23].

light arriving at $\theta_1 = 0$, it is easy to see that, for the air-glass interface, $R \approx 0.04$, which means 4% of the light is lost to reflection, which is not an insignificant amount. This same effect occurs at every boundary, so the back side of the lens exhibits reflection, as do any boundaries associated with additional lenses. This can cause spurious reflections to bounce around inside an optical system, greatly degrading the image contrast and causing undesirable noise spots in the image.

To reduce this effect, antireflective (AR) coatings have been developed that incorporate one or more thin films (where the thickness of the coating is typically on the order of $\lambda/4$). Destructive interference of the reflected light (and constructive interference of the transmitted light) greatly reduces R and allows more light to be transmitted to the photodetectors. The effectiveness of an AR coating is dependent on both the wavelength and the angle of incidence. The term *coated optics* is sometimes used to describe AR optics. Essentially all optical components (lenses, beam splitters, turning mirrors, etc.) are available with AR coatings, and their use is highly recommended. See Chapter 12 on biomimetic AR coatings.

1.2.2 Fourier Optics Approach

A particularly powerful and practical method of dealing with design considerations such as apertures, lenses, photodetector size, and spatial sampling is called *Fourier optics*. The Fourier approach can even be continued (with a change of domains from space to time) to the electronics associated with obtaining an image from a given sensor system. The classic reference for Fourier optics is the excellent book by Goodman [10], although Wilson [11] is also very helpful; a succinct treatment can be found in Hecht [7, Ch. 11]. The method is very similar to the Fourier approach to the design and analysis of circuits and systems that is familiar

to electrical engineers. The treatment given here is brief and is not intended to be rigorous but rather to merely provide a few practical techniques.

1.2.2.1 Point-Spread Function

For the purposes of optical design, the two domains linked by the Fourier transform $\mathcal{F}\{\ \}$ are the spatial domain (i.e., distance) and the spatial frequency domain (i.e., cycles per unit distance), where the distance coordinates are usually assumed to be measured transverse to the optical axis, usually at the focal plane. Recall from the theory of Fourier transforms that convolution in one domain is equivalent to multiplication in the other domain; this will be useful. Every optical component (aperture, lens, etc.) has a point-spread function (PSF) defined in the spatial domain at the focal plane, which describes how an infinitesimally small (yet sufficiently bright) point of light (the optical equivalent of a Dirac delta function $\delta(x_o, y_o)$ at the object plane) is spread (or smeared) by that component. A perfect component, in the absence of both aberrations and diffraction, would pass the point unchanged. The PSF of an optical component is convolved in the spatial domain with the incoming light. This means that a perfect component would require a PSF that was also a delta function $\delta(x, y)$; how much the PSF deviates from $\delta(x, y)$ determines how much it smears each point of light. Since diffraction is always present, it provides the limit on how closely a PSF can approach $\delta(x, y)$; any aberrations simply make the PSF deviate even further from the ideal.

Assume that light enters the sensor system though an aperture and lens, and that the lens focuses an image at the focal plane. The amplitude transmittance of the aperture can be described mathematically by a simple aperture function $A(x_a, y_a)$, which is an expression of how light is transmitted through or is blocked by the aperture at the aperture plane. For example, an ideal circular aperture with a radius of r could be expressed as

$$A(x_a, y_a) = \begin{cases} 1, & \sqrt{x_a^2 + y_a^2} \leqslant r, \\ 0, & \sqrt{x_a^2 + y_a^2} > r, \end{cases} \quad (1.6)$$

where (x_a, y_a) are spatial coordinates at the aperture plane. The aperture function sets the limits of the field distribution (which is usually determined using the Fraunhofer diffraction approximation [10]).

Previously, we saw that the intensity pattern at the focal plane of a lens, due only to diffraction of a circular aperture, resulted in an Airy disk. It turns out that this intensity pattern at the focal plane is proportional to the squared magnitude of the Fourier transform of the aperture function. That is, $h_A(x,y) \propto |\mathcal{F}\{A(x_a, y_a)\}|^2$, where $h_A(x,y)$ is the PSF at the focal plane of the circular aperture. Normalized plots of $h_A(x,y)$ for a circular aperture are shown in Figure 1.7; compare that figure to Figures 1.3 and 1.4. If a point of light $\delta(x_o, y_o)$ from the object plane passes through this aperture to the focal plane, then the aperture PSF $h_A(x,y)$ is convolved with $\delta(x,y)$, and by the sifting property of delta function the result is $h_A(x,y)$. Thus, the smallest possible blur spot due to the aperture is the same Airy disk as we found before, only we can now use the power of the Fourier transform and linear systems theory to extend the analysis.

If the lens is nonideal, it will also contribute (through convolution) a PSF $h_L(x,y)$ that deviates from a delta function. The PSF $h_L(x,y)$ is determined primarily by the various lens aberrations that are present. The combined PSF of the aperture and the lens is thus $h_{AL}(x,y) = h_A(x,y) * h_L(x,y)$, where the * symbol denotes convolution. The combined PSF $h_{AL}(x,y)$ is convolved with an ideal image (from purely geometrical optics) to obtain the actual image at the focal plane. If multiple lenses are used, they each contribute a PSF via convolution in the same way (unless arranged in such a way as to compensate for each other's aberrations, as previously discussed). Many other image degradations such as misfocus, sensor vibration (relative movement), and atmospheric turbulence can also be modeled with an approximate PSF that contributes, through convolution, to the overall PSF. If specific details regarding the degradation are known, it can sometimes be mitigated through careful image processing, depending on the noise in the image [18].

1.2.2.2 Optical Transfer Function, Modulation Transfer Function, and Contrast Transfer Function

The Fourier transform of the PSF yields the *optical transfer function* (OTF). That is, $H(u,v) = \mathcal{F}\{h(x,y)\}$, where (u,v) are the spatial-frequency coordinates at the focal plane. The PSF is in the spatial domain, and the OTF is in the spatial frequency domain, both at the focal plane. For most incoherent imaging systems, we are most interested in the magnitude of the OTF, called the *modulation transfer function* (MTF). See Figure 1.8 for a normalized plot of the MTF due only to the PSF of a circular aperture (e.g., the circular aperture PSF shown in Figure 1.7).

In this context, modulation m is a description of how a sinusoidal pattern of a particular spatial frequency at the object plane can be resolved. It quantifies the contrast between the bright and dark parts of the pattern, measured at the image plane. Specifically, $m = (\max - \min)/(\max + \min)$. From Figure 1.8, you can see that as the spatial frequency increases, the ability of the optical system to resolve the pattern decreases, until at some frequency the pattern cannot be discerned. Note the MTF in Figure 1.8 is zero at $u \geqslant D/\lambda$; thus, D/λ is called the *cutoff frequency* f_c. However, no real-world optical system can detect a sinusoidal pattern all the way out to f_c; the practical contrast limit (sometimes called the *threshold modulation*) for the MTF is not zero but more like 2%, 5%, or higher, depending on the system and the observer.

The *goodness* of an MTF relates to how high the modulation level remains as frequency increases,

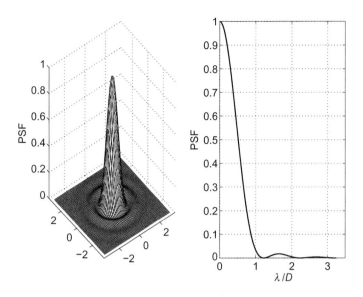

FIGURE 1.7 The normalized point-spread function (PSF) of a circular aperture is an Airy disk. Angular units of the horizontal axes (x, y directions) are radians; to convert from angular to linear units, multiply by the focal length of the lens.

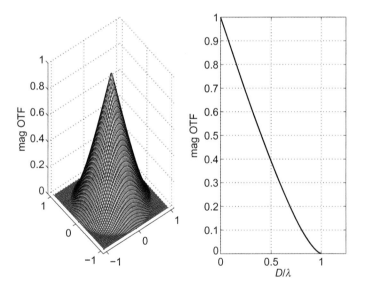

FIGURE 1.8 The normalized magnitude of the OTF of a circular aperture. This magnitude is typically abbreviated MTF. Spatial frequency units of the horizontal axes (u, v directions) are cycles/radian.

so some authors use the area under the MTF curve as a figure of merit. However, this is often too simple an approach, and a more application-specific comparison of MTFs may be warranted.

Each PSF in the spatial domain for each component of an optical system is associated with an MTF in the spatial frequency domain, and the MTFs are all combined into an overall MTF through multiplication. To empirically measure the overall MTF of an optical system, sinusoidal patterns of various spatial frequencies could be imaged, and the value of *m* could be determined for each of those frequencies. This piecewise data could be used to estimate the MTF. If desired, an inverse Fourier transform of this empirically derived MTF could then be used to estimate the overall PSF of the optical system.

Note that optical imaging systems are often specified in terms of units such as lp/mm, which stands for *line pairs per millimeter*, instead of specifying some maximum frequency sinusoidal pattern that can be detected. A *line pair* is a white stripe and black stripe, and it is a far easier pattern to produce than an accurate sinusoidal pattern. However, a line-pair pattern is not sinusoidal, so measurements using this pattern do not directly yield the MTF; instead, they yield something called the *contrast transfer function* (CTF). But a line-pair pattern is to a sinusoidal pattern as a square wave is to a sine wave, so the conversion between CTF and MTF is well known [24]. Shown in one dimension[7] for compactness, the relationship is

$$M(u) = \frac{\pi}{4}\left[C(u) + \frac{C(3u)}{3} - \frac{C(5u)}{5} + \frac{C(7u)}{7} + \frac{C(11u)}{11} - \frac{C(13u)}{13} \cdots\right] \quad (1.7)$$

where $M(u)$ is the MTF and $C(u)$ is the CTF at spatial frequency u. Note that the existence and sign of the odd harmonic terms are irregular; see Coltman [24] for details. While the relationship is an infinite series, using just the six terms explicitly shown in Eq. (1.7) will usually provide sufficient fidelity.

Figure 1.9 compares the normalized MTF to the normalized CTF, due only to a circular aperture. Note that the CTF tends to overestimate the image contrast compared to the true MTF. Other

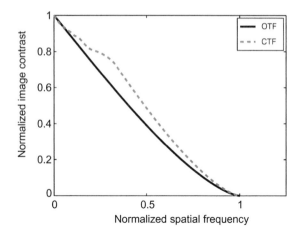

FIGURE 1.9 Comparison of an MTF with a CTF of a circular aperture.

methods, such as the use of laser speckle patterns [25], have been developed for empirically obtaining the MTF of optics and detector arrays[25].

1.2.2.3 Aberrations

When the optics include significant aberrations, the PSF and OTF can be complex-valued and asymmetrical. In general, aberrations will broaden the PSF and consequently narrow the OTF. An excellent treatment of these effects can be found in Smith [8]. Aberrations can reduce the cutoff frequency, cause contrast reversals, cause zero-contrast bands to appear below the cutoff frequency, and generally reduce image quality. A proposed quantitative measure of aberrated image quality is the *Strehl ratio*, which is the ratio of the volume integral of the aberrated two-dimensional MTF to the volume integral of the associated diffraction-limited MTF [8]. In discussing aberrations, it is important to recall from earlier that optical components that are not separated by a diffuser of some sort may compensate for the aberrations of each other—hence the term

---

[7] To simplify the discussion at some points, we assume that two-dimensional functions are separable in Cartesian coordinates. This is not exactly true, but the errors caused by this assumption are typically very small [15].

corrected optics. Otherwise, the MTFs of individual system components are all cascaded by multiplication. The concept of aberration tolerance should be considered: How much aberration can be considered acceptable within the system requirements? Smith [8] advises that most imaging systems can withstand aberration resulting in up to one-quarter wavelength of optical path difference from a perfect (i.e., ideal) reference wavefront without a noticeable effect on image quality. High-quality optics typically achieve this goal [9].

1.2.2.4 Detector Arrays

With a scene imaged at the focal plane, the next step is to use some photosensitive devices to convert the light energy to electrical energy. One of the most common techniques is to use a rectangular, planar array of photodetectors (typically CCD or CMOS) at the focal plane; this is often called a *focal plane array* (FPA). Use of the FPA introduces two more effects on the image: the spatial integration of light over the finite photosensitive area

of each photodetector, and the spatial sampling of the continuous image. For compactness, a one-dimensional approach is used where appropriate for the explanation that follows.

Each photodetector of the FPA must have a large enough photosensitive area to capture a sufficient number of photons to obtain a useable signal above the noise floor. This finite area results in spatial integration that produces a blurring or smearing effect. In the most common case, the sensitivity of the photodetector is relatively constant over the entire photosensitive area, so the effective PSF of the FPA is just a rectangular (or *top-hat*) function with a width determined by the size of the photosensitive area in the given dimension for the individual photodetectors. The OTF of the FPA is the Fourier transform of a rectangular function, which is the well-known sinc function. Thus, the MTF of FPA is the magnitude of the associated sinc:

$$\text{MTF}_{\text{FPA}} = \left| \frac{\sin(\pi x_d u)}{\pi x_d u} \right| = |\text{sinc}(x_d u)|, \quad (1.8)$$

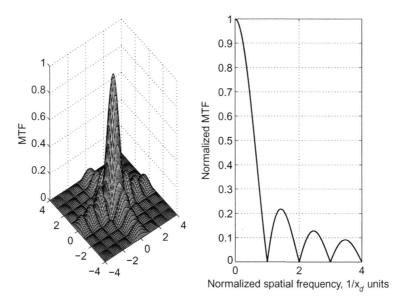

FIGURE 1.10 Normalized MTF of a typical focal plane array with a square photosensitive area, where $x_d = y_d$. The spatial frequency units of the horizontal axes (u, v directions) are the reciprocal of the spatial dimension of a single photodetector in that direction, as in $1/x_d$ or $1/y_d$, respectively.

where x_d is the size of the photosensitive area in the x direction (Figure 1.10). While the magnitude of a sinc function extends to infinity (so technically there is no associated cutoff frequency with this MTF), it is common to consider the MTF of an FPA only up to the first zero of the sinc, which occurs at $1/x_d$. Thus, a larger photosensitive area may gather more light but results in the first zero occurring at a lower spatial frequency, which induces more blurring in the image. Note that the units of spatial frequency in Eq. (1.8) for the MTF of an FPA are linear units (e.g., cycles/meter); when optical MTFs were previously discussed, we followed the common convention of using angular units. Angular spatial frequency divided by the focal length of the optical system equals linear spatial frequency at the focal plane, assuming the small angle approximation holds.

A commonly used mathematical description of image sampling is to convolve the point-spread function of all the nonideal aspects of the imaging system (optics, sensor array, and associated electronics) with an ideal continuous image source, followed by sampling via an ideal basis function such as a train of delta functions [15, 26]. Thus, for an image source sampled in the x direction,

$$g[n] = [h(x) * f(x)] \sum_{n \in z} \delta(x - nX_s), \quad (1.9)$$

where $g[n]$ is the discrete space–sampled image result, $h(x)$ is the combined point-spread function of all the nonideal effects, $f(x)$ is the ideal continuous space image, and X_s is the center-to-center spacing of the photodetectors.[8] Sampling in the y direction has a similar form. The top-hat PSF due the FPA would contribute to the overall $h(x)$ in Eq. (1.9). But what of the effect of sampling? Since X_s is the spatial sampling interval,

then $F_s = 1/X_s$ is the spatial sampling frequency. From the sampling theorem, we can conclude that any spatial frequencies sampled by the FPA that are higher than one-half the sampling frequency, or $F_s/2 = 1/2X_s$, will be aliased as described by Eq. (1.3). With no dead space between detectors (i.e., a fill factor of 1), the first zero of the FPA's MTF occurs at F_s, so the FPA will respond to spatial frequencies well above $F_s/2$, which means aliasing is likely. Lower fill factors exacerbate the potential for aliasing, since a smaller detector size moves the first zero of the MTF to a higher spatial frequency.

This may or may not present a problem, depending on the application. Monochrome aliasing tends to be less objectionable to human observers than color aliasing, for example.[9] Real-world images are often bandlimited only by the optical cutoff frequency of the optics used to form the image. This optical cutoff frequency is often considerably higher than $F_s/2$, and in that case aliasing will be present. However, some FPAs come with an optical low-pass filter in the form of a birefringent crystal window on the front surface of the array.

Note that the spatial sampling can be implemented in various ways to meet the requirements of the application, as depicted in Figure 1.11, but the treatment of spatial integration (which leads to the MTF) and the spatial sampling (with considerations of aliasing) remain the same. Ideal sampling (as shown in Figure 1.11a) is a mathematical construct, useful for calculations [as in Eq. (1.9)] but not achievable in practice. The most common type of sampling is top-hat sampling on a planar base, as shown in Figure 1.11b, where the fill factor implied by the figure is 50%. The MTF associated with top-hat sampling was given in Eq. (1.8). Some FPAs do

---

[8] If $X_s = x_d$ then the *fill factor* FF in the x direction is 1, or 100%, meaning there is no dead space between photosensitive areas in the x direction. In general, the two-dimensional fill factor is FF $= (x_d y_d)/(X_s Y_s)$.

[9] Color images are often formed with a combination of a single FPA and a filter array such as a Bayer mosaic. This results in a different F_s for different colors; typically green has an F_s twice that of blue or red, but half of the F_s is implied by just the pixel count.

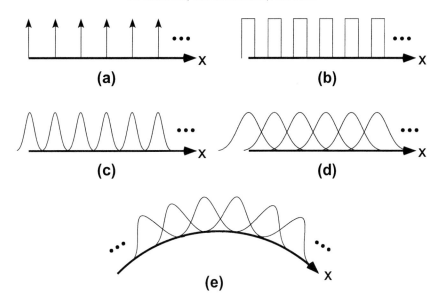

FIGURE 1.11 Various types and geometries of spatial sampling for vision sensors: (a) ideal, (b) top-hat, (c) Gaussian, (d) overlapping Gaussian, and (e) overlapping Gaussian with a nonplanar base.

not exhibit constant sensitivity over the photosensitive area of each photodetector, with the most common variation being an approximation to the Gaussian shape, as shown in Figure 1.11c. The MTF of this type of array is Gaussian, since the Fourier transform of a Gaussian is a scaled Gaussian. Figures 1.11d and e show Gaussian sampling with an intentional overlap between adjacent samples, on a planar and on a spherical base; these specific variations will be discussed further in the case study describing a biomimetic vision sensor based on *Musca domestica*, the common housefly.

1.2.2.5 Image Acquisition Electronics

The Fourier approach is commonly used in the design and analysis of electronic circuits and systems. The terminology is only slightly different between optical and electronic systems: The point spread function is similar to the impulse response, and the optical transfer function is similar to the transfer function. Optical system designers look mainly at the magnitude of the optical transfer function; electronic system designers usually are concerned with both the magnitude and the phase of the transfer function. Electronics operate in the time domain, whereas optics operate in the spatial domain.[10] If it is desirable to maintain the link to the spatial domain as one analyzes the associated image acquisition electronics, then one can map time to space. For example, assume the FPA data is read out row by row (i.e., horizontal readout). The spatial samples from the FPA are sent to the readout electronics at a certain temporal rate T_s. Knowing the center-to-center distance between detectors

[10] In addition to the spatial domain, optics and imaging systems have an implied time-domain aspect in terms of how an image changes over time. An obvious example of this is video, which is a time sequence of static images taken at some *frame rate*. Any motion detection or object tracking also implies that the time axis must be considered.

on a row of the FPA, one can map across the row the spatial distance associated with one temporal sample. The readout then shifts down to the next row, and that distance is the center-to-center distance between detectors on a column of the FPA. Whether this mapping is useful depends on the application.

Though many modern camera and imaging systems use a digital interface to pass data to computer, recording, and/or display devices, there are still many older (and some newer special-purpose) systems that make use of an analog interface at some point in the signal chain. This requires special consideration. A common example is the relatively inexpensive analog video camera (e.g., RS-170, RS-330, NTSC, CCIR, PAL, or SECAM) that processes (i.e., modulates) the output from the FPA into a specific analog video format to be carried by a cable to a computer, whereupon a specialized analog-to-digital (A/D) converter (called a *frame grabber*)[11] turns the analog image data into discrete digital pixel data. Two issues are predominant: bandwidth and resampling. The analog bandwidth B allocated to the video signal[12] puts a limit on the horizontal resolution the camera can provide (regardless of the number of photodetectors on the FPA), such that the varying analog voltage level in the video signal cannot have more than $2B$ independent values per second [27]. The resampling that occurs (the first sampling was at the FPA, the second at the frame grabber) almost always means that a pixel of digital image data has no direct correspondence to a particular photodetector location on the FPA. For some applications, this can have serious ramifications to the design.

How the interface between optical analysis and electronic analysis is handled is up to the designer and the particular application, but the process warrants significant thought to avoid erroneous conclusions.

1.2.3 Recommended Approach

This section has presented a very brief overview of optics and photodetectors in a practical way for the purpose of either designing or analyzing a vision sensor. Repetitive calculations can be made easily using a numerical analysis program such as MATLAB®, which can also provide insightful plots. More extensive optical analysis, simulation, and design can be achieved with a program designed specifically for optics, such as Zemax®. Tools such as MATLAB and Zemax are extremely valuable for this purpose, can save a great deal of time, and help avoid dead ends for potential design approaches. A particularly handy figure of merit for an imaging system that is easy to calculate and takes into account both optics and the detector array is $F\lambda/d$, where F is the f-number, λ is the wavelength under consideration, and d is the detector size of one element of the FPA in the given direction, as discussed earlier [12, 13]. With this figure of merit, $F\lambda/d < 1$ results in a detector-limited system, and $F\lambda/d > 1$ results in an optics-limited system. When $F\lambda/d \geqslant 2$, there is no aliasing possible, but this condition may result in too much image blur. See Holst [12, 13] for more detail.

1.3 BIOMIMETIC APPROACHES TO VISION SENSORS

Essentially all vision sensors developed by humans in the past and present are in some way biomimetic vision sensors. That is, humans studied how various animals *see* the world around them, then applied known principles of optics and light detection to mimic certain aspects of how animal vision systems evolved and thus created artificial vision sensors.[13]

---

[11] There are optional frame grabbers for digital cameras, without the A/D circuitry.

[12] The video signal specifics, such as scan rate, blanking interval, and bandwidth, must be known.

[13] Two optical techniques, zoom lenses and Fresnel lenses, have not yet been found in nature.

There are at least 10 known variants of animal eyes [1, 28]. These 10 types can be grouped as either noncompound eyes or compound eyes. Of the noncompound eyes, the most frequently mimicked type for vision sensors is the *refractive cornea eye*, which is often called a *camera eye*, since it is very similar to how most cameras operate. This is the type of eye found in almost all mammals, birds, reptiles, and most other terrestrial vertebrates. For completeness, we discuss this type of eye briefly in the next section. Of the compound eyes, the three most commonly mimicked types for vision sensors are the *apposition eye*, the *optical superposition eye*, and the *neural superposition eye*. We cover these types of eyes in more detail, since some of the most recent work in biomimetic vision sensors is based on them.

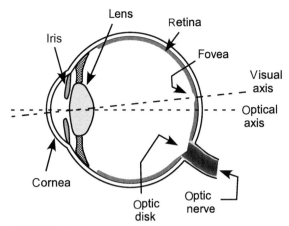

FIGURE 1.12 Simplified cross-sectional diagram of the human eye.

1.3.1 Camera Eye

Nearly all mammals, including humans, have camera eyes. As an example, Figure 1.12 is a highly simplified diagram of the human eye. The region between the cornea and the lens (the anterior chamber) is filled with a waterlike substance called the *aqueous humor*; the region behind the lens (the posterior chamber) is filled with an optically clear but somewhat gelatinous substance called the *vitreous humor* (or *vitreous body*). In this type of eye, the primary refractive power is due to the air/cornea optical interface.[14] An additional refractive effect is sometimes provided by an internal lens, such as the variable-shape crystalline lens that humans use to accommodate focus for close objects.

The use of significant refractive power allows the use of a relatively large aperture in the camera eye, permitting good light gathering and keeping the blur spot acceptably small

(necessary for good static acuity) in the short focal distance required of a compact vision organ.[15] An artificial vision sensor based on a camera eye typically uses a single large-aperture lens or lens system (mimicking the cornea and lens) combined with a relatively large, high-resolution focal plane array of photodetectors (mimicking the photoreceptors in the retina).

The human eye is a highly complex sensor that responds to electromagnetic stimuli at wavelengths of approximately 400–700 nm [29, 30]; for obvious reasons this band is called the *visible wavelength*. Ambient light enters the cornea and through the anterior chamber (containing the aqueous humor), through the pupil opening of the iris (which determines the effective aperture size), and through the crystalline lens and then passes through the vitreous humor before striking the retina. The retina consists of many layers of neural tissue that contains, among other things, the photoreceptors (rods, cones, and nonimaging photosensitive ganglion

---

[14] Aquatic mammals must use a lens as the primary refractive element, since the index of refraction of water is very close to the indices of refraction of the cornea and the aqueous humor.

[15] Due to aberrations in the human eye, the actual PSF is somewhat larger than the diffraction-limited Airy disk and has an approximately Gaussian shape. Aberrations are greater, and therefore the PSF is wider, for larger pupil diameters.

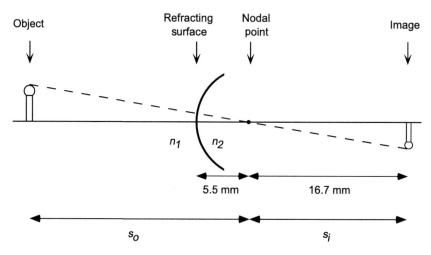

FIGURE 1.13 The reduced eye model for simplified calculations of human retinal distances.

cells). Note that the centered optical axis is slightly different from the physiological visual axis; although both pass through the nodal point (optical center) of the eye, the latter is referenced to the fovea, which is the area of highest acuity vision on the retina.

To simplify practical calculations related to the human eye, a method of estimating spatial distances on the curved surface of the retina with sufficient accuracy is helpful. A straightforward technique to convert between angular span and spatial distance on the retina makes use of a simplified version of Hemholtz's schematic eye called the *reduced eye model*, shown in Figure 1.13 [31, 32]. Compare Figure 1.13 to Figure 1.1. In the reduced eye model, there is a single refractive surface having a 5.5 mm radius of curvature and a posterior nodal distance of 16.7 mm. Since a ray passing through the nodal point is by definition an undeviated ray, an angular span in degrees can be equated to a spatial distance on the retina by

$$16.7 \text{ mm} \times \tan (1°) = 291.5 \text{ μm/ deg}. \quad (1.10)$$

As an example, the optic disk[16] in a typical human eye is roughly circular and has a mean diameter of 1,800 μm [33]. This value has been found to be nearly constant in all humans regardless of race, sex, or age and is often used as a yardstick in fundus photographs [34]. From Eq. (1.10), we see that the optic disk subtends an arc of roughly 6.2°.

Note that because the human eye is not truly spherical, these distance calculations become less accurate as one moves away from the region of the fundus near the optic disk and fovea often referred to as the *posterior pole* [35]. However, the peripheral regions of the retina have no critical vision anatomy and so are not of great interest to designers of biomimetic vision sensors. Therefore, the approximation of Eq. (1.10) will usually suffice.

It may be desirable to compare the spatial resolution of a biomimetic vision sensor to that of the human eye. An often-quoted resolution limit for the human eye (based only on the diffraction-limited point-spread function of the pupil) is 1 min of arc (equivalent to 4.9 μm at the posterior

---

[16] The optic disk is the nearly circular *blind spot* where the optic nerve exits the orb of the eye; no photodetectors exist inside the area of the optic disk.

pole); this assumes a 2 mm pupil and illumination wavelength of 550 nm [7]. A *normal* (i.e., emmetropic) human eye that measures 20/20 on a standard eye chart sees with a resolution of approximately 1 min of arc. Perhaps not coincidentally, the minimum separation of individual photoreceptors in the fovea [32] and the sampling theorem requirement of at least two samples per cycle of spatial frequency [27] together yield a physiological resolution limit for the eye of just over 1 min of arc [22], closely matching the theoretical optical resolution limit due to the diffraction-limited cutoff frequency of a 2 mm pupil in daylight. Certain types of acuity, such as vernier acuity, have been shown to exceed this theoretical resolution, most likely due to higher-level processing in the brain's visual cortex [32].

As mentioned earlier, the camera eye, of which the human eye is a particularly good example, is the basis on which nearly all standard cameras and vision sensors are based. Some optical system is used to bring an image to focus on an FPA, which then spatially samples the image. The FPA typically is rigidly attached to the sensor frame and cannot move. The focal length is also typically fixed (except for zoom lenses). With reference to Figure 1.1, if some particular object distance s_o is desired, the optical center (nodal point) of the lens or lens system is moved axially to ensure that the image distance s_i remains equal to the distance between the nodal point and the fixed FPA. This simple camera-eye model can be used to analyze or design a wide variety of vision sensors. However, in search of capabilities beyond that of standard cameras and vision sensors, much of the recent work in the area of biomimetic vision sensors has turned to models of the compound eye.

1.3.2 Compound Eye

Many insects and other species have compound eyes. In this section we provide a basic description of compound eyes followed by three basic configurations as described by Land in a seminal work [36]. We then narrow our focus to the common housefly, *M. domestica*, and describe its vision processes in some detail, including the feature of motion hyperacuity. A review of physical sensors and navigation systems inspired by the insect world is then provided. We conclude with potential applications for these specialized sensors. This section is based in part on several previous journal papers on this topic [37–40].

Compound eyes have been around for some time. Trilobites featured compound eyes, and they existed from approximately 500 million to 250 million years ago [41]. Compound eyes are also quite common in insects. Insect compound eyes are quite mobile relative to the head. This allows advanced features such as a wide field of view and depth perception by employing stereo vision [41].

Depending on specific species, a compound eye includes hundreds to tens of thousands of individual hexagonal-shaped facet lenses. Each individual lens is composed of a chitinous-type material that serves as the cornea for the primary modular vision unit of the compound eye, called an *ommatidium*. The lens ranges in size from 15 to 40 μm, again depending on species [41].

The ommatidia form a repeatable pattern across the curved surface of the compound eye. The angle between adjacent ommatadia, called the *interommatidial angle*, ranges from 1 to 2 degrees, depending on species. As we shall see, this angle has a large impact on determining the resolving power of the eye [41].

After light enters the facet lens, it passes through and is focused by the crystalline cone. The cone is made up of a transparent solid material and focuses the impinging light on the proximal end of the rhabdom. The rhabdom channels the light to the photosensitive receptors called *rhabdomeres* [41].

1.3.2.1 *Types of Insect Compound Eyes*
Land described three basic configurations of insect vision: apposition, superposition, and neural superposition compound eyes [36].

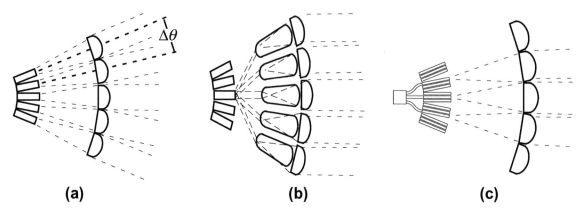

(a)   **(b)**   **(c)**

FIGURE 1.14 The three primary insect vision configurations: (a) apposition eye, (b) superposition eye, and (c) neural superposition eye. Adapted from Ref. 36.

These three configurations are illustrated in Figure 1.14. The primary difference between the configurations is the way light is routed and processed. Because of these differences, each compound eye configuration has its own inherent advantages and disadvantages. Insect species are equipped with a specific configuration that is best suited for the role the insect has in nature and the activities it accomplishes.

In the *apposition* compound eye, the rhabdom is in direct contact with the apex of the crystalline cone. Each cone and rhabdom are insulated by light-absorbing pigments such that light leakage to lateral, adjacent structures is significantly reduced [41]. The individual light-gathering contributions from each rhabdomere are pooled. The spatial acuity of the apposition compound eye is primarily determined by the interommatidial angle ($\Delta\theta$) described by

$$\Delta\theta = D/R, \qquad (1.11)$$

where D is the diameter of the facet lens, and R is the local radius of curvature of the eye [36, 42]. As shown in Figure 1.14a, $\Delta\theta$ describes the angular displacement between adjacent ommatidia, thus the name *interommatidial angle*.

The optical *superposition* eye pools light from adjacent ommatidia as shown in Figure 1.14b. Because there is no pigment between adjacent

crystalline cones, light exiting multiple cones may fall on the same rhabdom. This effectively enhances the light-gathering capability of this insect vision configuration, but it reduces the effective acuity due to the blurring effect of spatial superposition [41].

In the *neural superposition* eye, illustrated in Figure 1.14c, one rhabdomere in seven adjacent ommatidia shares an overlapped field of view with one another. This results in overlapping, Gaussian sensitivity patterns for the individual light-sensitive cells. These overlapped fields of view provide a motion resolution greater than that implied by the photoreceptor spacing of the retinal array, a phenomenon known as *motion hyperacuity* [43].

1.3.3 Visual Processing

Let us now review the vision-processing mechanisms of various biological species, followed by a more in-depth view of the common housefly vision system. It is important to study these processes because they inspire the development of sensors and the processing of their respective outputs. We look at early processes that occur in the first several cellular levels and at more complex processing. Our coverage of these topics is brief. The literature provides a rich heritage of this extensive body of work.

In the early vision processes, there is considerable visual information to be gleaned from the photoreceptors and their interaction with the first several cellular layers. There is enough evidence in the literature that fairly significant vision processes may take place early in the vision system (prior to the brain). For example, several researchers studying single-unit recordings in early vision processes of various organisms found visual neurons that had sensitivity to moving images [44–46]. Specifically, they noted the impulse rates of specific cells were modified by changes in the visual stimulus direction. Marr also posited that primitive forms of object recognition may take place on the retina. He noted that single neurons may perform more complex processing tasks than had been previously thought, including the ability to detect pattern elements and to discriminate the depth of objects [2]. Marr based his conclusions on the experiments conducted on frogs by Barlow, who observed that the frog's selective stimulation of neurons at the retinal layer served as bug detectors. That is, the neurons were providing a primitive form of object recognition on the retina [2, 44].

1.3.3.1 Optical Flow

Additional work by Nakayama and Loomis further supported sophisticated processing at the retinal level. These researchers postulated that center-surround motion-detection neurons exist and are directionally sensitive to different stimuli [47]. Several motion-detector cells, with different orientation sensitivities, feed a higher-order convexity cell to produce optical flow. *Optical flow* is the apparent motion of surfaces or objects in a scene, resulting from the motion difference between the observer and a scene. An example of optical flow, as discussed in Chapter 9 by Chahl and Mizutani, is evident when driving a vehicle: when we looking straight ahead, objects appear to be stationary; as we shift our gaze gradually to the side, objects in the

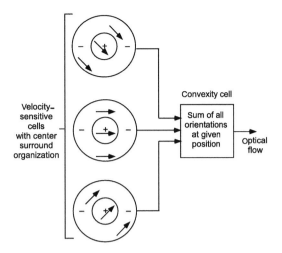

FIGURE 1.15 Convexity cell for extracting optical flow. Adapted from Ref. 47.

peripheral field of view appear to move faster. Sensors that are specifically sensitive to optical flow have been designed [48, 49].

The elegance of the Nakayama–Loomis theory is the ability to compute optical flow globally by employing local, simple neuronal networks, as depicted in Figure 1.15. Work by O'Carroll on dragonflies further noted that these and other related insects possess neurons that are tuned for detecting specific pattern features such as oriented line edges and moving spots [50].

1.3.3.2 Motion Processing

The literature includes a considerable body of work that has been accomplished to describe vision and motion processing at higher-order process regions of the vision system. There are four broad categories of motion-processing models [43, 51]:

- Differential- or gradient- based models employing first and second derivatives to determine velocity;
- Region- or feature-based matching to determine movement between adjacent temporal image scenes (frames);

- Phased-based models employing an array of band-pass filters that parse the incoming signal according to scale, speed, and orientation; and
- Energy- or frequency-based methods that quantify the output energy from velocity-tuned Gabor filters [52–54].

Nakayama provided a generalized model of motion processing divided into several stages, as shown in Figure 1.16. This model conceptually describes the four different broad categories of motion-processing models. Nakayama's generalized model is divided into three stages: an input receptor field stage, a directionally sensitive processing stage, and an integration stage. The input receptor field stage consists of a number of receptors sensitive to spatial position and frequency on the retinal image. This stage feeds the directionally sensitive stage that provides as outputs directionally sensitive and velocity-specific outputs. This stage combines, by addition or multiplication, the time-delayed input from an input receptor field with the immediate output from a physically displaced (Δs) input receptor field. The final stage spatially and temporally integrates the outputs from the directionally sensitive stages to provide a motion signal [43].

Space does not permit a detailed discussion of each motion model; we concentrate here on one of the best-known models, developed by Hassenstein and Reichardt. Borst reported that Hassenstein and Reichardt met as young men during World War II and decided to combine their talents in biology and physics [55,56]. They collaboratively developed a motion model based on their observations of the optomotor response of the beetle. Borst and Egelhaaf provided an excellent description of this motion-detection algorithm [56, 57]. An illustration, adapted from their work, is provided in Figure 1.17.

To detect motion, a stimulus must be viewed by two different photoreceptors that are displaced from one another by a distance Δs. If an object is moving from left to right, it will first be seen by photoreceptor A and then some time later by photoreceptor B. As show in Figure 1.17,

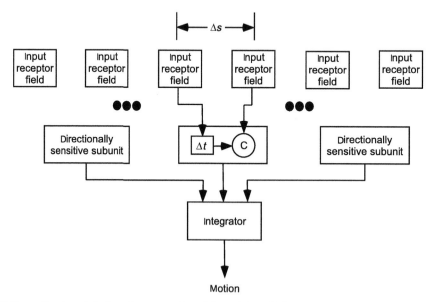

FIGURE 1.16 Generalized model of motion processing. Adapted from Ref. 43.

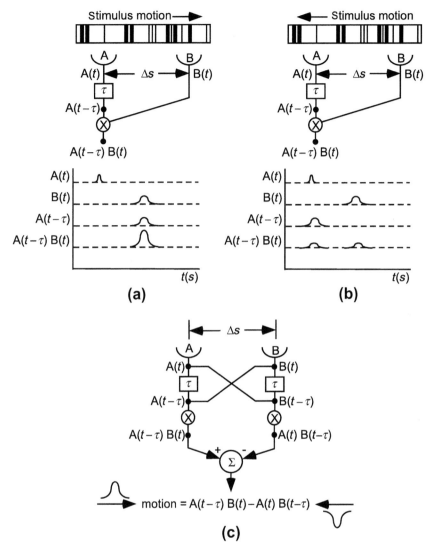

FIGURE 1.17 Hassenstein–Reichardt model of motion processing: (a) preferred direction, (b) null direction, and (c) motion director. Adapted from Refs. 55–57.

the view as seen by photoreceptor A is time delayed by τ and then multiplied by the immediate response from photoreceptor B. If the two photoreceptors have seen the same stimulus, there will be a high correlation, as indicated by a large response from the multiplier. The left-to-right motion is said to be in the *preferred direction* for this arrangement see Figure 1.17a. On the other hand, a stimulus traveling from right to left (termed the *null direction*) will have a minimal response from the multiplier stage; see Figure 1.17b.

To detect motion in both directions for a given pair of photoreceptors, a mirror-image configuration of hardware is added; see Figure 1.17c. This results in a large positive response in the preferred direction and a large negative response in the null direction.

In a physical sensor system based on this algorithm, the motion detector may be tuned to specific speeds of interest by varying the time delay and the physical displacement between a pair of sensing elements. An array of such detectors could be developed to detect a wide range of motion velocities and directions.

In addition to motion processing, other vision processes of interest include light adaptation and lateral inhibition.

1.3.3.3 Light Adaptation

Many vision systems must be able to view scenes with illumination levels that span multiple orders of magnitude. That is, the system must be able to operate from very low-light conditions to very high levels of illumination. This requires some mechanism for adaptation over a wide range of illuminance levels. Different species use a variety of *light adaption* mechanisms, including photomechanical mechanisms such as pupil changes, intensity-dependent summation mechanisms in both space and time, photochemical processes, and neuronal responses [58].

In *M. domestica*, two different processes are employed: chemical mediation of the photoreceptor response and membrance modification of the rhabdomeres to adjust the amount of light reaching the photoreceptors [58]. The fly further conditions the light via a *log transform-subtraction-multiplication* cellular-based algorithm. The signal is first logarithmically compressed. The average of surrounding ommatidia is then subtracted from the log-compressed signal, which removes the mean background illumination. The final multiplication step provides a gain to match the signal to the dynamic range of the monopolar cells [58]. Dean has used a special implementation of this algorithm to develop a sensor system that operates over a wide dynamic illumination range [59–61].

1.3.3.4 Lateral Inhibition

Another important vision process is *lateral inhibition*. Lateral inhibition allows an excited cell to mediate the response of its neighbors. This provides for an overall improvement of response. The concept of lateral inhibition was investigated by Hartline *et al.* in the horseshoe crab (*Limulus polyphemus*) [62].

The horseshoe crab is equipped with compound eyes with approximately 800 ommatidia per eye. The visual axis of the ommatidia diverge, with a slight overlap between adjacent pairs. Hartline *et al.* reported that each ommatidium functions as a single receptor unit. That is, each ommatidia has its own single nerve fiber. There is also an extensive system of cross-connecting strands of nerve fibers [62].

In a series of experiments, Hartline *et al.* illuminated a single ommatidium to elicit a response in its corresponding nerve fiber. The nerve fiber provided an output of approximately 65 impulses per second. Ommatidia near the original were then illuminated to determine their effect on the response. Hartline *et al.* found that when the surrounding light was activated, it inhibited the response and decreased the number of pulses. When the surrounding light was removed, the original ommatidial response returned to its previous value [62]. For this investigative work in lateral inhibition, Ragnar Granit, Haldan Hartline, and George Wald received the 1967 Nobel Prize for Physiology or Medicine. Recently, Strube has modeled lateral inhibition in an array of fly-inspired biomimetic ommatidia and showed a significant improvement in edge detection capability [63]. Also, Petkov *et al.* have used the response of Gabor filters to model lateral inhibition [64].

1.3.3.5 Navigation

Srinivasan has studied the honeybee in great detail and developed a hypothesis supported by experimental observations that have implications for the development of bio-inspired navigation systems for robots and unmanned aerial vehicles. Through these observations, Srinivasan has concluded that bees demonstrate complex behavior derived from optical flow [65]. Specifically, bees:

- Gauge distance traversed en route by integrating the apparent motion of the visual scene. This integration appears to be independent of the contrast and spatial frequency content (structure) of the scene.
- Maintain equidistant separation from tunnel walls by balancing the images as seen by each eye as the tunnel is traversed. This is accomplished by holding the average image velocity constant between the two eyes.
- Maintain a controlled landing on a horizontal surface by maintaining a constant average image velocity for the surface approach. This is accomplished by maintaining a forward speed that is approximately proportional to altitude.

For a closer investigation of the visual system, in the next section, we concentrate on a single species: *M. domestica*.

1.4 CASE STUDY: MUSCA DOMESTICA VISION SENSOR

Dipterans, such as *M. domestica*, are flying insects equipped with two wings and two balancers; they exhibit a highly parallel, compartmentalized, analog vision system of the neural superposition type [42,66]. The primary visual system of the housefly consists of two compound eyes equipped with approximately 3000 ommatidia per eye. As previously discussed, an ommatidium is the major modular structural unit of the compound eye. Each ommatidium of the housefly is equipped with a cutinous, hexagonal, $25\,\mu m$ diameter facet lens and a cone-shaped

lens with a complement of photoreceptors, as shown in Figure 1.18 [41, 66–68].

Figure 1.18 shows how six photoreceptors (R1–R6) surround two coaxially aligned photoreceptors (R7, R8) in an irregular pattern; R1–R6 terminate in the neurons of the lamina, whereas R7 and R8 bypass the lamina and connect directly to the medulla [41, 66–68]. The photoreceptors function as transducers to convert light energy into ionic current, and are thought to be sensitive to both magnitude and angle of the impinging light. The angular sensitivity of each photoreceptor has a profile that is approximately Gaussian [41, 69–74], as depicted in Figure 1.18.

Six photoreceptors (R1–R6) contribute to the neural superposition effect; the remaining two photoreceptors (R7, R8) bypass the lamina and are therefore not directly involved in neural superposition. For the purposes of this discussion of neural superposition, we will concentrate on R1–R6. As previously described for a neural superposition compound eye, each individual photoreceptor (R1–R6) is connected in the lamina with five other common view photoreceptors from adjacent ommatidia. This is depicted in the cross-sectional view of six adjacent ommatidia on the far-left side of Figure 1.18. Pick carefully studied this arrangement and reported that the axis of the receptors converge at a distance of 3–6 mm in front of the corneal surface. This provides for a substantial gain in light sensitivity. It also results in a slight blurring of the image, but, as Pick noted, it provides the fly with the ability to determine the distance from an object by sensing the imbalance of response from the photoreceptor grouping [75]. The resulting overlapped Gaussian profile responses are depicted in Figure 1.18.

Photoreceptor signals from R1–R6 are then combined and processed in the lamina by monopolar cells L1 and L2 [76]. There is some disagreement concerning the function of the L1 and L2 cells. Some hypothesize the cells could

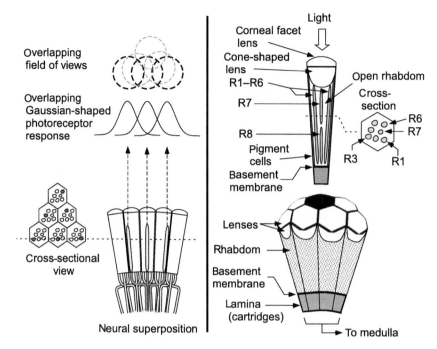

FIGURE 1.18 Schematic diagrams of ommatidial structure and the Gaussian-shaped photoreceptor sensitivity. Monopolar cells such as L1, L2, and L4 exist in the lamina.

be encoding signals with position information [77, 78]; others have concluded that the responses of L1 and L2 to motion and flicker are indistinguishable at equivalent contrast frequencies and are therefore not motion-specific [79]. Juusola concluded that the large monopolar cells contribute to the light adaption mechanisms previously described [74]. Although the exact roles of the L1 and L2 monopolar cells remain elusive, there is a rich body of work contained in the literature to inspire sensor feature development.

Another monopolar neuron in the lamina, designated L4, is connected distally to the L1 and L2 monopolar cells. The parent cartridge L4 cell is also proximally connected to two neighboring cartridges. However, Strausfeld has shown that L4 meets components of six other L4 cells from six neighboring cartridges. This interconnected network of L4 cells

provides a regular and orderly array of cellular interconnections across the fly's eye. Many of the vision processes previously described (e.g., motion, flow, lateral inhibition, etc.) require connections to adjacent ommatidia. The L4 network provides such a mechanism for these connections [79].

Nakayama defined hyperacuity as visual localization better than that which would be predicted purely by photoreceptor spacing [43]. The common housefly, equipped with a neural signal from each photoreceptor that projects into a cartridge, shares a common visual axis and thus views an overlapped sample of the same point in space [75, 77, 78, 80]. As previously mentioned, Pick reported that photoreceptors do not share precisely the same visual axis and are slightly misaligned with one another [75]. Some have hypothesized that the fly would not maintain such a misalignment if the eye was not enjoying some advantage with the disparate photoreceptor axes [77, 78]. Though the photoreceptor response in *M. domestica* is Gaussian-shaped, hyperacuity may also be achieved with other continuous, nonlinear functions.

Luke *et al.* studied the requirements for motion hyperacuity in detail. *Motion hyperacuity* is defined as the ability of an imaging system to detect object motion at a much finer resolution than photoreceptor spacing would suggest. It should not be confused with *static hyperacuity* such as the subpixel resolution of line pairs. It is interesting to note that the fly has relatively poor resolution in terms of static hyperacuity of a random, unknown visual scene [40].

A traditional imaging system is compared to a motion hyperacuity–capable system in Figure 1.19. A traditional imaging system, shown in only the x direction, consists of a series of pixel elements with a rectangular function response profile. To detect motion (Δx), sufficient movement must occur such that there is a detectable change in pixel response. With the flat, rectangular profile, considerable movement relative to the pixel width must occur before a detectable response occurs. On the other hand, a motion hyperacuity system has overlapping Gaussian-shaped profiles that detect movement immediately in the primary pixel and in those adjacent to it. The limit of detectability is the noise floor of the system, or the contrast limit. Luke *et al.* have shown that careful control of both preblurring of each pixel and the pixel spacing can optimize motion-detection capability of a biomimetic vision sensor [40].

1.5 BIOMIMETIC VISION SENSOR DEVELOPMENTS

A brief review of some past and recent developments in biomimetic vision sensors is presented in this section. Pointers to the literature are provided to allow readers to obtain additional information on specific projects.

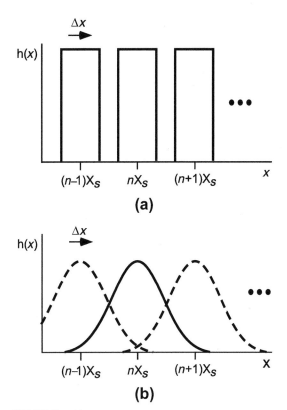

FIGURE 1.19 Traditional pixel imaging (a) compared to a motion hyperacuity-capable imaging system (b). Adapted from Luke *et al.* [40].

Many sensors based on the compound eye have been developed. In 1979, Angel developed an imaging system for an X-ray telescope based on the eye of macruran crustaceans such as the lobster. The lobster eye consists of a series of small, rectangular-shaped tubes with reflective internal surfaces. The tubes are arranged on a spherical surface with their axes radiating from the center of the eye. The reflective nature of the tubes coupled with their orientation produce a focusing effect identical to a reflective-type telescope [81].

Several sensors have been developed based on the apposition compound eye. Currin developed a point-tracking system using three gradient index (GRIN) lenses with overlapping fields of view. The system was designated the Multi-Aperture Vision System (MAVS) [82].

Bruckner *et al.* [83] developed an array of sensors based on the apposition compound eye. In their design, an array of pinhole aperture photosensors was coupled with a microlens array in a layered arrangement. The pitch of the two arrays was dissimilar, which enabled different viewing directions of the separate optical channels. The Gaussian overlapped response of neighboring sensors in the array provided for the localization of a point source with hyperacuity.

Ogata *et al.* [84] also employed a layered sensor technique to develop an apposition-style 8×8 sensor. The sensor consisted of microlens, pinhole, and photodiode arrays in tightly coupled layers. Jeong *et al.* [85] developed techniques to manufacture a biologically inspired three-dimensional, spherical compound eye employing microlens technology. They noted that a light from a distant point source impinging on the omnidirectional array would have a different coupling efficiency with each ommatidium [85].

Tanida *et al.* [86–88] developed a compact image-capturing system designated TOMBO (Thin Observation Module for Bound Optics). The TOMBO system employs compound-eye imaging optics to capture a series of rendered images to obtain the object image.

Hoshino *et al.* [89] developed an insect-inspired retina chip that integrates a microlens array, a photodiode array, and an electrostatically driven scanning slit. This system can image a contrast grating with high temporal resolution.

Considerable work has been devoted to the development of a sensor based on neural superposition compound eyes [37, 39, 40]. In neural superposition eyes, the overlapping Gaussian photoreceptor acceptance profiles provide for motion hyperacuity. To achieve the overlapped response, a variety of optical configurations have been employed, including optical fibers equipped with ball lens, off-the-shelf photodiodes, and optical fibers equipped with small lensets [40]. Each of these configurations is depicted schematically in Figure 1.20, with the corresponding physical sensor prototypes shown in Figure 1.21. Wilcox *et al.* developed a VLSI-based array of neural superposition sensors that demonstrated hyperacuity [77, 78].

Several research groups have developed a number of bio-inspired processors that provide for optic flow and aerial vehicle navigation. Harrison *et al.* compiled a noteworthy body of work based on the fly's flow-field processes. After studying the fly's system in detail, Harrison rendered silicon-chip flow-field generators. He developed a single-chip analog VLSI sensor that detects imminent vehicle collisions by measuring radially expanding optic flow based on the delay-and-correlate scheme similar to that first proposed by Reichardt [90, 91]. Pudas *et al.* also developed a bio-inspired optic flow-field sensor based on low-temperature co-fired ceramics (LTCC) technology. The process provides reliable, small-profile optic-flow sensors that are largely invariant to both contrast and spatial frequency [92].

Netter and Franceschini have demonstrated the ability to control a model unmanned aerial vehicle (UAV) with biologically inspired optical flow processes [93], based on earlier work by Aubépart and Franceschini [48]. A model UAV was equipped with a 20-photoreceptor linear array. The photoreceptor outputs were processed by 19 analog elementary motion detectors (EMDs). Each of the EMDs detects motion in a particular direction within a limited field of view. The overall output from the EMD is a pulse of which the voltage is proportional to the detected speed. Terrain-following capability was achieved in the model setup by varying thrust such that the measured optical flow was adjusted to the reference optical flow. Other approaches to hardware sensors that are specifically sensitive to optical flow have been designed by researchers such as Chahl and Mizutani [49], discussed further in Chapter 9.

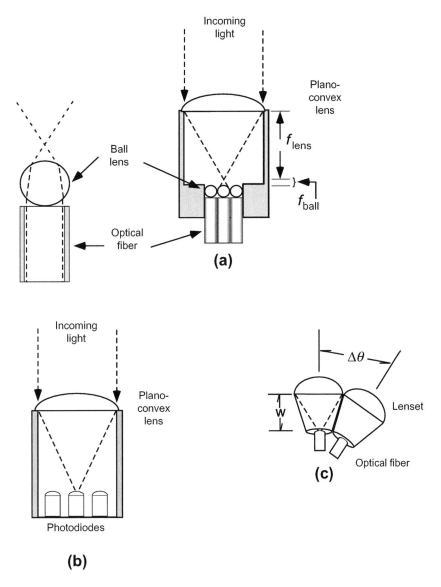

FIGURE 1.20 Optical front-end schematics of neural-superposition-based sensors capable of motion hyperacuity. Adapted from Ref. 40.

O'Carroll and Brinkworth developed a detailed model for optic flow coding based on the insect vision system. The motion detector employed in the project is based on the Hassenstein–Reichardt algorithm. They tested the model under a variety of conditions and concluded that accurate and robust detection of global motion is possible using low-resolution optics and simple mathematical operations that may be implemented in digital or analog hardware [51].

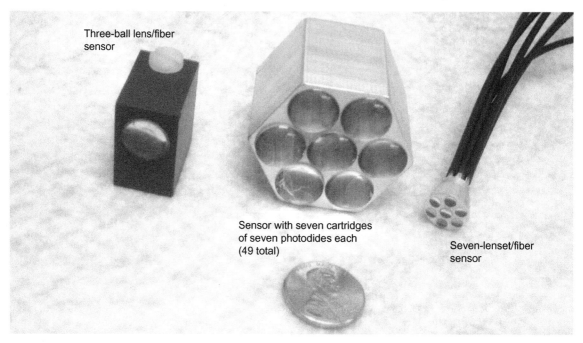

FIGURE 1.21 Prototypes of neural-superposition-based sensors capable of motion hyperacuity. Adapted from Ref. 40.

1.6 CONCLUDING REMARKS

Biomimetic vision sensors are inspired by what we have learned about various methods of vision that have evolved in the animal kingdom, and they may be employed in a wide variety of applications. Common configurations of digital cameras and other traditional vision sensors were shown to be based on the mammalian camera eye. Less-traditional biomimetic sensors have been based on the compound-eye configurations found in many insects, and these insect-eye sensors can provide capabilities not matched by traditional vision sensors. Research is ongoing to employ fly-eye types of biomimetic vision systems in small unmanned aerial systems, aerial system obstacle avoidance, and high-speed inspection in the manufacturing and transportation industries. Also, they have been employed in robotic applications as discussed in Chapters 5 and 9. Although they are not, in general, intended to supplant traditional imaging systems and vision sensors, they can provide enhanced results for specific applications. Thus, the greatest anticipated use of nontraditional biomimetic sensors is in a hybrid design, where both camera-eye and compound-eye designs can complement each other.

References

[1] M.F. Land and D. Nilsson, *Animal eyes*, Oxford University Press, New York, NY, USA (2002).
[2] D. Marr, *Vision*, W.H. Freeman, New York, NY, USA (1982).
[3] R. Szema and L.P. Lee, Biologically inspired optical systems, in *Biomimetics: biologically inspired technologies*, (Y. Bar-Cohen, ed.), CRC Press, Boca Raton, FL (2006), 291–308.
[4] Y. Bar-Cohen (ed.), *Biomimetics: biologically inspired technologies*, CRC Press, Boca Raton, FL, USA (2006).
[5] Y. Bar-Cohen and C. Breazeal, *Biologically inspired intelligent robots*, SPIE Press, Bellingham, WA, USA (2003).

[6] K. Kirschfeld, The resolution of lens and compound eyes, in *Neural principles in vision* (F. Zettler and R. Weiler, eds.), Springer-Verlag, New York, NY, USA (1976), 354–370.

[7] E. Hecht, *Optics*, 4th ed., Addison-Wesley, San Francisco, CA, USA (2002).

[8] W.J. Smith, *Modern optical engineering*, 2nd ed., McGraw-Hill, New York, NY, USA (1990).

[9] C.L. Wyatt, *Electro-optical system design*, McGraw-Hill, New York, NY, USA (1991).

[10] J.W. Goodman, *Introduction to Fourier optics*, 2nd ed., McGraw-Hill, New York, NY, USA (1996).

[11] R.G. Wilson, *Fourier series and optical transform techniques in contemporary optics*, Wiley, New York, NY, USA (1995).

[12] G.C. Holst, Imaging system performance based upon $F\lambda/d$, *Opt Eng* **46** (2007), 103204.

[13] G.C. Holst, Imaging system fundamentals, *Opt Eng* **50** (2011), 052601.

[14] J. Nakamura, *Image sensors and signal processing for digital still cameras*, CRC Press, Boca Raton, FL, USA (2006).

[15] R.H. Vollmerhausen, D.A. Reago, Jr., and R.G. Driggers, *Analysis and evaluation of sampled imaging systems*, SPIE Press, Bellingham, WA, USA (2010).

[16] G.C. Holst, *Electro-optical imaging system performance*, 3rd ed., SPIE Press, Bellingham, WA, USA (2003).

[17] B.E.A. Saleh and M.C. Teich, *Fundamentals of photonics*, Wiley, New York, NY, USA (1991).

[18] R.C. Gonzalez and R.E. Woods, *Digital image processing*, 3rd ed., Prentice-Hall, Upper Saddle River, NJ, USA (2008).

[19] R.N. Bracewell, *The Fourier transform and its applications*, 3rd ed., McGraw-Hill, New York, NY, USA (2000).

[20] N. van Hulst, Many photons get more out of diffraction, *OSA optics and photonics focus*, vol. 4, Story 1. http://www.opfocus.org/index.php?topic=volume&v=4 (accessed 10 April 2013).

[21] M. Bass, E.W. van Stryland, D.R. Williams, and W.L. Wolfe (eds.), *Handbook of optics*, 2nd ed., vol. 1, McGraw-Hill, New York, NY, USA (1995).

[22] V. Ronchi, *Optics: the science of vision*, Dover, New York, NY, USA (1991).

[23] M. Zghal, H.-E. Bouali, Z.B. Lakhdar, and H. Hamam, The first steps for learning optics: Ibn Sahl's, Al-Haytham's and Young's works on refraction as typical examples, *Proceedings of the 10th SPIE conference on education and training in optics and photonics*. http://spie.org/etop/etop2007.html (accessed 10 April 2013).

[24] J.W. Coltman, The specification of imaging properties by response to a sine wave target, *J Opt Soc Am* **44** (1954), 468–471.

[25] A.D. Ducharme and S.P. Temple, Improved aperture for modulation transfer function measurement of detector arrays beyond the Nyquist frequency, *Opt Eng* **47** (2008), 093601.

[26] M. Unser, Sampling—50 years after Shannon, *Proc IEEE* **88** (2000), 569–587.

[27] C.E. Shannon, A mathematical theory of communication, *Bell Syst Tech J* **27** (1948), 623–656.

[28] M.F. Land and R.D. Fernald, The evolution of eyes, *Annu Rev Neurosci* **15** (1992), 1–29.

[29] W.F. Ganong, *Review of medical physiology*, 16th ed., Appleton & Lange, Norwalk, CT, USA (1993).

[30] D.G. Vaughan, T. Asbury, and P. Riordan-Eva, *General ophthalmology*, 13th ed., Appleton & Lange, Norwalk, CT, USA (1992).

[31] Y. LeGrand and S.G. El Hage, *Physiological optics*, Springer-Verlag, New York, NY, USA (1980).

[32] R. Sekuler and R. Blake, *Perception*, 3rd ed., McGraw-Hill, New York, NY, USA (1994).

[33] H.A. Quigley, A.E. Brown, J.D. Morrison, and S.M. Drance, The size and shape of the optic disk in normal human eyes, *Arch Ophthalmol* **108** (1990), 51–57.

[34] A.M. Mansour, Measuring fundus landmarks, *Invest Ophthalmol Vis Sci* **31** (1990), 41–42.

[35] N. Drasdo and C.W. Fowler, Non-linear projection of the retinal image in a wide-angle schematic eye, *Brit J Ophthalmol* **58** (1974), 709–714.

[36] M.F. Land, Visual acuity in insects, *Annu Rev Entomol* **42** (1997), 147–177.

[37] D.T. Riley, W.M. Harmann, S.F. Barrett, and C.H.G. Wright, *Musca domestica* inspired machine vision sensor with hyperacuity, *Bioinsp Biomim* **3** (2008), 026003.

[38] J.D. Davis, S.F. Barrett, C.H.G. Wright, and M.J. Wilcox, A bio-inspired apposition compound eye machine vision sensor system, *Bioinsp Biomim* **4** (2009), 046002.

[39] R.S. Prabhakara, C.H.G. Wright, and S.F. Barrett, Motion detection: a biomimetic vision sensor versus a CCD camera sensor, *IEEE Sens J* **12** (2012), 298–307.

[40] G.P. Luke, C.H.G. Wright, and S.F. Barrett, A multi-aperture bio-inspired sensor with hyperacuity, *IEEE Sens J* **12** (2012), 308–314.

[41] G.A. Mazokhin-Porshnyakov, *Insect vision*, Plenum Press, New York, NY, USA (1969).

[42] M.F. Land, Optics and vision in invertebrates, in *Handbook of sensory physiology*, vol. VII/6B (H. Autrum, ed.), Springer-Verlag, New York, NY, USA (1981), 472–592.

[43] K. Nakayama, Biological image motion processing: a review, *Vision Res* **25** (1985), 625–660.

[44] H.B. Barlow, Summation and inhibition in the frog's retina, *J Physiol* **199** (1953), 69–88.

[45] J.Y. Lettvin, R. Maturana, W.S. McCulloch, and W.H. Pitts, What the frog's eye tells the frog's brain, *Proc Inst Radiat Eng* **47** (1959), 1940–1951.

[46] D.H. Hubel and T.N. Weisel, Brain mechanisms of vision, *Sci Am* **241** (3) (March 1979), 150–162.

[47] K. Nakayama and J.M. Loomis, Optical velocity patterns velocity sensitive neurons, and space perception: a hypothesis, *Perception* **3** (1974), 63–80.

[48] F. Aubépart and N. Franceschini, Bio-inspired optic flow sensors based on FPGA: application to micro-air-vehicles, *Microproc Microsys* **31** (2007), 408–419.

[49] J. Chahl and A. Mizutani, Biomimetic attitude and orientation sensors, *IEEE Sens J* **12** (2012), 289–297.

[50] D.C. O'Carroll, Feature-detecting neurons in dragon-flies, *Nature* **362** (1993), 541–543.

[51] R.S.A. Brinkworth and D.C. O'Carroll, Robust models for optic flow coding in natural scenes inspired by insect biology, *PLoS Comput Biol* **5** (2009), 1–14.

[52] J.G. Daugman, Uncertainty relation for resolution in space, spatial frequency, and orientation optimized by two-dimensional visual cortical filters, *J. Opt. Soc. Am. A* **2** (1985), 1160–1169.

[53] J.G. Daugman, Complete discrete 2-D Gabor transforms by neural networks for image analysis and compression, *IEEE Trans Acoust Speech Signal Process* **36** (1988), 1169–1179.

[54] E.H. Adelson and J.R. Bergen, Spatiotemporal energy models for the perception of motion, *J Opt Soc Am A* **2** (1985), 284–299.

[55] A. Borst, Models of motion detection, *Nat Neurosci Suppl* **3** (2000), 1168.

[56] J. Haag, W. Denk, and A. Borst, Fly motion vision is based on Reichardt detectors regardless of the signal-to-noise ratio, *Proc Natl Acad Sci* **101** (2004), 16333–16338.

[57] A. Borst and M. Egelhaaf, Principles of visual motion detection, *Trends Neurosci* **12** (1989), 297–306.

[58] S.B. Laughlin and R.C. Hardie, Common strategies for light adaptation in the peripheral visual systems of fly and dragonfly, *J Comp Physiol* **128** (1978), 319–340.

[59] B.K. Dean, C.H.G. Wright, and S.F. Barrett, The design of an analog module for sensor adaptation to changes in ambient light, *ISA Biomed Sci Intrum* **45** (2009), 185–190.

[60] B.K. Dean, C.H.G. Wright, and S.F. Barrett, Advances in sensor adaptation to changes in ambient light: a bio-inspired solution, *ISA Biomed Sci Intrum* **46** (2010), 20–25.

[61] B.K. Dean, C.H.G. Wright, and S.F. Barrett, Preliminary tests of a possible outdoor light adaptation solution of a fly inspired visual sensor: a biomimetic solution, *ISA Biomed Sci Intrum* **47** (2011), 147–152.

[62] H.K. Hartline, H.G. Wagner, and F. Ratliff, Inhibition in the eye of the limulus, *J Gen Physiol* **39** (1956), 651–673.

[63] K.H. Strube, *Musca domestica inspired vision sensor cartride modeling*, Master's thesis, University of Wyoming (2010).

[64] N. Petkov, T. Lourens, and P. Kruizinga, Lateral inhibition in cortical filters, *Proceedings of the international conference on digital signal processing and international conference on computer applications to engineering systems* (1993), 122–129.

[65] M.V. Srinivasan, S.W. Zhang, M. Lehrer, and T.S. Collett, Honeybee navigation en route to the goal: visual flight control and odometry, *J Exp Biol* **199** (1996), 237–244.

[66] R. Hardie, Functional organization of the fly retina, in *Progress in sensory physiology*, vol. V (D. Ottoson, ed.), Springer-Verlag, New York, NY, USA (1985), 1–79.

[67] V. Braitenberg and P. Debbage, A regular net of reciprocal synapses in the visual system of the fly *Musca domestica*, *J Comput Physiol* **90** (1974), 25–31.

[68] V. Braitenberg, Patterns of projection in the visual system of the fly: I. retina-lamina projections, *Exp Brain Res* **3** (1967), 271–298.

[69] A.W. Snyder, Acuity of compound eyes: physical limitations and design, *J Comp Physiol* **116** (1977), 161–182.

[70] A.W. Snyder, D.G. Stavenga, and S.B. Laughlin, Spatial information capacity of compound eyes, *J Comp Physiol* **116** (1977), 183–207.

[71] A.W. Snyder, Physics of vision in compound eyes, in *Handbook of sensory physiology*, vol. VII/6A (H. Autrum, ed.), Springer-Verlag, New York, NY, USA (1979), 225–313.

[72] L.A. Popp, E.S. Tomberlin, S.F. Barrett, and C.H.G. Wright, *Musca domestica* lamina monopolar cell response to visual stimuli and their contribution to early motion detection, *ISA Biomed Sci Intrum* **43** (2007), 134–139.

[73] E.S. Tomberlin, *Musca domestica's large monopolar cell responses to visual stimuli*, Master's thesis, University of Wyoming (2004).

[74] M. Juusola and A.S. French, Visual acuity for moving objects in first- and second-order neurons of the fly compound eye, *J Neurophysiol* **77** (1997), 1487–1495.

[75] B. Pick, Specific misalignments of rhabdomere visual axes in the neural superposition eye of dipteran flies, *Biol Cybern* **26** (1977), 215–224.

[76] V. Braitenberg and H. Hauser-Holschuh, Patterns of projection in the visual system of the fly: II. Quantitative aspects of second-order neurons in relation to models of movement perception, *Exp Brain Res* **16** (1972), 184–209.

[77] M. Wilcox, D. Thelen Jr., G. Peake, S. Hersee, K. Scott, and C. Abdallah, An analog model of the retinal information processing in the eye of the fly, *Proceedings of the 6th NASA symposium on VLSI design* (1997), 3.4.1–3.4.15.

[78] M.J. Wilcox and D.C. Thelen Jr., A retina with parallel input and pulsed output, extracting high-resolution information, *IEEE Trans Neural Networks* **10** (1999), 574–583.

[79] N. Strausfeld and J. Campos-Ortega, The L4 monopolar neuron: a substrate for lateral interaction in the visual system of the fly, *Brain Res* **59** (1973), 97–117.

[80] D.W. Arnett, Spatial and temporal integration properties of units in first optic ganglion of dipterans, *J Neurophysiol* **35** (1972), 429–444.

[81] J. Angel, Lobster eyes as x-ray telescopes, *Astrophys J* **233** (1979), 364–373.

[82] M.S. Currin, *Design aspects of multi-apertures vision system point trackers that use apposition eyelets*, Master's thesis, University of Memphis (1994).

[83] A. Bruckner, J. Duparre, and A. Brauer, Artificial compound eye applying hyperacuity, *Opt Express* **14** (2006), 12076–12084.

[84] S. Ogata, J. Ishida, and T. Sasano, Optical sensor array in an artificial compound eye, *Opt Eng* **33** (1994), 3649–3655.

[85] K.-H. Jeong, J. Kim, and L.P. Lee, Biologically inspired artificial compound eyes, *Science* **312** (2006), 557–561.

[86] J. Tanida, T. Kumagai, K. Yamada, S. Miyatake, K. Ishida, T. Morimoto, N. Kondou, D. Miyazaki, and Y. Ichioka, Thin observation module by bound optics (TOMBO), *Appl Opt* **40** (2001), 1806–1813.

[87] J. Tunida, Color imaging with an integrated compound imaging system, *Opt Express* **11** (2003), 2109–2117.

[88] R. Horisaki, S. Irie, Y. Ogura, and J. Tanida, Three-dimensional information acquisition using a compound imaging system, *Opt Rev* **14** (2007), 347–350.

[89] K. Hoshino, F. Mura, and I. Shimoyama, A one-chip scanning retina with an integrated micro-mechanical scanning actuator, *J Micromech Syst* **10** (2001), 492–497.

[90] R.R. Harrison and C. Koch, A robust analog VLSI Reichardt motion sensor, *Analog Integr Circuits Signal Process* **24** (2000), 213–229.

[91] R.R. Harrison, A biologically inspired analog IC for visual collision detection, *IEEE Trans Circuits Sys I* **52** (2005), 2308–2318.

[92] M. Pudas, S. Viollet, F. Ruffier, A. Kruusing, S. Amic, S. Leppavuori, and N. Franceschini, A miniature bio-inspired optic flow sensor based on low temperature co-fired ceramics (LTCC) technology, *Sens Actuat A: Phys* **133** (2007), 88–95.

[93] T. Netter and N. Franceschini, A robotic aircraft that follows terrain using a neuromorphic eye, *Proceedings of the 2002 IEEE/RSJ international conference of intelligent robots and systems* (October 2002), 129–134.

ABOUT THE AUTHORS

Cameron H. G. Wright, PhD, PE is Associate Professor and Associate Department Head, Department of Electrical and Computer Engineering, University of Wyoming, Laramie, WY. He was previously Professor and Deputy Department Head, Department of Electrical Engineering, United States Air Force Academy, and served as an R&D engineering officer in the U.S. Air Force for over 20 years. He received the BSEE (summa cum laude) from Louisiana Tech University (1983), the MSEE from Purdue University (1988), and PhD from the University of Texas at Austin (1996). His research interests include signal and image processing, real-time embedded computer systems, biomedical instrumentation, and engineering education. He is a member of ASEE, IEEE, SPIE, BMES, NSPE, Tau Beta Pi, and Eta Kappa Nu, and is a registered PE in Wyoming and California.

Steven F. Barrett, PhD, PE, BS, Electronic Engineering Technology, University of Nebraska at Omaha, 1979; MEEE, University of Idaho at Moscow, 1986; PhD, University of Texas at Austin, 1993. He serves as Professor of Electrical and Computer Engineering and Associate Dean for Academic Programs, University of Wyoming. He is a member of IEEE (senior) and Tau Beta Pi (chief faculty advisor, WY A). His research interests include image processing, computer-assisted laser surgery, and embedded systems. He is a registered PE in Wyoming and Colorado. He co-wrote with Dr. Daniel Pack several textbooks on microcontrollers and embedded systems.

CHAPTER

2

Noise Exploitation and Adaptation in Neuromorphic Sensors

Thamira Hindo and Shantanu Chakrabartty

Department of Electrical and Computer Engineering, Michigan State University, East Lansing, MI 48824, USA

Prospectus

Even though current micro-nano fabrication technology has reached integration levels at which ultra-sensitive sensors can be fabricated, the sensing performance (bits per Joule) of synthetic systems are still orders of magnitude inferior to those observed in neurobiology. For example, the filiform hair in crickets operates at fundamental limits of noise and energy efficiency. Another example is the auditory sensor in the parasitoid fly *Ormia ochracea* that can precisely localize ultra-faint acoustic signatures in spite of the underlying physical limitations. Even though many of these biological marvels have served as inspirations for different types of neuromorphic sensors, the main focus of these designs has been to faithfully replicate the biological functions, without considering the constructive role of *noise*. In manmade sensors, device and sensor noise are typically considered nuisances, whereas in neurobiology *noise* has been shown to be a computational aid that enables sensing and operation at fundamental limits of energy efficiency and performance. In this chapter, we describe some of the important noise exploitation and adaptation principles observed in neurobiology and how they can be systematically used for designing neuromorphic sensors. Our focus is on two types of noise exploitation principles, namely, (a) stochastic resonance and (b) noise shaping, which are unified within a framework

called $\Sigma\Delta$ learning. As a case study, we describe the application of $\Sigma\Delta$ learning for the design of a miniature acoustic source localizer, the performance of which matches that of its biological counterpart (*O. ochracea*).

Keywords

Acoustic sensors, Adaptation, Localization, Neuromorphic sensors, Neurons, Noise exploitation, Noise shaping, Signal-to-noise ratio, Spike-time-dependent plasticity (STDP), Stochastic resonance, Synapse

2.1 INTRODUCTION

Over the last decade significant research effort has been expended in designing systems inspired by biology. Neuromorphic engineering constitutes one such discipline in which the objective has been to design sensors and systems that mimic or model the physical principles observed in neurobiology [1–3]. The key motivation behind this effort has been to reduce the performance gap that exists between neurobiological sensors and their synthetic counterparts. For instance, it has been known that biology, in spite of its physical and fundamental limitations,

Engineered Biomimicry

http://dx.doi.org/10.1016/B978-0-12-415995-2.00002-7

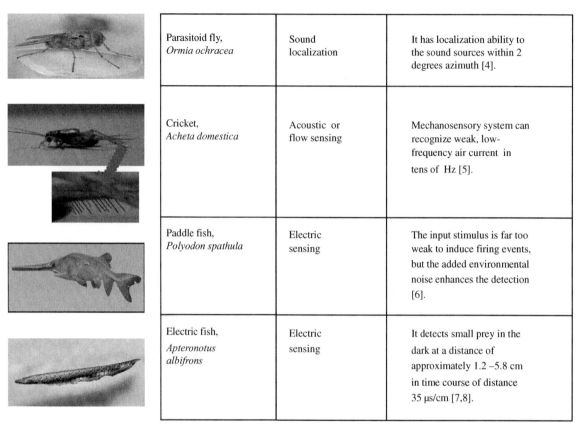

Parasitoid fly, *Ormia ochracea*	Sound localization	It has localization ability to the sound sources within 2 degrees azimuth [4].	
Cricket, *Acheta domestica*	Acoustic or flow sensing	Mechanosensory system can recognize weak, low-frequency air current in tens of Hz [5].	
Paddle fish, *Polyodon spathula*	Electric sensing	The input stimulus is far too weak to induce firing events, but the added environmental noise enhances the detection [6].	
Electric fish, *Apteronotus albifrons*	Electric sensing	It detects small prey in the dark at a distance of approximately 1.2–5.8 cm in time course of distance 35 µs/cm [7,8].	

FIGURE 2.1 Efficient sensing in organisms of sound localization, flow sensing, and electric sensing. (Images from Wikipedia and Encyclopedia of Life.)

can overcome constraints imposed by noise and environmental artifacts to achieve remarkable performance (illustrated in Figure 2.1). For example, the parasitoid fly (*Ormia ochracea*) can localize the sound of crickets with a remarkable accuracy of less than 2°. Even though the fly's eardrums are spaced less than 0.5 mm apart, it can successfully overcome the physical constraints due to sound propagation and resolve localization cues of less than 50 ns [4]. Another marvel of biology is the filiform hair in the cricket (*Acheta domestica*) that exhibits exquisite vibration-sensing capabilities. These mechanoreceptive sensory hairs have been shown to be capable of sensing vibrations as low as tens of

µm/s flow speed, even in the presence of large ambient noise [5]. Yet another example is the electro sensory receptors in the paddlefish (*Polyodon spathula*) that the fish uses to sense and localize plankton, even in the presence of ambient aquatic turbulence [6]. A similar mechanism is also found in the electric fish (*Apteronotus albifrons*) where the performance of the receptors has been shown to supersede the state-of-the-art manmade impedance spectroscopy systems. Using neurobiological techniques, the organs of the electric fish can monitor microvolt-level voltage perturbations caused by surrounding objects and detect small prey at distances greater than 25 mm [7, 8].

To date neuromorphic systems have attempted to mimic the functionalities of these sensory organs, among many other neurobiological structures, such as the cochlea and retina [9]. Except for a few cases, the performance of most neuromorphic architectures falls far short of that of their biological counterparts in terms of sensing capabilities and energy efficiency. One of the primary reasons for this performance gap could be that the design principles focus mainly and only on enhancing the system's signal-to-noise ratio (SNR) by alleviating system artifacts like *noise*, nonlinearity, and sensor imperfections. In biology, however, *noise* and nonlinearity play a constructive role where by sensing and signal detection are in fact enhanced due to these system artifacts. For instance, it has been shown that the ability of the paddlefish (*P. spathula*) to localize and capture plankton is significantly enhanced when different amplitudes of random noise are intentionally added to its environment [6]. In biology, learning and adaptation also play key roled in noise exploitation by shaping the system signal and system noise in the frequency domain.

The purpose of this chapter is to describe some of the important noise exploitation and adaptation principles observed in neurobiology and show how they can be used for designing neuromorphic sensors. The chapter is organized as follows: Section 2.2 briefly introduces the organization of a typical neurobiological sensory system and includes a brief overview of the structure of a neuron, synapses, and different types of neural coding. Section 2.3 provides an overview of two noise exploitation principles: (a) stochastic resonance (SR) and (b) noise shaping. Section 2.4 describes the adaptation mechanisms in neural systems with a focus on plasticity and learning. Section 2.5 presents a case study of a neuromorphic acoustic source sensor and localizer that emulates the neurobiological principles observed in the parasitoid fly (*Ormia ochracea*). The chapter concludes in Section 2.7 by discussing open problems and challenges in this area.

2.2 ORGANIZATION OF NEUROBIOLOGICAL SENSORY SYSTEMS

The typical structure of a neurobiological sensory system is shown in Figure 2.2. The system consists of an array of sensors (mechanoreceptors, optical, or auditory) that are directly coupled to a group of sensory neurons, also referred to as afferent neurons. Depending on the type of sensory system, the sensors (skin, hair, retina, cochlea) convert input stimulus such as sound, mechanical, temperature, or pressure into electric stimuli. Each of the afferent neurons could potentially receive electrical stimuli from multiple sensors (as shown in Figure 2.2), an organization that is commonly referred to as the sensory *receptive field*.

For example, in the electric fish, the electrosense receptors distributed on the skin detect a disruption in the electric field (generated by the fish itself) that corresponds to the movement and identification of the prey. The receptive field in this case corresponds to electrical intensity spots that are then encoded by the afferent neurons using spike trains [10].

The neurons are connected with each other through specialized junctions known as *synapses*. While the neurons (afferent or non-afferent) form the core signal-processing unit of the sensory system, the synapses are responsible for adaptation by modulating the strength of the connection between two neurons. The dendrites of the neurons transmit and receive electrical signals to and from other neurons, and the soma receives and integrates the electrical stimuli. The axon, which is an extension of the soma, transmits the generated signals or spikes to other neurons and higher layers.

The underlying mechanism of a spike or action-potential generation is due to unbalanced movement of ions across a membrane, as shown in Figure 2.3, which alters the potential difference between the inside and the outside of the neuron. In the absence of any stimuli to the

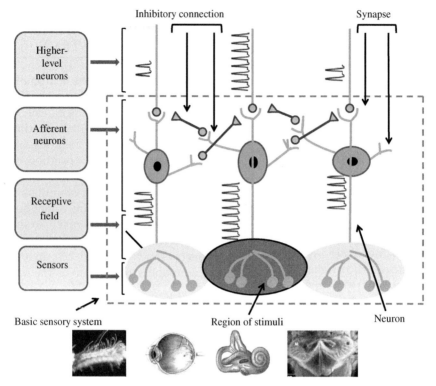

FIGURE 2.2 Organization of a generic neurobiological sensory system. Images adapted from Wikipedia and Ref. 11.

neuron, the potential inside the membrane with respect to the potential outside the membrane is about −65 mV, also referred to as the *resting potential*. This potential is increased by the influx of sodium ions (Na^+) inside the cell, causing depolarization, whereas the potential is decreased by the efflux of potassium ions (K^+) outside the cell, causing hyper polarization Once the action potential is generated, the Na^+ ion channels are unable to reopen immediately until a built-up potential is formed across the membrane. The delay in reopening the sodium channels results in a time period called the *refractory period*, as shown in Figure 2.3, during which the neuron cannot spike.

The network of afferent spiking neurons can be viewed as an analog-to-digital converter, where the network faithfully encodes different features of the input analog sensory stimuli using

a train of spikes (that can be viewed as a binary sequence). Note that the organization of the receptive field introduces significant redundancy in the firing patterns produced by the afferent neurons. At the lower level of processing, this redundancy makes the encoding robust to noise, but as the spike trains are propagated to higher processing layers this redundancy leads to degradation in energy efficiency. Therefore, the network of afferent neurons self-optimizes and adapts to the statistics of the input stimuli using inhibitory synaptic connections.

The process of inhibition (among the same layer of neurons) is referred to as *lateral inhibition,* where by the objective is to optimize (reduce) the spiking rate of the network while faithfully capturing the information embedded in the receptive field. This idea is illustrated in Figure 2.2, where the afferent neural network emphasizes

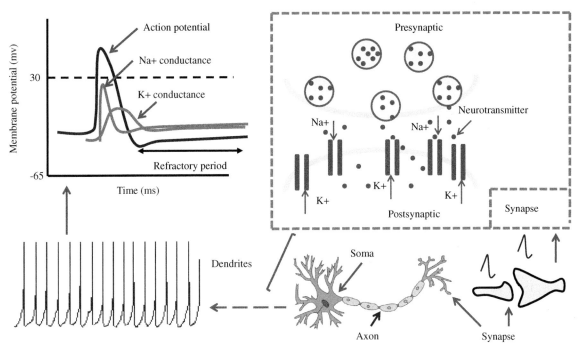

FIGURE 2.3 Mechanism of spike generation and signal propagation through synapses and neurons. (Images from Wikipedia.)

the discriminatory information present in the input spike trains while inhibiting the rest. This not only reduces the rate of spike generation at the higher layer of the receptive field (leading to improved energy efficiency), but it also optimizes the information transfer that facilitates real-time recognition and motor operation. Indeed, later inhibition and synaptic adaptation are related to the concept of noise shaping. Before we discuss the role of noise in neurobiological sensory systems, let us introduce some mathematical models that are commonly used to capture the dynamics of spike generation and spike-based information encoding.

2.2.1 Spiking Models of Neuron and Neural Coding

As a convention, the neuron transmitting or generating a spike and incident onto a synapse

is referred as the *presynaptic neuron*, whereas the neuron receiving the spike from the synapse is referred as the *postsynaptic neuron* (see Figure 2.3). Also, there are two types of synapses typically encountered in neurobiology: excitatory synapses and inhibitory synapses. For excitatory synapses, the membrane potential of the *postsynaptic neuron* (referred to as the *excitatory postsynaptic potential*, or EPSP) increases, whereas for inhibitory synapses, the membrane potential of the post-synaptic neuron (referred to as the *inhibitory postsynaptic potential*, or IPSP) decreases.

It is important to note that the underlying dynamics of EPSP, IPSP, and the action potential are complex and several texts have been dedicated to discuss the underlying mathematics [12]. Therefore, for the sake of brevity, we only describe a simple integrate-and-fire neuron model that has been extensively used for the design of neuromorphic sensors [9] and is sufficient to explain the noise exploitation techniques described in this chapter.

We first define a spike train $\rho(t)$ using a sequence of time-shifted Kronecker delta functions as

$$\rho(t) = \sum_{m=1}^{\infty} \delta(t - t_m), \qquad (2.1)$$

where $\delta(t)=0$ for $t \neq 0$ and $\int_{-\infty}^{+\infty} \delta(\tau)d\tau = 1$. In the above Eq. (2.1), the spike is generated when t is equal to the firing time of the neuron t_m. If the somatic (or membrane) potential of the neuron is denoted by $v(t)$, then the dynamics of the integrate-and-fire model can be summarized using the following first-order differential equation:

$$\frac{d}{dt}v(t) = -v(t)/\tau_m - \sum_{j=1}^{N} W_j[h(t)*\rho_j(t)] + x(t), \qquad (2.2)$$

where N denotes the number of presynaptic neurons, W_j is a scalar transconductance representing the strength of the synaptic connection between the jth presynaptic neuron and the postsynaptic neuron, τ_m is the time constant that determines the maximum firing rate, $h(t)$ is a presynaptic filtering function that filters the spike train $\rho_j(t)$ before it is integrated at the soma, and $*$ denotes a convolution operator. The variable $x(t)$ in Eq. (2.2) denotes an extrinsic contribution to the membrane current, which could be an external stimulation current.

When the membrane potential $v(t)$ reaches a certain threshold, the neuron generates a spike or a train of spikes. Again, different chaotic models have been proposed that can capture different types of spike dynamics. For the sake of brevity, specific details of the dynamical models can be found in Ref. 13. We next briefly describe different methods by which neuronal spikes encode information.

The simplest form of neural coding is the rate-based encoding [13] that computes the instantaneous spiking rate of the ith neuron $R_i(t)$ according to

$$R_i(t) = \frac{1}{T} \int_{t}^{t+T} \rho_i(t)dt, \qquad (2.3)$$

where $\rho_i(t)$ denotes the spike train generated by the ith neuron and is given by Eq. (2.1), and T is the observation interval over which the integral or spike count is computed. Note that the instantaneous spiking rate $R(t)$ does not capture any information related to the relative phase of the individual spikes, and hence it embeds significant redundancy in encoding. However, at the sensory layer, this redundancy plays a critical role because the stimuli need to be precisely encoded and the encoding have to be robust to the loss or temporal variability of the individual spikes.

Another mechanism by which neurons improve reliability and transmission of spikes is through the use of *bursting*, which refers to trains of repetitive spikes followed by periods of *silence*. This method of encoding has been shown to improve the reliability of information transmission across unreliable synapses [14] and, in some cases, to enhance the SNR of the encoded signal. Modulating the bursting pattern also provides the neuron with more ways to encode different properties of the stimulus. For instance, in the case of the electric fish, a change in bursting signifies a change in the states (or modes) of the input stimuli, which could distinguish different types of prey in the fish's environment [14].

Whether bursting is used or not, the main disadvantage of rate-based encoding is that it is intrinsically slow. The averaging operation in Eq. (2.3) requires that a sufficient number of spikes be generated within T to reliably compute $R_i(t)$. One possible approach to improve the reliability of rate-based encoding is to compute the rate across a population of neurons where each neuron is encoding the same stimuli.

The corresponding rate metric, also known as the population rate $R(t)$, is computed as

$$R(t) = \frac{1}{N} \sum_{i=1}^{N} R_i(t), \qquad (2.4)$$

where N denotes the number of neurons in the population. By using the population rate, the stimuli can now be effectively encoded at a signal-to-noise ratio that is $N^{1/2}$ times higher than that of a single neuron [15].

Unfortunately, even an improvement by a factor of N is not efficient enough to encode fast-varying sensory stimuli in real time. Later, in Section 2.4, we show that lateral inhibition between the neurons would potentially be beneficial to

enhance the SNR of a population code by a factor of N^2 [16] through the use of noise shaping.

We complete the discussion of neural encoding by describing other forms of codes: time-to-first spike, phase encoding, and neural correlations and synchrony. We do not describe the mathematical models for these codes but illustrate the codes using Figure 2.4d.

The time-to-spike is defined as the time difference between the onset of the stimuli and the time when a neuron produces the first spike. The time difference is inversely proportional to the strength of the stimulus and can efficiently encode the real-time stimuli compared to the rate-based code. Time-to-spike code is efficient since

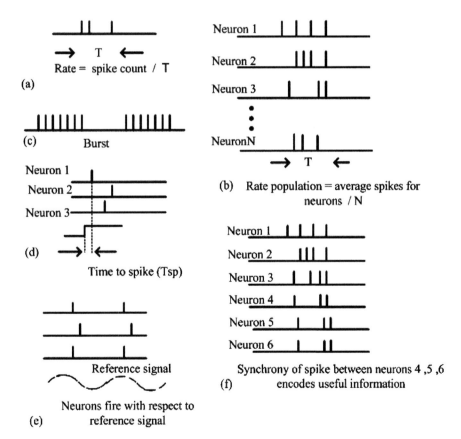

FIGURE 2.4 Different types of neural coding: (a) rate, (b) population rate, (c) burst coding, (d) time-to-spike pulse code, (e) phase pulse code, and (f) correlation and synchrony-based code. Adapted from Ref. 13.

most of the information is conveyed during the first 20–50 ms [17, 18]. However, time-to-first-spike encoding is susceptible to channel noise and spike loss; therefore, this type of encoding is typically observed in the cortex, where the spiking rate could be as low as one spike per second.

An extension of the time-to-spike code is the phase code that is applicable for a periodic stimulus. An example of phase encoding is shown in Figure 2.4e, where the spiking rate is shown to vary with the phase of the input stimulus. Yet another kind of neural code that has attracted significant interest from the neuroscience community uses the information encoded by correlated and synchronous firings between groups of neurons [13]. The response is referred to as *synchrony* and is illustrated in Figure 2.4f, where a sequence of spikes generated by neuron 1, followed by neuron 2 and neuron 3, encodes a specific feature of the input stimulus. Thus information is encoded in the trajectory of the spike pattern and so can provide a more elaborate mechanism of encoding different stimuli and its properties [19].

2.3 NOISE EXPLOITATION IN NEUROBIOLOGY

As mentioned in Section 2.1, noise plays a constructive role in neurobiology. A single neuron, by its very nature, acts as a noisy and crude (less than 3 bits accurate) computational unit [20–23]. It is not only affected by intrinsic *noise* (e.g., thermal noise in the ion channels) and extrinsic *noise* (e.g., noise due to the neurotransmitters present in the synaptic junctions), but it is also affected by noise in the sensor [24–27]. For example, the photoreceptor cells in the retina generate thermal and quantum noise due to the photons impinging on the retinal membrane. Thus, the spike train generated by a neuron not only exhibits a significant amount of jitter and drift but is also severely limited in its dynamic range and bandwidth (less than 500 Hz) due to its refractory period. In spite

of these limitations, networks of spiking neurons are remarkably accurate and are able to process large-bandwidth (much higher than 500 Hz) analog sensory signals with very high precision (greater than 120 dB) [28]. Through evolution, neurobiological systems have evolved to exploit noise as a computational resource rather than a hindrance. A study reported in Ref. 29 demonstrated the increase in the reliability of neuronal firings with the addition of noise. In yet another study [30], it was shown that noise facilitates reliable synchronization of the firing patterns in a population of neurons.

In this section, we describe two types of noise exploitation techniques commonly observed in neurobiology: (a) stochastic resonance (SR) and (b) noise shaping. In stochastic resonance, the addition of random noise enhances the detection of a weak, periodic signal, the amplitude of which is smaller than the firing threshold of the neuron. Noise-shaping principles apply to a population of neurons where the SNR of the network is enhanced by shifting the intrinsic noise out of the frequency bands where the signals of interest are present.

2.3.1 Stochastic Resonance

Figure 2.5 shows the basic principle of signal enhancement using stochastic resonance. The threshold of an integrate-and-fire neuron is denoted by V_{th}, whereas $v(t)$ is the membrane potential driven by a periodic stimulus. When *noise* or random perturbation is absent, as shown in Figure 2.5a, the neuron does not fire because the amplitude of the membrane potential $v(t)$ is below the threshold V_{th}. When noise (extrinsic or intrinsic) is added to the system, as shown in Figure 2.5b, there exists a finite probability that the membrane potential $v(t)$ will cross the threshold V_{th}, which would result in the generation of spikes. The rate of spikes would therefore be proportional to the level of the noise and to the amplitude of the membrane potential or input stimulus. However, when the

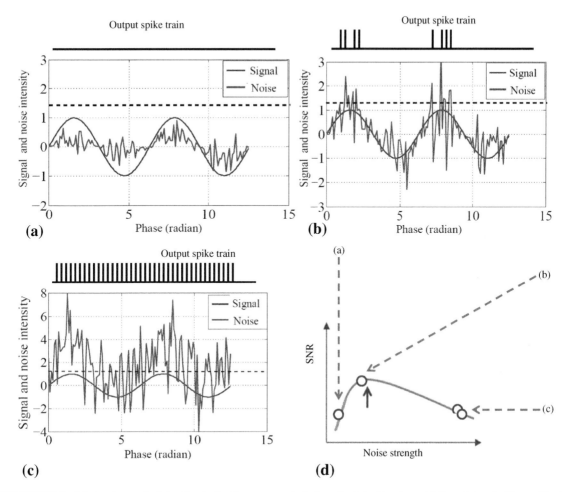

FIGURE 2.5 Mechanism underlying the stochastic resonance technique for detecting sinusoidal input signal with amplitude lower than the spiking threshold of the neuron: (a) at low levels of random noise, (b) at optimal level of random noise, (c) at large magnitudes of random noise, and (d) signal-to-noise ratio for the output spike trains corresponding to the condition (a), (b), or (c).

magnitude of the noise is significantly large, as shown in Figure 2.5c, then the spike is generated even without the presence of the stimuli. Thus, the SNR of the system exhibits a resonance-like phenomenon for which the peak is determined by the level of additive noise and by the amplitude of the input stimuli.

Stochastic resonance has been extensively studied in literature, and numerous mathematical models exist that capture the resonance phenomenon under various stimuli in the presence of different noise statistics [29, 31, 32]. The existence of SR in neurobiology was first reported in the mechanoreceptor hair cells of the crayfish (*Procambarus clarkii*). The cells were shown to use stochastic resonance to enhance the detection of small vibrations (caused by planktons) in aquatic environments. Stochastic resonance was also observed in mechanosensory systems of crickets, where it was used to recognize weak, low-frequency acoustic signatures emitted by the wing beats of a predator wasp [33]. Also, it was shown that adding noise to a weak stimuli improves the timing precision in neuronal activity and that the cells are able to *adapt* their intrinsic threshold values to the overall input signal power. We defer our discussion of the role of *adaptation* in SR until Section 2.4.

An organism in which stochastic resonance is exploited for enhanced electrical sensing is the paddlefish (*P. spathula*). Its electro-sensory receptors use stochastic resonance to detect and localize low-frequency electrical fields (0.5–20 Hz) emanated by planktons (*Daphnia*). In this case, the source of noise is due to the prey themselves, which in turn increases the sensitivity of the paddlefish electro-sensory receptors [6].

2.3.2 Noise Shaping

In Section 2.2, we described population rate encoding or averaging the firing activity across multiple neurons as a method for achieving higher dynamic range. Unfortunately, the simple averaging of noise across independent neurons in the network is suboptimal because the SNR improves only as a square root of the number of neurons [30]. It would therefore require an extraordinary number of neurons to achieve the SNR values (greater than 120 dB) typically observed in biological systems.

It has been proposed that a possible mechanism behind the remarkable processing acuity achieved by neuronal networks is *noise shaping*, a term that refers to the mechanism of shifting the energy contained in noise and interference out of the regions (spectral or spatial) where the desired information is present. It has been argued by Mar *et al.* [30] that inhibitory connections between neurons could lead to noise-shaping behavior and that the SNR improves directly as the number of neurons, a significant improvement over simple averaging techniques.

In this section, we describe the noise-shaping mechanism using the integrate-and-fire model described *via* Eq. (2.2). Consider a neuronal network consisting of N integrate-and-fire neurons.

Each neuron is characterized by its intrinsic voltage $v_i(t)$, $i \in [1, N]$, and the neuron fires whenever v_i exceeds a threshold V_{th}. Between consecutive firings, the dynamics of the membrane potential can be expressed using the integrate-and-fire model as [30]

$$\frac{d}{dt}v_i(t) = -v_i(t)/\tau_m - \sum_{j=1}^{N} W_{ij} \exp(-(t - t_j^m)/\tau_s)$$
$$+ \alpha x_i(t). \qquad (2.5)$$

Here, t_j^m are the set of firing times of the jth neuron and τ_m denotes the time constant of the neuron, capturing the leaky nature of integration denoted by the leaky potential of the membrane $v_i(t)/\tau_m$. The exponential term and the related time constant τ_s model the presynaptic filter $h(t)$ in Eq. (2.2) and the time constant of the presynaptic spike train. The parameter set W_{ij} denotes the synaptic weights between the ith and jth neurons and denotes the set of learning parameters for this integrate-and-fire neural network.

To show how the synaptic weights W_{ij} influence noise shaping, consider two specific cases as described in Ref. 30: (a) when $W_{ij}=0$, implying there is no coupling between the neurons and each neuron fires independently of the other; and (b) when $W_{ij}=W$, implying that the coupling between the neurons is inhibitory and constant. For a simple demonstration, τ_m is set to 1 ms and N is set to 50 neurons.

For the case in which the input $x_i(t)$ is constant, the raster plots indicating the firing of the 50 neurons for the uncoupled case and for the coupled case, as in Figures 2.6a and b, respectively. The bottom trace in each panel shows the firing pattern of the neuronal population that has been obtained by combining the firings of all the neurons. For the uncoupled case, the population firing shows clustered behavior where multiple neurons fire could fire in close proximity, whereas for the coupled case, the firing rates are uniform, indicating that the inhibitory coupling reduces the correlation between the neuronal firings.

To understand the implication of the inhibitory coupling for noise shaping, a sinusoidal input at frequency $f_0 = 1\,kHz$ was applied to all the neurons and the population firing rates are analyzed in frequency domain using a short-time Fourier transform. Figure 2.6c shows a comparison of the power spectrum for a single neuron, a neuron in a population of a coupled and an uncoupled network, respectively. The spectrum for a single neuron shows that it is unable to track the input signal since its bandwidth (1 kHz)

is much larger than the firing rate of the neuron, whereas for the uncoupled/coupled neurons in a population case, the input signal can be easily seen. For the uncoupled case, the noise floor, however, is flat, whereas for the coupled case, the noise floor from the signal band is shifted in the higher-frequency range, as shown in Figure 2.6c. The shaping of the in-band noise floor enhances the SNR ratio of the network for a large network, and the improvement is directly proportional to the number of neurons [30].

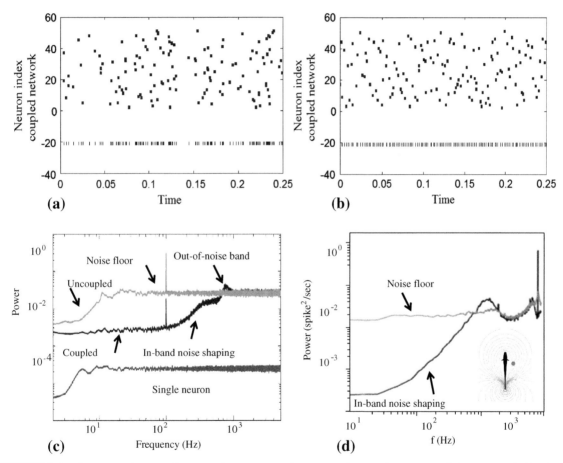

FIGURE 2.6 Illustration showing the noise-shaping principle in a population of integrate-and-fire neurons [30] and in electric fish [14, 34]: (a) spiking patterns generated when no inhibitory coupling exist, (b) spiking patterns generated when inhibitory coupling exists between the neurons that make the firing more uniform compared to the uncoupled case, (c) comparison in the spectral domain that clearly shows the connection between inhibitory coupling and noise shaping, and (d) noise shaping observed in electric fish. Adapted from Refs. 14 and 34.

Noise shaping has been observed in the sensory system of the electric fish that detects the perturbation of the ambient electric field. The intensity of the receptive field directly modulates the firing rate of the afferent neurons. The firing rates are synchronized where spikes with long interspike intervals follow spikes with short interspike intervals. This correlation between the interspike intervals leads to a noise-shaping effect, as shown in Figure 2.6d, where the noise is shifted out of the input stimuli [34].

The role of synaptic weights W_{ij} in noise shaping is not yet understood. These network parameters have to be learned and their values are critical for the successful exploitation of noise shaping and stochastic resonance. Synaptic learning and adaptation are the topics of the next section.

2.4 LEARNING AND ADAPTATION

The primary mechanism of learning and adaptation in the nervous system is the spike-time-dependent plasticity (STDP). Although different models have been reported for describing STDP, all of them agree on the causal relationship between the pre- and postsynaptic spikes. The increase or decrease in the strength of the synapse (W) depends on whether the presynaptic spike arrives before or after the postsynaptic spike and the time duration between the pre- and postsynaptic spikes as illustrated in Figure 2.7. If the presynaptic spike were generated at time t_j and the postsynaptic spike were generated at time t_i, then one form of STDP can be mathematically expressed as

$$\Delta W_{ij}(\Delta t) \propto \text{sgn}(\Delta t) e^{|\Delta t|/\tau}, \qquad (2.6)$$

where $\Delta t = t_j - t_i$, sgn(·) denotes the sign of its argument, whereas A and τ refer to the amplitude and time parameters of the synapse, respectively. Depending on the retention time of the synapse, increase of weight is referred as the short-term potentiation (STP) or long-term potentiation (LTP). Similarly, the decrease of weight is referred to as the short-term depression (STD) or long-term depression (LTD).

Equation (2.6) is the time-domain representation of the Hebbian rule

$$\Delta W_{ij}[n] = \eta\, y_i[n]\, x_j[n], \qquad (2.7)$$

where $x_i[n]$ and $y_j[n]$ are the pre- and postsynaptic signal amplitudes, respectively, and n denotes the learning-rate parameter. The causality in

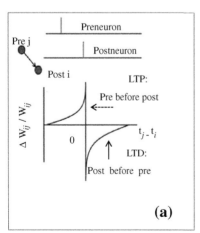

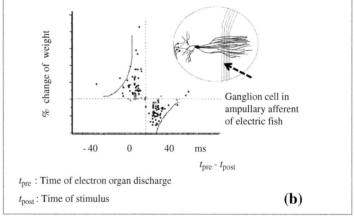

FIGURE 2.7 (a) STDP and (b) plasticity in electric fish. Part (b) is adapted from Ref. 35.

STDP and the time-domain Hebbian rule arises from the fact that, when the spike generated by a presynaptic neuron causes the post-synaptic neuron to fire as shown in Figure 2.7, the relative strength of the synapse connecting them should strengthen. However, anti-Hebbian responses have also been observed in synapses in organisms like the mormyrid electric fish (*Gnathonemus petersii*), in which negative correlation between the pre- and postsynaptic spike trains increases the strength of the connection [11].

At the sensory level, the interaction between two sets of neurons shows a specific pattern of synaptic plasticity. The first set of neurons comprises those in the receptive field that receive spikes from the environment; the second set of neurons are those that emit spikes to the surrounded field. The time difference between the spikes in the sets is processed in the sensory system to find the change of weight needed for learning. Adjustment of the weight (plasticity and learning) in the neural sets at the sensory level is used to eliminate the internal and external noise from masking the important sensory information [11, 35]. In fact, anti-Hebbian synapses are more important to achieve noise shaping, a evident from the case study presented in the next section.

2.5 CASE STUDY: NEUROMORPHIC ACOUSTIC SOURCE LOCALIZER

In this section, we present a systematic approach to designing neuromorphic sensors that incorporates the noise exploitation and adaptation principles. We apply the technique for the design of a miniature, neuromorphic acoustic source localizer as a case study.

Localization of acoustic sources using miniature microphone arrays poses a significant challenge due to fundamental limitations imposed by the physics of sound propagation [36]. When the spacing between the microphones is much smaller than the wavelength of the acoustic signal, the resolution of acute localization cues becomes difficult due to precision artifacts, intrinsic noise, and extrinsic noise. However, insects such as *O. ochracea* can overcome these fundamental limitations and demonstrate remarkable localization abilities (with accuracy better than 2°), comparable to the localization ability demonstrated by humans [4, 11, 36]. The mechanism underlying this remarkable ability is the coupling between the miniature differential eardrums (separated by less than 0.5 mm) of the fly and its neuronal circuitry (which acts as an equivalent analog-to-digital converter).

Figure 2.8a shows the structure of the acoustic sensor composed of two ears that are mechanically coupled with membranes that are connected across the middle by a flexible cuticular lever that provides a basis for directional sensitivity. The direction of the acoustic signal incident on the two sensory levers is shown in Figure 2.8b. This system acts as an amplifier that magnifies the acoustic cues such as the interaural time difference (ITD) by about 50 times, as shown in Figure 2.8c. The ITD amplification occurs at two levels. The first amplification is caused by the mechanical differentiation of the acoustical signal; the second amplification is caused by signal processing and adaptation at the neuronal level.

The differential-sensing-and-signal-processing mechanism is further elaborated using a simple far-field acoustic model. Consider the differential cantilever that is modeled using two point masses (shown in Figure 2.8a) vibrating about their center of mass and separated by a distance that is much smaller than the wavelength of sound. In this case, the signals exciting each of the cantilever sensors can be approximated using a far-field wave-propagation model [37, 38] where by the acoustic wavefront can be assumed to be planar, as shown in Figure 2.8b. For acoustic signals with a frequency range of 100 Hz–20 kHz, this distance is typically less than $\lambda/10 \approx 3.4$ cm, where λ is the wavelength of the audio signal in air. Also, the distance to the source from the center G of the array is assumed to be larger than

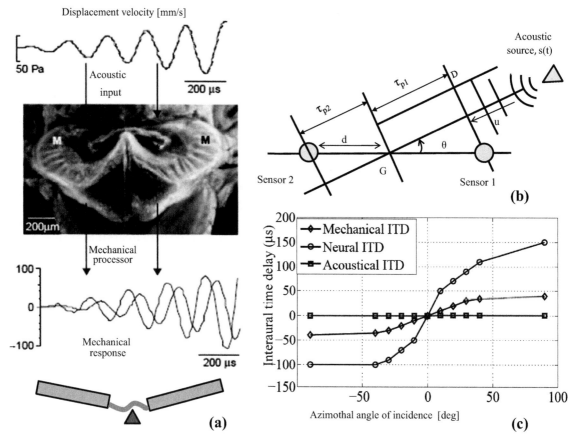

FIGURE 2.8 Sensory mechanism in the parasitoid fly *Ormia ochracea*: (a) structure of the mechanically coupled eardrums modeled as a vibrating cantilever, (b) a conceptual far-field model where by the acoustic wavefront impinges the sensors at an angle θ, and (c) amplification of the ITD ($\tau_{p1} + \tau_{p2}$) at the acoustic, mechanical, and the neuronal levels. Adapted from Ref. 11.

the inter-element distance. Therefore, the acoustic signal wavefront is considered planar as it reaches the sensor, as shown in Figure 2.8b. The signal $x(\mathbf{p}_j, t)$ recorded at the jth element (location specified by a three-dimensional position vector $\mathbf{p}_j \in \mathcal{R}^3$ with respect to the center $\mathbf{G}$ of the array) can be expressed as a function of the bearing θ that is the angle between the position vector $\mathbf{p}_j$ and the source vector $\mathbf{u}$. The signal $x(\mathbf{p}_j, t)$ is written as

$$x(\mathbf{p}_j, t) = a(\mathbf{p}_j)s(t - \tau(\mathbf{p}_j)), \quad (2.8)$$

where $a(\mathbf{p}_j)$ and $\tau(\mathbf{p}_j)$ denote the attenuation and delay for the source $s(t)$, respectively, measured relative to the center of the microphone array. Equation (2.8) is expanded using the Taylor theorem as

$$x(\mathbf{P}_j, t) = a(\mathbf{p}_j) \sum_{k=0}^{\infty} \frac{(-\tau(\mathbf{p}_j))^k}{k!} s^{(k)}(t), \quad (2.9)$$

where $s^{(k)}(t)$ is the kth temporal derivative of $s(t)$. Under far-field conditions, $a(\mathbf{p}_j)$ is approximately constant and we set $a(\mathbf{p}_j) \approx 1$ without sufficient loss of generality. Also, for far-field conditions the time delay $\tau(\mathbf{p}_j)$ can be assumed to be a linear projection between the vector $\mathbf{p}_j$ and the unit vector u oriented toward the direction of arrival of the acoustic wavefront (see Figure 2.8). Thus,

$$\tau(\mathbf{p}_j) = \frac{1}{c}\mathbf{u}^T \cdot \mathbf{p}_j, \qquad (2.10)$$

where c is the speed of sound waves in air. Ignoring the higher-order terms in the series expansion (under far-field assumptions), we simplify Eq. (2.9) as

$$x(\mathbf{p}_j, t) \approx s(t) - \tau(\mathbf{p}_j)s^{(1)}(t), \qquad (2.11)$$

thereby implying that under far-field conditions, the signals recorded at the microphone array are linear with respect to the bearing parameter $\tau(\mathbf{p}_j)$. Thus, the differential signal is given by

$$\Delta x(t) = x(\mathbf{p}_2, t) - x(\mathbf{p}_1, t) = 2s^{(1)}(t)\frac{d}{c}\cos\theta, \quad (2.12)$$

whose envelope is a function of the bearing angle θ, and $s^{(1)}(t)$ is the first-order temporal derivative of $s(t)$. The fly's neural circuitry uses this differential signal to accurately extract the amplitude and hence estimate of the bearing. Adaptation and noise exploitation play a critical role in this process and directly affect the precision of the estimated bearing angle. Thus, the signal-measurement process is integrated with the statistical learning and adaptation process that alleviates the effects of sensor artifacts and noise.

This integration served as an inspiration for a novel online learning framework called ΣΔ *learning* [39] that integrates the noise-shaping principles with the statistical learning and adaptation process. The framework exploits the structural similarities between an integrate-and-fire neuron and a ΣΔ modulator for which stochastic resonance and noise shaping have been demonstrated in literature. Let us now show how ΣΔ learning can be used to resolve acute bearing cues in subwavelength microphone arrays.

2.6 ΣΔ LEARNING FRAMEWORK

ΣΔ learning is based on an optimization framework that integrates ΣΔ modulation with statistical learning [40]. Given a random input vector $\mathbf{x} \in \mathcal{R}^D$ and a vector $\mathbf{v} \in \mathcal{R}^M$ (where each element of the vector $\mathbf{v}$ is the membrane potential of each neuron), a ΣΔ learner estimates the parameters of the synaptic-weight matrix $\mathbf{W} \in \mathcal{R}^D \times \mathcal{R}^M$ according to the following optimization criterion:

$$\max_{\mathbf{W} \in \mathcal{C}}(\min_{\mathbf{v}} f(\mathbf{v}, \mathbf{W})), \qquad (2.13)$$

where $\mathcal{C}$ denotes a constraint space of the transformation matrix $\mathbf{W}$ and.

$$f(\mathbf{v}, \mathbf{W}) = ||\mathbf{v}||_1 - \mathbf{v}^T \mathcal{E}_x\{\mathbf{W}^T \mathbf{x}\}. \qquad (2.14)$$

Here, $\mathcal{E}_x\{.\}$ denotes an expectation operator with respect to the random vector $\mathbf{x}$ (or rate encoding of $\mathbf{x}$).

The term $||\mathbf{v}||$ bears similarity to a regularization operator that is extensively used in machine learning algorithms to prevent over-fitting [41, 42]. However, the L_1 norm in Eq. (2.14) forms an important link in connecting the cost function in Eq. (2.13) to spike generation. This is illustrated in Figure 2.9a, which shows an example of a one-dimensional regularization function $||\mathbf{v}||$. The piecewise behavior of $||\mathbf{v}||$ leads to discontinuous gradient sgn($\mathbf{v}$), as shown in Figure 2.9b. The signum is a Boolean function that indicates whether a spike is generated or not. The minimization step in Eq. (2.13) ensures that the membrane potential vector $\mathbf{v}$ is correlated with the transformed input signal $\mathbf{Wx}$ (signal-tracking step), and the maximization step in Eq. (2.13) adapts the parameters of $\mathbf{W}$ such that it minimizes the correlation (decorrelation step), similar to the lateral inhibition observed in afferent neurons. The uniqueness of the proposed approach, compared to other optimization techniques to solve Eq. (2.13), is the use of bounded gradients to generate SA limit cycles about a

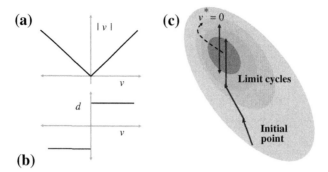

FIGURE 2.9 Mathematical concepts of $\Sigma\Delta$ learning: (a) a one-dimensional L_1 regularizer and (b) its derivative that leads to a binary (Boolean) function, (c) limit cycles due to $\Sigma\Delta$ learning about the minima that leads to spike generation.

minimum $\mathbf{v}^*$, as shown in Figure 2.9c. This limit-cycle behavior will capture the spiking dynamics of the system. It can be verified that, if $\|\mathbf{w}\|_\infty \le 1$, then $\mathbf{v}^*=0$ is the location of the minimum with $f(\mathbf{v}^*, \mathbf{W})=0$.

The link between the optimization criterion of Eq. (2.13) and $\Sigma\Delta$ modulation (or integrate-and-fire neuron) is through a stochastic gradient minimization [43] of the cost function given by Eq. (2.14). If the input random vector $\mathbf{x}$ is assumed to be stationary and if the probability density function of $\mathbf{x}$ is assumed to be well behaved (i.e., the gradient of the expectation operator is equal to the expectation of the gradient operator), the stochastic gradient step with respect to $\mathbf{v}$ for each iteration n yields

$$\mathbf{v}[n] = \mathbf{v}[n-1] - \left.\frac{\partial f(\mathbf{v}, \mathbf{W})}{\partial \mathbf{v}}\right|_{(n-1)}, \quad (2.15)$$

$$\mathbf{v}[n] = \mathbf{v}[n-1] + \mathbf{W}[n-1]^T\mathbf{x}[n-1] - d[n], \quad (2.16)$$

where $d[n]=\mathrm{sgn}(\mathbf{v}[n-1])$ denotes a Boolean function indicating whether a spike is generated. Solution of the discrete-time recursion Eq. (2.16) leads to

$$\frac{1}{N}\sum_{n=1}^{N}d[n] = \frac{1}{N}\sum_{n=0}^{N-1}\mathbf{W}[n]^T\mathbf{x}[n] + \frac{1}{N}(\mathbf{v}[n]-\mathbf{v}[0]). \quad (2.17)$$

The bounded property of $\mathbf{w}[n]$ then leads to the asymptotic property:

$$\mathcal{E}_n\{d[n]\} \stackrel{n\to\infty}{\longrightarrow} \frac{1}{\lambda}\mathcal{E}_n\{\mathbf{W}[n]^T\mathbf{x}[n]\}, \quad (2.18)$$

where $\mathcal{E}_n\{\cdot\}$ denotes an empirical expectation with respect to time index n. Thus, the recursion Eq. (2.16) produces a binary (spike) sequence, the mean of which asymptotically encodes the transformed input at infinite resolution. This is illustrated in Figure 2.9c, which shows a two-dimensional optimization contour. The objective of the $\Sigma\Delta$ learning is to follow the trajectory from an initial condition to the minimum and induce limit cycles about the minimum $\mathbf{v}^*$. The dynamics of the limit cycles then encode the shape of the optimization contour and hence also encode the estimation parameters.

The maximization step (decorrelation) in Eq. (2.13) yields updates for matrix $\mathbf{W}$ according to:

$$\mathbf{W}[n] = \mathbf{W}[n-1] + \left.\xi\frac{\partial f(\mathbf{v}, \mathbf{W})}{\partial \mathbf{W}}\right|_{n-1}, \quad (2.19)$$

which leads to

$$\mathbf{W}[n] = \mathbf{W}[n-1] - \xi\mathbf{v}[n-1]\mathbf{x}[n-1]^T; \mathbf{W}[n] \in \mathcal{C}. \quad (2.20)$$

The adaptation step in Eq. (2.20) can be expressed by its digital equivalent as

$$\mathbf{W}[n] = \mathbf{W}[n-1] - \xi[n]\mathrm{sgn}(\mathbf{x}[n-1])^T; \mathbf{W}[n] \in \mathcal{C}, \quad (2.21)$$

where we have used the relationship $d[n] = \text{sgn}(\mathbf{v}[n-1])$. Note that the update rate ξ does not affect the stability of learning. The constraint space $\mathcal{C}$ is restricted to transforms represented by lower-triangular matrices with diagonal elements set to unity that can be expressed as $W_{ij} = 0;\ \forall i < j;\ W_{ii} = 1$. Thus, to satisfy the constraint $\mathbf{W} \in \mathcal{C}$, only the lower-diagonal elements are updated in Eq. (2.21). It can be seen from Eq. (2.21) that if $||\mathbf{W}||_\infty \le B > 0$ is true, then the recursion (21) will asymptotically lead to

$$\mathcal{E}_n\{d[n]\text{sgn}(\mathbf{x}[n])^T\} \overset{n\to\infty}{\longrightarrow} 0 \qquad (2.22)$$

for $\mathbf{W}_\infty \in \mathcal{C}$. Equation (2.22) shows that the proposed ΣΔ learning algorithm produces binary (spike) sequences that are mutually uncorrelated to a nonlinear function of the input signals.

2.6.1 ΣΔ Learning for Bearing Estimation

The design of the bio-inspired acoustic source localizer based on ΣΔ learning has been presented by Gore et al. [39]. Figure 2.10 shows the localizer constructed using an array of off-the-shelf electret microphones interfacing with a custom-made application-specific integrated circuit (ASIC). Similar to the mechanical cantilever model in a parasitoid fly, the differential electrical signals recorded by the microphones can be approximated by

$$\left.\begin{array}{l} \Delta x_1(t) = -s^{(1)}(t)\frac{d}{c}\cos\theta \\ \Delta x_2(t) = -s^{(1)}(t)\frac{d}{c}\sin\theta \\ \Delta x_3(t) = +s^{(1)}(t)\frac{d}{c}\cos\theta \\ \Delta x_4(t) = +s^{(1)}(t)\frac{d}{c}\sin\theta \end{array}\right\}, \qquad (2.23)$$

which bear similarity to the differential signals observed for the parasitoid fly. Figure 2.10 also shows a scope trace of the sample differential output produced by the microphone array when a 1 kHz tone is played from a standard computer speaker. The scope trace clearly shows that the differential signal $\Delta x_2(t)$ is 90°

out of phase with respect to $\Delta x_1(t)$, and the differential signal $\Delta x_3(t)$ is 180° out of phase with respect to $\Delta x_1(t)$.

To apply ΣΔ learning for bearing estimation, three of the four differential signals in Eqs. (2.23) are chosen as inputs, and the synaptic matrix $\mathbf{W}$ is chosen to be of the form

$$\mathbf{W} = \begin{bmatrix} W_{11} & 1 & 0 \\ W_{21} & W_{22} & 1 \end{bmatrix}. \qquad (2.24)$$

When applied to the three differential signals of the microphone array modeled by Eq. (2.23), the ΣΔ recursions in Eq. (2.16) lead to

$$\begin{aligned} v_1[n] &= v_1[n-1] &&+ \Delta x_2[n] + W_{11}[n]\Delta x_1[n] - d_1[n], \\ V_2[n] &= V_2[n-1] &&+ \Delta x_3[n] + W_{21}[n]\Delta x_1[n], \\ & &&+ W_{22}\Delta x_2[n] - d_2[n], \qquad (2.25) \end{aligned}$$

with $d_1[n] = \text{sgn}(v_1[n])$ and $d_2[n] = \text{sgn}(v_2[n])$. The adaptation steps for the parameters W_{11}, W_{21}, and W_{22} based on Eq. (2.21) can be expressed as

$$\begin{aligned} W_{11}[n] &= W_{11}[n-1] - \xi d_1[n]\text{sgn}(\Delta x_1[n]), \\ W_{21}[n] &= W_{21}[n-1] - \xi d_2[n]\text{sgn}(\Delta x_1[n]), \\ W_{22}[n] &= W_{22}[n-1] - \xi d_2[n]\text{sgn}(\Delta x_2[n]). \end{aligned}$$
$$(2.26)$$

One of the implications of ΣΔ adaptation steps in Eq. (2.26) is that, for the bounded values of W_{11}, W_{21}, and W_{22}, the following asymptotic result holds:

$$\lim_{N\to\infty} \frac{1}{N} \sum_{n=1}^{N} d_{1,2}[n] \to 0. \qquad (2.27)$$

To demonstrate how Eqs. (2.25) and (2.26) can be used for bearing estimation, consider two different cases based on the quality of common-mode cancellation.

Case I: Perfect common-mode cancellation. The differential microphone is assumed to completely suppress the common-mode signal $x_{\text{cm}}(t)$ in Eq. (2.23). Also, the bearing of the source is assumed to be located in the positive

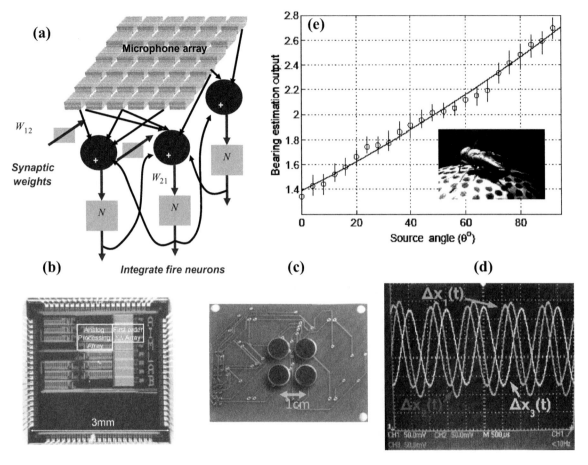

FIGURE 2.10 A neuromorphic acoustic source localizer mimicking the sensory mechanism in the parasitoid fly [11]: (a) architecture of the localizer interfacing with a microphone array, (b) micrograph of the localizer IC fabricated in a 0.5 μm CMOS processor, (c) localizer system board integrating the microphone array and the localizer IC, (d) outputs from the microphone array showing the differential amplification of the ITD when the input is a tone, and (e) output of the localizer demonstrating a bearing estimation better than 2° resolution.

quadrant $90° > \theta > 0°$. Inserting Eq. (2.23) into Eqs. (2.25) and (2.26), we obtain the following expressions:

$$W_{11}[N] = -\xi \sum_{n=1}^{N} d_1[n]\mathrm{sgn}(s^{(1)}[n]),$$

$$d_1[n] = \dot{s}[n]\sin\theta + W_{11}[n]s^{(1)}[n]\cos\theta - (v_1[n] - v_1[n-1]). \tag{2.28}$$

Combining the two Eqs. (2.28), we get

$$W_{11}[N] = -\xi \sum_{n=1}^{N} |s^{(1)}[n]|(\sin\theta + W_{11}[n]\cos\theta)$$

$$+ \xi \sum_{n=1}^{N} (v_1[n] - v_1[n-1])\mathrm{sgn}(s^{(1)}[n]). \tag{2.29}$$

Let W_{11}^* denote the converged value such that $W_{11}[n] \xrightarrow{n\to\infty} W_{11}^*$. Eq. (2.29) then leads to

$$\frac{1}{N}\sum_{n=1}^{N}|s^{(1)}[n]|(\sin\theta + W_{11}^{*}\cos\theta)$$

$$-\frac{1}{N}\sum_{n=1}^{N}(v_1[n] - v_1[n-1])\mathrm{sgn}(s^{(1)}[n]) \xrightarrow{N\to\infty} 0.$$

$$(2.30)$$

The second part on the right side of Eq. (2.30) converges to zero, thereby implying that

$$\lim_{N\to\infty}\frac{1}{N}\sum_{n=1}^{N}|s^{(1)}[n]|(\sin\theta + W_{11}^{*}\cos\theta) = 0.$$

$$(2.31)$$

The solution to Eq. (2.31) is given by $W_{11}^{*} = -\tan\theta$. Eqs. (2.27) and (2.25) then lead to

$$\frac{1}{N}\sum_{n=1}^{N}(\Delta x_2[n] + W_{11}[n]\Delta x_1[n]) \xrightarrow{N\to\infty} 0, \quad (2.32)$$

and hence,

$$\frac{1}{N}\sum_{n=1}^{N}(s^{(1)}[n]\sin\theta + W_{11}[n]s^{(1)}[n]\cos\theta) \xrightarrow{N\to\infty} 0.$$

$$(2.33)$$

Equation (2.33) then produces the bearing estimate

$$\hat{\theta} = \arctan(-W_{11}^{*}). \qquad (2.34)$$

In a similar fashion, W_{21} and W_{22} can also be used to estimate the bearing of the source according to

$$\hat{\theta} = \arctan\left(\frac{1 - W_{21}}{W_{22}}\right). \qquad (2.35)$$

Both the estimates given by Eqs. (2.34) and (2.35) use ΣΔ learning to estimate the correlation between differential microphone signals. However, the correlation-based approach does not yield robust results when the common-mode cancellation is imperfect, which leads to the second case.

Case II: Imperfect common-mode cancellation. In practice, microphone artifacts and mismatch in analog circuits limit the precision of common-mode cancellation. Suppose the common-mode signal $s_c(t)$ is assumed to be the same for all the microphones in the array and Eqs. (2.23) is modified as follows:

$$\left.\begin{array}{l}\Delta x_1(t) = s_c(t) - \dot{s}(t)\frac{d}{c}\cos\theta \\ \Delta x_2(t) = s_c(t) - \dot{s}(t)\frac{d}{c}\sin\theta \\ \Delta x_3(t) = s_c(t) + \dot{s}(t)\frac{d}{c}\cos\theta \\ \Delta x_4(t) = s_c(t) + \dot{s}(t)\frac{d}{c}\sin\theta\end{array}\right\}. \qquad (2.36)$$

The underlying assumption is valid, since sensor mismatch only changes the location of the point of reference according to which the measurements are made.

Then, using similar assumptions and approach as in *Case I*, we have verified that the estimate of the bearing is given by:

$$\hat{\theta} = \arctan\left(\frac{W_{21} - 2W_{21} - W_{11}W_{22} - 1}{W_{21} - W_{11}W_{22} + 1}\right). \qquad (2.37)$$

Figure 2.10 shows the prototype of the neuromorphic acoustic source localizer, consisting of a microphone array and a custom-made integrated circuit that implements the integrate-and-fire neural network. The microphone array is formed with the omnidirectional electret condenser microphones. This condenser microphone has the sensitivity of $-42\,$dB at 1 kHz. The SNR of this microphone is 55 dB. The four-microphone array was designed with the condenser microphones 1 cm apart. The far-field effect is generated by keeping the acoustic source at a distance of 1 m from the microphone array. Because ΣΔ learning involves manipulation of analog parameters, we implemented the algorithms using a custom-made integrated circuit.

A 1 kHz tone was played through a standard loudspeaker, and the synaptic parameters W_{11}, W_{21}, and W_{22} on the integrated circuit were continuously adapted. In all the experiments, the position of the speaker was fixed and the microphone array was rotated in steps of the desired angle. For each angular orientation of the source, 10 sets of measurements (W_{11}, W_{21}, and W_{22}) were recorded every 80 ms using a field-programmable gate array. These sets were then used to estimate the bearing. The measured response shown in Figure 2.10 demonstrates a resolution of $2°$ that is similar to or better than the localization capability of the parasitoid fly. For all the experiments, the miniaturized microphone array consumes only a few microwatts of power, which is also comparable to the energy efficiency of the parasitoid fly's localization apparatus.

2.7 CONCLUSIONS

In this chapter, we described two important noise exploitation and adaptation mechanisms observed in neurobiology: (a) stochastic resonance and (b) noise shaping. We argued the importance of synaptic plasticity in its role in noise exploitation and signal enhancement. The concepts of stochastic resonance, noise shaping, and adaptation (synaptic plasticity) have been integrated into a unified algorithmic framework called $\Sigma\Delta$ learning. As a case study, we described the application of the $\Sigma\Delta$ learning framework toward the design of a miniature acoustic source localizer that mimics the response of a parasitoid fly (O. ochracea). This case study illustrates one specific example, and the future challenge lies in extending the framework to more complex sensory tasks such as recognition and perception. In this regard, the objective will be to apply the noise exploitation and adaptation concepts to emerging devices such as the memristor, which provides a compact apparatus for storing synaptic weights. Furthermore, extensive simulation and emulation studies need to be conducted that can verify the scalability of the $\Sigma\Delta$ learning framework.

References

[1] C. Mead, *Analog VLSI and neural systems*, Addison-Wesley, Boston, USA (1989).

[2] K. Boahen, Neuromorphic microchips, *Sci Am* **292** (5) (May 2005), 56–63.

[3] P. Lichtsteiner, C. Posch, and T. Delbruck, A 128 × 128 120 db 15 µs latency asynchronous temporal contrast vision sensor, *IEEE J Solid State Circ* **43** (2008), 566–576.

[4] A.C. Mason, M.L. Oshinsky, and R.R. Hoy, Hyperacute directional hearing in a microscale auditory system, *Nature* **410** (2001), 686–690.

[5] G. Krijnen, A. Floris, M. Dijkstra, T. Lammerink, and R. Wiegerink, Biomimetic micromechanical adaptive flow-sensor arrays, *Proc SPIE* **6592** (2007), 65920F.

[6] D.F. Russell, L.A. Wilkens, and F. Moss, Use of behavioural stochastic resonance by paddle fish for feeding, *Nature* **402** (1999), 291–294.

[7] M. Nelson, Smart sensing strategies: Insights from a biological active sensing system, *Proceedings of the 5th International workshop on advanced smart materials and smart structures technology*, Boston, MA (July 2009).

[8] J.B. Snyder, M.E. Nelson, J.W. Burdick, and M.A. Maciver, Omnidirectional sensory and motor volumes in electric fish, *PLoS Biol* **5** (2007), e301.

[9] S.-C. Liu and T. Delbruck, Neuromorphic sensory systems, *Curr Opin Neurobiol* **20** (2010), 288–295.

[10] M.E. Nelson, Biological smart sensing strategies in weakly electric fish, *Smart Struct Syst* **8** (2011), 107-117.

[11] D. Robert and M.C. Göpfert, Novel schemes for hearing and orientation in insects, *Curr Opin Neurobiol* **12** (2002), 715–720.

[12] E.M. Izhikevich, *Dynamical systems in neuroscience: The geometry of excitability and bursting computational neuroscience*, MIT Press, Cambridge, MA, USA (2006).

[13] W. Gerstner and W.M. Kistler, *Spiking neuron models: Single neurons, populations, plasticity*, Cambridge University Press, Cambridge, UK (2002).

[14] R. Krahe and F. Gabbiani, Burst firing in sensory systems, *Nat Rev Neurosci* **5** (2004), 13–23.

[15] R.B. Stein, E.R. Gossen and K.E. Jones, Neuronal variability: noise or part of the signal? *Nat Rev Neurosci* **6** (2005), 389–397.

[16] G.B. Ermentrout, R.F. Galán, and N.N. Urban, Reliability, synchrony and noise, *Trends Neurosci* **31** (2008), 428–434.

[17] S. Thorpe, D. Fize, and C. Marlot, Speed of processing in the human visual system, *Nature* **381** (1996), 520–522.

[18] M.J. Tovee and E.T. Rolls, Information encoding in short firing rate epochs by single neurons in the primate temporal visual cortex, *Vis Cogn* **2** (1995), 35–58.

[19] M. Abeles, Firing rates and well-timed events, *Models of neural networks 2* (E. Domany, K. Schulten, and J.L. van Hemmen (eds.)), Springer, NY, USA (1994), 121–140

[20] Z.F. Mainen and T.J. Sejnowski, Reliability of spike timing in neocortical neurons, *Science* **268** (1995), 1503–1506.

[21] R. Lyon and C. Mead, An analog electronic cochlea, *IEEE Trans Acoust Speech Signal Process* **36** (1988), 1119–1134.

[22] H.E. Derksen and A.A. Verveen, Fluctuations of resting neural membrane potential, *Science* **151** (1966), 1388–1389.

[23] A. Destexhe, M. Rudolph, and D. Pare, The high-conductance state of neocortical neurons in vivo, *Nat Rev Neurosci* **4** (2003), 739–751.

[24] E.T. Rolls and G. Deco, *The noisy brain: Stochastic dynamics as a principle of brain function*, Oxford University Press, Oxford, UK (2010).

[25] A.A. Faisal, Noise in the nervous system, *Nat Rev Neurosci* **9** (2010), 292–303.

[26] H. Markram and M. Tsodyks, Redistribution of synaptic efficacy between neocortical pyramidal neurons, *Nature* **382** (1996), 807–810.

[27] J.A. White, J.T. Rubinstein, and A.R. Kay, Channel noise in neurons, *Trends Neurosci* **23** (2000), 131–137.

[28] W. Gerstner, Population dynamics of spiking neurons: fast transients, asynchronous states, and locking, *Neural Comput* **12** (1999), 43–89.

[29] L. Gammaitoni, P. Hnggi, P. Jung, and F. Marchesoni, Stochastic resonance, *Rev Mod Phys* **70** (1998), 223–287.

[30] D.J. Mar, C.C. Chow, W. Gerstner, R.W. Adams, and J.J. Collins, Noise shaping in populations of coupled model neurons, *Proc Natl Acad Sci* **96** (1999), 10450–10455.

[31] A. Longtin, F. Moss, and K. Bulsara, Time-interval sequences in bistable systems and the noise-induced transmission of information by sensory neurons, *Phys Rev Lett* **67** (1991), 656–659.

[32] O. Rosso and C. Masoller, Detecting and quantifying stochastic and coherence resonances via information-theory complexity measurements, *Phys Rev E* **79** (2009), 040106.

[33] J.E. Levin and J.P. Miller, Broadband neural encoding in the cricket cereal sensory system enhanced by stochastic resonance, *Nature* **380** (1996), 165–168.

[34] M.J. Chacron, B. Lindner, L. Maler, A. Longtin, and J. Bastian, Experimental and theoretical demonstration of noise shaping by interspike interval correlations, *Proc SPIE* **5841** (2005), 150–163.

[35] C.C. Bell, V.Z. Han, Y. Sugawara, and K. Grant, Synaptic plasticity in a cerebellum-like structure depends on temporal order, *Nature* **387** (1997), 278–281.

[36] R.N. Miles and R.R. Hoy, The development of a biomedically inspired directional microphone for hearing aids, *Audiol Neurol* **11** (2006), 86–94.

[37] M.N. Do, Toward sound-based synthesis: the far-field case, *Signal Process* **2** (2004), 601–604.

[38] M. Stanacevic and G. Cauwenberghs, Micropower gradient flow acoustic localizer, *IEEE Trans Circuit Sys* **152** (2005), 2148–2157.

[39] A. Gore, A. Fazel, and S. Chakrabartty, Far-field acoustic source localization and bearing estimation using learners, *IEEE Trans Circuit Sys* **I** (2010), 783–792.

[40] A. Gore and S. Chakrabartty, Large margin multi-channel analog-to-digital conversion with applications to neural prosthesis, *Proceedings of the advances in neural information processing system (NIPS 2006)*, Vancouver, British Columbia, Canada (2006), 497–504.

[41] V.V. Vapnik, *The nature of statistical learning theory*, Springer-Verlag, Heidelberg, Germany (1995).

[42] F. Girosi, M. Jones, and T. Poggio, Regularization theory and neural networks architectures, *Neural Comput* **7** (1996), 219–269.

[43] S. Boyd and L. Vandenberghe, *Convex optimization*, Cambridge University Press, Cambridge, UK (2004).

ABOUT THE AUTHORS

Thamira Hindo received BE and ME degrees from the School of Control and Systems Engineering, University of Technology, Baghdad, Iraq, in 1984 and 1990, respectively. Her ME thesis was related to the design of personal-based instruments for diagnosing electrocardiogram (ECG) arrhythmia, for diagnosing and implementing pulmonary function tests, and for measuring the body temperature. From 1990 to 2008, she worked as a lecturer in the School of Control and Systems Engineering, teaching undergraduate courses in digital techniques, electrical circuits, assembly language, and computer architecture. Since 2009 she has been a PhD candidate in the department of Electrical and Computer Engineering at Michigan State University. Her current research interests are in the area of neuromorphic engineering and in the design of asynchronous pulse-mode computer architectures.

Shantanu Chakrabartty received his BTech degree from Indian Institute of Technology, Delhi, in 1996 and his MS and PhD degrees in electrical engineering from Johns Hopkins University, Baltimore, in 2002 and 2005, respectively. He is currently an associate professor in the Department of Electrical and Computer Engineering at Michigan State University (MSU). From 1996 to 1999 he worked with Qualcomm Inc., San Diego, and during 2002 he was a visiting researcher at the University of Tokyo. His work covers a variety of aspects of analog computing, in particular non-volatile circuits, and his current research interests include energy-harvesting sensors and neuromorphic and hybrid circuits and systems. He was a Catalyst Foundation fellow from 1999 to 2004 and is a recipient of the National Science Foundation's CAREER Award, the University Teacher-Scholar Award from MSU, and the 2012 Technology of the Year Award from MSU Technologies. He is a senior member of the Institute of Electrical and Electronics Engineers (IEEE) and is currently serving as the associate editor for *IEEE Transactions of Biomedical Circuits and Systems*, associate editor for the *Advances in Artificial Neural Systems* journal, and a review editor for the *Frontiers in Neuromorphic Engineering* journal.

Biomimetic Hard Materials

Mohammad Mirkhalaf, Deju Zhu, and Francois Barthelat

Department of Mechanical Engineering, McGill University,
Montreal, QC H3A 2K6, Canada

Prospectus

Materials such as bone, teeth, and seashells possess remarkable combinations of properties despite the poor structural quality of their ingredients (brittle minerals and soft proteins). Nacre from mollusk shells is 3,000 times tougher than the brittle mineral it is made of, a level of *toughness amplification* currently unmatched by any engineering material. For this reason, nacre has become the model for bioinspiration for novel structural materials. The structure of nacre is organized over several length scales, but the microscopic brick-and-mortar arrangement of the mineral tablets is prominent. This staggered structure provides a *universal* approach to arranging hard building blocks in nature and is also found in bone and teeth. Recent models have demonstrated how an attractive combination of stiffness, strength, and toughness can be achieved through the staggered structure. The fabrication of engineering materials that duplicate the structure, mechanics, and properties of natural nacre still present formidable challenges to this day.

Keywords

Aragonite, Biological materials, Composite materials, Fracture toughness, Freeze casting, Layer-by-layer assembly, Mechanical properties, Mineralization, Nacre, Stiffness, Strength, Template-assisted fabrication

3.1 INTRODUCTION

Research in to structural materials seeks new pathways to achieve novel and attractive combinations of mechanical properties. For example, toughness (resistance to the initiation and growth of a crack) and stiffness are both desirable properties for structural application, but high stiffness and high toughness are difficult to combine in traditional engineering materials [1,2]. Steel is tough but not as stiff as ceramics, which suffer from low toughness. Current research seeks to overcome this type of limitation by, for example, incorporating microstructural features to increase the toughness of ceramics.

There are three main approaches to achieving novel combinations of properties. Manipulating and tailoring the fundamental chemistry of the material (new alloys or new polymers) is one of them. With a fixed composition, the microstructure of the materials can be altered (e.g., austenite and martensite phases of steel) and optimized to obtain desired properties by thermomechanical processing (e.g., quenching). A third approach consists of combining two or more immiscible

http://dx.doi.org/10.1016/B978-0-12-415995-2.00003-9

materials with distinct and complementary mechanical properties, leading to hybrid materials. Carbon-fiber-reinforced polymers enter this category.

The concept of hybrid materials has been elegantly described by Ashby's equation: hybrid material = material A + material B + configuration + scale [1, 3], i.e., combination of materials A and B (or more materials) in a controlled shape, configuration, and scale. A or B can be theoretically any material, including a gas or liquid (for the case of cellular solids). Hybrid materials are powerful alternatives to traditional monolithic materials and offer a promising pathway to push the property envelope of traditional materials. Cellular materials with re-entrant cells [4] and anisotropic fibrous composites [5, 6] can display unusual properties such as negative Poisson's ratio or negative coefficient of thermal expansion [7].

Hybrid materials are abundant in nature, with mollusk shells, bone, tooth, tendon, and glass-sponge skeleton as only a few examples of such materials. In addition to specific chemistries, the structures of many of these hierarchical materials display specific features over distinct length scales. These materials boast remarkable combinations of properties, in some cases unmatched in engineering materials. For example, mollusk shell and bone are stiff and at the same time tough and flaw-tolerant, meaning that their strength is not compromised by the presence of initial flaws (Figure 3.1a). These attributes arise due to the combination of the high strength of inorganic minerals and the ductility of organic macromolecules through specific architectures at different length scales [8]. In contrast, hard and stiff engineering materials like ceramics usually display low levels of toughness, and tough flaw-tolerant materials like metals are not usually very stiff (Figure 3.1b).

In addition to attractive combinations of structural properties, natural composites such as bone possess remarkable capabilities such as self-healing, damage sensing, and self-repair; see Chapter 13 on self-organization, self-sealing, and self-healing in many natural systems. It has therefore become very attractive to fabricate materials that duplicate the structure and mechanics of these natural materials, in line with a paradigm called *biomimetics* [2, 9, 10].

Natural composites are composed of minerals and polymers arranged in intricate, compact structures. The hard inorganic minerals come in the form of inclusions of finite size embedded in a matrix of organic polymers. The organic polymers can be seen as interfaces between the hard

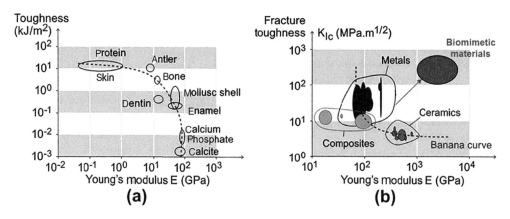

FIGURE 3.1   Stiffness-toughness charts for (a) biological materials and (b) engineering materials. Adapted from Ref. 40.

phases, and as such how they deform and fail plays an important role in the overall behavior of natural composites [11–13]. The roles of interfaces in natural composites include: (i) transferring stresses between inclusions; (ii) improving the fracture resistance of materials containing high amounts of brittle minerals [14]; (iii) acting as bridges to connect two highly different materials through functionally graded properties, exemplified by a gradient in mechanical properties along the muscle byssus connecting the soft muscle to the hard substrate (e.g., a rock) [15]; (iv) allowing large deformation and energy dissipation [16]; and (v) acting as actuators that can change shape or/and exert force upon the application of a stimulus [17].

The interfaces in natural composites are composed of several different types of polymers that provide a modular elongation that may convey toughness to the material [18]. Interfaces in nacre, for example, are composed of diverse groups, mainly including β-chitin, glycine, and alanine-rich proteins. Understanding the mechanisms associated with the interfaces in natural composites is key to the improvement of biomimetic materials [19,20].

Nacre inspires many biomimetic studies because: (i) its toughness is three orders of magnitude higher than that of its main constituent, (ii) it has less complex structure compared to other biological materials, and (iii) its structure is mainly optimized for mechanical functions (protecting living organisms) rather than biological functions (such as sensing, cell reproduction, etc.). Nacre, a layered nanocomposite found in mollusk shells, is composed of 95 %w/w of aragonite (a crystalline from of $CaCO_3$) and 5 %w/w of organic macromolecules. The aragonite layers are about 0.5 μm thick and are bonded by 20–30 nm-thick organic layers [18, 21]. Hard biological composites, such as nacre, incorporate a large amount of stiff mineral through a process called *biomineralization*. In this process, minerals nucleate and grow into pre-self-assembled organic scaffolds [19,22].

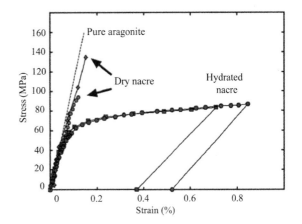

FIGURE 3.2 Tensile stress-strain behavior of nacre. Adapted from Ref. 14.

The mineralization of these organic scaffolds improves stiffness and compressive strength. Figure 3.2 shows the tensile stress-strain behavior of dry nacre, hydrated nacre, and pure aragonite. Dry nacre acts like monolithic aragonite and fails in a brittle fashion at low levels of strain (almost 0.1%). The response is linear elastic with Young's modulus $E \sim 90$ GPa. Hydrated nacre, however, acts initially as a linear elastic material ($E \sim 70$ GPa) followed by a large inelastic deformation before failure. Hydrated nacre fails at almost 1% strain, which is high for a material composed of 95 %w/w mineral. This high level of strain is a result of progressive spread of deformation through the material, stemming from the structural properties of nacre, i.e., tablet waviness [14].

3.2 DESIGN GUIDELINES FOR BIOMIMETIC HARD MATERIALS

The remarkable performance of materials like bone and nacre is due to their sophisticated microstructure, in which high-aspect-ratio mineral tablets are bonded between layers of organic materials and arranged in a brick-wall structure

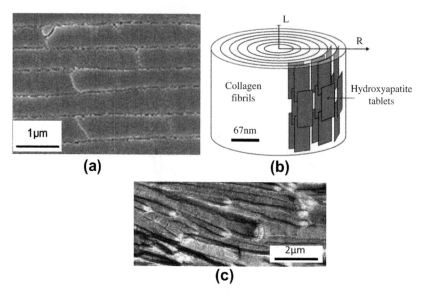

FIGURE 3.3 Staggered arrangement of mineral tablets. (a) scanning electron micrograph of nacre, (b) schematic of mineralized collagen fibrils in bone (adapted from Ref. 25), and (c) scanning electron micrograph of tooth dentin. Adapted from Ref. 26.

known as a *staggered arrangement* [14,23]; see Figure 3.3. This staggered arrangement of mineral tablets is also found in other biological materials such as teeth, collagen fiber, spider silk, and cellulose fiber and is now considered a universal pattern [20], providing attractive combinations of stiffness, strength, and toughness [23,24]. Several analytical and numerical models have therefore been proposed to predict the behavior of staggered structures [25,26]. These models for stiffness, strength, and toughness are briefly reviewed next.

3.2.1 Stiffness

The tensile behavior of staggered structures has been investigated by a simple shear-tension chain model. The model is based on a small representative volume element (RVE) of the structure shown in Figure 3.4 [25]. The tablets have length L, thickness t, and overlap $L/2$.

These tablets are bonded between organic layers of thickness t_i.

Assuming an elastic, perfectly plastic behavior for the interface and a constant shear field along the interface, Kotha *et al.* [27] derived the following expression for the tensile modulus E of staggered composites:

$$\frac{1}{E} = \left(1 + \frac{t_i}{t}\right)\left[\frac{1}{E_m} + 2\frac{t}{L}\frac{t_i\gamma}{G_i}\left(\frac{1 + \cosh(\gamma L)}{\sinh(\gamma L)}\right)\right],$$

$$(3.1)$$

where

$$\gamma = \sqrt{\frac{G_i}{E_m}\frac{1}{t_i t}},$$

$$(3.2)$$

E_m is the Young's modulus of the mineral, and G_i is the shear modulus of the interface.

A reasonable assumption for biological hard materials is that the organic interfaces are much

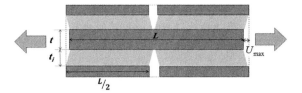

FIGURE 3.4 Representative volume element of a staggered composite when loaded in tension (U_{max} represents the maximum cohesive displacement).

softer than the mineral tablets ($G_i \ll E_m$) [14]. Equation (3.2) can then be rewritten as [28,29]

$$\frac{1}{E} \approx \frac{1}{\varphi E_m} + 4 \frac{1}{\rho^2} \frac{1-\varphi}{\varphi^2} \frac{1}{G_i}, \tag{3.3}$$

where ϕ is the volume fraction of mineral tablets and $\rho = L/t$ is the tablet aspect ratio. This expression reveals the effects of microstructural parameters on the modulus. Thus, for constant G_i, the elastic modulus of the material converges to ϕE_m when the mineral volume fraction ϕ increases to values near 1. Therefore, for high mineral concentrations, the modulus of staggered composites reaches its theoretical limit associated with a Voigt (uniform strain) composite.

3.2.2 Strength

Under tensile loading, staggered structures fail either at the interfaces (tablet pull-out fracture mode) or through the tablets (brittle fracture). The latter should be prevented so that the tablets slide on each other and energy is dissipated through inelastic deformation at the interfaces. Assuming tablet pull-out fracture mode and using a simple shear-tension load-transfer chain, we see that the strength of the composite σ_s can be written as [30]

$$\sigma_s = \frac{1}{2} \rho \tau_s, \tag{3.4}$$

provided that $t_i \ll t$. This expression shows that the strength of the composite is controlled by the shear strength τ_s of the interfaces and the aspect ratio ρ of tablets. Although strength increases with the aspect ratio, very high aspect

ratios may result in brittle fracture of the tablets, which is a detrimental failure mode. This limit to the aspect ratio of tablets is discussed in Section 3.2.5.

3.2.3 Toughness

Toughness is the most remarkable property of hard biological materials [31,32]. Nacre, as an example, shows a toughness that is 3,000 times higher than that of its main constituent (aragonite). Several experimental studies have identified the main toughening mechanisms of these materials [14,33]. Recently, it has also been theoretically demonstrated that the toughness achieved through the staggered arrangement of nacre is far greater than the toughness of both the mineral and organic mortar [26,29].

Several powerful mechanisms exist in nacre to resist crack propagation and increase toughness. Crack deflection, crack bridging, and viscoplastic-energy dissipation in volumes of material around cracks (process zone) are the dominant toughening mechanisms. Bridging develops as a crack advances and occurs when mineral tablets are not completely pulled out. The shear stresses between tablets therefore apply closure forces to the crack faces. The process zone, where the tablets slide onto one another, consists of two parts: the frontal zone and the wake. The frontal zone is the area in front of the crack tip experiencing inelastic deformation at the interfaces. Once the crack is advanced through the frontal zone, the stresses are released and some of the deformation is recovered, leaving a wake behind the crack tip, as depicted schematically in Figure 3.5. The energy is therefore dissipated through loading and unloading of inelastic interfaces. The effect of moisture on this inelastic behavior is also crucial; increasing the moisture plasticizes the organic molecules and increases their deformability [9].

When a crack advances across the direction of the tablets, as shown in Figure 3.5, the effect

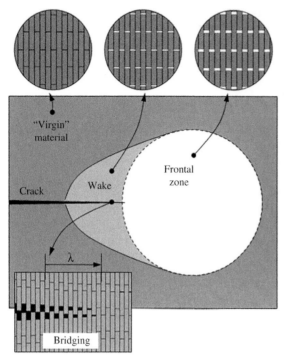

FIGURE 3.5 Schematic of a crack advancing in a staggered composite, where λ represents the bridging length and *a* represents the crack advance. Adapted from Ref. 30.

of bridging and process zone can then be estimated for the steady-state case; thus [29]

$$J = \frac{\rho}{2.5 - (U_{max}/L)(1/\rho)(E/\tau_s)} J_i, \quad (3.5)$$

where J is the mode-I fracture toughness of the composite, U_{max} is the maximum cohesive displacement as depicted in Figure 3.4, and J_i is the toughness of the interface. In this model, debonding is assumed to happen after the ultimate shear strain of the organic material at the interface is exceeded. The *ultimate shear strain* is the maximum shear strain that the organic material at interfaces can withstand.

The values for toughness predicted by this model were found to be in good agreement with experimental values for red-abalone nacre [29]. This shows that the toughness of nacre can be analytically explained by the effect of bridging

and process zone. Equation (3.5) also shows that: (i) staggered arrangement amplifies the toughness of interfaces, (ii) increasing the aspect ratio positively affects the toughness, and (iii) composites made of stiff inclusions and soft interfaces have enhanced toughness (term E/τ_s) [29]. These models for stiffness, strength, and toughness can greatly help the designers choose the best materials and microstructural parameters in order to tailor and optimize the performance of biomimetic staggered composites.

3.2.4 Strain Hardening at the Interfaces

The foregoing models show the effects of shear strength and shear modulus of interfaces on the behavior of staggered composites. However, in natural composites like nacre, the shear strength of the interfaces is not constant and increases with increases in shear strain. This strain hardening at the interfaces causes progressive tablet sliding (Figure 3.6a), which is one of the most important deformation mechanisms of biological hard materials like nacre and is the origin of toughening mechanisms such as viscoelastic energy dissipation at process zone.

Progressive sliding prevents strain localization and spreads the deformation through large volumes of material, thereby providing high levels of strain and therefore improving the energy absorption properties of the material (because this energy is the area under the stress–strain curve). Figures 3.6b and c show how the incorporation of wavy tablets improves the load transfer [14]. In the case of flat tablets, the load is transferred between the tablets only by shear stresses. For wavy tablets, tablet sliding generates transverse tensile and compression stresses, which contribute to the load transfer, increase the resistance over sliding, and generate hardening. The organic material itself generates hardening if the shear resistance of organic material increases with shear strain. The choice of organic material is therefore crucial for

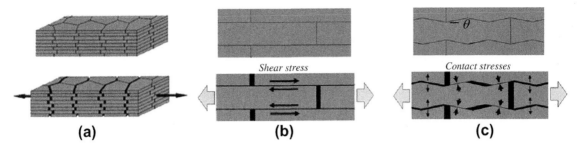

FIGURE 3.6 (a) Schematic of progressive tablet sliding in staggered composites. Load-transfer mechanisms in (b) flat tablets and (c) wavy tablets (θ is the dovetail angle).

improving the load transfer in biomimetic materials.

3.2.5 Size Effects

In staggered composites, the flow of stress is such that the interfaces are under shear while mineral inclusions are under tension (tension-shear chain). Therefore, the mineral inclusions should resist high levels of tensile stress in order to prevent brittle fracture. Brittle materials are sensitive to initial flaws, which, for example, include organic molecules embedded in the mineral crystals during the biomineralization process [34,35]. These organic molecules are much softer than the mineral and act as cracks within the material.

For a cracked brittle inclusion, the condition for failure is governed by the Griffith criterion:

$$\sigma_m^f = \alpha E_m \phi, \quad \phi = \sqrt{\frac{\gamma_s}{E_m h}}, \tag{3.6}$$

where σ_m^f is the fracture strength of the mineral, γ_s is the surface energy, h is the thickness of the mineral tablet, and the parameter α depends on the geometry of the crack. Based on the Griffith criterion, Gao *et al.* [26] showed that the strength of the inclusions increases when they are made smaller, because they can only contain small defects. In theory, inclusions smaller than a critical size of 30 nm [26] have a strength

approaching the theoretical strength of the material. Interestingly, the size of mineral inclusions in hard biological materials like bone and tooth is on the same order [36,37]. This suggests that nanometer inclusions in these materials maximize their fracture resistance.

Although in nacre the tablets are in the micrometer range, their small size still confers on them high strength. For example, Bekah *et al.* [30] found that the aspect ratio must be small enough so that an assumed edge crack extending halfway through the tablet is prevented from propagating further. This condition is given by:

$$\rho < 0.56 \frac{K_{IC}}{\tau_s \sqrt{t}}, \tag{3.7}$$

where K_{IC} is the mode-I fracture toughness of the tablets. This expression suggests that by decreasing the thickness of the tablets, the maximum allowable aspect ratio in the structure increases. Increasing the aspect ratio is desirable because it improves the performance of materials with staggered structure, as indicated by Eqs. (3.1)–(3.3).

Bekah *et al.* [30] also argued that junctions in the staggered composites act as crack-like features when the material is loaded in tension. Thus, decreasing the tablet thickness results in a decrease in the size of these crack-like features and therefore decreases the resulting stress-intensity factor K_I. Computing this stress-intensity factor and setting the condition $K_I < K_{IC}$ gives the *soft-wrap* condition for preventing the fracture of tablets:

$$\frac{K_{IC}}{\tau_s \sqrt{t}} > 0.58. \tag{3.8}$$

The *soft-wrap* condition also shows that decreasing the thickness of the tablets improves the fracture resistance of the structure. Rather than improving the fracture resistance of the tablets, reducing the size of building blocks also increases the number of interfaces and therefore increases the energy dissipation during loading and unloading of organic material, improving fracture resistance. These recent developments in the understanding of design aspects of these materials can greatly assist the optimization of biomimetic hard and stiff materials.

3.3 BIOMIMETIC HARD MATERIALS AT THE MACROSCALE

3.3.1 Fabrication

The fabrication of nacre-like synthetic composites that duplicate its structure, mechanisms, and properties has been and remains a formidable challenge. To this day there is no fabrication technology that can duplicate the highly regular brick-and-mortar structure of nacre at the microscale. One possible approach is to *relax* the constraints on small-length scales to fabricate millimeter-size structures. Although the advantage of small-length scales on the strength of the inclusion is lost, working at larger scale means that highly accurate fabrication techniques can be used. The structure and composition of nacre in term of organic/inorganic materials ratios can be duplicated, and the resulting materials show interesting mechanisms that are similar to natural nacre.

Clegg *et al.* [38] showed that a laminated composite material made of silicon-carbide (20 mm × 20 mm × 200 μm) sheets with graphite interfaces (Figure 3.7a) displays stiffness and strength values comparable to those of monolithic silicon carbide and toughness values four times higher. Following the same idea of fabrication of biomimetic materials at macro scales, Mayer [39] developed a segmented nacre-like composite made of almost 90 %v/v of alumina tablets that are bonded with adhesive tapes (Figure 3.7b). This material shows toughness six times higher than that of monolithic alumina.

The effects of various arrangements of tablets, different glues at the interface, and different mineral concentrations were researched by Mayer, whose conclusions are summarized as follows: (i) using weaker glues at the interfaces improves the toughness of the structure, which agrees well with Eq. (3.5); (ii) composites made of continuous or segmented layers with 82 %v/v ceramic undergo a catastrophic failure, whereas the composite made of segmented layers with 89 %v/v ceramic undergo significant deformation before failure, as shown in Figure 3.7b; and (iii) the toughness amplification of these macroscale artificial nacres is a result of considerable crack deflection observed when the material is loaded in bending [39].

Espinosa *et al.* [40] utilized a rapid prototyping method to fabricate artificial nacres composed of acrylonitrile butadiene styrene (ABS, a common prototyping material) inclusions as the stiffer phase and BGEBA (diglycidyl ether of bisphenol-A) epoxy as the softer phase. The tablets were made wavy by incorporating dovetails at both edges in order to generate hardening in the materials; see Figure 3.8a. This study showed that the incorporation of wavy inclusions results in the spread of the deformation through the whole material so that catastrophic failure is prevented. The section labeled C–D in Figure 3.8a shows that the presence of dovetail-like features in the tablets results in interlocking and hardening in the material.

Barthelat and Zhu utilized traditional machining techniques to fabricate wavy tablets

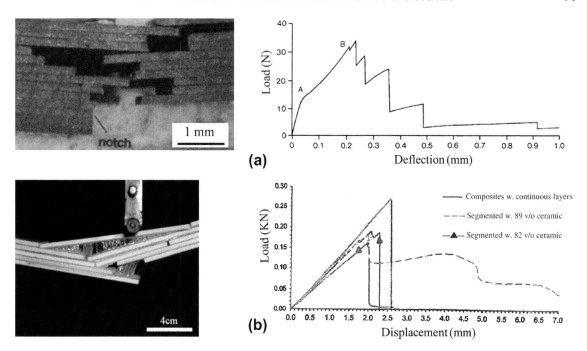

FIGURE 3.7 (a) Fractured laminated composite (SiC sheets bonded with graphite) and its load-deflection curve in bending (adapted from Ref. 38), and (b) artificial nacre made of segmented alumina tablets bonded with adhesive tape and its load versus its deflection curve in bending. Adapted from Ref. 39.

(inspired from the waviness observed in red-abalone shell [41]) to develop an artificial nacre at macroscale (Figure 3.8b) [42]. This artificial nacre is composed of PMMA wavy tablets held together by transverse fasteners, which play the role of mineral bridges in natural nacre. Using machining techniques for fabrication of tablets enables the designer to duplicate the structural features of natural composites into synthetic counterparts. This material duplicates the strain hardening, from progressive tablet locking observed in nacre and undergoes almost 10% strain before failure. This high strain is a result of using wavy tablets, which prevent strain localization, generate hardening, and embed efficient load transfer throughout the material [42]. The shape of the tensile stress-strain curve shown in Figure 3.8b resembles that of nacre, with an initial small linear elastic region

followed by large strains with pronounced hardening. Significantly, the behavior of this composite can be tuned by changing the initial compression on the layers exerted by transverse fasteners. No glue was used in this material, and shear cohesion and energy dissipation are provided by Coulomb dry friction. Analytical and finite element models were developed in order to predict, optimize, and design this type of material [42].

3.4 BIOMIMETIC HARD MATERIALS AT THE MICRO- AND NANOSCALES

As discussed in previous sections, reducing the size of the inclusions while keeping the volume constant improves mechanical properties,

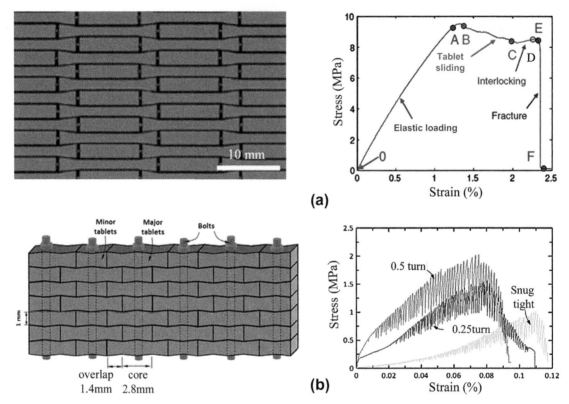

FIGURE 3.8 (a) Schematic illustration of artificial nacre composed of flat and wavy tablets and its tensile behavior (adapted from Ref. 40), and (b) schematic illustration of artificial nacre composed of wavy tablets and its tensile behavior. Adapted from Ref. 42.

particularly toughness. Significant research efforts are currently underway to develop artificial biomimetic materials with structures on the micrometer and nanometer scales by using a wide variety of fabrication techniques [16].

Nature uses biomineralization to build sophisticated hard and stiff materials. Although excellent qualitative investigations have been reported in natural biomineralization, some aspects of this complex process are still unknown [43–45]. Harvesting and controlling mineralization to fabricate complex and highly controlled mineral structures is still a research challenge. Material scientists, chemists, and engineers have therefore proposed alternative approaches consisting of innovative and unconventional techniques. In order for a methodology to successfully

duplicate the mechanisms observed in nature, it should (i) guarantee a microstructure design that conveys the main structural characteristic of natural composites, and (ii) utilize a fabrication strategy that is able to implement the design in several length scales so that large-scale structures with tailored microstructure can be fabricated from small inclusions. This type of microstructure increases the number of interface interactions, which ultimately results in efficient load transfer and improves mechanical performance.

The design of biomimetic hard materials starts by selecting the ingredients, which of course do not need to be of the same chemical composition as the natural original. For the hard and stiff inclusions, materials such as hydroxyapatite, calcium carbonate, glass, graphite, montmorillonite

(MTM) clay, silicon carbide, aluminum oxide, mica, and talc have been utilized [9, 46, 47].

The selection criteria for the organic phase are more complicated. The organic phase should show a strong adhesion to the mineral phase in order to prevent delamination at low levels of stress. Concurrently, the organic phase is required to be ductile and deformable so that it can resist high levels of strain. Different types of polymers, including polyelectrolytes (PEs), polyvinylalcohol (PVA), polycarbonate (PC), and poly(methyl methacrylate) (PMMA) have been used as the organic phase in biomimetic materials [46, 48–50]. Some of these polymers are attractive for chemical assembly processes. Thus, PVA has a good layering ability and forms hydrogen bonds with some minerals [48, 49], and PEs contain electric charges that facilitate the assembly processes [51].

Several innovative approaches were developed recently to assemble the chosen ingredients [52]. These methods can be classified as: (i) the freeze-casting method, (ii) layer-by-layer (LBL) assembly, (iii) direct-deposition techniques, and (iv) centrifugation, sedimentation, shearing, and gel casting. Other methods such as template-assisted fabrication have been utilized to develop bone-like materials. A brief review of these techniques along with their aptitudes and limitations is included in the remainder of this section.

3.4.1 Freeze-Casting Method

The freeze-casting method has recently been utilized to develop layered nacre-like composites. This innovative technique is based on the anisotropic growth of ice crystals by controlled freezing. The process starts with preparation of a suspension of ceramic particles in water. This suspension is then frozen in a controlled fashion so that flat ice structures with thickness of several micrometers are generated. As the ceramic particles are expelled from the forming ice, they assemble to form ceramic layers constrained between the planes of ice (Figure 3.9a).

Sublimating the ice by freeze-drying followed by a sintering step produces a porous scaffold composed of distinctive ceramic layers (Figure 3.9b). This scaffold is then filled with a tougher second phase (Figure 3.9c) such as a polymer, a metal, or a metallic alloy, resulting in a layered ceramic/polymer (or ceramic/metal) composite [50, 53, 54]. The layered composite can then be pressed so that the ceramic phase breaks into tablets, resulting in segmented layered composite resembling nacre (Figure 3.9d). Figure 3.9e shows the bending stress-strain behavior of an alumina/PMMA composite fabricated by the freeze-casting technique compared to that of hydrated red-abalone nacre. It shows that the strength and area under the strain-stress curve are higher for the alumina/ PMMA composite. However, it is noteworthy that the constituents and the composition are also different for the two materials.

The advantages of this method are as follows:

(i) Bulk-layered hybrid composites with proper control over the thickness of layers can be developed.

(ii) The success of the process is not dependent on the interfacial compatibility of different phases, so that a large variety of materials can be used.

(iii) Well-controlled interface behavior can be induced either by changing the surface roughness of phases or by improving chemical bonds between phases (grafting) [54].

This method yields materials with microstructures and deformation behavior resembling those of natural nacre, but it cannot control the overlap between segmented minerals at neighboring layers. Munch et al. [54] used this strategy to fabricate artificial nacre, composed of 80 %v/v of Al_2O_3 and 20 %v/v PMMA, which showed 1.4% strain to failure, a strength higher than the prediction of the rule of mixtures [55], and toughness of two orders of magnitude higher than that of Al_2O_3.

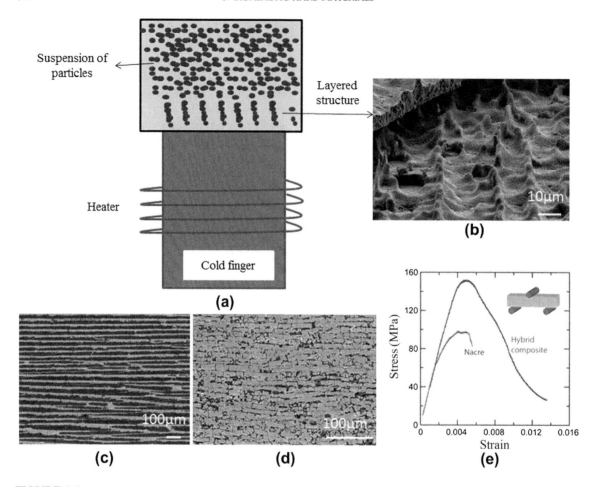

FIGURE 3.9 Freeze casting method to develop artificial nacres. (a) Schematic of the set-up, (b) aluminum oxide layers after the ice layers are sublimated, (c) layered composite made of alumina layers (the lighter phase) filled with PMMA (the darker phase), (d) the segmented layered structure resulted from compressing the layered composite, and (e) flexural stress-strain behavior of the composite compared to that of nacre. Adapted from Ref. 54.

3.4.2 Layer-by-Layer Assembly

Layer-by-layer (LBL) assembly is based on sequential deposition of nanometers-thick layers of substances from oppositely charged compounds [56]. The substances can be PEs, carbon nanotubes, charged nanoparticles, biological macromolecules, etc. [56–62].

A substrate is coated with a sacrificial layer and then is sequentially dipped into solutions of oppositely charged substances with intermediate rinsing steps (Figure 3.10a). In each dipping step, a layer of substances that have opposite charge to the substrate is coated on the surface by electrostatic interactions. For the case of artificial nacre, these substances can be nanoclays and different polymers. After the layers are coated, the sacrificial layer is dissolved so that a stable freestanding film is developed. This film is stabilized by ionic and hydrogen bonds, which

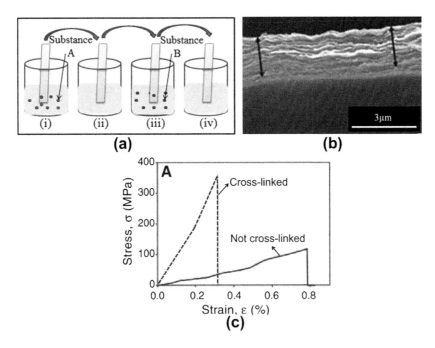

FIGURE 3.10   (a) The process plan for LBL assembly consists of sequential (i) dipping in substance A, (ii) rinsing, (iii) dipping in substance B, and (iv) rinsing. (b) A cross-sectional scanning electron micrograph of a multilayered nano particle/polymer composite fabricated by LBL assembly. (c) Tensile behavior of a PVA/MTM composite with and without cross-linking. Adapted from Ref. 48.

lead to an effective load transfer [48, 58, 59]. The polymer can be cross-linked to incorporate covalent bonds at the interface.

In addition to the possibility of manipulating the interface behavior, this process has advantages such as: (i) nanometer control over the thickness of layers, (ii) ability to fabricate homogenous films with little or no phase separation, and (iii) incorporation of antimicrobial properties and biocompatibility with the human osteoblast for biomedical applications [51]. However, for the case of artificial nacre, the mechanical properties of the film degrade for mineral concentrations higher than 15 %v/v. Other main disadvantages are that it is a slow process and there are practical limitations to the maximum number of layers that can be deposited.

The most successful artificial nacre fabricated by this method is a PVA/MTM layered composite, which shows a strength of 10 times higher than the organic phase (PVA) and an average ultimate tensile strain of 0.7% (Figure 3.10b) [48,63]. Figure 3.10c shows the tensile behavior of the composite with and without cross-linking. Covalent cross-linking at the interfaces increases stiffness and strength at the expense of reducing the extensibility of the material.

Another innovative technique is based on sequential deposition of inorganic monolayers formed at a liquid-gas interface followed by a polymer coating stage. Submicrometer surface-modified platelets are attracted to each other by hydrophobic interactions at a water-air interface to form an inorganic monolayer (Figure 3.11a) [64]. This monolayer is then deposited on a substrate and coated with a polymer so that a bilayer of inorganic/organic materials is produced. This process is repeated several times to fabricate a multilayered composite with total thickness on the order of few tens of micrometers (Figure 3.11b). A free-standing composite film is then obtained by peeling this multilayer off with a razor blade [46].

Bonderer *et al.* [65] used this strategy to fabricate an alumina/chitosan hybrid material with mineral concentration of almost 15 %v/v, which shows a high level of inelastic deformation (17% strain to failure). The tensile stress-strain curves of four sets of composites containing different mineral concentrations are shown in Figure 3.11c, indicating that increasing the mineral concentration up to 15 %v/v results in overall improved tensile behavior. The large number of hydrogen bonds between the hydrophobic surface of the inorganic platelets and chitosan results in highly efficient load transfer between the inclusions and the matrix.

3.4.3 Direct-Deposition Techniques

Deposition techniques such as sequential sputtering [66], inkjet deposition [67], and spin coating [68] have been utilized for fabrication of artificial nacres. Sequential ion-beam sputtering of metal/oxide layers will result in a layered structure with individual layer thickness on the order of several nanometers. Using this technique, He *et al.* [66] reported a layered TiN/ Pt nanocomposite, which is harder than the

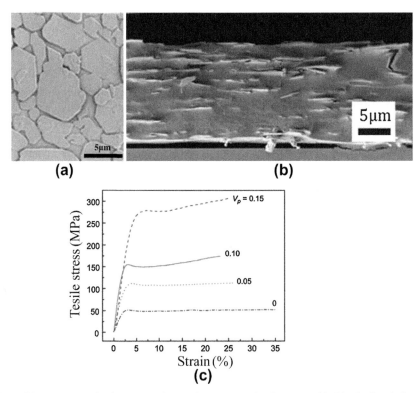

FIGURE 3.11   A SEM micrograph of one monolayer of inorganic platelets assembled by hydrophobic interactions and deposited on a substrate, (b) a cross-section of the multilayered composite (scale bar=5 μm), and (c) tensile behavior of the composites with different mineral concentration. Adapted from Ref. 46.

TiN phase (the harder phase in the structure). This technique yields multilayered structures with sharp interfaces. However, it is a very slow process, as discussed in Chapter 15 by Martín-Palma and Lakhtakia, and the load-transfer mechanics are only based on molecular interaction at interfaces. The restricted choice of materials is another disadvantage.

Another novel technique that can be categorized as either an LBL technique or as a deposition technique is sequential inkjet deposition of different materials on a substrate. This promising technique demonstrates nanometer control over the thickness of constituents and well-controlled mineral concentration. It has two advantages over traditional LBL techniques: (i) there is no need to go through several intermediate rinsing steps, as required for LBL techniques, and (ii) direct patterning on the substrate is possible, thereby eliminating the need for subsequent etching steps. Andres et al. [67] recently utilized this technique to develop stable multilayered nanocomposite films. A combination of this technique and the traditional LBL technique, has also been used for several purposes, including selectively activating a multilayered surface for metal plating or modifying the surface to enhance adhesion [69,70].

3.4.4 Centrifugation, Sedimentation, Shearing, and Gel Casting

These processes usually start with coating nano-platelets with polymers. The nanoplatelets can then be assembled using a variety of techniques, including centrifugation [71], sedimentation [49,72], and shearing [73]. These techniques are relatively faster than other bottom-up assembly techniques such as LBL assembly. Using a simple sedimentation technique, Walther et al. [49] recently fabricated a PVA/nanoclay composite with a tensile strength of 170 MPa and ultimate tensile strain of 0.014. The results are consistent with the results obtained earlier using the

LBL technique [48]. However, the sedimentation technique is much faster than LBL and could lead to high-throughput nanocomposite fabrication.

Gel casting can also be used to fabricate polymer-reinforced platelets. As the first step, the platelets are well dispersed in a solution that turns to a gel upon cooling, so that the particles are locked in their place. The solvent is then vaporized to make the structure denser. This polymer-reinforced platelet is then hot-pressed to further align the platelets. Bonderer et al. [74] showed that, for platelet concentration up to 40 % v/v, the mechanical behavior of these composites can be improved by increasing the platelet content but degrades for higher concentrations. This composite has strength and stiffness values of, respectively, 83% and 13 times higher than the pure polymer matrix.

3.4.5 Template-Assisted Fabrication

Template-assisted fabrication methods are defined here as the methods in which surface-modified particles are assembled into a mechanically or chemically modified template. This template controls the attraction, growth, and morphology of the resulting structures [75]. For example, use of cavities with the same size as the colloidal particles would direct the particles to self-assemble into the cavities in a lock-and-key manner [76,77]. In many cases, a two-dimensional patterned template is used to control the growth and morphology of colloidal crystals [78,79]. One-dimensional linear templates like DNA or single-walled carbon nanotubes (SWNTs) are also used to direct the organization of nanoparticles [80,81] or polymers [82,83].

Template-assisted self-assembly has found wide applications in bone mimicry and reconstruction [84,85]. Zhao et al. reported the use of chemically functionalized SWNTs as templates to grow bone-like materials [85]. Hydroxyapatite particles are nucleated and grown into these chemically patterned SWNTs in a well-aligned platelet form. The thickness of these hydroxyapatite layers is controlled by the coating time and can reach values as high as $3\,\mu$m.

3.5 CONCLUSION AND OUTLOOK

Despite significant efforts to duplicate the structural and functional properties of biological hard and stiff materials, only a few successful implementations of these mechanisms in biomimetic materials have been reported [48, 54]. Also, none of these studies have been able to develop micro-/nanocomposites with the high level of structural organization observed in natural composites. Finally, the high level of mineral concentration in hard biological materials (e.g., 95 %w/w for nacre, 99 %w/w for enamel) has not been achieved in their synthetic counterparts.

Experimental investigations show that high mineral concentrations result in poor mechanical properties, particularly toughness, in biomimetic materials [46,86,87] whereas models for fracture toughness of biological composites predict the opposite trend [29]. This moderate success of biomimetic materials can be explained by the following limitations:

(1) In biomimetic materials, the mineral tablets are not organized in a controlled fashion in the polymer matrix [88], whereas the mineral tablets in natural composites like bone and nacre are arranged in a well-organized, staggered structure. This lack of configuration in biomimetic materials necessitates the use of highly directed fabrication methods rather than simple mixing of the phases [89]. This also restricts fabrication of these materials to higher-length scales so that controlled arrangements of tablets can be more easily achieved.

(2) The bonding between mineral and polymer phases is not as efficient as it is in natural composites. In nacre, the sacrificial ionic bonds at the interface, which can reform, are one of the keys to its superior mechanical properties. This encourages fundamental research to incorporate such bonding in biomimetic materials. The presence of these bonds prevents catastrophic failure as the broken bonds at the interfaces can reform so that the integrity of the structure is maintained, even at high levels of strains. The challenging question of whether this break-reform fashion can be engaged in other bonds such as covalent bonds remains open [87].

Meanwhile, composites with well-organized structures made of macroscale inclusions have been developed relatively easily at the expense of losing the advantages of using inclusions of small size. Therefore, although reducing the size of the inclusions is beneficial according to the design guidelines for staggered composites, it may not result in the expected high performance due to the limitations of the fabrication processes at small scales, i.e., poor structural organization. There is therefore a trade-off between the size of inclusions and scalability of well-designed structure. State-of-the-art biomimetic studies, however, aim to explore innovative and promising fabrication methods to develop structures with high levels of structural organization made of micro/nano inclusions.

References

[1] M.F. Ashby, Hybrids to fill holes in material property space, *Phil Mag* **85** (2005), 3235–3257.

[2] F. Barthelat, Biomimetics for next-generation materials, *Phil Trans R Soc Lond A* **365** (2007), 2907–2919.

[3] M.F. Ashby and Y.J.M. Bréchet, Designing hybrid materials, *Acta Mater* **51** (2003), 5801–5821.

[4] R. Lakes, Advances in negative Poisson's ratio materials, *Adv Mater* **5** (1993), 293–296.

[5] M. Miki and Y. Murotsu, The peculiar behavior of the Poisson's ratio of laminated fibrous composites, *JSME Int J I, Solid Mech Strength Mater* **32** (1989), 67–72.

[6] C.T. Herakovich, Composite laminates with negative through-the-thickness Poisson's ratios, *J Compos Mater* **18** (1984), 447–455.

[7] R. Lakes, Cellular solid structures with unbounded thermal expansion, *J Mater Sci Lett* **15** (1996), 475–477.

[8] A.P. Jackson, J.F.V. Vincent, and R.M. Turner, The mechanical design of nacre, *Proc R Soc Lond B* **234** (1988), 415–440.

[9] G. Mayer, Rigid biological systems as models for synthetic composites, *Science* **310** (2005), 1144–1147.

[10] G.M. Luz and J.F. Mano, Mineralized structures in nature: examples and inspirations for the design of new composite materials and biomaterials, *Compos Sci Technol* **70** (2010), 1777–1788.

[11] J.W.C. Dunlop and P. Fratzl, Biological composites (D.R. Clarke, M. Ruhle, and F. Zok, eds.), *Annu Rev Mater Res* **40** (2010), 1–24.

[12] P. Fratzl, I. Burgert, and H.S. Gupta, On the role of interface polymers for the mechanics of natural polymeric composites, *Phys Chem Chem Phys* **6** (2004), 5575–5579.

[13] J.W.C. Dunlop, R. Weinkamer, and P. Fratzl, Artful interfaces within biological materials, *Mater Today* **14** (3) (March 2011), 70–78.

[14] F. Barthelat, H. Tang, P.D. Zavattieri, C.M. Li, and H.D. Espinosa, On the mechanics of mother-of-pearl: a key feature in the material hierarchical structure, *J Mech Phys Solids* **55** (2007), 306–337.

[15] M.J. Harrington and J.H. Waite, How nature modulates a fiber's mechanical properties: mechanically distinct fibers drawn from natural mesogenic block copolymer variants, *Adv Mater* **21** (2009), 440–444.

[16] R.O. Ritchie, The conflicts between strength and toughness, *Nat Mater* **10** (2011), 817–822.

[17] G.E. Fantner, T. Hassenkam, J.H. Kindt, J.C. Weaver, H. Birkedal, L. Pechenik, J.A. Cutroni, G.A.G. Cidade, G.D. Stucky, D.E. Morse, and P.K. Hansma, Sacrificial bonds and hidden length dissipate energy as mineralized fibrils separate during bone fracture, *Nat Mater* **4** (2005), 612–616.

[18] B.L. Smith, T.E. Schäffer, M. Vlani, J.B. Thompson, N.A. Frederick, J. Klndt, A. Belcher, G.D. Stuckyll, D.E. Morse, and P.K. Hansma, Molecular mechanistic origin of the toughness of natural adhesives, fibres and composites, *Nature* **399** (1999), 761–763.

[19] S. Mann, Molecular recognition in biomineralization, *Nature* **332** (1988), 119–124.

[20] M.J. Buehler, S. Keten, and T. Ackbarow, Theoretical and computational hierarchical nanomechanics of protein materials: deformation and fracture, *Prog Mater Sci* **53** (2008), 1101–1241.

[21] I.A. Aksay, M. Trau, S. Manne, I. Honma, N. Yao, L. Zhou, P. Fenter, P.M. Eisenberger, and S.M. Gruner, Biomimetic pathways for assembling inorganic thin films, *Science* **273** (1996), 892–898.

[22] M.A. Meyers, P.Y. Chen, A.Y.M. Lin, and Y. Seki, Biological materials: structure and mechanical properties, *Prog Mater Sci* **53** (2008), 1–206.

[23] X. Guo and H. Gao, *Bio-inspired material design and optimization*, Springer-Verlag, Heidelberg, Germany, New York (2006).

[24] H. Gao, Application of fracture mechanics concepts to hierarchical biomechanics of bone and bone-like materials, *Int J Fracture* **138** (2006), 101–137.

[25] I. Jager and P. Fratzl, Mineralized collagen fibrils: a mechanical model with a staggered arrangement of mineral particles, *Biophys J* **79** (2000), 1737–1746.

[26] H. Gao, B. Ji, I.L. Jäger, E. Arzt, and P. Fratzl, Materials become insensitive to flaws at nanoscale: lessons from nature, *Proc Natl Acad Sci* **100** (2003), 5597.

[27] S. Kotha, Y. Li, and N. Guzelsu, Micromechanical model of nacre tested in tension, *J Mater Sci* **36** (2001), 2001–2007.

[28] S.P. Kotha, Y. Li, and N. Guzelsu, Micromechanical model of nacre tested in tension, *J Mater Sci* **36** (2001), 2001–2007.

[29] F. Barthelat and R. Rabiei, Toughness amplification in natural composites, *J Mech Phys Solids* **59** (2011), 829–840.

[30] S. Bekah, R. Rabiei, and F. Barthelat, Structure, scaling, and performance of natural micro- and nanocomposites, *BioNanoSci* **1** (2011), 1–9.

[31] J.C. Weaver, G.W. Milliron, A. Miserez, K. Evans-Lutterodt, S. Herrera, I. Gallana, W.J. Mershon, B. Swanson, P. Zavattieri, and E. DiMasi, The stomatopod dactyl club: a formidable damage-tolerant biological hammer, *Science* **336** (2012), 1275–1280.

[32] K.E. Tanner, Small but extremely tough, *Science* **336** (2012), 1237–1238.

[33] R. Rabiei, S. Bekah, and F. Barthelat, Failure mode transition in nacre and bone-like materials, *Acta Biomater* **6** (2010), 4081–4089.

[34] A. Sellinger, P.M. Weiss, A. Nguyen, Y. Lu, R.A. Assink, W. Gong, and C.J. Brinker, Continuous self-assembly of organic-inorganic nanocomposite coatings that mimic nacre, *Nature* **394** (1998), 256–260.

[35] J.H.E. Cartwright and A.G. Checa, The dynamics of nacre self-assembly, *J R Soc Interf* **4** (2007), 491–504.

[36] J.Y. Rho, L. Kuhn-Spearing, and P. Zioupos, Mechanical properties and the hierarchical structure of bone, *Med Eng Phys* **20** (1998), 92–102.

[37] W. Tesch, N. Eidelman, P. Roschger, F. Goldenberg, K. Klaushofer, and P. Fratzl, Graded microstructure and mechanical properties of human crown dentin, *Calcified Tissue Int* **69** (2001), 147–157.

[38] W.J. Clegg, K. Kendall, N.M. Alford, T.W. Button, and J.D. Birchall, A simple way to make tough ceramics, *Nature* **347** (1990), 455–457.

[39] G. Mayer, New classes of tough composite materials: Lessons from natural rigid biological systems, *Mat Sci Eng C* **26** (2006), 1261–1268.

[40] H.D. Espinosa, J.E. Rim, F. Barthelat, and M.J. Buehler, Merger of structure and material in nacre and bone: Perspectives on de novo biomimetic materials, *Prog Mater Sci* **54** (2009), 1059–1100.

[41] F. Barthelat, Nacre from mollusk shells: a model for high-performance structural materials, *Bioinsp Biomim* **5** (2010).

[42] F. Barthelat and D. Zhu, A novel biomimetic material duplicating the structure and mechanics of natural nacre, *J Mater Res* **26** (2011), 1203–1215.

[43] N.M. Alves, I.B. Leonor, H.S. Azevedo, R.L. Reis, and J.F. Mano, Designing biomaterials based on biomineralization of bone, *J Mater Chem* **20** (2010), 2911–2921.

[44] Y. Oaki and H. Imai, Hierarchically organized superstructure emerging from the exquisite association of inorganic crystals, organic polymers, and dyes: a model approach towards suprabiomineral materials, *Adv Funct Mater* **15** (2005), 1407–1414.

[45] T.E. Schaffer, C. Ionescu-Zanetti, R. Proksch, M. Fritz, D.A. Walters, N. Almqvist, C.M. Zaremba, A.M. Belcher, B.L. Smith, G.D. Stucky, D.E. Morse, and P.K. Hansma, Does abalone nacre form heteroepitaxial nucleation or by growth through mineral bridges? (Vol. 9, p. 1731, 1997), *Chem Mater* **10** (1998), 946–946.

[46] L.J. Bonderer, A.R. Studart, and L.J. Gauckler, Bioinspired design and assembly of platelet reinforced polymer films, *Science* **319** (2008), 1069–1073.

[47] T.H. Lin, W.H. Huang, I.K. Jun, and P. Jiang, Bioinspired assembly of surface-roughened nanoplatelets, *J Colloid Interf Sci* **344** (2010), 272–278.

[48] P. Podsiadlo, A.K. Kaushik, E.M. Arruda, A.M. Waas, B.S. Shim, J. Xu, H. Nandivada, B.G. Pumplin, J. Lahann, A. Ramamoorthy, and N.A. Kotov, Ultrastrong and stiff layered polymer nanocomposites, *Science* **318** (2007), 80–83.

[49] A. Walther, I. Bjurhager, J.M. Malho, J. Pere, J. Ruokolainen, L.A. Berglund, and O. Ikkala, Large-area, lightweight, and thick biomimetic composites with superior material properties via fast, economic, and green pathways, *Nano Lett* **10** (2010), 2742–2748.

[50] M.E. Launey, E. Munch, D.H. Alsem, E. Saiz, A.P. Tomsia, and R.O. Ritchie, A novel biomimetic approach to the design of high-performance ceramic-metal composites, *J R Soc Interf* **7** (2010), 741–753.

[51] P. Podsiadlo, S. Paternel, J.M. Rouillard, Z. Zhang, J. Lee, J.W. Lee, E. Gulari, and N.A. Kotov, Layer-by-layer assembly of nacre-like nanostructured composites with antimicrobial properties, *Langmuir* **21** (2005), 11915–11921.

[52] K. Liu and L. Jiang, Bio-inspired design of multiscale structures for function integration, *Nano Today* **6**(2) (April 2011), 155–175.

[53] S. Deville, E. Saiz, R.K. Nalla, and A.P. Tomsia, Freezing as a path to build complex composites, *Science* **311** (2006), 515–518.

[54] E. Munch, M.E. Launey, D.H. Alsem, E. Saiz, A.P. Tomsia, and R.O. Ritchie, Tough, bio-inspired hybrid materials, *Science* **322** (2008), 1516–1520.

[55] D. Hull and T. Clyne, *An introduction to composite materials*, Cambridge University Press, Cambridge, UK (1996).

[56] R.K. Iler, Multilayers of colloidal particles, *J Colloid Interf Sci* **21** (1966), 569–594.

[57] O. Mermut, J. Lefebvre, D.G. Gray, and C.J. Barrett, Structural and mechanical properties of polyelectrolyte multilayer films studied by AFM, *Macromolecules* **36** (2003), 8819–8824.

[58] C.A. Wang, B. Long, W. Lin, Y. Huang, and J. Sun, Poly(amic acid)-clay nacrelike composites prepared by electrophoretic deposition, *J Mater Res* **23** (2008), 1706–1712.

[59] Z. Tang, N.A. Kotov, S. Magonov, and B. Ozturk, Nanostructured artificial nacre, *Nat Mater* **2** (2003), 413–418.

[60] P. Podsiadlo, M. Qin, M. Cuddihy, J. Zhu, K. Critchley, E. Kheng, A.K. Kaushik, Y. Qi, H.S. Kim, S.T. Noh, E.M. Arruda, A.M. Waas, and N.A. Kotov, Highly ductile multilayered films by layer-by-layer assembly of oppositely charged polyurethanes for biomedical applications, *Langmuir* **25** (2009), 14093–14099.

[61] P. Lavalle, J.C. Voegel, D. Vautier, B. Senger, P. Schaaf, and V. Ball, Dynamic aspects of films prepared by a sequential deposition of species: perspectives for smart and responsive materials, *Adv Mater* **23** (2011), 1191–1221.

[62] A.A. Mamedov, N.A. Kotov, M. Prato, D.M. Guldi, J.P. Wicksted, and A. Hirsch, Molecular design of strong single-wall carbon nanotube/polyelectrolyte multilayer composites, *Nat Mater* **1** (2002), 190–194.

[63] P. Podsiadlo, M. Michel, J. Lee, E. Verploegen, N.W.S. Kam, V. Ball, Y. Qi, A.J. Hart, P.T. Hammond, and N.A. Kotov, Exponential growth of LBL films with incorporated inorganic sheets, *Nano Lett* **8** (2008), 1762–1770.

[64] U. Abraham, *An introduction to ultrathin organic films: from Langmuir-Blodgett to self-assembly*, Academic Press, London, UK (1946).

[65] L.J. Bonderer, A.R. Studart, J. Woltersdorf, E. Pippel, and L.J. Gauckler, Strong and ductile platelet-reinforced polymer films inspired by nature: microstructure and mechanical properties, *J Mater Res* **24** (2009), 2741–2754.

[66] J.L. He, J. Wang, W.Z. Li, and H.D. Li, Simulation of nacre with TiN/Pt multilayers and a study of their

mechanical properties, *Mater Sci Eng B* **49** (1997), 128–134.

[67] C.M. Andres and N.A. Kotov, Inkjet deposition of layer-by-layer assembled films, *J Am Chem Soc* **132** (2010), 14496–14502.

[68] T. Kato, Polymer/calcium carbonate layered thin-film composites, *Adv Mater* **12** (2000), 1543–1546.

[69] S. Limem, D. Li, S. Iyengar, and P. Calvert, Multi-material inkjet printing of self-assembling and reacting coatings, *J Macromol Sci A: Pure Appl Chem* **46** (2009), 1205–1212.

[70] T.C. Wang, B. Chen, M.F. Rubner, and R.E. Cohen, Selective electroless nickel plating on polyelectrolyte multilayer platforms, *Langmuir* **17** (2001), 6610–6615.

[71] H.B. Yao, Z.H. Tan, H.Y. Fang, and S.H. Yu, Artificial nacre-like bionanocomposite films from the self-assembly of chitosan-montmorillonite hybrid building blocks, *Angew Chem Int Ed* **49** (2010), 10127–10131.

[72] A. Walther, I. Bjurhager, J.M. Malho, J. Ruokolainen, L. Berglund, and O. Ikkala, Supramolecular control of stiffness and strength in lightweight high-performance nacre-mimetic paper with fire-shielding properties, *Angew Chem Int Ed* **49** (2010), 6448–6453.

[73] N. Almqvist, N.H. Thomson, B.L. Smith, G.D. Stucky, D.E. Morse, and P.K. Hansma, Methods for fabricating and characterizing a new generation of biomimetic materials, *Mater Sci Eng C* **7** (1999), 37–43.

[74] L.J. Bonderer, K. Feldman, and L.J. Gauckler, Platelet-reinforced polymer matrix composites by combined gel-casting and hot-pressing. Part I: polypropylene matrix composites, *Compos Sci Technol* **70** (2010), 1958–1965.

[75] B.A. Grzybowski, C.E. Wilmer, J. Kim, K.P. Browne, and K.J.M. Bishop, Self-assembly: from crystals to cells, *Soft Matter* **5** (2009), 1110–1128.

[76] C.J. Hernandez and T.G. Mason, Colloidal alphabet soup: monodisperse dispersions of shape-designed LithoParticles, *J Phys Chem C* **111** (2007), 4477–4480.

[77] T.D. Clark, R. Ferrigno, J. Tien, K.E. Paul, and G.M. Whitesides, Template-directed self-assembly of 10-μm-sized hexagonal plates, *J Am Chem Soc* **124** (2002), 5419–5426.

[78] A. Van Blaaderen, R. Ruel, and P. Wiltzius, Template-directed colloidal crystallization, *Nature* **385** (1997), 321–323.

[79] P. Schall, I. Cohen, D.A. Weitz, and F. Spaepen, Visualization of dislocation dynamics in colloidal crystals, *Science* **305** (2004), 1944–1948.

[80] J.D. Le, Y. Pinto, N.C. Seeman, K. Musier-Forsyth, T.A. Taton, and R.A. Kiehl, DNA-templated self-assembly of metallic nanocomponent arrays on a surface, *Nano Lett* **4** (2004), 2343–2347.

[81] R.J. Macfarlane, M.R. Jones, A.J. Senesi, K.L. Young, B. Lee, J. Wu, and C.A. Mirkin, Establishing the design rules for DNA-mediated programmable colloidal crystallization, *Angew Chem Int Ed* **49** (2010), 4589–4592.

[82] Y. Shao, Y. Jin, and S. Dong, DNA-templated assembly and electropolymerization of aniline on gold surface, *Electrochem Commun* **4** (2002), 773–779.

[83] H. Li, S.H. Park, J.H. Reif, T.H. LaBean, and H. Yan, DNA-templated self-assembly of protein and nanoparticle linear arrays, *J Am Chem Soc* **126** (2004), 418–419.

[84] G. Atlan, O. Delattre, S. Berland, A. LeFaou, G. Nabias, D. Cot, and E. Lopez, Interface between bone and nacre implants in sheep, *Biomaterials* **20** (1999), 1017–1022.

[85] B. Zhao, H. Hu, S.K. Mandal, and R.C. Haddon, A bone mimic based on the self-assembly of hydroxyapatite on chemically functionalized single-walled carbon nanotubes, *Chem Mater* **17** (2005), 3235–3241.

[86] L.J. Bonderer, K. Feldman, and L.J. Gauckler, Platelet-reinforced polymer matrix composites by combined gel-casting and hot-pressing. Part II: Thermoplastic polyurethane matrix composites, *Compos Sci Technol* **70** (2010), 1966–1972.

[87] P. Podsiadlo, *Layer-by-layer assembly of nanostructured composites: mechanics and applications in chemical engineering*, PhD dissertation, University of Michigan (2008).

[88] X.L. Xie, Y.W. Mai, and X.P. Zhou, Dispersion and alignment of carbon nanotubes in polymer matrix: a review, *Mater Sci Eng R* **49** (2005), 89–112.

[89] Y. Zhang and J.R.G. Evans, Approaches to the manufacture of layered nanocomposites, *Appl Surf Sci* **258** (2012), 2098–2102.

ABOUT THE AUTHORS

Mohammad Mirkhalaf was born in Isfahan, Iran, in 1985. He received his BS and MEng in mechanical engineering at Isfahan University of Technology (IUT) and Nanyang Technological University (NTU), respectively. He is now a PhD candidate in mechanical engineering at McGill University. His main research interests include structure and performance of biological materials, design and development of biomimetic materials, and micro/nano fabrication methods. Mr. Mirkhalaf has won several scholarships and awards, namely A*STAR scholarship at NTU, the McGill Engineering Doctoral Award (MEDA), and the James McConnell Fellowship at McGill.

Deju Zhu is a postdoctoral fellow in the Department of Mechanical Engineering at McGill University, Montreal. He received his BS and MS in civil engineering from Northeast Forestry University (China) in 2001 and 2004, respectively, and his PhD in civil engineering from Arizona State University in 2009. He is a member of the American Concrete Institute and the Society for Experimental Mechanics. His research interests include structure and mechanics of natural materials and novel bio-inspired materials, dynamic behavior of fabrics and textile-reinforced cement composites, experimental techniques of impact and high strain-rate testing, and computational simulation of woven fabrics subjected to ballistic impact.

Francois Barthelat is an associate professor in the Department of Mechanical Engineering at McGill University (Montreal, Canada). In 2006 he created the Biomimetic Materials Laboratory, which focuses on the structure and mechanics of natural materials and on the design, fabrication, and testing of engineering materials inspired from nature. Projects include mechanics of seashells, bone, natural interfaces and adhesives, and fish scales. Professor Barthelat and his team have also developed new nacre-like materials and bio-inspired fibers for composites. He has won a number of awards, including the Hetenyi Award for the best research article in the journal *Experimental Mechanics* (2003), the Award for the Best Paper by a Young Researcher at the 12th International Conference on Fracture (2009), and the Best Paper Award at the 2011 Society for Experimental Mechanics Annual Conference (Biological Systems and Materials Division). He currently serves on the Editorial Boards of *Bioinspiration and Biomimetics* and *Experimental Technique*.

Biomimetic Robotics

Ranjan Vepa

School of Engineering and Materials Science, Queen Mary, University of London,
London E1 4NS, United Kingdom

Prospectus

Some basic features of biomimetic robotics and the technologies that are facilitating their development are discussed in this chapter. The emergence of smart materials and structures, smart sensors and actuators capable of mimicking biological transducers, bio-inspired signal-processing techniques, modeling and control of manipulators resembling biological limbs, and the shape control of flexible systems are the primary areas in which recent technological advances have taken place. Some key applications of these technological developments in the design of morphing airfoils, modeling and control of anthropomorphic manipulators and muscle activation modeling, and control for human limb prosthetic and orthotic applications are discussed. Also discussed, with some typical examples, are the related developments in the application of nonlinear optimal control and estimation, which are fundamental to the success of biomimetic robotics.

Keywords

Biomimetic robots, Electromyography, Electro-active polymer, Electrorheological fluids, Kalman filter, Magnetorheological fluids, Morphing, Nernst equation, Nonlinear optimal control, Orthotics, Prosthetics, Shape control, Shape-memory alloys, Smart structures

4.1 INTRODUCTION TO BIOMIMICRY

Biomimicry refers to the process of mimicking living plants and animals in the way in which they solve problems or tackle tasks confronting them. Thus, biomimicry aims to imitate biological processes or systems.

Research into biomimetic robotics is currently being conducted in many research institutions worldwide with the aim of developing a class of biomimetic robots that exhibit much greater competence and robustness in performance in unstructured environments than the robots that are currently in use. Biomimetically designed and empowered robots are expected to be more compliant and stable than current robots. Designers of such robots are taking advantage of new developments in materials, fabrication technologies, sensors, and actuators to provide a wide range of capabilities that allow these robots to mimic biological processes and systems.

Biomimetic robots are particularly suited to perform autonomous or semi-autonomous tasks, such as reconnaissance and exploration

Engineered Biomimicry

http://dx.doi.org/10.1016/B978-0-12-415995-2.00004-0

of new environments, both on Earth and in space. For the details of biomimetic robotics, readers are referred to a recent book [1].

It is important to mention at the outset that biomimicry is facilitated by the presence of mirror neurons that are fundamental in giving humans the ability to empathize. Developing empathy is also a way of learning, and biomimicry facilitates the design and building of self-learning robots. In this chapter we briefly review the technologies that facilitate biomimicry, such as smart materials and structures, biomimetic sensors and actuators, biomimetically inspired signal-processing and analysis techniques, novel mechanisms and manipulators, and the technology of shape changing and controlled morphing. Typical application examples are also discussed.

4.2 TECHNOLOGIES FACILITATING BIOMIMETIC ROBOTICS

4.2.1 Smart Materials and Smart Structures

Smart materials are a special class of materials of which the material properties can be influenced in a significant or novel way by changing the environmental conditions in which they operate. There are many examples of smart materials, such as electro-active polymers (EAPs) [2], which are a class of polymers of which the shape or size can be influenced by exposing them to an electric field. The novel feature of these materials is that, while undergoing high deformations due to the influence of the electric field, they are able to sustain large forces.

Metallic alloys called *shape-memory alloys* (SMAs) furnish examples of another type of smart material [3]. When an SMA is thermally activated by heating it beyond the activation temperature, there is a phase transformation and a rearrangement of the molecules.

Consequent to this phase change, the material changes its shape. However, when it is cooled back to its original temperature, the material reverts to the original phase and the shape of the material is recovered. Certain alloys of (1) nickel and titanium; (2) copper, zinc, and aluminum; and (3) copper, aluminum, and nickel exhibit the shape-memory property. The primary phases involved are generally the martensite phase at low temperatures and the austenite phase at high temperatures [4]. Because of the fact these SMAs are biologically compatible with human physiological organs, they are used for a variety of applications associated with robotic surgery.

Another class of smart materials is constituted by electrorheological and magnetorheological fluids [5], which are, respectively, controllable by the application of electric and magnetic fields. They consist of colloidal suspensions of particles that form structural chains parallel to the applied field. The resulting structure has a yield point that is determined by the applied field. Beyond the yield point, any applied shear stress results in an incremental rate of shear. Thus, the resistance to motion can be controlled by varying the applied field.

Smart structures may be considered to be a class of hybrid or composite structures that are capable of sensing and responding to a stimulus in a controllable and predictable way by the integration of various embedded smart materials or elements for sensing, actuation, autonomous power generation, signal processing, or any dedicated filtering task. Furthermore, like an ordinary structure, they are able to sustain mechanical loads while also performing additional functions such as isolating vibrations, reducing acoustic noise transmission, monitoring their own state or the environment, automatically performing precision alignments, or changing their shape or mechanical properties on command.

The concept of a smart or intelligent structure was mooted based on the stimulus-response characteristics of living organisms. A smart structure can be considered to be biomimetic when it is a smart structural system containing multifunctional parts that can perform sensing, control, and actuation as well as a primitive analog of a biological organism. Typically, a biomimetic smart structure is an integration of sensors, actuators, and a control system that is able to mimic a biological organism in performing many desirable functions, such as synchronization with environmental changes, self-repair of damage, etc.

A typical example of a smart structure is a composite structure with embedded piezoceramic layers capable of acting as piezoelectric actuators [6]. The piezoelectric effect is a property of certain materials including certain ceramics, such as lead zirconate titanate (PZT), and polymers, such as polyvinylidene flouride (PVDF), that cause it to generate a stress in response to an applied voltage. The stress generated in turn causes a particular type of strain that may then be employed either to bend or to alter the shape of a structure. Moreover, one is able to minutely deform a structure at specific locations in a controlled and relatively fast manner, thereby facilitating their use in high-precision positioning applications. Although these materials they exhibit the property of hysteresis, methods of compensation have been developed so they can be controlled in a predictive way. For detailed exposition of the dynamics of smart structures, see Kim and Tadokoro [2] and Vepa [7].

Another typical example of a smart material that can be embedded in a structure is furnished by a class of electrostrictive materials that are ferroelectric materials such as lead-magnesium-niobate (PbMgN). Electrostrictive and magnetostrictive materials are capable of directly transforming an electric or magnetic stimulus into a small mechanical displacement response with little or no hysteresis. Consequently, these materials are used in high-speed and high-precision positioning applications. However, their response to the stimulus is not linear and the presence of even a small amount of hysteresis is difficult to compensate. Yet, they are used in high-temperature applications because they are able to perform beyond the Curie temperature in a temperature domain where piezoelectric actuation is not possible. They are particularly suitable for acoustic noise or sound generators and for generating large forces with low applied voltages.

4.2.2 Biomimetic Sensors and Actuators

Biomimicry refers to mimicry of life—that is, to imitate biological systems. The term is derived from *bios*, meaning life, and *mimesis*, meaning imitation. Biomimicry concerns the design of biologically inspired mechanical devices. One is then interested in isolating the special features associated with biomimetic sensors. At first sight, sensors associated with human physiology, such as the eye, motion sensors within the inner ear, and touch sensors, seem to be like conventional pinhole cameras, accelerometers and gyroscopes, and pressure sensors, respectively. And yet a closer examination of the eye reveals that it is able to alter the focal length of the lens, whereas a pinhole camera cannot. Thus, the human eye can be considered a compliant sensor.

The saccule and the utricle in the inner ear function as inclinometers, which are essentially accelerometers measuring a component of the acceleration in a preferred direction. Yet, because the saccule and the utricle are immersed in a fluid medium, they function as integrating accelerometers over a high-frequency band. The saccule is primarily located in a vertical plane, while the utricle is primarily oriented horizontal with respect to the horizon. Both organs have sensory membranes called the *maculae*. The maculae are covered by the otolithic membrane. The primary sensors in these organs are the macular hair cells that project into the otolithic membrane. These hair cells are oriented along a centerline that curves through the center of the macula. On each side of this centerline, the tips of the hair cells are oriented in opposite directions, facing the centerline in the utricle and turning away from the centerline in the saccule. Hair cells of particular type are located in abundance near the centerline. Due to the opposite orientation of the tips of the hair cells and the curvature of the centerline in both otolith organs, they are sensitive to multiple directions of linear acceleration over a center-frequency bandwidth. Yet, because they are immersed in a fluid medium, they function as integrating accelerometers over high frequencies. This feature, which represents a form of compliance, provides these sensors with a relatively large operational bandwidth.

Located within the inner ear three semicircular canals, which are fluid-filled membranous ovals oriented in three different planes, function like rate gyroscopes. At the base of each is a region called the *ampulla*, which contains hair cells called the *crista* that are affected by movement. These hair cells also respond to movement in a manner similar to a set of integrating accelerometers. Again, the primary feature of the semicircular canals is their compliance.

Another form of compliance is also a primary feature in the case of touch sensors. In fact, on observing many biological sensors, we find that a primary underlying feature is that they are essentially compliant in some way. When robotic manipulators are performing noncontact or unconstrained tasks, there is no need for compliance. The need for compliance arises when contact tasks that impose severe constraints due to the geometry of the local environment are being performed. When the robot attempts to meet these constraints, there is generally a need for it to be compliant with the geometry or other constraints imposed by the local environment.

For example, a minimally invasive approach to robotic surgery uses three-dimensional imaging techniques coupled with tiny robotic manipulators to perform delicate urological surgery, including prostatectomy (removal of the prostate gland to treat cancer). It relies on making

microscopic incisions with accuracy and control and subsequently cuts and manipulates tissue through the tiny holes. Thus, successful completion of the tasks depends on the robot being able to meet the compliance requirements, which in turn requires that the compliance be monitored and controlled.

Since the compliance requirements depend to a large extent on the forces and moments generated by contact, there is a need to monitor and control the forces and moments generated by contact. Sensors with the ability to self-monitor and control the forces and moments generated by contact can therefore be considered biomimetic sensors.

There is yet another important smart feature associated with certain biological sensors. Every biological organism generates noise, but only some species such as bats have highly sensitive sensors capable of sensing the self-generated noise [8]. Furthermore, because these organisms are endowed with memory, they can compare the sensed noise with a record of the noise sensed at an earlier instance. This ability to compare and correlate the measurements empowers these organisms with an interesting kind of perception. Thus, bats have acoustic sensors that act like sonar and allow them to perceive an obstacle in the dark. The key features are the sensors' sensitivity, the presence of memory, and an element of built-in self-signal processing, a topic that is broadly discussed in Section 4.2.3.

We are now in a position to consider the special features associated with biomimetic actuators. Probably the best examples of biological actuators are the muscles in the human body. Biological actuators such as muscles are complex systems involving a variety of feedback mechanisms and pathways that facilitate both voluntary and involuntary responses to stimuli. Moreover, they generally incorporate a class of biomimetic sensors and so possess all the features of biomimetic sensors discussed earlier in this section. Muscles work in pairs, called *agonists* and *antagonists*, to facilitate the coordinated movement of a joint.

Muscles can only exert a force while contracting. On the other hand, a muscle can allow itself to be stretched, although it cannot exert a force when this happens. Thus, there is a need for a pair of muscles, the agonist providing the force by contraction while the antagonist stretches in acting against the agonist. Thus, the antagonist is responsible for moving the body part back to its original position.

An example of this kind of muscle pairing is furnished by the biceps brachii and triceps brachii. When the biceps are contracting, the triceps are stretching back to their original position. The opposite happens when the triceps contract.

The agonist and antagonist muscles are also referred to as (1) extensors when the bones move away from each other as the muscle contracts (triceps), and (2) flexors when the bones move toward each other as the muscle contracts (biceps). Combined with the concept of controlled compliance, the resulting biomimetic actuators generally feature controllable stiffness. The control of stiffness and, consequently, the control of compliance can be achieved by means of passive components or by employing feedback and active components.

A hybrid combination of both active and passive compliance is referred to as *semi-passive compliance* and generally features the advantages of both actively compliant and passively compliant actuators. Whereas actively compliant actuators use an unlimited power source and passively compliant actuators use no external power source, the semi-passive compliant actuators use a limited power source, thus facilitating a limit variation of the stiffness. These actuators are designed using smart materials such as EAPs and SMAs, whereas their active counterparts use electrohydraulic and electropneumatic actuators. On the other hand, passively compliant actuators use passive components only, and the variation of stiffness in these actuators is achieved by increasing the mechanical complexity of their design.

4.2.3 Biomimetic and Bio-Inspired Signal Processing

Probably the most widely used biomimetic methods of signal processing are artificial neural networks, which are mathematical models of the manner in which the biological- and cortical-based neural networks function in the human body and the brain, respectively. These are a well-known class of tools that have been applied to a wide range of engineering problems well beyond the field of robotics. Another class of methods is based on the concept of fuzzy logic, which was proposed in the 1960s to capture aspects of human possibilistic reasoning. It must be said at the outset that, in spite of these developments, researchers are still interested in biomimetic and bio-inspired signal-processing methods because most biological methods still excel in real-time and real-world perception and control applications, input data processing from distributed arrays of sensors, and holistic pattern recognition, compared with several advanced functionally like-methods in engineering systems.

From the concepts of learning and evolution in biological species, the field of genetic algorithms, capturing a host of Darwinian and Lamarckian evolutionary mechanisms, has not only arisen but is used in many engineering applications. See Chapter 17 by Banzhaf on evolutionary approaches in computation. From the collective behaviors of elementary entities, new concepts of swarm intelligence have been isolated, leading to new methods of genetic optimization such as particle-swarm optimization and ant-colony optimization.

Typically, when a biological sensor such as the ear or the eye encounters a signal, the signal goes through a range of processing steps before the information is acted upon. In an engineered system also, the signal undergoes a host of signal-processing steps. These steps are initiated by acquisition of the signal and followed by either a sequence of steps leading to signal identification or activity detection. *Activity detection* refers to the process of segmenting an isolated word out of the continuous data stream. This step is followed by feature set extraction and dimensionality reduction. Reducing the dimensionality of the data involves finding a finite set of uncorrelated variables and establishing a transformation relating them to a large number of correlated variables describing the observed data. Feature extraction is the process of reducing the dimensionality of the data to facilitate subsequent classification. A standard approach to reduction of the dimensionality of the data is to use *principal component analysis* (PCA) [9]. PCA reduces a large number of original variables into a smaller number of uncorrelated *components* that represent most of the variance in the original data. At this stage the vast quantities of data generated are generally classified. A variety of data analysis paradigms have been developed for feature extraction and classification, including multivariate statistical methods and linear discriminant analysis (LDA) [10] and support vector machines (SVMs) [11]. Following an assessment of the classified data, a command signal is generated and a control function is executed.

4.2.4 Modeling and Control of Anthropomorphic Manipulators

Modeling of human limb dynamics by equivalent robot manipulators and establishing a one-to-one equivalence of links and joints offers the opportunity to create a new generation of prosthetic limbs for both healthy and disabled people.

Humans use naturally developed algorithms for control of limb movement, which are limited only by the muscle strength. Furthermore, whereas robotic manipulators can perform tasks requiring large forces, the associated control algorithms do not provide the flexibility to perform tasks in a range of environmental conditions while preserving the same quality of performance as humans. Thus, interfacing a robot manipulator

to the human body so that the manipulator is controlled by electromyographically sensed signals will allow the manipulator to be controlled by the human brain, as well as provide the ability to perform tasks requiring large forces.

Alternately, an exoskeleton robot, serving as an assistive device, may be worn by a human as an orthotic device. If such a robot is to be properly matched to assist the human, its joints and links must correspond to those of the human body, and its actuators must share a portion of the external load on the operator. For this reason, it is important to design and build robotic manipulators that are able to emulate human limbs. The aim then is to model the human arm so as to emulate its degrees of freedom, kinematics, and dynamics.

To model the human forearm, it may be observed that at three key nodal points—the shoulder joint, the elbow, and the wrist—the human arm is characterized by a minimum of three, two, and three degrees of freedom, respectively. These are adduction/abduction, flexion/extension, and interior/exterior rotation at the shoulder; flexion/extension, rotation, and supination/pronation at the elbow; and flexion/extension and ulnar/radial deviation at the wrist and link twist at the wrist end.

To emulate these joints, the human arm is modeled by four rigid bodies, where the first one has two degrees of freedom—the joint angle about a vertical axis and the link twist about a horizontal axis—and the remaining three links each have two degrees of freedom characterized by the joint angles and link twist. The first and last links are relatively small in length, and the first two joints along with the first link constitute the shoulder joint. The joint between the second and third links constitutes the elbow; the joint between the third and fourth links and the fourth link constitute the wrist, which is assumed to be attached to the hand.

In Section 4.3.1.3, the performance of the nonlinear controller of such a robot manipulator (Figure 4.1) based on a complete dynamic model

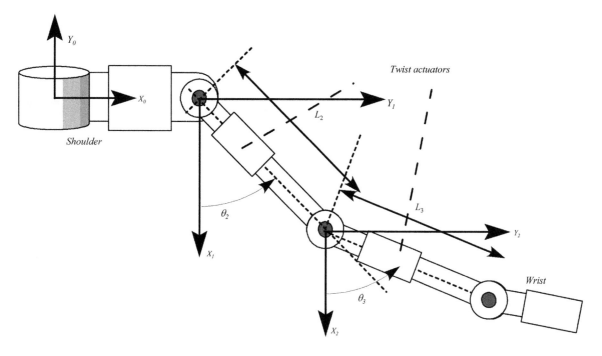

FIGURE 4.1 Mechanical model of the forearm.

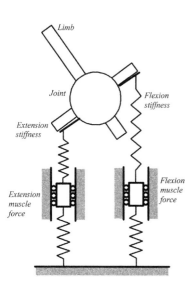

FIGURE 4.2 Mechanical model of a joint driven by an antagonistic pair of actuators.

is discussed. Each joint can be modeled as a mechanical system, as shown in Figure 4.2, where the dampers are omitted since these are generally parallel to the springs. Each joint in the serial manipulator is assumed to be driven by an antagonistic pair of actuators.

4.2.5 Shape and Morphing Control

Recent studies conducted by Weisshaar [12] and Bae *et al.* [13] have indicated that morphing of the airfoil shape has several potential benefits

FIGURE 4.3 Airfoil actuated by an assemblage of rigid-link actuators.

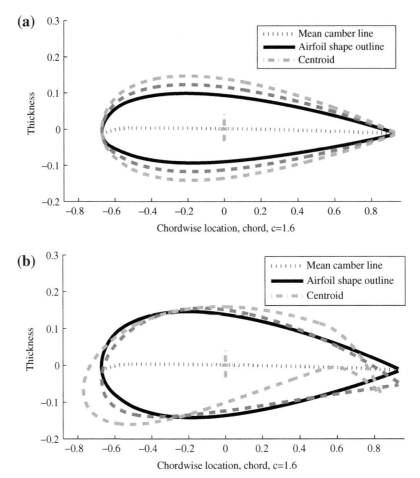

FIGURE 4.4   (a) Thickness control of a morphing airfoil, and (b) camber control of the morphing airfoil.

in generating lift on a wing. There are methods of affecting and altering shape. Typically, these involve the use of (1) multiple serial manipulators with rigid links, (2) multiple serial manipulators with elastic links, and (3) dynamically controlled morphing elastic/smart structures. Figure 4.3 illustrates a typical airfoil actuated by a truss-like assemblage of rigid-link actuators [14]. The upper and lower surfaces are actuated by independent serial manipulators while the thickness is controlled by a set of extendable actuators.

The morphing schedule is defined by a series of standard airfoil shapes. For example, a typical morphing schedule of NACA airfoil shapes is 1112, 1115, 1118, 4418, and 8818. NACA 1112 is a slightly cambered 12% thick airfoil. The thickness and camber morphing sequences are illustrated in Figure 4.4. First, as shown in Figure 4.4a, the thickness is increased to 15% (NACA 1115) and then to 18% (NACA 1118). Then, as shown in Figure 4.4b, the thickness is held at 18% and the maximum camber is increased from 1 to 4 and then to 8 (the first digit; the second digit defines the location of maximum camber). To secure the benefits of morphing, one would ideally like to morph from a low-speed airfoil such as NACA 0012 to a supercritical airfoil at transonic speeds such as NASA SC(2) 0712

NACA0012

FIGURE 4.5 The NASA SC(2) 0712 supercritical airfoil overlaid on top of a symmetric NACA 0012 airfoil in the background.

(Figure 4.5). The net increase of the maximum lift coefficient could be almost as high as 34% with a slight reduction in the stall angle, the airfoil's angle of attack beyond which the airfoil experiences a marked loss of lift.

To obtain a smoother shape, one could in principle use flexible or compliant links that could be either passively or actively controlled [15–17]. In the latter case, one could precisely control the shape of the airfoil as each elastic segment or link in the manipulator is also actively controlled by a suitable actuator. The key to achieving shape control is the availability of typical actuators that can affect the shape with the expenditure of minimum energy. This aspect has been researched by several investigators [13, 18–21]. Apart from conventional actuators, one could use PZT-based piezoceramic or SMA actuators.

In the remainder of this section, let us consider the derivation of the general equations describing the deformation and morphing dynamics of a structure [22]. We seek to change the radius of curvature of a curved beam from one continuously varying function to another by the simultaneous application of longitudinal and transverse forces as well as a distributed bending moment along the track of the beam.

Let us consider a uniform, curved beam with variable radius of curvature and its dynamics under the action of (1) external distributed control forces acting both tangentially and along the radials as well as (2) an external distributed bending, control moment acting about the sectional neutral axes. The fundamental assumptions made to simplify the analysis are as follows:

1. A plane section of the cross-section before bending remains planar after bending.
2. Every cross-section of the beam is symmetrical with respect to the loading plane.
3. The constitutive equations relating the stresses and strains are defined by Hooke's law.
4. Deformations are assumed to be dynamic and slowly varying, so all stresses other than those used to define the in-plane and the bending stress resultants do not change.

The geometry of a uniform symmetric curved beam with a variable radius of curvature $R(\alpha)$ is illustrated in Figure 4.6. The span of the beam, maximum rise, height of the middle surface, and inclination to the x axis are, respectively, denoted by L, h, $y(x)$, and α. The mass per unit length of the beam is denoted by m, and the section mass moment of inertia about an axis transverse to the plane of bending and passing through the section

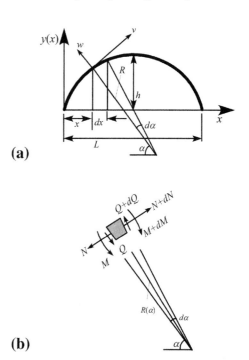

(a)

(b)

FIGURE 4.6 (a) Geometry of uniform curved beam, and (b) in-plane and shear forces and bending moment acting on an element of the beam.

center of mass is denoted by I_m. The displacements in the direction of the positive radial and the tangent to the curve are assumed to be w and v, respectively. The rotation of the cross-section is assumed to be ψ. The in-plane, axial stress resultant, and bending moment stress resultant are denoted by N and M, respectively. The external distributed torque acting is denoted as T. The shear force acting at any section is given by Q, and the external longitudinal and transverse forces are denoted by F and P, respectively.

The displacements in the direction of the positive radial and the tangent to the curve w and v, and the rotation of the cross-section ψ, are related to each other by the relation

$$dw/d\alpha \equiv w' = v + R\psi, \qquad (4.1)$$

where the prime denotes differentiation with respect to α, and $R = R(\alpha)$ is the local radius curvature. On considering a typical element of the beam as shown in Figure 4.7, the conditions of force and moment equilibrium are as follows:

$$dF + dN + Q\,d\alpha = mR\,d\alpha\,\frac{d^2}{dt^2}(w' - R\psi),$$
$$(4.2a)$$

$$dP + dQ - N\,d\alpha = mR\,d\alpha\,\frac{d^2}{dt^2}w, \quad (4.2b)$$

$$dT - dM - QR\,d\alpha = I_m R\,d\alpha\,\frac{d^2}{dt^2}\psi. \quad (4.2c)$$

Dividing both sides of Eqs. (4.2a)–(4.2c) by $R\,d\alpha$, we obtain

$$\frac{1}{R}\frac{dF}{d\alpha} + \frac{1}{R}\frac{dN}{d\alpha} + \frac{Q}{R} = m\frac{d^2}{dt^2}(w' - R\psi),$$
$$(4.3a)$$

$$\frac{1}{R}\frac{dP}{d\alpha} + \frac{1}{R}\frac{dQ}{d\alpha} - \frac{N}{R} = m\frac{d^2}{dt^2}w, \quad (4.3b)$$

$$\frac{1}{R}\frac{dT}{d\alpha} - \frac{1}{R}\frac{dM}{d\alpha} + Q = I_m\frac{d^2}{dt^2}\psi. \quad (4.3c)$$

The bending moment stress resultant is related to the transverse displacement by the relation

$$M = -\frac{EI_a}{R^2}(w'' + w). \qquad (4.4)$$

The in-plane stress resultant is

$$N = \frac{EA}{R}(w + v') - \frac{M}{R}$$
$$= \frac{EA}{R}\left[w + \frac{d}{d\alpha}(w' - R\psi)\right] - \frac{M}{R}. \qquad (4.5)$$

In Eqs. (4.4) and (4.5), E is the Young's modulus of the material of the beam that is assumed to be homogeneous, A is the area of cross-section of the beam, and I_a is the section area moment of inertia about an axis transverse to the plane of bending and passing through the section's geometric center. Eliminating M and N from Eqs. (4.3a)–(4.3c), we obtain three partial differential equations that must be solved to obtain w, ψ, and Q. Alternately, they may also be expressed in terms of w, v, and Q.

Given the equations of motion and the control inputs, one could design an appropriate control for morphing the structure. In flight, one would need to include the unsteady loads. However, the downside of morphing structures is their susceptibility to divergence and flutter instabilities, which must be avoided.

4.3 ENGINEERING APPLICATIONS

In this section several key applications are considered in some detail.

4.3.1 Modeling and Control of Robotic Manipulators

4.3.1.1 Introduction

Motion control of a robot manipulator is a fundamental problem that must be addressed at the design stage. Two categories of motion-control problems may be identified during the use of robotic manipulators: (1) point-to-point motion control, and (2) motion control with prescribed path tracking. Kahn and Roth [23] were the first to propose an optimal control for a robotic manipulator that belonged to the point-to-point control category and was based on the local linearization of the robot dynamics.

The motion-control problem with prescribed path tracking requires a path to be specified and is related to the general problem of path planning. Both motion-control approaches can be implemented by the use of optimal control methods, which could also be classified into two broad groups: (1) methods based on the solution of the Hamilton–Jacobi–Bellman equation or nonlinear programming, and (2) methods based on reduction of the solution to a two-point boundary value problem. The latter class of optimal controllers can yet again be classified into two groups based on the nature of the control: (2a) on-off or bang-bang control or time-optimal control, or (2b) continuous feedback quadratic cost-optimal control.

Time-optimal control methods were initiated in the mid-1980s by Bobrow *et al.* [24] and Shin and McKay [25], and efficient algorithms for bang-bang (time-optimal) control were proposed by Pfeiffer and Johanni [26] and Shiller [27]. Chen and Desrochers [28] were able to prove that that the time-optimal control problem is really a bang-bang solution. Bobrow *et al.* [29], Bien and Lee [30], and Chen [31] dealt with multiple manipulator arms. More recent efforts have been focused on the application of state and control constraints.

The application of evolutionary optimal methods and learning and adaptive strategies to time-optimal control have also been extensively pursued. For example, Arimoto *et al.* [32] introduced the concept of iterative learning control; Berlin and Frank [33] applied the concept of model predictive control to a robotic manipulator. There are also several other evolutionary techniques based on the use of fuzzy sets, genetic algorithms, and neural networks

that seek control laws that ensure optimal performance. The application of dynamic programming methods to robotic manipulators was also initiated in the 1980s, although the design of optimal trajectories was done in the 1970s.

One of the most well-known robot control schemes that emerged in the 1980s is the *computed-torque control* method, which involves the computation of the appropriate control torques and forces based on the robot dynamics, using (1) the sensed and estimated values of the generalized coordinates and velocities, and (2) the estimated values of the generalized accelerations. When the robotic manipulator dynamics and the loads are precisely known and if the sensors and actuators are error-free and the environment is noise-free, the computed-torque control method assures that the trajectory error goes to zero. Gilbert and Ha [34] have shown that the computed-torque control method is robust to a small modeling error. Moreover, the control law has the structure of a nonlinear-feedback control law.

The computed-torque control method naturally led to the use of quadratic cost-optimal control coupled with sensor-based estimation to generate the feedback control torques, with or without the use of a prescribed trajectory. This also led to the development of independent sensor-based quadratic cost-optimal control algorithms for robot manipulators [35].

The influence of flexibility on robot dynamics was first considered in the late 1970s and early 1980s (see, e.g., Book [36]). The inclusion of the effects of the dynamics of flexibility into the controller design process began shortly thereafter. The paper by Spong [37] is a good example. Many control schemes that are extensions of the control schemes for rigid manipulators, such as the computed-torque control method [38], optimal feedback control [39], and optimal and robust control [40], have been successfully proposed to tackle the modeling and control problems of flexible manipulators. Several new approaches to modeling [41–44] as well as a number of controller synthesis techniques tailored to flexible robots driven by control actuators [45–47] have also been published. The problem of motion control of flexible multilink manipulators has recently been considered by Wang *et al.* [48]. In the early 1990s, papers on the influence of nonlinear elastic effects first began to appear. Fenili *et al.* [49] developed the governing equations of motion for a damped beamlike slewing flexible structure driven by a DC motor and presented discussions of nonlinear effects.

Optimal control techniques [50] based on optimization of H_2 and H_∞ norms of a cost functional have emerged as reliable and computationally fast methods for the synthesis of feedback control laws for systems that are essentially described by a set of linear differential equations. However, most engineering systems, including robotic manipulators, particularly those involving some form of energy conversion or systems associated with biological processes, are patently nonlinear and must be dealt with in their natural form. The primary advantage of a nonlinear regulator is that closed-loop regulation is much more uniform and, consequently, the response is always closer to the set point than a comparable linear controller. Although the roots of nonlinear optimal control theory were established a long time ago when Bellman [51] formulated his dynamic programming approach and characterized the optimal control as a solution of the Hamilton–Jacobi–Bellman equation [52], computational methods for rapid computation of the feedback control laws are not easily available.

Several other methods, such as the use of control Lyapunov functions and model predictive control, have also emerged [53, 54]. One approach, first suggested by Cloutier *et al.* [55], is nonlinear regulation and nonlinear H_∞ control. A related technique due to Vlassenbroeck and van Dooren [56] is based on an orthogonal series expansion of the dynamical equations followed by a sequential numerical solution of the conditions for optimality. Another approach is based on replacing the nonlinear optimal control problem by sequence of linear quadratic

optimal control problems. Several researchers have addressed the issue of the synthesis of full-state feedback control laws when all states are measurable [57–60]. However, there are relatively fewer examples of the application of nonlinear H_∞ control [61, 62] to systems where only a limited number of states are available for measurement.

4.3.1.2 Modeling of a Multilink Serial Manipulator

Let us apply the H_∞ controller synthesis algorithm to the position control of a flexible, three-link serial manipulator. The traditional H_∞ controller for a linear system consists of a full-state estimate linear feedback control law and a H_∞ state estimator. The feedback control law is synthesized by solving an algebraic Riccati equation [63], whereas the estimator is similar in form to a *Kalman filter* (KF) [64]. The nonlinear controller is obtained by synthesizing a *frozen estimated state* optimal control law and replacing the linear estimator by a nonlinear filter known as the *unscented Kalman filter* (UKF) that is obtained by applying a nonlinear transformation known as the *unscented transformation* [65–67] while propagating the estimates and covariance matrices. The resulting controller is a nonlinear controller.

The unscented H_∞ estimator, a nonlinear estimator that bears the same relationship to the linear H_∞ estimator as the relationship of the UKF to the linear KF, is constructed by the same process as the UKF, by employing a weighted combination of estimates and process covariance matrices evaluated at a finite set of sigma points.

The modeling of a multilink manipulator can be done by adopting the Lagrangian formulation [1]. With the correct choice of reference frames, the dynamics can be reduced to a standard form. A typical three-link serial manipulator used to represent a human prosthetic limb is illustrated in Figure 4.7. The Euler–Lagrange equations may be expressed as a standard set of three coupled second-order equations for the joint variables θ_i that are defined in Figure 4.7. The nonlinear

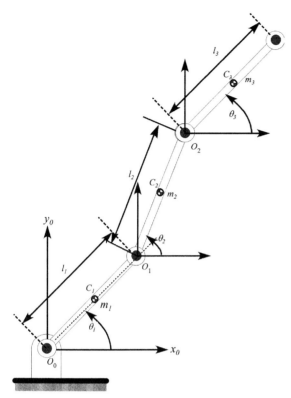

FIGURE 4.7 Typical three-link manipulator showing the definitions of the degrees of freedom.

described by nonlinear strain–displacement equations and a linear stress–strain relation, which results in an approximate nonlinear equation [68–70] for beam vibrations. The generalized *external* nonlinear forces due to stretching of the midplane may then be determined. The complete equations of motion may then be expressed in standard state-space form.

Expressions for the additional coefficient matrices in the dynamics due to the presence of flexibility and midplane stretching may be obtained from the expressions for the additional kinetic and potential energies due to flexibility and the additional generalized work done by the nonlinear force distribution and structural damping forces. For brevity, the details are not presented here. The state space is now augmented to include the amplitudes of the assumed flexible displacement modes, and the corresponding equilibrium and perturbation equations may be found. The nonlinear H_∞ controller synthesis problem is solved with the plant assumed to be nonlinear in terms of state-space description, but with the measurements and the performance assessment outputs assumed to be linear in terms of the states and inputs. The adopted nonlinear H_∞ controller synthesis method is based on the linear H_∞ controller synthesis method [71–75].

It should be mentioned at this stage that both nonlinear H_2 control and the UKF are special cases of the corresponding H_∞ controller and the unscented H_∞ estimator. To get the benefits of both H_2 and H_∞ control, one design approach is to choose the free parameters in the H_∞ controller design within appropriate limits. Such an approach will let the designer allow for the fact that the nonlinear closed-loop system, with the H_∞ controller in place, is generally a lot less robust than the linear closed system with the linear H_∞ controller in place.

4.3.1.3 Application to the Position Control of a Three-Link Flexible Serial Manipulator

The three-link flexible serial manipulator considered in this section serves as a benchmarking

perturbation dynamics about the equilibrium may be obtained from these equations.

In the perturbation equations, the external moments acting on the links are the sum of the control torques and the disturbance torques. The nonlinear equations can be written in state-space form by adopting standard techniques for defining an augmented state vector.

The additional kinetic and potential energies due to elastic transverse (normal) displacement normal to the link are used to determine the equations of motion of a flexible manipulator. A beam that is simply supported at its edges experiences a midplane stretching when deflected. The influence of this geometric nonlinearity or stretching on the response increases with increasing amplitude of motion. The dynamics may be example. The nonlinear H_∞ controller synthesis method is applied to this problem, although there are more effective methods of controlling the plant, such as controller partitioning. Thus, the feasibility of the nonlinear control technique and its limitations can be assessed. One of the major advantages of the use of the nonlinear controller synthesis methodology is the availability of well-defined algorithms for solving algebraic Riccati equations and for unscented Kalman filtering, which could be routinely adapted to solve the nonlinear controller synthesis problem.

The measurements were assumed to be the three angular link positions measured at the pivots. All of the process and measurement disturbances were modeled as white noise or delta-correlated random inputs. The measurement disturbances were assumed to be relatively very small compared with the process noise. Each of the process disturbance torques was assumed to be

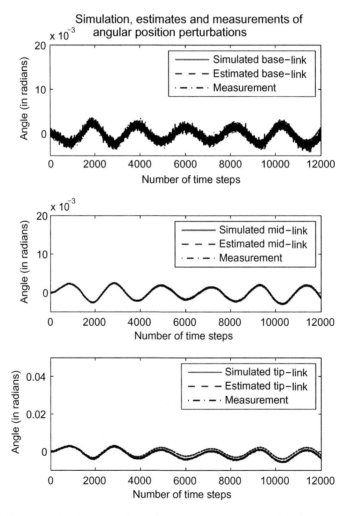

FIGURE 4.8 Simulated, estimated and measured angular positions, in radians, of the base-link, mid-link and tip-link in the controlled 3-link flexible manipulator obtained by using the improved state estimator to compute the control gains.

less than 0.05 nm in magnitude. The measurement disturbances were assumed to be less than 0.5°. The masses, joint stiffnesses, and lengths of the three links are assumed to be equal in magnitude. The three links were assumed to be uniform, with the centers of mass located at the middle of the links. Only the first two modes were included to represent the elastic effects in each of the three links. As a result, in the example considered, a total of six modal amplitudes are included in the model, whereas the total number of states is 18.

Only a typical result of the simulated, estimated, and measured positions of the three links obtained by using the frozen estimated state approximation to compute the control gains is shown in Figure 4.8. The time step for the calculation is $dt = 0.0002s$ and the timeframe for the calculation is 12,000 time steps. The manipulator was commanded to move from a configuration with all three angles equal to zero, to a configuration where the bottom, middle, and top links were at the angles 24°, 18°, and 12° to the horizontal. On examining the results, it is observed that the estimator convergence in the closed loop is the most important requirement for long-term stable performance. Thus, the estimator was tuned by tuning the unscented transformation. All other parameters in the unscented filter algorithm are maintained the same. The control gains are again computed using the frozen estimated state approximation and the simulations repeated. Although the very small amplitude oscillations ($< 0.2°$) and small steady-state pointing errors were unavoidable, the closed-loop system was stable in the long term. Increasing the number of time steps provided no difficulties, and the unscented H_∞ estimator continued to converge to the measurement in spite of small variations in angular positions about the desired set point because of the presence of the noise.

The results presented here indicate that in the case of the closed-loop nonlinear H_∞ control, the estimator tracks the angular positions of a flexible multilink serial manipulator to within $\pm 3\%$ of the set point (the initial error was 2%), and this value increases considerably for other cases that are not optimal. Further reductions in the tracking error are possible if one relaxes the bounds on the maximum permissible magnitude of the control input, which have been implicitly set by scaling the inputs. Moreover, the influence of nonlinear flexibility was relatively small, indicating that the estimator's performance did not degrade due to the presence of small structural nonlinearities in the dynamic model. Like its linear counterpart, the nonlinear H_∞ controller seeks to maintain a balance between disturbance rejection and stability. When the H_∞ estimator is combined with methods involving the control law synthesis based on replacing the nonlinear optimal control problem by sequence of linear optimal control problems, a powerful computational tool may be established for synthesizing control laws for a variety of nonlinear controller synthesis applications in biomimetic robotics.

4.3.2 Muscle Activation Modeling for Human Limb Prosthesis and Orthotics

4.3.2.1 Muscle Activation Dynamics Modeling

Let us consider a representation of a human limb as an articulated linkage system of intercalated bony segments. The bones, articulations, and some of the ligaments (bone-to-bone joining tissues) and tendons (muscle-to-bone joining tissues) involved in flexion and extension of a limb serve as the template for generic modeling of limbs. *Flexion* is defined as the bending of, or decrease in angle between, articulated body parts. *Extension* is the opposite of flexion, resulting in a straightening of anatomic links. In regard to limbs, *adduction* and *abduction* are defined, respectively, as motion away from or toward the center line of the limb— that is, the vertical axis of the limb. Limb motion generated solely by muscle activation inputs is referred to as *active* movement, whereas motion generated without reference to muscle activation inputs is referred to as *passive*. The terms *proximal* and *distal* are anatomical descriptors meaning, respectively, closer to and farther from the central reference point of the limb in question.

Connective tissues (collagen), which bind the parallel fibers that make up skeletal muscles, join together at the ends of a muscle to form tendons. For intrinsic muscles, these elastic tendons terminate on the limb component's bones. In the case of the extrinsic muscles, the tendons extend from remotely located muscles, spanning multiple joints in the limb. They are tethered to intermediate bones by fibrous tunnels that smoothly guide the tendons to maintain their relative positions to the neighboring smaller bones instead of assuming straight paths during flexion or extension. The organization, networking, and dynamics of connecting tendons primarily determine the range of motion of a limb.

Tendon connections are responsible for the forces generated by bundles of muscle fibers to be translated into torques about the joints. If the resulting moment exceeds an opposing moment from an antagonistic group of muscles or a distribution of external loads, then the limb will rotate about the joint. Because muscles provide power only during contraction, complementarily oriented muscles must command tendons to the bones they are connected to. These muscles are referred to as *flexors* and *extensors*. A flexor signal is said to supinate the limb, and an extensor signal is said to pronate the limb. Control of the limb is the result of these and other muscles acting together on different sides of a rotational joint axis. The redundancy of the tendons enables muscles to contract either antagonistically and/or synergistically and thereby optimally tune the loading of articulated joints for different tasks.

Since muscles can pull but not push, to produce a moment or rotation in both directions (flexion/extension) one will need a pair of muscles opposing each other (antagonistic muscles). However, the stiffness and damping contributions of the muscles add while the exerted moments subtract. Thus, by carefully coordinating muscle activities, the net joint torque may be held constant

while the net impedance increases. Impedance is primarily provided by the damping of the joint's relative motion, whereas the damping within a joint is provided by ligaments. It is assumed that the ligaments also provide for impedance control. But the impedance must also meet stability considerations, and the provision of a redundant set of tendons and ligaments satisfies this requirement. Thus, it is not only extremely important to be able to model the ligament damping, the tendon dynamics, and the generation of both forces by a single muscle but also the moments and the dynamics of an antagonistic pair of muscles.

Continuously commanded control inputs to the muscles are assumed to provide neural signals that make their way into the muscle fiber via neuromuscular junctions. These signals can be measured and constitute the *electromyogram* (EMG) signals. The main features of EMG signals, when used as continuous control command signals for prostheses, are their nonlinear and nonstationary characteristics. To develop an EMG controller for a prosthesis, one approach is to mimic neuromuscular control models of human limbs. Modeling of the biological limb dynamics is expected to result in a more responsive controller for prosthetic limbs. Our objective is to develop a generic approach to neuromuscular modeling of the dynamics of human limbs. There are four aspects to consider:

1. Modeling of the neuromuscular action potentials,
2. Modeling of the muscle activation dynamics,
3. Modeling of the muscle moment generation, and
4. Modeling of the tendon length dynamics.

4.3.2.2 *Modeling of the Neuromuscular Action Potentials*

Modeling of the neuromuscular action potentials that mimic the measurements of EMG signals can be done using either surface-mounted electrodes or thin-wire electrodes embedded within a muscle. These measurements are aggregate models of delayed weighted summations of neural signals in the form of action potentials emerging from the nerve cells embedded in each muscle fiber within a pool of muscle fibers. It is these weight summations of several single fiber action potentials that constitute a typical EMG measurement. The EMG signal is assumed to be an amplitude-modulated *carrier* with multiplicative noise (to model the firing and recruitment of motor units near the sensors when there is change in the muscle force level) and additive noise (arising from distal motor units) of the activation inputs to a neuromuscular model of the limb dynamics. It is assumed that the activation signals are extracted following several steps that include pattern recognition and/or blind signal analysis.

4.3.2.3 Modeling of the Muscle Activation Dynamics

Sensory and motor nerve fibers enter the muscle in one or two nerve branches. Most muscle fibers are supplied by alpha motoneurons [1; see Section 4.2.4.5], while gamma motoneurons [1; see Section 4.2.4.5] supply the muscle fibers within the muscle spindle. The nerve fiber enters the muscle at the motor endplate, and the whole of the neuromuscular junction at the nerve fiber constitutes a motor unit. Depolarization of the nerve fiber activates the muscle fiber at the motor unit and is responsible for the activation that in turn increases the concentration of calcium ions, which in turn switches on the muscle contraction. The activation state is a function of calcium ion concentration.

Concentration dynamics can be modeled by the Nernst equation, but a simplified model is used for the gating function of the calcium ions. Zajac [76] defined it by

$$\frac{dq}{dt} = \frac{[\beta + (1 - \beta)u]}{\tau_{act}} q + \frac{1}{\tau_{act}} u, \quad (4.6)$$

where u is the depolarizing input from a motor unit, q is the activating state, β is the fraction of the activating state that is not influenced by the external control, and τ_{act} is the time constant of

the order of 10 ms for fast twitch fibers and 60 ms for slow tonic fibers.

4.3.2.4 Modeling of the Muscle Moment Generation

To model the forces within a muscle, it is essential to consider several recruitment schemes of multiple motor units for different fibers. The sudden recruitment of a motor unit at its initial firing rate causes a step increase of muscle force; the size of that step is determined by the fractional physiological cross-sectional area of that unit, which is a function of the number of motor units. Upon recruitment, the lumped motor unit modulates its frequency according to an effective recruitment signal that is proportional to the amount of muscle recruited. The effective recruitment signal is characterized by a rise and fall time constant that is determined by the first-order dynamics of the exchange of calcium between the nerve cells and muscle fiber within the motor unit [77]. The level of effective activation of each fiber results from a linear combination of multiple motor unit activations weighted by their respective fractional physiological cross-sectional area. The differences between tonic (or slow) and twitch (or fast) fiber types are reflected in rise and fall time constants of the excitation dynamics that model the sagging or yielding properties, the activation frequency that represents the calcium dynamics, and the muscle force-length and muscle force-velocity properties. Thus, the muscle force is represented as

$$F = F_0 \sum_{i=1}^{J} W_i \times A_{fi} \times \left[F_{iV}(V) F_{iL}(L) q_i + F_{pi}(L) \right],$$

$$(4.7)$$

where F_0 is the maximum force generated, which could itself be dynamic and could be modeled by a first-order dynamic model, although it is assumed constant here; W_i are a set of weights; A_{fi} is the frequency of activation, which primarily depends on the dynamics of calcium ion concentration; F_{iV} is the force-velocity function, which is obtained from a model like the Hill-type model [78, 79], as $(F_{iV} + a)(V + b) = \text{constant} = b(F_{iV0} + a)$, where F_{iV0} is the isometric muscle force (in practice, the Hill-type function is sometimes substituted by a sigmoid function); F_{iL} is the force-length function that is modeled by a parabolic function; and F_{ip} is the passive force-length function that is modeled as a fraction of the total force or by a sigmoid function. The actual calculation of the forces is done exactly as by Csercsik [58, 80, 81]. Thus, the muscle forces F are expressed as

$$F = F_{\max}f_L(\alpha, d_p, d_d)f_V(\alpha, \omega)q, \quad (4.8)$$

where $F_{\max}$ is the maximum muscle force, f_L is the parabolic muscle force of the muscle, d_p is the proximal distance between the joint and the origin of the muscle, d_d is the distal distance between the joint and the insertion point of the muscle, f_V is a Hill or sigmoid function representing the muscle force-velocity characteristic, and q is the muscle activation state.

4.3.2.5 Modeling of the Tendon Length Dynamics

To model the tendon length dynamics, the models of viscoelastic tissue discussed by Fung [82] are considered. These are the standard linear solid and the quasilinear viscoelastic solid models. The tendon length-time relation for a standard linear solid is given by

$$L_t(t) = L_{t0} \left[1 - \left(1 - \frac{\tau_\sigma}{\tau_\varepsilon}\right) e^{-t/\tau_\varepsilon} \right] \Bigg/ \left[1 - \left(1 - \frac{\tau_\sigma}{\tau_\varepsilon}\right) \right], \quad (4.9)$$

where τ_ε and τ_σ are the relaxation times for constant strain and constant stress, respectively, and L_{t0} is the initial tendon length. In the case of a quasilinear viscoelastic solid,

$$L_t(t) = L_{t0} \left[\left(1 + c \int_{\tau_1}^{\tau_2} e^{-t/\tau} d\tau/\tau \right) \Bigg/ \left(1 + c \int_{\tau_1}^{\tau_2} d\tau/\tau \right) \right]. \quad (4.10)$$

The integral may be replaced by a finite summation of exponential terms:

$$L_t(t) = L_{t0} \left[\left(1 + \sum_{i=1}^{I} c_i e^{-t/\tau_i} \right) \Bigg/ \left(1 + \sum_{i=1}^{I} c_i \right) \right]. \quad (4.11)$$

The resulting model is a higher-order generalization of that for a linear elastic solid. The model assumed by Csercsik [80] could be considered to be a special case of the model defined by Eq. (4.11). In our case, the tendon dynamics is modeled by the linear viscoelastic solid, although the provision for using the higher-order model is retained. In the case of the linear elastic solid, the tendon length satisfies the dynamic equation

$$\frac{dL_t}{dt} = -\frac{\left\{ L_t - L_{t0} \Big/ \left[1 - \left(1 - \frac{\tau_\sigma}{\tau_\varepsilon}\right)\right] \right\}}{\tau_\varepsilon} + w_t, \quad (4.12)$$

where w_t is the assumed Gaussian white noise process driving the rate of change of the tendon length. The complete set of state equations for the limb dynamics actuated by a flexor and an extensor pair of muscles may now be assembled and discretized by standard methods.

4.3.2.6 Application to Muscle Dynamic System Parameter Identification

In this section the adaptive method is coupled with the UKF developed by Julier and Uhlmann [65] and used to estimate the parameters of a bilinear model of a binary muscle-activation system that belongs to the Weiner type of block-oriented models. The details of the adaptive Kalman filter implementation may be found in the work of Vepa and Zhahir [83].

The adaptive UKF is applied to the problem of identification of a typical flexor–extensor muscle pair. A typical set of parameters was assumed for the muscle–limb dynamics and the

TABLE 4.1 Definitions and typical values assumed for the muscle and limb parameters.

Parameter	Description (Units)	Value
β	Fraction of activation state	0
τ_{act}	Activation time constant (s)	0.015
B_c	Coefficient of viscous damping in limb (Kg.m^2.s)	1×10^{-04}
I_c	Moment of inertia of limb (Kg.m^2)	2×10^{-05}
τ_ε	Relaxation time for constant strain (s)	0.1
τ_σ	Relaxation time for constant stress (s)	0.077
r_{te}	Ratio of final to initial extensor tendon length	1.3
ml_c	Limb unbalance (N.m)	0.005
α_s	Reference limb rotation angle	$\pi/2$

TABLE 4.2 Typical initial values assumed for the muscle and limb dynamic states.

State	Description (Units)	Initial Value
q_f	Flexor muscle activation state	0
q_e	Extensor muscle activation state	0
α	Limb rotation angle (rad)	0.2
ω	Limb rotation rate (rad/s)	0
L_{tf}	Flexor muscle tendon length (m)	0.1
L_{te}	Extensor muscle tendon length (m)	0.1
d_f	Flexor muscle force moment arm (m)	0.3
d_e	Extensor muscle force moment arm (m)	0.25
d_p	Proximal distance of muscle origin to joint (m)	0.2
d_d	Distal distance of muscle insertion point to joint (m)	0.3
F_{max}	Maximum muscle force (N)	0.01
L_{tf0}	Flexor muscle initial tendon length (m)	0.1
L_{te0}	Extensor muscle initial tendon length (m)	0.1

state response simulated by numerical integration of the equations in MATLAB. The assumed set of parameters is listed in Table 4.1. Table 4.2 lists the initial values of the states, that were estimated by the adaptive UKF estimator. Of these states, the last seven are assumed to be constants. All the equations are assembled and expressed in discrete form to facilitate the application of the adaptive UKF.

In the first instance, the UKF was implemented without any adaptation of the process noise or measurement noise covariance matrices. The first activation inputs used to determine the limb response over a 10-second timeframe corresponding to 10,000 time steps are obtained from real measured EMG data available on the Web [84]. The subject did a protocol of 30 seconds water cycling in a 100 bpm cadence. The second

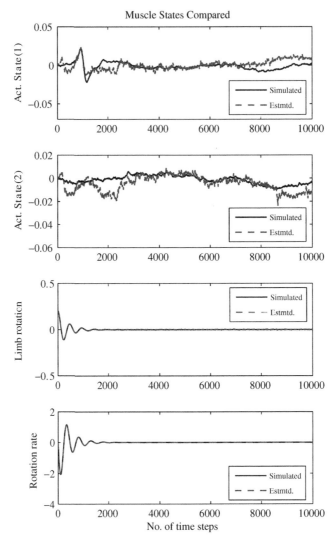

FIGURE 4.9 UKF estimated and simulated responses of the two activation states, the limb rotation, and the limb rotation rate.

activation input was scaled white noise. The estimated and simulated responses of the two activation states, the limb rotation, and the rotation rate are shown in Figure 4.9.

The case of the adaptive UKF-based state estimation and parameter identification, where the process noise covariance is continuously updated, was also considered. A key difference

between the results obtained by the use of the UKF and the adaptive UKF is in the estimates of the initial length of the tendon. For this reason a nonlinear model of the tendon length dynamics was also employed. The advantage of using the nonlinear model is that the tendon could not only be allowed to be slack in its rest state, but one is also allowed to incorporate nonlinearities such as dead band and hysteresis. However, both tendons cannot be allowed to be slack simultaneously if the limb is to maintain its orientation. When the nonlinear model was used only for the tendon that was less stressed, the results resembled the case of the UKF without adaptation of the process covariance.

In this section the successful application of the UKF and the adaptive UKF to the identification of model parameters of a typical nonlinear muscle-limb dynamic system is illustrated. It has been shown by example that UKF-based and adaptive UKF–based state estimation and parameter identification are an option that is particularly well suited for the dynamic systems associated with muscle-limb interactions. Although the generic case of a limb actuated by a pair of antagonistic muscles was considered, the identification method could be fine-tuned and applied to any specific pair of antagonistic muscles actuating a particular limb. The method is currently being employed to design nonlinear control laws that can be used to control smart prosthetic limbs.

An adaptive UKF is used to estimate the unmeasurable state variables and kinetic parameters of the muscle-limb model. Although the UKF has a simple structure, the tuning of these estimators is a relatively difficult task. The use of the adaptive approach eliminates the need for the tuning of the covariance parameters of the UKF estimator. However, the estimates of the process covariance matrices obtained adaptively can vary widely depending on the adaptation scheme adopted. For this reason, the adaptive algorithm is recommended to be used only in the initial stages as a tuning method.

One of the features of the parameter-identification problem considered here is the presence of internal parameters in the definition of the applied moment. These internal parameters, such as the distances to the two muscle distal and proximal anchor points from the center of rotation, d_d and d_p, constitute a set of weakly identifiable parameters since they may in some cases be inaccessible for observation. Although formal identifiability analysis of the model has not been carried out, the problem of weakly identifiable parameters has been dealt with by the use of optimal state estimation of nonlinear dynamic systems. Moreover, physically meaningful, coupled models of the internal tendons' length dynamics facilitate the identification of the weakly identifiable parameters.

4.4 PROGNOSIS FOR THE FUTURE

Biomimetic roboticists will continue to develop the field so as to mimic physiological dynamic models of human limbs and the manner in which they are controlled. These developments will lead to the routine use and implementation of sophisticated and complex systems. In particular, these developments are expected to be in the area of the design and implementation of compliant robot limbs and hands with exceptional sensory and control capabilities so that they can be interfaced to the human brain and controlled by it.

References

[1] R. Vepa, *Biomimetic robotics: mechanisms and control*, Cambridge University Press, New York, NY, USA (2009).

[2] K.J. Kim and S. Tadokoro (eds.), *Electroactive polymers for robotic applications: artificial muscles and sensors*, Springer-Verlag, New York, NY, USA (2010).

[3] K. Otsuka and C.M. Wayman, *Shape memory materials*, Cambridge University Press, London, UK (1998).

[4] D.R. Askeland and P.P. Phule, *The science and engineering of materials*, 5th ed., Thomson Canada, Toronto, Canada (2006).

[5] R.G. Larson, *The structure and rheology of complex fluids*, Oxford University Press, New York, NY, USA (1998), pp. 360–385.

[6] J.E. Segel (ed.), *Piezoelectric actuators*, Nova Science Publishing Inc., New York, NY, USA (2011).

[7] R. Vepa, *Dynamics of smart structures*, Wiley, Chichester, UK (2010).

[8] J. Weaver, Bats broaden sonar field of view to maneuver around obstacles, *PLoS Biol* **9** (2011), e1001147.

[9] M. Ringnér, What is principal component analysis? *Nat Biotechnol* **26** (2008), 303–304.

[10] T. Hastie, R. Tibshirani, and J.H. Friedman, *The elements of statistical learning: data mining, inference and prediction*, Springer-Verlag, Berlin, Germany (2001).

[11] B. Schökopf and A. Smola, *Learning with kernels: support vector machines, regularization, optimization and beyond*, MIT Press, Cambridge, MA, USA (2002).

[12] T.A. Weisshaar, Morphing aircraft technology—new shapes for aircraft design, NATO unclassified documents, *Proceedings of RTO-MP-AVT-141*, Neuilly-sur-Seine, France (2006), O1-1–O1-20.

[13] J.-S. Bae, N.-H. Kyong, T.M. Seigler, and D.J. Inman, Aeroelastic considerations on shape control of an adaptive wing, *J Intell Mater Syst Struct* **16** (2005), 1051–1056.

[14] A.Y.N. Sofla, D.M. Elzey, and H.N.G. Wadley, Shape morphing hinged truss structures, *Smart Mater Struct* **18** (2009), 065012.

[15] D.S. Ramrkahyani, G.A. Lesieutre, M. Frecker, and S. Bharti, Aircraft structural morphing using tendon actuated compliant cellular trusses, *Proceedings of the 45th AIAA/ASME/ASCE/AHS/ASC structures, structural dynamics and materials conference*, Palm Springs, CA, USA (April 19–22, 2004).

[16] N.M. Ursache, A.J. Keane, and N.W. Bressloff, Design of postbuckled spinal structures for airfoil camber and shape control, *AIAA J* **44** (2006), 3115–3124.

[17] K.-J. Lu and S. Kota, Design of compliant mechanisms for morphing structural shapes, *J Intell Mater Syst Struct* **14** (2003), 379–391.

[18] T.J. Lu, J.W. Hutchinson, and A.G. Evans, Optimal design of a flexural actuator, *J Mech Phys Solids* **49** (2001), 2071–2093.

[19] M. Amprikidis and J.E. Cooper, Development of smart spars for active aeroelastic structures, *AIAA Paper 2003-1799* (2003).

[20] K. Chandrashekhara and S. Varadarajan, Adaptive shape control of composite beams with piezoelectric actuators, *J Intell Mater Syst Struct* **8** (1997), 112–124.

[21] A. Achuthan, A.K. Keng, and W.C. Ming, Shape control of coupled nonlinear piezoelectric beams, *Smart Mater Struct* **10** (2001), 914–924.

[22] I. Karnovsky and O. Lebed, *Non-classical vibrations of arches and beams*, McGraw-Hill, New York, NY, USA (2004).

[23] M.E. Kahn and B. Roth, The near-minimum-time control of open-loop articulated kinematic chains, *ASME J Dyn Syst Meas Contr* **93** (1971), 164–172.

[24] J.E. Bobrow, S. Dubowsky, and J.S. Gibson, Time-optimal control of robotic manipulators along specified paths, *Int J Rob Res* **4** (1985), 3–17.

[25] K.G. Shin and N.D. McKay, Minimum-time control of robotic manipulators with geometric constraints, *IEEE Trans Autom Control* **30** (1985), 531–541.

[26] F. Pfeiffer and R. Johanni, A concept for manipulator trajectory planning, *IEEE J Rob Autom* **3** (1987), 115–123.

[27] Z. Shiller, On singular points and arcs in path constrained time optimal motions, *ASME Dyn Syst Contr Div* **42** (1992), 141–147.

[28] Y. Chen and A.A. Desrochers, Structure of minimum-time control law for robotic manipulators with constrained paths, *Proceedings of the IEEE international conference robotics automation*, Scottsdale, AZ, USA (1989), 971–976.

[29] J.E. Bobrow, J.M. McCarthy, and V.K. Chu, Minimum-time trajectories for two robots holding the same workpiece, *Proceedings of the 29th IEEE conference decision and control*, Honolulu, HI, USA (1990), 3102–3107.

[30] Z. Bien and J. Lee, A minimum-time trajectory planning method for two robots, *IEEE Trans Rob Autom* **8** (1992), 414–418.

[31] Y. Chen, Structure of the time-optimal control law for multiple arms handling a common object along specified paths, *IEEE Trans Autom Control* **37** (1992), 1648–1652.

[32] S. Arimoto, S. Kawamura, and F. Miyazaki, Bettering operation of robots by learning, *J Rob Syst* **1** (1984), 123–140.

[33] F. Berlin and P.M. Frank, Robust predictive robot control, *Proceedings of the 5th international conference on advanced robotics*, vol. 2, Pisa, Italy (19–22 June 1991), 1493–1496.

[34] E.G. Gilbert and I.J. Ha, An approach to nonlinear feedback control with application to robotics, *IEEE Trans Syst Man Cybern* **14** (1984), 879–884.

[35] D. Xiao, B.K. Ghosh, N. Xi, and T.J. Tarn, Sensor-based hybrid position/force control of a robot manipulator in an uncalibrated environment, *IEEE Trans Control Syst Technol* **8** (2000), 635–645.

[36] W.J. Book, Recursive Lagrangian dynamics of flexible manipulator arms, *Int J Rob Res* **3** (1984), 87–101.

[37] M.W. Spong, Modeling and control of elastic joint robots, *ASME J Dyn Syst Meas Contr* **109** (1987), 310–319.

[38] C.M. Pham, W. Khalil, and C. Chevallereau, A nonlinear model-based control of flexible robots, *Robotica* **11** (1992), 73–82.

[39] P.T. Kotnic, S. Yurkovich, and U. Ozguner, Acceleration feedback for control of a flexible manipulator arm, *J Rob Syst* **5** (1988), 181–196.

[40] J. Daafouz, G. Garcia, and J. Bernussou, Robust control of a flexible robot arm using the quadratic d-stability approach, *IEEE Trans Control Syst Technol* **6** (1998), 524–533.

[41] H.A. Attia, Dynamic modelling of a serial robot manipulator using point and joint coordinates, *Adv Model Anal* **58** (2003), 11–26.

[42] J.M. Martins, Z. Mohamed, M.O. Tokhi, J. Sa da Costa, and M.A. Botto, Approaches for dynamic modelling of flexible manipulator systems, *IEE Proc Control Theory Appl* **150** (2003), 401–411.

[43] B.C. Bouzgarrou, P. Ray, and G. Gogu, New approach for dynamic modelling of flexible manipulators, *Proc Inst Mech Eng, Part K: J Multi-body Dyn* **219** (2005), 285–298.

[44] M.M. Fateh, Dynamic modeling of robot manipulators in D-H frames, *World Appl Sci J* **6** (2009), 39–44.

[45] E. Garcia and D.J. Inman, Advantages of slewing an active structure, *J Intell Mater Syst Struct* **1** (1991), 261–272.

[46] M.K. Kwak, K.K. Denoyer, and D. Sciulli, Dynamics and control of slewing active beam, *J Guid Contr Dyn* **18** (1994), 185–189.

[47] Q. Sun, Control of flexible-link multiple manipulators, *ASME J Dyn Syst Meas Contr* **124** (2002), 67–76.

[48] Z. Wang, H. Zeng, D.W.C. Ho, and H. Unbehauen, Multi-objective control of a four-link flexible manipulator: a robust H_∞ approach, *IEEE Trans Control Syst Technol* **10** (2002), 866–875.

[49] A. Fenili, J.M. Balthazar, and R.M.L.R.F. Brasil, On the mathematical modeling of beam-like flexible structure in slewing motion assuming nonlinear curvature, *J Sound Vib* **282** (2004), 543–552.

[50] A.E. Bryson and Y.-C. Ho, *Applied optimal control*, Hemisphere, New York, NY, USA (1975).

[51] R. Bellman, The theory of dynamic programming, *Proc Natl Acad Sci* **38** (1952), 360–385.

[52] R. Bellman, *Dynamic programming*, Dover Publications, New York, NY, USA (2003).

[53] J.A. Primbs, *Nonlinear optimal control: a receding horizon approach*, Ph.D. dissertation, California Institute of Technology (1999).

[54] J.A. Primbs, V. Nevistić, and J.C. Doyle, Nonlinear optimal control: a control Lyapunov function and receding horizon perspective, *Asian J Control* **1** (1999), 14–24.

[55] J.R. Cloutier, C.N. D'Souza, and C.P. Mracek. Nonlinear regulation and nonlinear H_∞ control via the state-dependent Riccati equation technique, *Proceedings of the 1st international conference on nonlinear problems in aviation and aerospace*, Daytona Beach, FL, USA (1996).

[56] J. Vlassenbroeck and R. Van Dooren, A Chebyshev technique for solving nonlinear optimal control problems, *IEEE Trans Autom Control* **33** (1988), 333–340.

[57] D. Georges, C.C. de Wit, and J. Ramirez, Nonlinear H_2 and H_∞ optimal controllers for current-fed induction motors, *IEEE Trans Autom Control* **44** (1999), 1430–1435.

[58] P.A. Frick and D.J. Stech, Solution of the optimal control problems on parallel machine using epsilon method, *Optim Control Appl Methods* **16** (1995), 1–17.

[59] H. Jaddu and E. Shimemura, Computation of optimal control trajectories using Chebyshev polynomials: parametrization and quadratic programming, *Optim Control Appl Methods* **20** (1999), 21–42.

[60] J. Tamimi and H. Jaddu, Nonlinear optimal controller of three-phase induction motor using quasi-linearization, *Proceedings of the second international symposium on communications, control and signal processing*, Marrakech, Morocco (March 13–15, 2006).

[61] A.J. van der Schaft, L_2-gain analysis of nonlinear systems and nonlinear state feedback H_∞ control, *IEEE Trans Autom Control* **37** (1992), 770–784.

[62] A.J. van der Schaft, L_2-gain and passivity techniques in nonlinear control, Springer-Verlag, Berlin, Germany (1996).

[63] J.B. Burl, Linear optimal control: H_2 and H_∞ methods, Addison Wesley Longman, Reading, MA, USA (1999).

[64] R.B. Brown and P.Y.C. Hwang, *Introduction to random signals and applied Kalman filtering*, 3rd ed., Wiley, New York, NY, USA (1997).

[65] S.J. Julier and J. Uhlmann, Unscented filtering and nonlinear estimation, *Proc IEEE* **92** (2000), 401–422.

[66] S.J. Julier, J. Uhlmann, and H.F. Durrant-Whyte, A new method for the nonlinear transformation of means and covariances in filters and estimators, *IEEE Trans Autom Control* **45** (2000), 477–482.

[67] S.J. Julier, The scaled unscented transformation, *Proceedings of the American control conference*, vol. 6 (2002), 4555–4559.

[68] F.C. Moon, *Chaotic vibrations*, Wiley, New York, NY, USA (1987).

[69] A.H. Nayfeh, D.T. Mook, and S. Sridhar, Nonlinear analysis of the forced response of structural elements, *J Acoust Soc Am* **55** (1974), 281–291.

[70] A.H. Nayfeh and D.T. Mook, *Nonlinear oscillations*, Wiley, New York, NY, USA (1979).

[71] K. Glover and J.C. Doyle, State space formulae for all stabilizing controllers that satisfy an H_∞-norm bound and relations to risk sensitivity, *Syst Control Lett* **11** (1988), 167–172.

[72] J.C. Doyle, K. Glover, P.P. Khargonekar, and B. Francis, State-space solutions to the standard H_2 and H_∞ control problems, *IEEE Trans Autom Control* **34** (1989), 831–847.

[73] P. Gahinet and P. Apkarian, A linear matrix inequality approach to H_∞ control, *Int J Robust Nonlinear Control* **4** (1994), 421–448.

[74] P.P. Khargonekar and M.A. Rotea, Mixed H_2/H_∞ control: a convex optimization approach, *IEEE Trans Autom Control* **36** (1991), 824–837.

[75] K. Zhou, J.C. Doyle, and K. Glover, *Robust and optimal control*, Prentice-Hall, Upper Saddle River, NJ, USA (1995).

[76] F.E. Zajac, Muscle and tendon: properties, models, scaling and application to biomechanics and motor control, *CRC Crit Rev Biomed Eng* **17** (1989), 359–411.

[77] D. Song, G. Raphael, N. Lan, and G.E. Loeb, Computationally efficient models of neuromuscular recruitment and mechanics, *J Neural Eng* **5** (2008), 175–184.

[78] J.M. Winters, Hill-based muscle models: a systems engineering perspective, in *Multiple muscle systems: biomechanics and movement organization* (J.M. Winters and S.L. Woo, eds.), Springer-Verlag, New York, NY, USA (1990).

[79] D.G. Thelen, Adjustment on muscle mechanics model parameters to simulate dynamic contractions in older adults, *Trans ASME J Biol Eng* **125** (2003), 70–77.

[80] D. Csercsik, *Analysis and control of a simple nonlinear limb model*, Diploma thesis, Budapest University of Technology and Economics (2005).

[81] D. Csercsik, *Modern control methods of a simple nonlinear limb model*, Diploma thesis, Budapest University of Technology and Economics (2007).

[82] Y.C. Fung, *Biomechanics: mechanical properties of living tissue*, Springer-Verlag, New York, NY, USA (1993).

[83] R. Vepa and A. Zhahir, High-precision kinematic satellite and doppler aided inertial navigation system, *R Inst Navig, J Navig* **64** (2011), 91–108.

[84] PLUX repository of physiologic signals. www.opensignals.net (accessed 25 March 2011).

ABOUT THE AUTHOR

Ranjan Vepa is a Lecturer with the School of Engineering and Material Science at Queen Mary, University of London. Prior to joining Queen Mary he was with NASA Langley Research Center, where he was awarded a National Research Council Fellowship. Earlier, he was with the National Aeronautical Laboratory, India and the Indian Institute of Technology, Madras, India.

His research interests are in modelling and simulation of aero-servo-elasticity and smart structures with applications to robotics, aircraft and energy systems. His books *Biomimetic Robotics* and *Dynamics of Smart Structures* were published in 2009 and 2011 respectively.

Bioinspired and Biomimetic Microflyers

Jayant Sirohi

Department of Aerospace Engineering and Engineering Mechanics,
University of Texas at Austin, Austin, TX 78712, USA

Prospectus

This chapter describes recent developments in the area of manmade microflyers. The design space for microflyers is described, along with fundamental physical limits to miniaturizing mechanisms, energy storage, and electronics. Aspects of aerodynamics at the scale of microflyers are discussed. Microflyer concepts developed by a number of researchers are described in detail. Because the focus is on bioinspiration and biomimetics, scaled-down versions of conventional aircraft, such as fixed-wing micro air vehicles and micro-helicopters, are not addressed. Modeling of the aeromechanics of flapping wing microflyers is described with an illustrative example. Finally, some of the sensing mechanisms used by natural flyers are discussed.

Keywords

Aeroelasticity, Cycloidal rotor, Flapping wing, Low reynolds number, Mechanical samara, Micro air vehicle (MAV), Microflyer, Nano air vehicle (NAV), Unsteady aerodynamics.

5.1 INTRODUCTION

For centuries, humans have been inspired by the flight of birds, bats, and insects. Early attempts at building aircraft by replicating the shape of bird and bat wings, without understanding the underlying principles of aerodynamics that govern flight, resulted in failure. It was only systematic observations of bird flight and morphology, in conjunction with experiments on models, that led to successful glider designs by pioneers such as Otto Lilienthal, finally culminating in the powered flight by the Wright brothers.

The aircraft that we see around us today have little in common with biological flyers except for the general shape and basic aerodynamic principles. There are no aircraft in production today that incorporate flapping wings. There are numerous aircraft that bear little resemblance to the planform of birds in flight. Modern aircraft have far exceeded biological flyers in many aspects of performance, such as maximum speed and payload. However, they still cannot compete with biological flyers in other aspects such as maneuverability, gust recovery, and autonomy. In general, modern aircraft development has seen an increase in maximum gross weight, payload fraction, and maximum speed of the aircraft. However, over the past decade, there has been considerable interest in miniaturizing

http://dx.doi.org/10.1016/B978-0-12-415995-2.00005-2

aircraft to create a class of extremely small, remotely piloted vehicles with a gross weight on the order of tens of grams and a dimension on the order of tens of centimeters. These are collectively referred to as *micro-aerial vehicles*, or *micro air vehicles* (MAVs).

The concept of using miniaturized, remotely piloted aerial vehicles for covert surveillance is not new. In the 1970s, the Central Intelligence Agency (CIA) developed an insect-sized, mechanical dragonfly to carry a miniature listening device [1]. The flapping wings of the dragonfly were actuated by a miniature engine powered by a liquid propellant. This MAV was designed to be steered using a laser beam. Due to difficulties in controlling the dragonfly, the project was not pursued beyond the fabrication of a flying prototype, which is now on display in the CIA museum.

In the early 1990s, a research project at Los Alamos National Laboratory theoretically investigated the feasibility of microrobots fabricated using microlithographic techniques for military uses [2] such as intelligence gathering and sensing or disruption of a variety of environmental stimuli (electrical, mechanical, and chemical). The small size of these systems would have made them difficult to detect, and the intent was to increase the probability of mission success by deploying a large number of microrobots. Mass production, similar to the process used to fabricate integrated circuits, was expected to keep the costs of each individual robot low. During the course of this research, the conceptual design of a rotary-wing vehicle was explored using the smallest commercially available electromagnetic motors ($\sim$1.5 g in mass), with different rotor blades, thin-film batteries, and miniaturized video cameras, acoustic sensors, and communications chips. Four different flying vehicle configurations were also investigated: fixed-wing, rotary-wing, microairship, and a passive, autorotative device based on a maple seed. The conceptual fixed-wing vehicle had a total mass of 4 g, with 1 g of sensors and a cruising speed of 900 cm/s. The conceptual rotary-wing vehicle had a counter-rotating rotor

configuration with a sensor mass of 1 g, a cruising speed of 200 cm/s, and an endurance of 5 min. The conceptual microairship had a total mass of 1.8 g, featured an almost transparent film envelope filled with hydrogen, and had a cruising speed of 200 cm/s. The conceptual autorotating device had a total mass of 0.3 g, with a wing area of 1.5 cm $\times$ 5 cm. These were designed to remain aloft for the maximum time possible after being deployed by a stealthy mother vehicle.

In late 1992, the RAND Corporation conducted a workshop for the Advanced Research Projects Agency (ARPA) on "Future Technology-Driven Revolutions in Military Operations" [3]. The objective of this workshop was to identify breakthrough technologies that could revolutionize future military operations. The identified applications included a "fly on the wall," or a miniature fly-sized vehicle carrying sensors, navigation, processing, and communication capabilities. The vehicle design featured the ability to move around by flying, crawling, or hopping. Another application involved the addition of a "stinger" on the vehicle that was intended to disable enemy systems. For conceptual design, a vehicle mass on the order of 1 g, with a size on the order of 1 cm, was selected. The power required for hovering and for forward flight was estimated, using momentum theory, to be $\sim$30 mW/g and $\sim$45 mW/g. In comparison, the hovering power requirement of large insects ranges from $\sim$9 mW/g to 19 mW/g, and for hummingbirds, $\sim$19 mW/g to 26 mW/g. Based on using a 530 J thin-film lithium polymer battery, this was calculated to yield an estimated hover time of 4.9 h and a flight time of 3.3 h, covering 80 km.

In the late 1990s, the Defense Advanced Research Projects Agency (DARPA) released a solicitation for MAVs that would have a dimension no larger than 15 cm, a mass of about 100 g (with a payload of 20 g), and a mission endurance of around 1 h.

These vehicles were intended to be man-portable robots that could fly to a target and

TABLE 5.1 Key MAV design requirements.

Maximum dimension	<15.24 cm
Take-off mass	<100 g
Range	Up to 10 km
Endurance (loiter time)	60 min
Payload mass	20 g
Maximum flight speed	15 m/s

relay video and audio information back to the operator. In this way, the MAVs would enhance the situational awareness of the soldier. Other possible civilian applications of MAVs included sensing of biological/chemical agents in an accident zone without risk to the human operator, fire rescue, traffic monitoring, mobile communications links, and civil structure inspection. Responses to this solicitation included several fixed-wing MAVs such as the MicroStar by Lockheed Martin and the Black Widow by Aerovironment. Rotary-wing MAVs included the LuMAV by Lutronix Inc. In general, it was observed that the fixed-wing MAVs outperformed the rotary-wing MAVs in terms of cruise speed, range, and endurance. However, the major advantage of the rotary-wing MAVs is their hover and low speed capability, which is very useful for surveillance indoors or in confined areas. The key design requirements for a MAV as described by this solicitation are listed in Table 5.1. Note that these specifications are very stringent; over the years, the term *MAV* in published literature has been used to refer to unmanned aerial vehicles with a range of dimensions, from palm-sized to meter-sized.

Recently, DARPA released specifications for the *nano air vehicle* (NAV) program [4]. The goal of this program was to develop a vehicle even smaller than the MAV specifications. The gross mass of the NAV was specified as 10 g, with a payload of 2 g. Configurations that were selected for Phase 1 of this program were a coaxial helicopter, a flapping-wing vehicle inspired by a hummingbird, a flapping-wing vehicle inspired

by a cicada, and a single-bladed rotary-wing vehicle (typically called a *monocopter*) inspired by a maple seed. The size, mass, and performance requirements of the NAV were intended to push the limits of aerodynamics, propulsion systems, and electronics. The vehicle based on the hummingbird was selected for further development in Phase 2 of the program. The final prototype was capable of stable, controllable flight indoors as well as outdoors, with an onboard camera and a fuselage fairing that made it look like a real hummingbird. The prototype met all of the original specifications except for the gross mass. However, all the components were commercially available and it is expected that developing components specifically optimized for this application will enable a significant reduction in gross mass.

The Air Force Research Laboratory has also released a future-vision plan that describes a fully autonomous robotic bird by the year 2015 and a fully autonomous robotic insect by 2030. Several research groups are currently investigating a variety of issues related to such vehicles, specifically focusing on flapping-wing aerodynamics, wing aeroelasticity, gust response, stability, and control as well as autonomous flight.

Bioinspiration and biomimetics form a common theme of many of these micro and nano air vehicles (referred to as *microflyers*), for two key reasons. The first reason is the belief that a microflyer performing a surveillance mission can remain undetected by looking like a real bird or insect and literally *hiding in plain sight*. The second reason is that by virtue of their size, microflyers fall in a size regime that is naturally populated by large insects and small birds. It is believed that by copying several of the characteristics of these natural fliers, man-made microflyers can improve several aspects of their performance such as flight endurance, maneuverability, and gust tolerance. However, it is important to caution against blindly copying biological systems without properly understanding their function. It is quite tempting to conclude that if a certain feature exists on a bird wing, and the bird flies well, then that feature is essential for flight. An example of this reasoning is to conclude that feathers on birds, by virtue of their beneficial aerodynamic properties, must have evolved to enable flight. However, it is now a widely accepted fact that birds evolved from theropod dinosaurs, and feathers evolved for several reasons before the ancestors of birds could fly. Some of these reasons include thermal insulation, water repellancy, and coloration to attract a mate. Numerous fossils have confirmed that feathers existed in nonavian dinosaurs. These early feathers reflect the stages of feather development predicted by theoretical reasoning based on evolutionary developmental biology [5]. It has been stated that "proposing that feathers evolved for flight now appears to be like hypothesizing that fingers evolved to play the piano" [6].

This chapter describes recent developments in the area of manmade microflyers, along with fundamental limits to their performance. Because the focus is on biomimicry, scaled-down versions of conventional aircraft, such as fixed-wing micro air vehicles and micro-helicopters, are not discussed.

5.2 DESIGN SPACE FOR MICROFLYERS

Manmade microflyers typically have dimensions on the order of 10 cm and a gross mass on the order of 100 g or less. Based on conventional fixed-wing aircraft that range in size from single-passenger light aircraft to large civil transport aircraft such as the Boeing 747, it is possible to develop scaling laws for the size and performance of an aircraft of given dimension. These parameters are broadly governed by the square-cube law. That is, the mass of the aircraft, and other parameters related to the mass, vary directly with the volume of the aircraft, which is proportional to the cube of its representative dimension. The surface area of the aircraft, and other parameters related to the area, vary directly with the square of the representative dimension.

As a result, the wing loading W/S of an aircraft, which is the ratio of its weight W to the area of its wings S, varies approximately linearly with the representative dimension l as

$$W/S \propto \frac{l^3}{l^2} \propto l \propto W^{1/3}. \qquad (5.1)$$

The cruise speed V is related to the wing loading by

$$W/S \propto V^2. \qquad (5.2)$$

As a result, it can be seen that a heavier aircraft tends to have a higher cruising speed.

A number of trends can be deduced using similar scaling arguments. Tennekes [7] discussed several of these scaling laws and developed the Great Flight Diagram (Figure 5.1), which plots a number of natural as well as manmade flyers on the basis of their weight, cruising speed, and wing loading. It is quite remarkable that manmade aircraft, birds, and insects all follow a common trend line quite closely. Specially developed aircraft—for example, solar-powered or human-powered aircraft—do not follow this trend because they have been engineered to achieve specific requirements. In a similar way, the spread of aircraft around the trend line is a result of specific mission or operational requirements. Note that the spread is the largest for insects, which perhaps have to satisfy numerous other requirements that are not considerations for birds or aircraft. Another remarkable feature of this diagram is that the trend line is the same regardless of the mechanism used for flight, that is, fixed-wing manmade aircraft follow the same trend as flapping-wing natural flyers. Although the diagram does not explicitly indicate bats, they fall within the range of small birds on the trend line.

Scaling laws and trends such as this are useful in developing conceptual designs of new

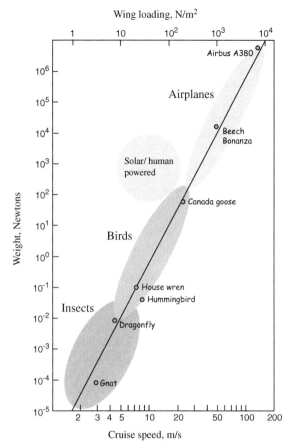

FIGURE 5.1   The Great Flight Diagram. Adapted from Tennekes [7].

wing planform shapes, flapping-wing mechanisms, high-lift aerodynamic devices, unsteady aerodynamic effects, morphing wings, and sensors for flight stabilization and navigation. These can be grouped together under the broad categories of airframes, performance enhancement mechanisms, and sensors.

5.3 PHYSICAL CHALLENGES AT SMALL SCALES

In addition to the performance limits imposed by the square-cube law, several other challenges appear as the size of an aircraft is decreased. These challenges are related to the behavior of mechanical assemblies, energy storage in batteries, miniaturization of electronics, and aerodynamics.

5.3.1 Mechanical Assemblies

Structures typically become relatively stiffer as their dimensions are reduced. For example, the bending stiffness of a cantilever beam varies inversely with the cube of its length. From this point of view, miniaturization of structural members does not result in increased deflections based on the applied loads. An exception is the deformation of wings and other aerodynamic surfaces. Due to the requirement for minimum thickness and minimum weight, wings at the microscale typically consist of membranes attached to a framework of stiffeners. The major consequence of such a construction is the low torsional stiffness of the wing. This can result in significant aeroelastic deformations during operation. However, appropriate deformation can be beneficial in terms of creating a passive pitching motion that would otherwise require a complicated set of hinges. The aeroelastic couplings inherent in insect and bird wings are believed to play an important role in flight efficiency as well as stability. These couplings are still poorly understood

aircraft. However, as the dimension of manmade microflyers is much smaller than the smallest light aircraft, it becomes difficult to extrapolate parameters based on conventional aircraft. Natural flyers such as birds and insects complement conventional aircraft by providing a vast number of data points at the small scale. Given the desired weight and dimension of microflyers, it can be seen that they fall into a region between large insects and small birds. Therefore, a majority of researchers have turned toward natural flyers for inspiration in developing microflyers. This inspiration includes a wide variety of areas such as vehicle configuration, and are the subject of active research. The structural arrangement of an insect or bird wing also accommodates specialized functions such as wing folding. Therefore, it is not straightforward to replicate the structure of a natural wing to obtain the desired structural couplings.

Another important consequence of scaling down a mechanical assembly is the effect on hinges, linkages, and bearings. Conventional hinges based on rotary joints suffer from frictional losses as well as loss of precision due to play between the fixed and rotary members. The relative losses increase as the size of the hinge decreases. Flexure hinges are ideally suited to small precision assemblies due to their lack of moving parts, repeatability, and absence of friction [8]. Several microflyers rely on flexure-based mechanisms to power their flapping flight. These often feature simplified kinematics to focus on specific degrees of freedom, rather than exactly copy mechanisms that occur in nature. As an example, the wing joint of a honeybee contains a number of extremely complicated shapes, linkages, and muscle attachments [9]. Although the functions of each of these features has been mapped out, it is very challenging to replicate the shape and dynamic characteristics of each of these components in a mechanical assembly.

For example, Wood [10] developed a robotic insect of 60 mg mass using a smart composite microstructure consisting of rigid carbon fiber reinforced prepegs sandwiching a thin polyimide layer that acts as a flexure. The wing joints included three degrees of freedom, out of which only one was controlled by an actuator and the other two responded passively. Similar flexure joints have also been used to fabricate the entire resonant thorax mechanism of microscale insects [11]. Compliant drive mechanisms for actuating wing flapping have also been fabricated using an injection-molding process that combines a soft flexural material with a stiff structural material [12].

5.3.2 Energy Storage

The majority of manmade microflyers rely on stored electrical energy for flight in the form of batteries. The energy storage capacity of batteries is often the major bottleneck in terms of flight endurance. Batteries have a significantly lower energy density than hydrocarbon fuels. In addition, batteries have typically been limited in terms of the continuous current that can be drawn from them, which constrains the maximum power that they can supply to the electric motors driving the microflyers. Recently, there have been large improvements in the energy density as well as maximum current draw of batteries. Specifically, the introduction of lithium polymer batteries has revolutionized the field of remotely piloted aircraft and has brought these vehicles within the reach of a vast number of hobbyists. Figure 5.2 shows a comparison of the energy density of different battery chemistries. Note that although LiIon batteries have the highest energy density, they are limited in terms of the current that they can supply. LiPoly batteries can supply several times their charging current and hence they have the highest power density, which makes them ideal for use in microflyers. Some other battery chemistries

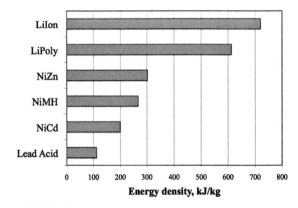

FIGURE 5.2 Nominal energy density of different battery chemistries. LiIon—lithium ion, LiPoly—lithium polymer, NiZn—nickel zinc, NiMH—nickel metal hydride, NiCd—nickel cadmium.

have been developed with higher power densities; however, to operate they require special conditions (such as high temperature) that make them unsuitable for microflyers.

As the size of electric motors decreases, the efficiency of converting electrical power to mechanical power also decreases. It is not uncommon for motors with a power output on the order of tens of Watts to have an efficiency of only around 50%. This means that the batteries that are installed on the microflyer must be sized for significantly higher power outputs than required to sustain flight.

5.3.3 Miniaturization of Electronics

The majority of microflyers are remotely piloted by a human pilot. These could either be in the line of sight or could be piloted in a *first-person view*, by the human pilot watching live video broadcast from cameras on the microflyer, thus giving the impression of being located inside the microflyer. The miniaturization of electronics has resulted in extremely small, lightweight microprocessors and associated sensor packages that are powerful enough to allow some degree of autonomy. This could range from on-board stability augmentation systems to autonomous take-off and landing algorithms. The ultimate goal would be to create an on-board sensing and processing system that replicates the nervous system and brain of natural flyers, enabling them to autonomously sense their environment, identify and avoid obstacles, recover from gusts, take off and land, and navigate to specific waypoints. Sensors based on microelectromechanical systems (MEMS) technology, such as accelerometers and gyros, are enabling significant sensing capability in a small package. However, the drawback of such sensors is a long-term drift, and techniques such as sensor fusion and Kalman filtering are required to correct errors. The size of computers decreases by a factor of 100 every ten years, and recently, computers with ultra-wideband transceivers having a volume of less

than 1 cc have been realized [13]. Further miniaturization of sensing, control, and communication electronics will enable a significant improvement in microflyer capabilities in the future.

5.3.4 Aerodynamics

A fundamental challenge at the scale of microflyers arises from the aerodynamics. The Navier–Stokes equations are fundamental physical relations that govern fluid flow. When the incompressible Navier–Stokes equations are nondimensionalized to make them independent of scale, they yield a dimensionless parameter called the *Reynolds number*. When the Reynolds number is kept constant between two flows at different scales or in different media, the fluid behavior is identical. The Reynolds number is given by

$$\mathrm{Re} = \frac{V l \rho}{\mu}, \qquad (5.3)$$

where V is the freestream velocity, l is a characteristic length, ρ is the density of the fluid, and μ is the dynamic viscosity of the fluid.

The Reynolds number represents the ratio between inertial forces and viscous forces. If the number is very high, then inertial effects dominate the flow and viscosity can be neglected. For conventional aircraft in cruise, the typical Reynolds number that they operate at is on the order of 1 million–10 million. However, at the scale of microflyers, the Reynolds number is on the order of 1–10,000. For small insects, the Reynolds number can be as low as several hundred. At these small Reynolds numbers, the flow is dominated by viscous effects, and the behavior of the flow can be quite different than at high Reynolds numbers.

As the Reynolds number is decreased, the maximum lift coefficient of an airfoil decreases, the profile drag coefficient increases, and the lift-to-drag ratio decreases. These effects are shown in Figures 5.3–5.5. Also shown for reference are the values for a flat plate; note that

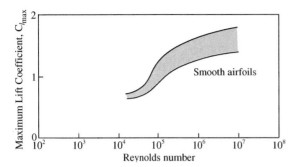

FIGURE 5.3 Variation of maximum lift coefficient $C_{l_{max}}$ as a function of Reynolds number for different airfoils. Adapted from Ref. 14.

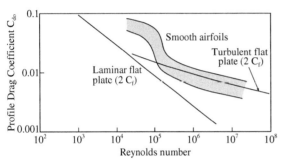

FIGURE 5.4 Variation of zero lift profile drag coefficient C_{do} as a function of Reynolds number for different airfoils (adapted from Ref. 14). The skin friction coefficient is C_f.

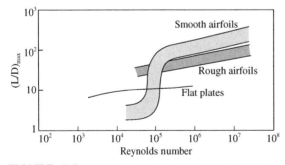

FIGURE 5.5 Variation of maximum lift-to-drag ratio as a function of Reynolds number for different airfoils. Adapted from Ref. 14.

insect wings typically operate at a Reynolds number of $\sim 10^4$. The variation of profile drag coefficient is very dependent on the skin friction

coefficient C_f, which can take on a range of values based on whether the surface is smooth or rough. This forms the basis of incorporating surface roughness elements such as turbulators, sandpaper, or trip wires on an airfoil to improve its performance at low Reynolds numbers.

A detailed review of flow physics at low Reynolds numbers was given by Carmichael [15]. He described twelve regimes of Reynolds number and corresponding natural or man-made flyers in each regime. For each regime, the basic flow physics and aerodynamic effects were described. The application of different types of devices to trip the boundary layer and improve aerodynamic performance at low Reynolds numbers was described. Schmitz [16] published a comprehensive database of aerodynamic performance of five different airfoils in the Reynolds number range of 42,000–420,000. These data were directed toward building model airplanes. Several other researchers have published data on airfoils at low Reynolds numbers for model airplanes as well as wind turbine applications [17–20].

Airfoils with rounded leading edges, which are quite efficient at high Reynolds numbers, perform poorly at low Reynolds numbers [14, 21]. Laitone [22, 23] measured the lift and drag on rectangular planform wings with different airfoil profiles and observed that thin cambered plates, circular arc airfoils, and airfoils with sharp leading edges have a significantly higher performance at low Reynolds numbers (less than 50,000) than conventional airfoils. In fact, he found that a reversed NACA0012 airfoil with flow incident on the trailing edge has a better lift-to-drag ratio than when the flow is incident on the leading edge. A circular arc airfoil with 5% camber and a thickness ratio of 1.3% was found to have the best lift-to-drag ratio and maximum lift coefficient at a Reynolds number of 20,700. The airfoils with a rounded nose were also found to be much more sensitive to freestream turbulence (Figure 5.6).

Figure 5.7 shows a comparison of the drag polars of a thin circular arc airfoil, and a

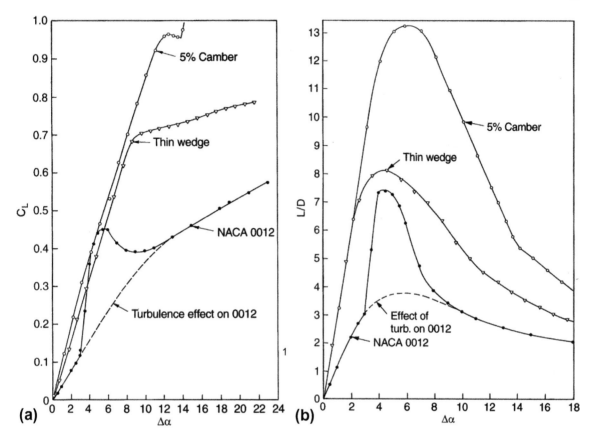

FIGURE 5.6　Aerodynamic performance of different profiles as a function of angle of attack $\Delta\alpha$, for a rectangular wing with aspect ratio=6 and Reynolds number=20,700 [23]. (a) Lift coefficient and (b) lift-to-drag ratio. With kind permission from Springer Science and Business Media, E.V. Laitone, Wind Tunnel Tests of Wings at Reynolds Numbers Below 70,000, Experiments in Fluids 23 (1997) 405–409.

conventional rounded-nose airfoil at a Reynolds number of 120,000. From this figure and Figure 5.6 it is seen that the Reynolds number has a relatively insignificant effect on thin airfoils with sharp leading edges, and has a significant effect on airfoils with a rounded nose. The circular-arc airfoil has the best performance at the lower Reynolds number, while the rounded-nose airfoil has the best performance at the higher Reynolds number, with substantially higher maximum lift coefficient.

At low Reynolds numbers (<50,000), the assumptions of potential flow break down and the

behavior of the airfoils is quite different than at the Reynolds numbers typical of full-scale aircraft (on the order of 10^6). For example, the Kutta condition (flow leaves the airfoil trailing edge smoothly) may not be satisfied at low Reynolds numbers. Also, the variation of lift with angle of attack is highly nonlinear at low Reynolds numbers, and the lift curve slope may be quite different than the potential flow prediction of 2π per radian. Figure 5.8 shows the lift coefficient as a function of angle of attack for a thin circular arc profile with a camber of 8% at a Reynolds number of 3.14×10^5. Although this is significantly

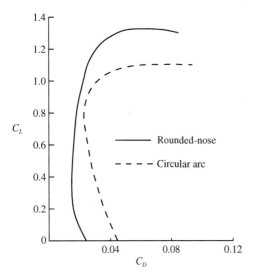

FIGURE 5.7 Comparison of drag polars of thin circular arc airfoil, and rounded-nose airfoil at Reynolds number of 120,000. Adapted from Ref. 21.

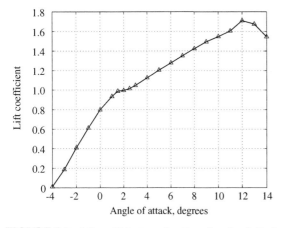

FIGURE 5.8 Lift coefficient as a function of angle of attack for an 8% cambered circular-arc profile when the Reynolds number $= 3.14 \times 10^5$. Adapted from Ref. 17.

higher than the Reynolds number typical of microflyers, it is still much lower than that of full-scale aircraft and shows significant nonlinear behavior. In general, at low Reynolds numbers, airfoils exhibit a lower maximum lift coefficient and a higher profile drag coefficient.

The best airfoils to use for microflyers and for propellers at the microscale are circular-arc profiles. However, these profiles still have a significant profile drag. Micro-helicopter rotors of diameter around 6 in., with blades having circular-arc airfoils and a tip Reynolds number on the order of 20,000, have a hovering efficiency of around half that of a full-scale helicopter rotor [24]. By modifying the planform in specific ways, sharpening the leading edge, and moving the maximum camber location forward of the airfoil mid-chord, the hovering efficiency can be improved to around 0.65 [25, 26]. In comparison, a modern full-scale helicopter rotor has a hovering efficiency of more than 0.8 [27, 28].

5.4 UNSTEADY AERODYNAMICS IN ANIMAL FLIGHT

The wings of birds, insects, and bats reflect the behavior of airfoils at low Reynolds numbers. Bird wings have a thin, cambered cross-section, which gives optimum performance at their flight Reynolds number. This fact was recognized early on by the pioneers of manmade flyers: Sir George Cayley, Otto Lilienthal, and the Wright brothers, who used thin, cambered airfoils for their airplane wings. However, as the flight speed of airplanes increased and their representative Reynolds number increased, thin cambered airfoils made way for the higher-performing, thicker, rounded-nose airfoils that are ubiquitous on airplanes today.

The wings of natural flyers continuously flap and deform, making their aerodynamic environment highly unsteady. The wing tips trace out complex patterns that change depending on flight speed and maneuvers. These paths are quite complex, and can be executed at a high frequency; this flapping frequency depends on the body mass and can range from around 900 Hz for a mosquito (mass ~ 1 mg) to around 1 Hz for a large bird such as a pelican (mass ~ 10 kg) [30].

In such a case, the forces generated can be quite different than in the steady case. Several researchers have observed that at the Reynolds numbers typical of bird and insect flight, steady aerodynamic forces are insufficient to sustain flight of the animal. Under steady conditions, the maximum lift coefficient of an airfoil is around 1.5 at these Reynolds numbers. For example, measurements on a gliding jackdaw yielded an estimate of 2.1 [29] for the lift coefficient.

Norberg [30] measured the wing-flapping kinematics of a hovering dragonfly and, using steady-state aerodynamics, calculated that the wings produce only 40% of the lift required to sustain the weight. Steady-state aerodynamics predicted lift coefficients between 3.1 and 6.4 for a hovering long-eared bat [31]. Weis-Fogh [32] studied the hovering flight of several species of insects and concluded that lift coefficients calculated from flight were far in excess of values predicted by steady-state aerodynamics. Therefore, he proposed several novel unsteady mechanisms for lift production, including significant elastic deformation of the wings, such as the clap-fling mechanism and the flip mechanism. In the clap-fling mechanism, the wings of the insect are clapped together at the end of the upstroke and are subsequently peeled apart during the beginning of the downstroke, as shown in Figure 5.9. The air rushing in to fill the space between the wings results in a large area of vorticity. This results in a significant transient increase in the lift on the insect. The flip mechanism occurs when the stroke reverses and the wing is rotated, capturing additional vorticity.

Ellington et al. [33] visualized the flow field around a scaled-up mechanical model of a flapping hawkmoth wing and discovered the presence of a three-dimensional leading-edge vortex stabilized by spanwise flow along the wing. This leading-edge vortex is believed to be responsible for the high lift measured on the hawkmoth wing, which cannot be explained by steady-state aerodynamics. Dickinson et al. [34] performed experiments on a scaled robotic flapping model of fruit-fly wings and described three unsteady mechanisms responsible for lift production in excess of steady-state values. One of these mechanisms, delayed stall, relies on the production of a leading-edge vortex similar to that observed during dynamic stall, while the wing is translating at a large angle of attack. In dynamic stall, a vortex is shed from the leading edge of the airfoil, which results in a transient lift significantly in excess of the maximum static lift. This vortex is subsequently convected downstream, causing a decrease in lift. The overall result of the dynamic stall phenomenon is to cause a hysteresis in the lift vs. angle of attack curve, accompanied by a large increase in the maximum lift in comparison to the static lift case. A detailed description of the dynamic stall phenomenon with experimental data for a NACA 0012 airfoil was given by Carr et al. [35].

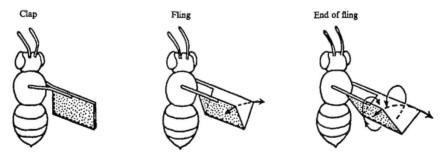

FIGURE 5.9 Clap-fling mechanism of lift production in hovering insects [32]. Adapted with permission from T. Weis-Fogh, Journal of Experimental Biology, 1973, 59(1), 169–230.

The other two mechanisms, rotational lift and wake capture, occur during stroke reversal. The flip and rotational lift mechanisms rely heavily on the aeroelastic deformation of the wing. Typically, the ribs in insect wings make them relatively stiff in bending but very flexible in torsion. Sane [36] reviewed the different mechanisms of lift production in insect flight, and Ellington [37] summarized these effects, including estimates of lift, power, and flight speed for potential application to microflyers.

The degree of unsteadiness in the flow is typically expressed in terms of the Strouhal number (St), given by

$$St = \frac{fA}{V_\infty}, \tag{5.4}$$

where f is the frequency of motion (such as flapping) in Hz (beats per second), A is the amplitude of motion, and V_∞ is the flight velocity or freestream velocity. In classical discussions of unsteady aerodynamics, the reduced frequency k is used as a measure of the unsteadiness of the flow [39, 40] and is closely related to the Strouhal number; here,

$$k = \frac{\omega c}{V_\infty}, \tag{5.5}$$

where ω is the frequency of the motion (in radians/s) and c is a chord length (typically parallel to the freestream). Many unsteady effects are directly related to the Strouhal number and reduced frequency. For example, Taylor *et al.* [41] found that flying and swimming animals, over a range of sizes, cruise at a Strouhal number between 0.2 and 0.4, which gives them the best propulsive efficiency. Classical aerodynamic theories involving the Theodorsen function or the Wagner function, for example, can be used to model the unsteady aerodynamics; however, they are only valid for attached flow [39, 40]. Consequently, these theories are often used to analyze the flapping flight of wings undergoing small motions. For higher-amplitude motion involving separated flow, vortex theories, indicial methods, and computational fluid dynamics are used to calculate the forces and power. Descriptions, analytical models, and reviews of the flight of different types of animals, in addition to their morphology, muscle energy consumption, and other physiological aspects, can be found in several references—for examples, Azuma [38], Pennycuick [42], Norberg [30, 43], Rayner [44].

5.5 AIRFRAMES

A wide variety of airframe configurations have been proposed for MAVs/NAVs. Two fixed-wing MAVs were developed under the DARPA-funded MAV project. The primary advantages of fixed-wing configurations are their relatively high lift-to-drag ratio (L/D) in cruise, mechanical simplicity, and high cruise speed. Their main disadvantage is their inability to hover. In addition, their high cruise speed makes it difficult to maneuver and avoid obstacles in indoor, cluttered environments. This can be seen in Figure 5.10 which plots the mass of several MAVs as a function of their endurance. The fixed-wing MAVs typically have a lower mass due to their mechanical simplicity and higher endurance, and their superior L/D, compared to rotary-wing MAVs. The DARPA specification is also indicated on this figure, which shows that the endurance requirement was very stringent, while the total mass specification was achievable. Subsequent versions of the fixed-wing MAVs were able to achieve significantly improved endurance by optimizing several of their subsystems. Also shown in the figure are vehicles that were designed to hover and transition to forward flight in fixed-wing mode (Hoverfly, Microcraft OAV). The penalty for this additional ability is an increased mass and marginal improvement in endurance. The parameters of these MAVs are summarized by Bohorquez and Pines [45], who also reviewed the state-of-the-art in MAVs and discussed the challenges for future development.

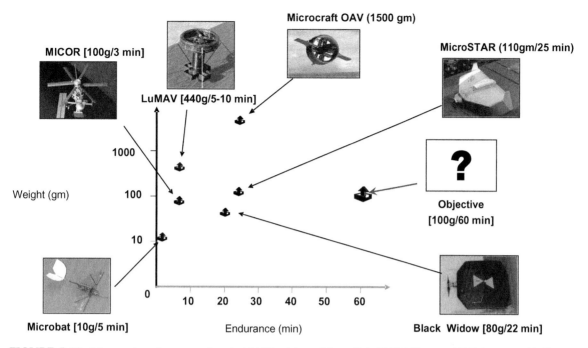

FIGURE 5.10 Mass and performance of typical MAVs. Adapted from Ref. 45. D.J. Pines and F. Bohorquez, Challenges facing future micro-air-vehicle development, Journal of Aircraft, Volume 43, 290-305, SPIE 2006.

The ability to hover as well as fly efficiently at low speeds is crucial for a microflyer designed for indoor surveillance. Accordingly, a large amount of research has been focused on rotary-wing and flapping-wing microflyer configurations. Some unconventional hover-capable configurations have also been proposed. The hover-capable microflyers are described in more detail in the following sections, considering that hovering flight itself is inspired by biological flyers.

5.5.1 Comparison of Rotary-Wing and Flapping-Wing Flyers

There has been a long-standing debate over whether rotary-wing flight or flapping-wing flight is more efficient at the microflyer scale. Many researchers have claimed that the existence of flapping-wing natural flyers in the size range of microflyers indicates that flapping-wing

flight is a solution favored by millenia of evolution and hence must be the most efficient. Other researchers have pointed out that it is impossible to realize in nature a high-speed rotary joint such as that required for a rotor shaft. This precluded the evolution of any natural rotary-wing flyers and therefore questions the superiority of flapping-wing flight.

Hall and Hall [46] developed a variational method to predict the circulation distribution on flapping wings that would yield the minimum power required for a given thrust and lift. This study concluded that the efficiency of flapping-wing flight is not necessarily greater than that of propeller-driven flight at low Reynolds numbers. Ellington and Usherwood [47] measured the performance of rotors at a Reynolds number range 10,000–50,000. They studied three types of blade planforms: a linear taper, the same planform as a hawkmoth wing, and a smooth outline with the same general shape as the hawkmoth wing but without any notches or discontinuities. They discovered that at these Reynolds numbers, under constant rotation speed and angle of attack, a leading-edge vortex periodically forms and breaks down, whereas for rotors at a lower Reynolds number and for the flapping motion of hawkmoth wings, a spanwise flow is developed that stabilizes the leading-edge vortex and enhances the lift.

Several hovering MAVs based on scaled-down single main rotor and coaxial helicopter configurations have been successfully built and flight tested [24, 25, 48, 49]. These MAV-scale rotors typically operate in the Reynolds number range from 10,000 to 100,000. Consequently, they experience much higher viscous drag than conventional helicopter rotors. As a result, MAV-scale rotors suffer from an inherent limitation in aerodynamic efficiency, which translates into poor endurance. By careful design of rotor blade geometric parameters such as solidity, twist, taper, camber, and tip shape, the maximum figure of merit achieved to date for a rotor of diameter 9 in. (22.86 cm) is around 0.64, and for a rotor of diameter 6 in. (15.24 cm) is around 0.55 [26], at a tip Reynolds number of 40,000. In comparison, a conventional helicopter rotor with a figure of merit of 0.64 is considered poor in terms of aerodynamic efficiency. In fact, most modern helicopter rotors have a maximum figure of merit of about 0.8 [27, 28].

5.5.2 Flapping-Wing Microflyers

Nowadays, there are a large number of remotely controlled, flapping-wing microflyers being sold as toys or hobby aircraft. Most of these microflyers are powered by an electric DC motor and feature simple wings with a rigid spar and thin membrane. Controls are incorporated in terms of a movable tail or by modifying the lift produced by each wing using a mechanism that changes the tension on the wing trailing edge.

These hobby flyers are available in several sizes; the smaller vehicles are suited for indoor flight, while the larger ones can fly outdoors. The largest ornithopter is a human-powered aircraft with a 32 m wingspan that was recently flown by a group at the University of Toronto Institute of Aerospace Studies.

A few research groups have focused on improving the performance of flapping-wing microflyers with the goal of achieving high-endurance, fully autonomous flight in the smallest possible dimensions. One of the earliest flapping-wing microflyers was the Caltech Microbat, developed as part of the DARPA MAV program [50], which had a total mass of around 11 g. Keennon et al. [51] developed a mechanical hummingbird as part of the DARPA NAV program. This microflyer has a wingspan of 16.5 cm and a total mass of 19 g. It can hover for around 4 min and can fly at a speed of 6.7 m/s. The flapping-wing mechanism is powered by DC motors and the wing-flapping frequency is 30 Hz. The remarkable feature of this microflyer is that it has a fuselage shaped and painted to make it look like a real hummingbird, which makes it ideal for covert operations. The electronics and control system were developed in-house and are enclosed within the body (Figure 5.11). All the control inputs are generated by varying the lift on the wings, and the vehicle does not rely on a tail for stability. The microflyer can fly stably outdoors under the control of a human pilot and transmits live video to a ground station.

deCroon et al. [52] developed a family of flapping-wing microflyers powered by DC electric motors. These consisted of the DelFly I with a 50 cm wingspan and a mass of 21 g, the DelFly II with a 28 cm wingspan and a mass of 16 g, and the DelFly Micro with a 10 cm wingspan and a mass of 3 g (Figure 5.12). The wings consisted of thin membranes attached to carbon fiber spars. Different types of tails were explored, and the main goal of these prototypes was to provide a stable platform for carrying a camera.

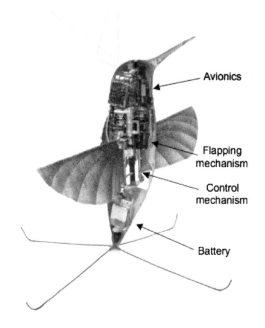

FIGURE 5.11 Nano hummingbird prototype, with fuselage cut away to show enclosed electronics [51]. By kind permission of M. Keennon.

FIGURE 5.12 DelFly Micro prototype [52]. The wing span is 10 cm and total mass is 3.07 g. Credits: G. C. H. E. de Croon, K. M. E. de Clerq, R. Ruijsink, B. Remes, and C. de Wagter. Design, Aerodynamics, and Vision-Based Control of the Delfly. International Journal of Micro Air Vehicles, 1(2):71–97, 2009.

The high power density of piezoelectric actuators (around 400 W/kg) compared to insect muscle (around 80 W/kg) [10] has motivated the development of piezoelectrically actuated flapping-wing mechanisms. Wood [10] developed a robotic insect with a wingspan of 3 cm and a mass of 60 mg. This insect was powered by piezoceramic bimorph actuators and had wings with 1.5 μm polyester membranes. The body of the insect, constructed from a laser-micromachined sandwich of carbon fiber and polymer, featured a flapping mechanism similar to that of an insect thorax. The robotic insect was powered from an external source and demonstrated a thrust greater than its weight.

The Micromechanical Flying Insect (MFI) [53] is another piezoelectrically actuated robotic insect with a body constructed from sandwiched composites and flexure hinges. The wingspan is 25 mm and the wing-flapping frequency is 275 Hz. Efforts are underway to increase the lift produced by this mechanism.

Cox *et al.* [54] developed several versions of a piezoelectrically driven flapping-wing mechanism based on four-bar and five-bar linkages. These had wingspans on the order of 15 cm and total mass around 7 g. Mechanisms based on piezoelectric actuators typically operate at resonance to obtain the largest amplitude of flapping. Although they have demonstrated good benchtop performance, the size of the power supply required for the piezoelectric actuators can be considerable, and no microflyer powered by piezoelectric actuators has achieved free flight to date.

5.5.3 Samara Type Microflyers

The term *samara* is a generic term for a winged seed. The seeds of many plants are dispersed by means of autorotation in wind, and there are several microflyers that have been inspired by this concept. The basic idea is to combine the simplicity of an autorotating samara with a source of thrust to sustain rotation, thus creating a simple, single-bladed helicopter. Azuma [38] provided a comprehensive review of several types of autorotating seeds along with their lift and drag characteristics. One of the earliest autorotating devices that was used in a submunition deployed from an airplane was described by Kline [55]. However, this was a completely passive device. As part of the DARPA NAV program, Lockheed Martin developed a vehicle based on the samara, called the Samarai [56, 57]. Early concepts of this vehicle featured a wing with a flap for control, driven by a fuel-powered pulsejet engine at the wing tip. More recently, this concept developed into a family of vehicles with a range of sizes from 17 cm to 72 cm, powered by electric motors driving propellers.

Ulrich *et al.* [58, 59] developed mechanical samaras incorporating a rapid-prototyped polymer body and a propeller driven by a DC brush- less motor. Three sizes of mechanical samara microflyers were developed, with a total mass of 75 g, 38 g, and 9.5 g, having a maximum dimension of 270 mm, 180 mm, and 75 mm (Figure 5.13). The largest of these had a flight time of around 20 min. Control is achieved by varying the angle of incidence of the wing with respect to the fuselage. It was envisaged that these microflyers would be deployed from

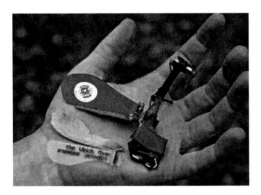

FIGURE 5.13 Mechanical samara microflyer, total mass 9.5 g, maximum dimension 75 mm [59]. Credits: E.R. Ulrich, D.J. Pines, J.S. Humbert, From Falling To Flying: The Path To Powered Flight Of A Robotic Samara Nano Air Vehicle, Bioinspiration & Biomimetics, 5, 045009, 2010.

a fixed-wing unmanned aerial vehicle and would then fly autonomously to execute their mission. Extensive experiments were performed characterizing the dynamics of this microflyer configuration and evaluating the effect of planform geometry [60].

5.5.4 Flap Rotors

There have been several attempts to harness unsteady mechanisms similar to those found in nature, such as flapping and pitching, to enhance the performance of conventional lift production mechanisms. For example, in a conventional helicopter rotor in hover, an airfoil at any spanwise location on the rotor blade experiences a steady aerodynamic environment. At higher levels of rotor thrust, as the airfoils operate close to their static stall angle, they experience a loss of lift and an increase in drag, resulting in a decrease in the hover efficiency of the rotor. It may be possible to improve this efficiency by creating an unsteady aerodynamic environment at the airfoil. The unsteady motion can be created mechanically in different ways. Due to the large forces involved in creating such motion, this approach is only feasible at the microscale.

Bohorquez and Pines [61] developed an active flapping and pitching mechanism for a 20 cm diameter, two-bladed helicopter rotor. The mechanism enabled the rotor blades to be actively pitched and flapped in an oscillatory fashion at a frequency independent of the rotor speed. The goal of the oscillatory pitching was to induce dynamic stall on the rotor blade, resulting in a large increase in lift coefficient. The goal of the flapping motion was to generate a radial flow along the rotor blade, which was expected to stabilize the leading-edge vortex created during the dynamic stall event. An appropriate combination of flapping and pitching amplitudes as well as frequencies was determined by experiments.

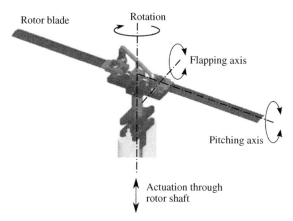

FIGURE 5.14 Schematic of flapping rotor operation. Adapted from Ref. 61. F. Bohorquez and D.J. Pines, Design and development of a biomimetic device for micro air vehicles, Volume 4701, 503–517, SPIE 2002.

A schematic of the rotor hub with articulation for flapping and pitching is shown in Figure 5.14. A scotch-yoke mechanism converts rotary input from a small electric motor into linear motion that is conveyed through the hollow rotor shaft into a lever mechanism on the rotor hub. These levers actuate the rotor blade in flapping and pitching about their respective axes. Large-amplitude flapping (total angle of 46°) and pitching ($\pm20°$) motion is possible, and the frequency of the motion depends on the rotational speed of the motor. A picture of the rotor with the blades at a large flap angle is shown in Figure 5.15.

The mechanism was tested by spinning the rotor at 2,000 rpm, keeping the flap motion fixed and prescribing a pitching motion of amplitude 6° at frequencies varying from 0.25 to 2 per revolution of the rotor. It was observed that while operating at mean pitch angles close to the static stall angle of the airfoil, the oscillatory pitching resulted in around 50% improvement in hover efficiency. As expected, there was negligible effect at lower angles of attack, where the

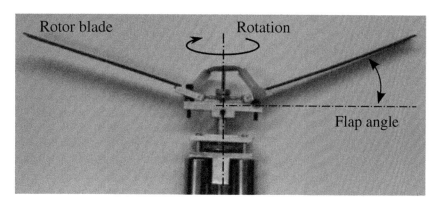

FIGURE 5.15 Rotor blades actuated to a high flap angle. Adapted from Ref. 61. F. Bohorquez and D.J. Pines, Design and development of a biomimetic device for micro air vehicles, Volume 4701, 503–517, SPIE 2002.

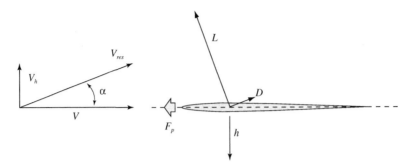

FIGURE 5.16 Schematic of thrust production by a plunging airfoil in a freestream.

oscillatory pitching does not cause dynamic stall to occur. In addition, the largest improvements were observed at low pitching frequencies (0.25–0.5 per revolution).

A further extension of this concept is to use oscillatory flapping of the rotor blades to reduce or eliminate the torque required to rotate the rotor. This idea is based on the Knoller–Betz effect: When an airfoil undergoes plunging motion in an incident freestream velocity, it can produce thrust, i.e., a force opposite the direction of the freestream velocity. Flyers with flapping wings utilize this effect to generate a propulsive force in flight. The effect is summarized in Figure 5.16. The airfoil is shown plunging in an incident freestream of velocity V. The plunging displacement of the airfoil is h, and the apparent velocity of the air is $V_h = dh/dt$. The resultant velocity incident on the airfoil is V_{res} at an angle of attack α. The lift L and drag D on the airfoil are perpendicular and parallel, respectively, to the resultant incident velocity. The thrust or propulsive force F_p is given by the summation of horizontal components of L and D, i.e.,

$$F_p = L\sin\alpha - D\cos\alpha. \qquad (5.6)$$

Therefore, based on a specific range of values of freestream velocity and plunging velocity, it is possible to create a positive propulsive force. This effect forms the basis of a unique microflyer developed by Jones and Platzer [62] in which the lift and propulsive force are generated by a pair of straight biplane-like wings located at the rear of the vehicle, flapping in opposition to each other. This gives the two wings an oscillatory pitching and plunging motion with respect to each other that results in both a lift force and a thrust force.

Heiligers *et al.* [63] developed a single-rotor helicopter, called the Ornicopter, with a mechanism that actively flapped the blades. The flapping resulted in the production of a propulsive force on the blades that created the torque required to spin the rotor. As a result, there was no reaction torque on the helicopter fuselage. A radio-controlled model helicopter was modified to accommodate the required flapping mechanisms. A series of experiments was performed to evaluate the yaw control authority and the optimum settings of rotational speed and flapping amplitude [64].

The rotor diameter was 1.5 m and the flapping was phased such that opposing pairs of blades on the four-bladed rotor flapped with the same phase and were out of phase with their neighboring blades. In this way, oscillatory inertial forces along the rotor shaft were eliminated. The prototype was tested at a rotational speed of 500 rpm, over a range of collective pitch settings and flapping amplitudes. Torque measurements indicated a range of settings over which thrust was produced at zero rotor torque. For example, at 4° collective pitch and a flapping angle of 8.3°, the rotor produced 8 N of thrust at zero torque. Yaw control was achieved by varying the

FIGURE 5.17 Bench test prototype of the Flotor, a micro rotor powered by blade flapping [65].

flapping amplitude at a given collective pitch setting. Note that rotor thrust was insensitive to flapping amplitude at a constant collective pitch. Future work should focus on increasing the thrust produced by the rotor by operating at higher rotational speed.

Fitchett and Chopra [65] developed a microscale rotor, called the Flotor, that was powered by blade flapping (Figure 5.17). A prototype was constructed and tested in three modes: pure flapping, pure rotation, and combined flapping/rotation. The geometry of the blades of the Flotor, as well as the rotational speed, was determined based on the wings of bats of similar size and their reduced frequency in cruise flight. The prototype rotor had two blades that were flapped in phase, a rotor radius of 80 mm, and a blade aspect ratio of 6.5. To ensure that pure flapping initiated rotor rotation in the correct direction, the blades were constructed with a main spar at the leading edge. This resulted in sufficient elastic twist in the blades to generate the appropriate propulsive forces at low rotational speed. The blades were constructed out of 0.25 mm thick mylar sheet and a carbon fiber framework.

In the pure flapping mode (passive rotation), a maximum disk loading of 10 N/m^2 was measured, which is low compared to conventional shaft-driven rotors. For the pure rotation and combined flapping/rotation tests, rigid blades with a circular arc profile were tested. Compared to the pure rotation cases, an increase in thrust of up to 20% and a decrease in torque of up to 30% were measured during combined flapping/rotation. Several recommendations were made for future research leading to a flight-capable prototype.

Although the concept of active blade flapping has been shown to enhance the performance of a conventional rotor, the main challenges to this approach are the mechanical complexity of the rotor hub, the inertial forces due to active blade-flapping, and the additional power required by the blade-flapping mechanism. These challenges must be addressed appropriately to enable flight testing of such a configuration.

5.5.5 Cycloidal Rotor

The cycloidal rotor is an unconventional lift-producing mechanism that has the potential to improve hover efficiency by harnessing unsteady aerodynamic effects. A cycloidal rotor consists of several blades that rotate about a horizontal axis

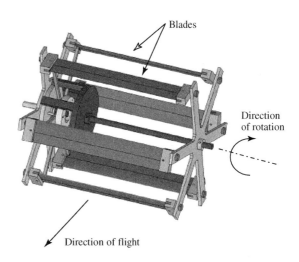

FIGURE 5.18 Cycloidal rotor configuration.

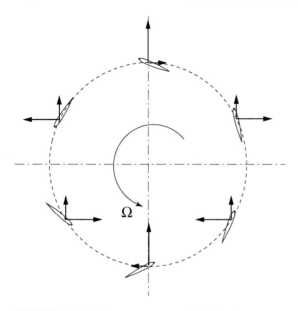

FIGURE 5.19   Thrust vectors at each blade cross-section.

that is perpendicular to the direction of flight (Figure 5.18). The blade span is parallel to the axis of rotation. As the blades rotate around the azimuth, their pitch angle is varied periodically, typically using a passive mechanism such as a four-bar linkage. Each spanwise blade element operates at about the same conditions—velocity, Reynolds number, angle of attack, centrifugal force—and thus can be designed to operate at its optimum efficiency.

Figure 5.19 shows a cross-section of a six-bladed cycloidal rotor rotating with an angular velocity Ω. Each of the blades produces a lift and a drag force. Blades at the top and bottom positions produce an almost vertical net force, while those at the sides produce small lateral forces because of their reduced angle of attack. The horizontal components of the forces cancel, resulting in a net vertical thrust. In addition, the amplitude and phase of the maximum blade pitch angle may be changed by modifying the configuration of the mechanical linkage. In this way, the magnitude and direction of the net

thrust vector of the rotor can be changed almost instantaneously.

The concept of cycloidal propulsion was first investigated in the 1920s by Kirsten [66] and in the 1930s by Wheatley [67, 68]. These early cycloidal rotors were intended for use in full-scale aircraft. Wind-tunnel tests were performed on 8 ft diameter cycloidal rotors and significant forces were obtained; however, due to incomplete theoretical knowledge of unsteady aerodynamic effects, it was not possible to accurately predict the performance of these devices. The cycloidal rotor can change the direction of its thrust vector almost instantaneously over a complete circle, i.e., over an angular range of 360°. Because of this unique ability, cycloidal rotors eventually made their way to marine systems, where they are used in tugboats to provide them with low-speed maneuverability. More recently, cycloidal rotors have made a reappearance in aircraft applications. They have been proposed for use on airships [69, 70] and on an UAV of gross weight 600 lb, where the wings are replaced by cycloidal rotors [71]. On a smaller scale, cycloidal rotors of span around 0.8 m have been investigated for VTOL UAVs of take-off mass around 50 kg [72, 73]. These rotors were able to demonstrate a power loading around 12 kg/HP at low thrust that asymptoted to 5 kg/HP at high thrust.

Due to the potential performance benefits of unsteady aerodynamic effects as well as the increased maneuverability afforded by the instantaneous change in thrust vector, cycloidal rotors have been explored for microflyers. Hwang et al. [74] designed a microscale cyclocopter with two cycloidal rotors of radius 0.2 m. Sirohi and Parsons [75] developed a six-bladed, six-inch-diameter cycloidal rotor for a micro-aerial vehicle. Experiments were performed on a prototype to measure the flowfield in the downwash of the rotor as well as the thrust and torque produced at rotational speeds up to 1,200 rpm. The rotor blades had a NACA0010 profile, and the amplitude of the oscillatory blade pitch angle could be set from 0° to 40°. This translated into a reduced frequency of around 0.167, which is considered highly unsteady. A time-domain formulation based on the Wagner's function, in conjunction with downwash predicted based on momentum theory, was used to predict the thrust and torque of the rotor. Good agreement with measured thrust was observed, but there was some discrepancy with measured torque. These discrepancies were attributed to an over-simplification of the flowfield, especially through the central part of the rotor. The power loading at low thrust settings was observed to be comparable to that of a conventional helicopter rotor of the same diameter and asymptoted to a lower value at high thrust settings.

Based on experimental results, a micro-aerial vehicle powered by two six-inch-diameter cycloidal rotors was designed (Figure 5.20). The total mass of this vehicle was around 250 g and the rotor speed was around 1,650 rpm.

Benedict *et al.* [76, 77] performed further experimental studies on a cycloidal rotor of the same size, with the goal of optimizing the performance of the cycloidal rotor MAV. The effects of number of blades (ranging from two to six), maximum pitching amplitude, and airfoil camber were investigated. Improved performance was achieved with a larger number of blades, higher pitching amplitude, and uncambered airfoils. Particle image velocimetry (PIV) measurements indicated a high degree of wake skewness as well as significant rotational flows inside the cycloidal rotor. Aeroelastic modeling of a cycloidal rotor using nonlinear finite elements and multibody simulations with different inflow models [78] indicated that the wake skewness and resulting side force arises from the mechanical linkage as well as a phase lag due to unsteady aerodynamic effects. Torsional deformations were shown to decrease the thrust produced. Further experimental studies on the blade airfoil profile and

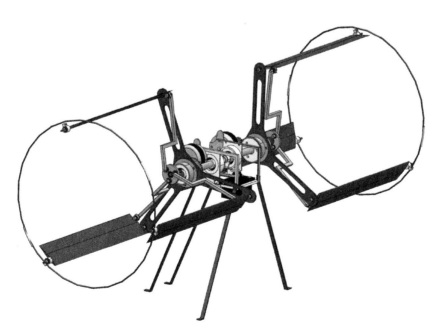

FIGURE 5.20 Conceptual twin cycloidal rotor MAV.

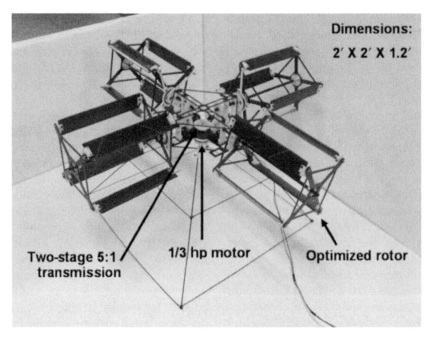

Dimensions:
2′ X 2′ X 1.2′

Two-stage 5:1 transmission   **1/3 hp motor**   **Optimized rotor**

FIGURE 5.21   Conceptual quad cycloidal rotor MAV [79]. By kind permission of M. Benedict.

location of pitching axis were performed, culminating in the design and successful hover flight of a micro-aerial vehicle with four cycloidal rotors [79], with a total mass of around 750 g and a power loading of 5.6 kg/HP (Figure 5.21).

Future work in this area is expected to focus on the forward flight capability of the cycloidal rotor as well as maneuverability and improvement of performance by further harnessing unsteady aerodynamic effects.

5.6 MODELING

Several analytical and computational models have been developed to calculate the forces generated by a flapping wing. Early studies did not include the effect of elastic wing deformations, while more recent studies feature coupled aeroelastic analyses. A few of these studies are described in this section, followed by a detailed discussion of a typical analysis based on strip theory.

Typically, the development of analytical tools has focused on the ability to accurately calculate dynamic loads generated by the flapping wings. The calculation of the coupled aeroelastic response of a flapping wing becomes increasingly challenging as the flexibility of the wing and the flapping frequency increase. Computational methods that incorporate detailed, coupled calculations of the structural deformations as well as the aerodynamic forces are required to accurately represent the dynamic behavior of flexible flapping wings. However, these methods are computationally expensive and are not suited to real-time control. In addition, they often yield less physical insight into the dynamics than simpler analytical models. Roget *et al.* [80] used a Reynolds-averaged Navier–Stokes computational solver with body-conforming, deformable grids to calculate the aerodynamic forces generated by a flexible flapping wing. The deformed shape of a wing measured in a parallel study by Harmon and Hubbard [81], using retroreflective markers in conjunction with motion-tracking algorithms, was input to the computational model. The computational model showed good agreement with measured forces at low flapping frequencies and was less accurate at higher flapping frequencies.

Two-dimensional strip theory is the approach most often adopted by researchers, in which the wings are discretized into chordwise strips distributed along the span. The flows over the strips are assumed to behave independently of each other, and so each strip is treated as a two-dimensional airfoil section. The forces and moments on each strip are calculated based on local flow velocities, angles, and airfoil characteristics, and the contributions of all the strips are summed to find the total forces on the flyer. A simplified unsteady aerodynamic analysis based on modified strip theory was developed by DeLaurier [82, 83] to model the flight performance of a harmonically flapping wing. In this analysis, the wing was assumed to be spanwise rigid in bending but flexible in torsion. A harmonic variation of pitching and flapping motion was assumed. A modified Theodorsen function was used to incorporate the unsteadiness of the flow as well as the finite aspect ratio of the wing. Post-stall characteristics were incorporated in the analysis in addition to a leading-edge suction force that account for the majority of forward-thrust production.

Figure 5.22 shows a schematic of a wing discretized into chordwise sections along its span. The wings flap about their axis of symmetry, at the mid-span location (only one wing is shown in the figure). The incident velocities, angles, and forces on the two-dimensional airfoil section are similar to that shown in Figure 5.16.

This analysis was used to find the performance of the flapping wings on an 18 ft span pterosaur model, which included a spanwise variation in airfoil chord as well as sweep, similar to that investigated by DeLaurier [82]. An empirical model was used to calculate the appropriate flapping frequency based on the total mass of the pterosaur (around 40 lbs). The flapping frequency f for any natural flyer was given by Pennycuick [84] as

$$f = m^{3/8} g^{1/2} b^{-23/24} S^{-1/3} \rho^{3/8}. \qquad (5.7)$$

Here, m is the mass of the bird, g is the acceleration due to gravity, b is the wingspan, S is the wing area, and ρ is the density of air. Using this relation, the flapping frequency was found to be 1.2 Hz. This is low in comparison to typical birds of today but can be attributed to the large wing span and low wing loading of pterosaurs. The calculations were performed at a flight speed of 44 ft/s, a flapping amplitude of 20°, and an angle of incidence of the flapping axis of 7.5°.

The average lift produced over one cycle as a function of the dynamic twist angle amplitude β_0 is shown in Figure 5.23a. It is seen that for values of $\beta_0 > 2.25°/$ft, the lift produced is about 42 lb, which is more than the weight of the pterosaur and hence sufficient to sustain flight.

The average thrust produced as a function of dynamic twist angle is plotted (Figure 5.23b). It is seen that the thrust produced peaks at a dynamic twist of around $\beta_0 = 2.25°/$ft. Beyond this value of β_0, the thrust rapidly decreases to around zero. This trend in thrust can be explained by the fact that upon increasing the dynamic twist to a larger value, the outboard sections of the wing become prone to stall, causing them to lose thrust.

A similar trend is seen in the propulsive efficiency curve (Figure 5.24), where a distinct maximum of 42% is reached at $\beta_0 = 2.25°/$ft. The propulsive efficiency is low compared to the

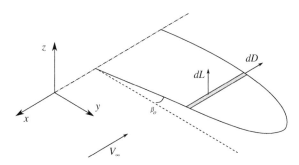

FIGURE 5.22 Schematic of flapping wing modeled using two-dimensional strip theory [82].

(a)

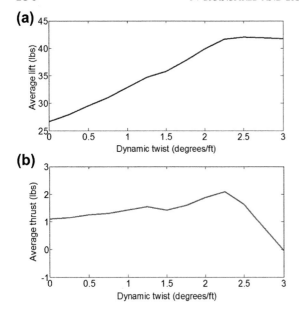

(b)

FIGURE 5.23 Average lift and average thrust produced over one flapping cycle as a function of dynamic twist amplitude. (a) Average lift and (b) average thrust.

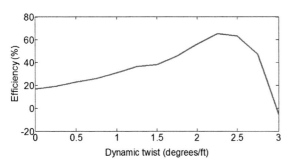

FIGURE 5.24 Propulsive efficiency as a function of dynamic twist.

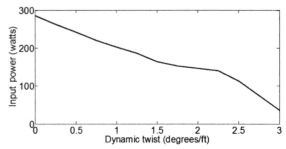

FIGURE 5.25 Input power as a function of dynamic twist.

less effort is needed to keep flapping. This means that with an increase in dynamic twist, more energy can be drawn from the airflow to produce thrust.

The variation of total lift and total thrust produced by the wing over a flapping cycle is shown in Figure 5.26a and b for a fixed dynamic twist amplitude of $\beta_0 = 2.25°/$ft. It is seen that the lift and drag have an approximately sinusoidal variation with flapping cycle angle. In the down stroke (cycle angle varying from 0° to 180°), the average lift and thrust produced per wing are high. In the up stroke (cycle angle varying from 180° to 360°), the lift produced is low and the thrust produced is negative. However, positive lift and thrust are produced over the entire cycle.

Refined structural models of flapping-wing flight include higher-order twist deformation in addition to bending deformations. In the case of insect wings having a lower aspect ratio, plate theories may be necessary to capture the appropriate dynamics. Nonlinearities may be introduced in terms of kinematic couplings or large deformations. Refined aerodynamic analyses rely on purely computational techniques to capture the complex, three-dimensional, unsteady flowfield. Larijani and DeLaurier [85] developed a nonlinear aeroelastic analysis to further investigate the aerodynamical and structural dynamical features of ornithopter flight. This analysis included a finite element structural model with

efficiency of modern propellers (70–80%). This is because the inherent mechanism of flapping consists of phases when lift and thrust are lost (during the down stroke) and the thrust generated is not constant over the flapping cycle.

The average input power, P_{in} (Figure 5.25) is also plotted against the dynamic twist angle. It is seen that with increase in dynamic twist, the input power required becomes lower and considerably

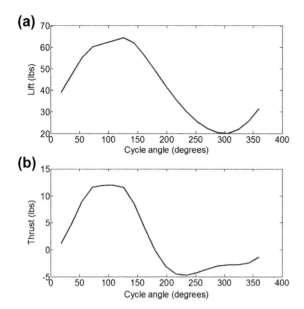

(a)

(b)

FIGURE 5.26 Variation of lift and thrust produced over one flapping cycle for spanwise dynamic twist amplitude $\beta_0 = 2.25°/\text{ft}$. (a) Instantaneous lift, and (b) instantaneous thrust.

lattice formulation that included vortex stretching and aging to model a plunging, pitching, twisting finite-span flapping wing. For simpler geometries, only a few vortex lattice rings were sufficient to characterize the flapping phenomenon, and these results were verified experimentally. Vest and Katz [89] developed a computational tool utilizing an unsteady potential flow-panel method. Singh and Chopra [90] developed a finite element-based structural model of an insect-based flapping wing in hover and included unsteady aerodynamics using indicial functions. They validated the analysis by performing experiments on a biomimetic flapping-wing mechanism and demonstrated the importance of aeroelastic twist deformation in producing lift. Recently, several researchers have used multibody mechanics to model flapping-wing micro air vehicles. Orlowski and Girard [91] described the nonlinear simulation of flapping-wing micro air vehicles and reviewed the state-of-art of dynamics modeling.

damping effects. Analytical predictions of average thrust, lift, spanwise bending moments, and wingtip twist angles for various ranges of flapping frequencies and airspeed were validated by a series of full-scale experiments. Grauer and Hubbard [86] developed a nonlinear multibody dynamics model of an ornithopter that treated the wings and body as rigid bodies having specific kinematic relationships.

Spedding *et al.* [87] measured the wake structure behind birds flying in the test section of an open test section wind tunnel using particle image velocimetry. The circulation measured in the wake was correlated with theoretical predictions; however, it was concluded that there still remained a significant gap in the understanding of how wing geometry affects wake structure.

To account for the complex, highly unsteady flowfield, several researchers have developed purely computational models. For example, Fritz and Long [88] used an unsteady vortex

5.7 SENSORS

Birds, bats, and insects have a wide variety of sensory mechanisms that are used for flight stabilization, navigation, and obstacle avoidance. These sensors are even more remarkable in insects due to their smaller size as well as higher bandwidth requirements associated with their higher flapping frequency and lower body inertia. Figure 5.27 shows a schematic of the different types of sensors on an insect. The compound eyes of the insect, shown in Figure 5.28 consist of a number of simple light sensors, each effectively operating like a single pixel in a digital camera.

Insects use the data from these simple light sensors in a number of different ways and can extract complicated patterns of information from them. For example, researchers have determined that honeybees use a technique called optical flow to measure distance flown and to

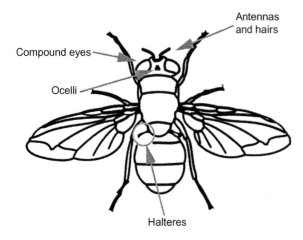

FIGURE 5.27   Different types of sensors on a typical insect [97].

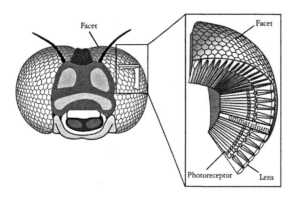

FIGURE 5.28   Compound eyes on an insect [97].

stabilize flight as well as avoid obstacles. *Optical flow* relies on the measurement of the rate of change of an image, which is directly related to the velocity of the image sensor with respect to the object being imaged. Consider a camera or an observer moving along a straight line at a constant velocity, imaging an object located along a direction perpendicular to the direction of motion. The closer the object is to the line of motion of the observer, the larger the angle it subtends. Therefore, closer objects appear to move faster through the field of view of the observer than objects that are farther away. For an observer moving at a constant velocity through a tunnel, the optical flow will be larger as the tunnel becomes narrower. Researchers observed that honeybees flying through tunnels to a food source significantly overestimated the distance flown, compared to when they flew through an open environment [92, 93]. It was concluded that the honeybees measure distance by integrating the optical flow across their eyes, and the higher optical flow created by the narrow tunnel walls resulted in an increased estimate of the distance flown.

In Chapter 9, Chahl and Mizutani discuss the use of optical flow for biomimetic sensing. The optical flow technique is being investigated by several researchers for application on microflyers, using CMOS sensors, special lenses, and dedicated electronics to minimize processing requirements [94]. For example, Barrows *et al.* [95] developed an optical flow sensor that they incorporated into a commercially available hobby indoor helicopter and demonstrated stable hover in a fixed location using feedback from the sensor. Garratt and Chahl [96] described an optical flow-based terrain-following system for an unmanned helicopter. They designed and constructed a system consisting of a downward-looking camera and hardware to compute the optical flow.

The system was installed on an 80 kg Yamaha RMAX helicopter as well as on a smaller 8 kg electric helicopter. Flight testing demonstrated that the system could accurately measure the height of the aircraft above ground; combined with global positioning system (GPS) measurements, the system was able to estimate height above terrain with an accuracy of 7.5% at a flight speed of 5 m/s. Other discussions of the physical principles that form the basis of optical flow and descriptions of several physical implementations can be found in Refs. [97, 98].

The halteres on insects are highly developed angular velocity sensors [99]. They vibrate up and down in resonance with the wing-flapping motion, and any angular velocity of the insect body, for example, in yaw, results in a bending moment on the halteres due to gyroscopic moments. Fine hair or other sensors at the root of the halteres measure this bending moment and provide feedback of the angular velocity to the insect. It has been observed that insects are unable to fly properly if these halteres are removed, and therefore they form an integral part of their flight control and stabilization system.

Researchers have also been investigating the sensory mechanisms involved in bat flight. In addition to the well-known acoustic sensing mechanisms such as echolocation, bats appear to employ several other mechanisms. Sterbing-D'Angelo et al. [100] investigated the function of bat hairs as aerodynamic sensors. Tactile receptors associated with the hairs on a bat's wings can sense airflow in a direction opposite to the hair growth, identifying the onset of stall and flow separation. The bat uses this information for control in various flight regimes. It was experimentally shown that shaving the hair from different areas of a bat's wings significantly altered their ability to maneuver and avoid obstacles and increased their flight speed. The exact mechanisms involved in this sensing as well as the processing of the vast amount of information received are still topics of active research.

Birds also feature several passive sensing mechanisms. The coverlets on birds, wings, located near the trailing edge, deploy automatically when the flow over the upper surface of the wing is stalled. These pop up locally in areas of separated flow and not only provide feedback to the bird, but also help in alleviating stall by increasing the post-stall lift. Bechert et al. [101] studied this phenomenon and measured the effect of passive coverlets in stall alleviation on an airfoil in a wind tunnel. Birds are known to sense the direction of light polarization as well as the Earth's magnetic field. Magnetometers are used in some microflyers for orientation; however,

these are not as sophisticated as the mechanisms used by birds. A large part of the stabilization and control mechanisms in birds and insects, from the point of view of both sensors and algorithms, is still a topic of active research.

5.8 FUTURE CHALLENGES

There remain a number of challenges to realizing a fully autonomous insect-sized or bird-sized microflyer. The fundamental limits imposed by the low Reynolds number flight regime necessitate harnessing unsteady aerodynamic mechanisms for achieving efficient flight. The unsteady mechanisms used by insects and birds have been studied for a number of years and are still topics of active research. A variety of computational tools are being developed to model these effects. These tools must be transitioned into design analyses to improve and optimize the performance of microflyers. In addition, the structural dynamic models of wings must be developed to account for aeroelastic behavior that forms a key part of the aeromechanics of natural flyers.

In terms of stability and control, natural flyers are highly maneuverable, which comes at the cost of stability. To replicate this kind of maneuverability in manmade microflyers requires high-bandwidth sensors and actuators as well as robust control algorithms. Each of these areas requires significant technical advances for realization of a microflyer with capabilities comparable to natural flyers. The enhanced maneuverability must not compromise the efficiency of flight. Actuators with higher power density must be developed to power the microflyers. The power density must be based not only on the actuator mass alone, but must include the power supply as well. The size as well as accuracy and drift behavior of sensors must be enhanced. In this area, biomimetic sensors such as optic flow-based sensors are very promising. Algorithms must be developed that require low processing capability so that on-board computers remain within mass and power budgets. Finally, natural flyers have a remarkable ability to withstand gusts and recover from mild collisions with objects. Practical man-made microflyers must also incorporate such characteristics if they are to carry out practical missions. Significant advances have been made in a number of these areas over the past decade, and these technical challenges will remain a fertile ground for researchers for the forseeable future.

Acknowledgments

The author thanks Anand Karpatne for his help in creating some of the figures. Support from the Cockrell School of Engineering is also gratefully acknowledged.

References

[1] http://www.cia.gov/about-cia/cia museum/cia-museum-tour/index.html (accessed 5.02.2013).

[2] J. Solem, The application of microrobotics in warfare. Technical Report LA- UR-96-3067, Los Alamos National Laboratory, Los Alamos, NM, USA (1991).

[3] R. Hundley and E.C. Gritton, Future technology-driven revolutions in military operations: results of a workshop. Technical Report DB-110-ARPA, RAND Corporation, Santa Monica, CA, USA (1994).

[4] T. Hylton, C. Martin, R. Tun, and V. Castelli, The DARPA nano air vehicle program. AIAA 2012–588, *50th AIAA Aerospace Sciences Meeting including the new horizons forum and aerospace exposition*, Nashville, TN, USA (January 9–12, 2012).

[5] R.O. Prum, Development and evolutionary origin of feathers, *Bioinsp Biomim* **285** (1999), 291–306.

[6] R.O. Prum and A.H. Brush, The origin and evolution of feathers, *Sci Am* **288** (3) (March 2003), 60–69.

[7] H. Tennekes, *The simple science of flight: from insects to jumbo jets*, MIT Press, Cambridge, MA, USA (2009).

[8] N. Lobontiu, *Compliant mechanisms: design of flexure hinges*, CRC Press, Boca Raton, FL, USA (2003).

[9] W. Nachtigall, A. Wisser, and D. Eisinger, Flight of the honeybee—VIII. Functional elements and mechanics of the flight motor and the wing joint- one of the most complicated gear-mechanisms in the animal kingdom, *J Comp Physiol B* **168** (1998), 323–344.

[10] R.J. Wood, The first takeoff of a biologically inspired at-scale robotic insect, *IEEE Trans Robot* **24** (2008), 341–347.

[11] B.M. Finio and R.J. Wood, Distributed power and control actuation in the thoracic mechanics of a robotic insect, *Bioinsp Biomim* **5** (2010), 045006.

[12] W. Bejgerowski, A. Ananthanarayanan, D. Mueller, and S.K. Gupta, Integrated product and process design for a flapping wing drive-mechanism, *ASME J Mech Des* **131** (2009), 061006.

[13] T. Nakagawa, M. Miyazaki, G. Ono, R. Fujiwara, T. Norimatsu, T. Terada, A. Maeki, Y. Ogata, S. Kobayashi, N. Koshizuka, and K. Sakamura, 1-cc computer using UWB-IR for wireless sensor network, *Design automation conference, 2008, ASPDAC 2008*, Seoul, Korea (March 2008), 392–397.

[14] J.H. McMasters and M.L. Henderson, Low speed single element airfoil synthesis, *Tech Soaring* **6** (1980), 1–21.

[15] B.H. Carmichael, Low Reynolds number airfoil survey, Volume I. Technical report, NASA-CR-165803, NASA (1981).

[16] F.W. Schmitz, Aerodynamics of the model airplane. Part 1: Airfoil measurements. Technical report, NASA-TM-X-60976, NASA (1967).

[17] R.A. Wallis, Wind tunnel tests on a series of circular arc airfoils, ARL Aero Note 74, CSIRO, Australia (1946).

[18] S.J. Miley, A catalog of low Reynolds number airfoil data for wind turbine applications. RFP-3387 VC-60, Rockwell International, US Department of Energy, Wind Energy Technology Division, Federal Wind Energy, Program, USA (1982).

[19] A. Bruining, Aerodynamic characteristics of a curved plate airfoil section at Reynolds numbers 60,000 and 100,000 and angles of attack from −10 to +90 degrees. Report Number VTHLR-281, Department of Aerospace Engineering, Technische Hogeschool, Delft, The Netherlands (1979).

[20] D. Althaus, ProfilePolaren für den Modellflug. Institut fur Aerodynamik und Gasdynamik der Universitat Stuttgart Neckar-Verlag, Villingen-Schwenningen, Germany (1980).

[21] S.F. Hoerner and H.V. Borst, *Fluid-dynamic lift*, Hoerner Fluid Dynamics, Bakersfield, CA, USA (1985).

[22] E.V. Laitone, Aerodynamic lift at Reynolds numbers below 7 x 10^4, *AIAA J* **34** (1996), 1941–1942.

[23] E.V. Laitone, Wind tunnel tests of wings at Reynolds numbers below 70,000, *Exp Fluids* **23** (1997), 405–409.

[24] F. Bohorquez, P. Samuel, J. Sirohi, D. Pines, L. Rudd, and R. Perel, Design, analysis and hover performance of a rotary wing micro air vehicle, *J Am Helicopter Soc* **48** (2003), 80–90.

[25] F. Bohorquez and D. Pines. Rotor and airfoil design for efficient rotary wing micro air vehicles, *Proceedings of the 61st Annual American Helicopter Society Forum*, Grapevine, TX, USA (June 2005).

[26] B.R. Hein and I. Chopra, Hover performance of a micro air vehicle: rotors at low Reynolds number, *J Am Helicopter Soc* **52** (2007), 254–262.

[27] R.W. Prouty, *Helicopter performance, stability and control*, Krieger, Malabar, FL, USA (1990).

[28] J.G. Leishman, *Principles of helicopter aerodynamics*, Cambridge University Press, New York, NY, USA (2000).

[29] M. Rosen and A. Hedenström, Gliding flight in a jackdaw: a wind tunnel study, *J Exp Biol* **204** (2001), 1153–1166.

[30] R.A. Norberg, Hovering flight of the dragonfly *Aeschna Juncea L*, in *Kinematics and aerodynamics*, vol. 2 (T.Y.-T. Wu, C.J. Brokaw, and C. Brennen, eds.), Plenum, New York, NY, USA (1975), 763–781.

[31] U.M. Norberg, Aerodynamics of hovering flight in the long-eared bat Plecotus Auritus, *J Exp Biol* **65** (1976), 459–470.

[32] T. Weis-Fogh, Quick estimates of flight fitness in hovering animals, including novel mechanisms for lift production, *J Exp Biol* **59** (1973), 169–230.

[33] C.P. Ellington, C. van den Berg, A.P. Willmott, and A.L.R. Thomas, Leading-edge vortices in insect flight, *Nature* **384** (1996), 626–630.

[34] M.H. Dickinson, F.O. Lehmann, and S.P. Sane, Wing rotation and the aerodynamic basis of insect flight, *Science* **284** (1999), 1954–1960.

[35] L.W. Carr, K.W. McAlister, and W. J. McCroskey, Analysis of the development of dynamic stall based on oscillating airfoil experiments. NASA TN-D-8382, NASA (1977).

[36] S.P. Sane, The aerodynamics of insect flight, *J Exp Biol* **206** (2003), 4191–4208.

[37] C.P. Ellington, The novel aerodynamics of insect flight: applications to micro-air vehicles, *J Exp Biol* **202** (1999), 3439–3448.

[38] A. Azuma, *The biokinetics of flying and swimming*, Springer-Verlag, Tokyo, Japan (1992).

[39] Y.C. Fung, *An introduction to the theory of aeroelasticity*, Wiley, New York, NY, USA (1955).

[40] R.L. Bisplinghoff, H. Ashley, and R.L. Halfman, *Aeroelasticity*, Addison-Wesley, Cambridge, MA, USA (1957).

[41] G.K. Taylor, R.L. Nudds, and A.L.R. Thomas, Flying and swimming animals cruise at a Strouhal number tuned for high power efficiency, *Nature* **425** (2003), 707–711.

[42] C.J Pennycuick, *Animal flight*, Arnold, London, UK (1972).

[43] U.M. Norberg, *Vertebrate flight*, Springer-Verlag, Berlin, Germany (1990).

[44] J. Rayner, Mathematical modelling of the avian flight power curve, *Math Method Appl Sci* **24** (2001), 1485–1514.

[45] D.J. Pines and F. Bohorquez, Challenges facing future micro-air-vehicle development, *J Aircraft* **43** (2006), 290–305.

[46] K.C. Hall and S.R. Hall, A rational engineering analysis of the efficiency of flapping flight, in *Fixed and flapping wing aerodynamics for micro air vehicle applications*, (T.J. Mueller, ed.), AIAA, Reston, VA, USA (2001), 249–274.

[47] C.P. Ellington and J.R. Usherwood, Lift and drag characteristics of rotary and flapping wings, in *Fixed and flapping wing Aerodynamics for micro air vehicle applications* (T.J. Mueller, ed.), AIAA, Reston, VA, USA (2001), 231–248.

[48] J. Sirohi, M. Tishchenko, and I. Chopra, Design and testing of a microaerial vehicle with a single rotor and turning vanes, *Proceedings of the 61st annual American Helicopter Society forum*, Grapevine, TX, USA (June 1–3, 2005).

[49] F. Bohorquez and D. Pines, Hover performance and swashplate design of a coaxial rotary wing micro air vehicle, *Proceedings of the 60th annual American Helicopter Society forum*, Baltimore, MD, USA (2004).

[50] T. Pornsin-Sisirak, S.W. Lee, H. Nassef, J. Grasmeyer, Y.C. Tai, C.M. Ho, and M. Keennon, MEMS wing technology for a battery-powered ornithopter, *Thirteenth IEEE International conference on micro electro mechanical systems*, vol. 122, Miyazaki, Japan, USA (2000), 23–27.

[51] M. Keennon, K. Klingebiel, and H. Won, Development of the nano hummingbird: a tailless flapping wing micro air vehicle, AIAA 2012-588, *50th AIAA Aerospace Sciences Meeting including the new horizons forum and aerospace exposition*, Nashville, TN, USA (January 9–12, 2012).

[52] G.C.H.E. de Croon, K.M.E. de Clerq, R. Ruijsink, B. Remes, and C. de Wagter, Design, aerodynamics, and vision-based control of the Delfly, *Internat J Micro Air Vehicles* **1** (2009), 71–97.

[53] E. Steltz, S. Avadhanula, and R.S. Fearing, High lift force with 275 Hz wing beat in MFI, *IROS 2007, IEEE/RSJ International conference on intelligent robots and systems*, (IEEE, 2007), 3987–3992.

[54] A. Cox, D. Monopoli, D. Cveticanin, M. Goldfarb, and E. Garcia, The development of elastodynamic components for piezoelectrically actuated flapping micro-air vehicles, *J Intell Mater Syst Struct* **13** (2002), 611–615.

[55] R. Kline and W. Koenig, Samara type decelerators, *8th aerodynamic decelerator and balloon technology conference*, Hyannis, MA April 2–4, Technical Papers (A84–26551 11–03), number AIAA-1984-807, pages 135141. American Institute of Aeronautics and Astronautics, New York, NY, USA (1984).

[56] K. Fregene and C.L. Bolden, Dynamics and control of a biomimetic single-wing nano air vehicle, *IEEE American Control Conference (ACC) 2010* (2010), 51–56.

[57] S. Jameson, K. Fregene, M. Chang, N. Allen, H. Youngren, and J. Scroggins, Lockheed Martin's SAMARAI nano air vehicle: challenges, research, and realization, AIAA-2012-584, *50th AIAA Aerospace Sciences Meeting including the new horizons forum and aerospace exposition*, Nashville, TN, USA (January 9–12, 2012).

[58] E.R. Ulrich, D.J. Pines, and S. Gerardi, Autonomous flight of a samara MAV, *American Helicopter Society 65th annual forum and technology display*, Grapevine, TX, USA (May 2009), 27–29.

[59] E.R. Ulrich, D.J. Pines, and J.S. Humbert, From falling to flying: the path to powered flight of a robotic samara nano air vehicle, *Bioinsp Biomim* **5** (2010), 045009.

[60] E.R. Ulrich and D.J. Pines, Effects of planform geometry on mechanical samara autorotation efficiency and rotational dynamics, *J Am Helicopter Soc* **57** (2012), 1–10.

[61] F. Bohorquez and D.J. Pines. Design and development of a biomimetic device for micro air vehicles, *Proc SPIE* **4701** (2002), 503–517.

[62] K.D. Jones and M.F. Platzer, Flapping-wing propulsion for a micro air vehicle, AIAA-2000-897, *38th aerospace sciences meeting and exhibit*, Reno, NV, USA (January 10–13, 2000).

[63] M.M. Heiligers, T. van Holten, and S.M. van den Bulcke, Test results of a radio-controlled ornicopter: a single rotor helicopter without reaction torque, AIAA-2006-820, *44th AIAA aerospace sciences meeting and exhibit*, Reno, NV, USA (January 9–12, 2006).

[64] D.J. Gerwen, M.M. Heiligers, and T. van Holten, *Ornicopter yaw control: testing a single rotor helicopter without reaction torq*ue, AIAA-2007-1253, *45th AIAA aerospace sciences meeting and exhibit*, Reno, NV, USA (8–11 January, 2007).

[65] B. Fitchett and I. Chopra, A biologically inspired flapping rotor for micro air vehicles, *Proceedings of the AHS international specialists meeting on unmanned rotorcraft*, Chandler, AZ, USA (January 23–25, 2007).

[66] F.K. Kirsten. Cycloidal propulsion applied to aircraft, *Trans ASME* 50(AER-50-12) (1928).

[67] J. Wheatley, Simplified aerodynamic analysis of the cyclogiro rotating-wing system, NACA-TN-467, National Advisory Committee for Aeronautics (1930).

[68] J. Wheatley and R. Windler, Wind tunnel tests of a cyclogiro rotor, NACA-TN-528, National Advisory Committee for Aeronautics (1935).

[69] R. Gibbens, Improvements in airship control using vertical axis propellers, AIAA-2003-6853, *Proceedings of the AIAA's 3rd annual aviation technology, integration, and operations forum*, Denver, CO, USA (November 17–19, 2003).

[70] M. Onda, K. Matsuuchi, N. Ohtsuka and Y. Kimura, Cycloidal propeller and its application to advanced LTA vehicles, AIAA-2003-6832, *Proceedings of the AIAA 3rd annual aviation technology, integration, and operations forum*, Denver, CO, USA (November 17–19, 2003).

[71] R. P. Gibbens, J. Boschma and C. Sullivan, Construction and testing of a new aircraft cycloidal propeller, AIAA-1999-3906, *Proceedings of the 13th AIAA lighter-than-air systems technology conference*, Norfolk, VA, USA (28 June-1 July, 1999).

[72] S. Kim, C. Yun, D. Kim, Y. Yoon, and I. Park, Design and performance tests of cycloidal propulsion systems. AIAA-2003-1786, *Proceedings of the 44th AIAA/ASME/ASCE/AHS structures, Structural dynamics, and materials conference*, Norfolk, VA, USA (April 7–10, 2003).

[73] C.Y. Yun, I. Park, H.Y. Lee, J.S. Jung, I.S. Hwang, S.J. Kim, and S.N. Jung. A new VTOL UAV cyclocopter with cycloidal blades system, *Proceedings of the 60th American Helicopter Society forum*, Baltimore, MD, USA (June 7–10, 2004).

[74] C.S. Hwang, I.S. Hwang, I.O. Jeong, S.J. Kim, C.H. Lee, Y.H. Lee, and S.Y. Min, Design and testing of VTOL UAV cyclocopter with 4 rotors, *Proceedings of the 62nd annual American Helicopter Society forum*, Phoenix, AZ, USA (2006).

[75] J. Sirohi, E. Parsons, and I. Chopra, Hover performance of a cycloidal rotor for a micro air vehicle, *J Am Helicopter Soc* **52** (2007), 263–279.

[76] M. Benedict, M. Ramasamy, and I. Chopra, Improving the aerodynamic performance of micro-air-vehicle-scale cycloidal rotor: an experimental approach, *J Aircraft* **47** (2010), 1117–1125.

[77] M. Benedict, M. Ramasamy, I. Chopra, and J.G. Leishman, Performance of a cycloidal rotor concept for micro air vehicle applications, *J Am Helicopter Soc* **55** (2010), 22002.

[78] M. Benedict, M. Mattaboni, I. Chopra, and P. Masarati, Aeroelastic analysis of a micro-air-vehicle-scale cycloidal rotor in hover, *AIAA J* **49** (2011), 2430–2443.

[79] M. Benedict, T. Jarugumilli, and I. Chopra, Experimental optimization of MAV-scale cycloidal rotor performance, *J Am Helicopter Soc* **56** (2011), 22005.

[80] B. Roget, J. Sitaraman, R. Harmon, J. Grauer, J. Hubbard, and S. Humbert, Computational study of flexible wing ornithopter flight, *J Aircraft* **46** (2009), 2016–2031.

[81] R. Harmon, J. Grauer, J. Hubbard, J. Humbert, B. Roget, J. Sitaraman, and J. Conroy, Experimental determination of ornithopter membrane wing shapes used for simple aerodynamic modeling, AIAA-2008-6237, *26th AIAA applied aerodynamics conference*, Honolulu, HI, USA (August 18–21, 2008).

[82] J.D. DeLaurier, An aerodynamic model for flapping-wing flight, *Aeronaut J* **97** (1993), 125–130.

[83] J.D. DeLaurier, The development of an efficient ornithopter wing, *Aeronaut J* **97** (1993), 153–1530.

[84] C. Pennycuick, Wingbeat frequency of birds in steady cruising flight: new data and improved predictions, *J Exp Biol* **199** (1996), 1613–1618.

[85] R.F. Larijani and J.D. DeLaurier, A nonlinear aeroelastic model for the study of flapping wing flight, *Prog Astronaut Aeronaut* **195** (2001), 399–428.

[86] J.A. Grauer and J.E. Hubbard, Multibody model of an ornithopter, *J Guid Control Dynam* **32** (2009), 1675–1679.

[87] G.R. Spedding, J. McArthur, M. Rosen, and A. Hedenstrom, Deducing aerodynamic mechanisms from near- and far-wake measurements of fixed and flapping wings at moderate Reynolds number, AIAA-2006-33, *44th AIAA aerospace sciences meeting and exhibit*, Reno, NV, USA (January 9–12, 2006).

[88] T. Fritz and L. Long, Object-oriented unsteady vortex lattice method for flapping flight, *J Aircraft* **41** (2004), 1275–1290.

[89] M. Vest and J. Katz, Unsteady aerodynamic model of flapping wings, *AIAA J* **34** (1996), 1435–1440.

[90] B. Singh and I. Chopra, Insect-based hover-capable flapping wings for micro air vehicles: experiments and analysis, *AIAA J* **46** (2008), 2115–2135.

[91] C.T. Orlowski and A.R. Girard, Modeling and simulation of nonlinear dynamics of flapping wing micro air vehicles, *AIAA J* **49** (2011), 969–981.

[92] M.V. Srinivasan, S. Zhang, M. Altwein, and J. Tautz, Honeybee navigation: nature and calibration of the odometer, *Science* **287** (2000), 851–853.

[93] H.E. Esch, S. Zhang, M.V. Srinivasan, and J. Tautz, Honeybee dances communicate distances measured by optic flow, *Nature* **411** (2001), 581–583.

[94] K. Weber, S. Venkatesh, and M. Srinivasan, Insect-inspired robotic homing, *Adaptive Behav* **7** (1999), 65–97.

[95] G. Barrows, T. Young, C. Neely and S. Humbert, Vision based hover in place, AIAA-2012-586, *50th AIAA aerospace sciences meeting including the new horizons forum and aerospace exposition*, Nashville, TN, USA (January 9–12, 2012).

[96] M.A. Garratt and J.S. Chahl, Vision-based terrain following for an unmanned rotorcraft, *J Field Robot* **25** (2008), 284–301.

[97] J.-C. Zufferey, *Bio-inspired flying robots—experimental synthesis of autonomous indoor flyers*, CRC Press, Boca Raton, FL, USA (2008).

[98] N. Franceschini, F. Ruffier, J. Serres, and S. Viollet, Optic flow based visual guidance: from flying insects to miniature aerial vehicles, in *Aerial Vehicles* (T.M. Lam, ed.), InTech, Vienna, Austria (2009).

[99] J.W.S. Pringle, The gyroscopic mechanism of the halteres of diptera, *Phil Trans R Soc B Lond* **233** (1948), 347–384.

[100] S. Sterbing-DAngelo, M. Chadha, C. Chiu, B. Falk, W. Xian, B. Janna, J.M. Zook, and C.F. Moss, Bat wing sensors support flight control, *Proc Natl Acad Sci* **108** (2011), 11291–11296.

[101] D.W. Bechert, M. Bruse, W. Hage, and R. Meyer, Biological surfaces and their technological application: laboratory and flight experiments on drag reduction and separation control, *28th fluid dynamics conference, AIAA-1997-1960*, Snowmass Village, CO, USA (June 29-July 2, 1997).

ABOUT THE AUTHOR

Jayant Sirohi is an Assistant Professor in the Department of Aerospace Engineering and Engineering Mechanics, at the University of Texas at Austin. He obtained his BTech degree in Aerospace Engineering from the Indian Institute of Technology, Chennai, India in 1996 and his MS and PhD degrees from the University of Maryland at College Park in 2002. He then worked as an Assistant Research Scientist in the Alfred Gessow Rotorcraft Center at the University of Maryland at College Park. In 2007, he joined the Advanced Concepts group in Sikorsky Aircraft Corporation, and in August 2008, he joined the faculty at the University of Texas at Austin.

He is interested in hovering micro-aerial vehicles, specifically in the effect of scaling and Reydesnolds number on the figure of merit of rotors and the reduced-order aeroelastic modeling of flapping wings. His present research activities include development of actuators and sensors for active helicopter rotors, aeroelasticity of flexible rotor blades, investigation of the efficiencies of different micro-aerial vehicle configurations, and measurement of rotor loads.

6

Muscular Biopolymers

Mohsen Shahinpoor

**Mechanical Engineering Department,
University of Maine, Orono, ME 04469, USA**

Prospectus

This chapter discusses properties and characteristics of ionic biopolymer-metal nanocomposites (IBMCs) as biomimetic multifunctional distributed nanoactuators, nanosensors, nanotransducers, and artificial muscles. After presenting some fundamental properties of biomimetic distributed nanosensing and nanoactuation of ionic polymer-metal composites (IPMCs) and IBMCs, the discussion extends to some recent advances in the manufacturing techniques and 3-D fabrication of IBMCs and some recent modeling and simulations, sensing and transduction, and product development. This chapter also presents procedures on how biopolymers such as chitosan and perfluorinated ionic polymers can be combined to make new nanocomposites with actuation, energy harvesting, and sensing capabilities. Chitin-based chitosan and ionic polymeric networks containing conjugated ions that can be redistributed by an imposed electric field and consequently act as distributed nanosensors, nanoactuators, and artificial muscles are also discussed. The manufacturing methodologies are briefly discussed, and the fundamental properties and characteristics of biopolymeric muscles as artificial muscles are presented. Two ionic models based on linear irreversible thermodynamics as well as charge dynamics of the underlying sensing and actuation mechanisms are also presented. Intercalation of biopolymers and ionic polymers and subsequent chemical plating of them with a noble metal by a reduction-oxidation (redox) operation is also reported and the properties of the new product are briefly discussed.

Keywords

Artificial muscles, Biopolymeric nanosensors, Chitin, Chitosan, Ion-containing macromolecular networks, Muscular biopolymers, Nanoactuators, Perfluorinated ionic polymers.

6.1 INTRODUCTION

Direct conversion of chemical to mechanical energy as occurs in biological muscles has been the focus of many scientists and researchers to achieve efficiencies as high as 50%. In comparison, most internal combustion engines and steam turbines have about 30% efficiency at best.

6.1.1 Brief Description of Mammalian Muscles

In order to reproduce similar properties of biological muscles in artificial counterparts, we have to fully understand mechanisms, behaviors, and properties of natural muscles. This section provides a brief summary of anatomical and

http://dx.doi.org/10.1016/B978-0-12-415995-2.00006-4

physiological characteristics of human or mammalian muscles.

Muscles provide the movement capability of the human body and form 40–45% of body weight. They provide a means of direct conversion of chemical to mechanical (movement, work, and force) energy. *Myology* deals with the scientific study of muscle.

There are three distinct types of muscle tissue that vary in microscopic anatomy and nervous controls. These are skeletal, cardiac, and smooth muscles.

Skeletal muscle tissues are primarily attached to bones and provide for the movement of the skeleton. They are striated; alternating light and dark bands are visible when the tissue is examined under a microscope. They are also classified as voluntary muscle tissue since they are under conscious control.

Cardiac muscle tissue is found in the heart only. It is also striated but involuntary, that is, its contraction is usually not under conscious control and uses a pacemaker included in the muscle to produce a heartbeat.

Smooth muscle tissue is found in the walls of hollow internal structures such as blood vessels, the stomach, the intestines, and most abdominal organs. It is nonstriated and involuntary.

There are three basic functions of muscle tissue by contraction or alternating contraction and relaxation: producing motion, providing stabilization, and generating heat.

There are four functional characteristics of the muscle as follows:

1. *Excitability*, or the ability to respond to certain stimuli by producing action potentials or impulses of electrical signal. The stimuli triggering action potentials are chemicals such as neurotransmitters released by neurons or hormones distributed by the blood.
2. *Contractility*, or the ability to contract and generate force to do work. Contraction is in response to one or more muscle action potentials.
3. *Extensibility*, or the ability of the muscle to be extended (stretched) without damaging the tissue. Most skeletal muscles are arranged in opposing or antagonistic pairs in which one muscle tissue contracts while the other one is relaxed or even stretched.
4. *Elasticity*, or the ability of the muscle tissue to return to its original shape after stimulation [1].

Nerves and blood vessels supply muscle in abundance. Motor neurons are responsible for stimulating muscle fibers. In order for muscle tissue to contract, it uses a good deal of adenosine triphosphate (ATP), an energy-rich molecule, and it has to produce ATP by inflow of nutrients and oxygen brought about by the blood capillaries (microscopic blood vessels) that are found in abundance in muscle tissue. Each muscle fiber (cell) is in close contact with one or more capillaries.

Connective tissue surrounds and protects muscle tissue. *Fascia* is a sheet of broadband fibrous connective tissue beneath the skin, around the muscles and other organs in the body. Deep fascia, a dense irregular connective tissue, lines the body wall and extremities, holds muscles together, and separates them into functional groups. Deep fascia allows free movement of muscles; carries nerves, blood, and lymphatic vessels; and fills spaces between muscles. Beyond deep fascia are three layers of dense, irregular connective tissues that further protect and strengthen skeletal muscle.

The outermost layer encircling the whole muscle is the *epimysium*. *Perimysium* then surrounds bundles (*faciculi* or *fascicles*) of 10–100 or more individual muscle fibers. Further penetrating the interior of each individual fascicle and separating muscle fibers from one another is the *endomysium*. Interested readers are invited to read Appendix A of Ref. 1.

A *motor neuron* delivers the stimulus that ultimately causes a muscle fiber to contract.

A motor neuron plus all the muscle fibers it stimulates is called a *motor unit*.

A typical skeletal muscle consists of hundreds or thousands of very long cylindrical cells called *muscle fibers*. The muscle fibers lie parallel to one another and range from 10 to 100 μm in diameter. While a typical length is 100 μm, some muscle fibers are up to 30 cm long.

The *sarcolemma* is a muscle fiber's plasma membrane, and it surrounds the muscle fiber's cytoplasm or sarcoplasm. Because skeletal muscle fibers arise from the fusion of many smaller cells during embryonic development, each fiber has many nuclei to direct synthesis of new proteins. The nuclei are at the periphery of the cell next to the sarcolemma, conveniently out of the way of the contractile elements. The *mitochondria* (energy packs) lie in rows throughout the muscle fiber, strategically close to muscle proteins that use ATP to carry on the contraction process. Within the muscle fibers are *myofibrils*, which are extended lengthwise in the sarcoplasm. Their prominent light and dark band colors, called *cross-striations*, make the whole muscle fiber appear striped or striated.

Myofibrils form the contractile element of the skeletal muscle. They are 1–2 μm in diameter and contain three types of smaller filaments called *myofilaments*. These are thin, thick, and elastic filaments, respectively. The thin filament is about 8 nm; the thick filaments are about 16 nm.

6.1.2 Muscle Contraction

In the mid-1950s, Jean Hanson and Hugh Huxley proposed the *sliding filament* mechanism of muscle contraction. They stated that skeletal muscle contraction was due to thick and thin filaments sliding past one another. During muscle contraction, myosin cross-bridges pull on the thin filaments, causing them to slide inward toward what is called the *H-zone*. As the cross-bridges pull on the thin filaments, the thin filaments meet at the center of the sarcomere. The myosin cross-bridges may even pull the thin filaments of each sarcomere so far inward that their ends overlap. As the thin filaments slide inward, structures called *Z-discs* come toward each other and the sarcomere shortens but the lengths of thick and thin filaments do not change. The sliding of the filaments and shortening of the sarcomere cause shortening of the whole muscle fiber and ultimately the entire muscle.

An increase in Ca^{2+} concentration in the sarcoplasm starts filaments sliding, whereas a decrease turns off the sliding process. When a muscle fiber is relaxed, the concentration of Ca^{2+} in the sarcoplasm is low. This is because the sarcoplasmic reticulum (SR) membrane contains Ca^{2+} active transport pumps that remove Ca^{2+} from the sarcoplasm. Ca^{2+} is stored or sequestered inside the SR. As a muscle action potential travels along the sarcolemma and into the transverse tubule system, Ca^{2+} release channels open in the SR membrane. The result is a flood of Ca^{2+} from within the SR into the sarcoplasm around the thick and thin filaments. The calcium ions released from the SR combine with troponin, causing it to change shape. This shape change slides the troponin–tropomyosin complex away from the myosin-binding sites on actin.

Muscle contraction requires Ca^{2+}. It also requires energy in the form of ATP. ATP attaches to ATP-binding sites on the myosin cross-bridges (heads). A portion of each myosin head acts as an ATPase, an enzyme that splits the ATP into ADP + $\underline{P}$, where $\underline{P}$ symbolizes the terminal phosphate group PO_4^{3-} and ADP is adenosine diphosphate, through hydrolysis reaction. This reaction transfers energy from ATP to the myosin head even before contraction begins. The myosin cross-bridges thus are in an activated (energized) state. Such activated myosin heads spontaneously bind to the myosin-binding sites on actin when the Ca^{2+} level rises and tropomyosin slides away from its blocking position. The shape change that occurs when myosin binds to actin produces the power stroke of contraction. During the power stroke, the myosin cross-bridge swivels toward the center of the sarcomere like the oars of a boat. This action draws the thin filaments past the thick filaments toward the H-zone. As the myosin heads swivel, they release ADP.

Once the power stoke is complete, ATP again combines with the ATP-binding sites on the myosin cross-bridges. As ATP binds, the myosin head detaches from actin. Again, ATP is split, imparting its energy to the myosin head, which returns to its original upright position. It is then ready to combine with another myosin-binding site further along the thin filament. The cycle repeats over and over.

The myosin cross-bridges keep moving back and forth like the cogs of a ratchet, with each power stroke moving the thin filaments toward the H-zone. At any instant, about half of the myosin cross-bridges are bound to actin and are swiveling. The other half are detached and preparing to swivel again.

Contraction is analogous to running on a nonmotorized treadmill. One foot (myosin head) strikes the belt (thin filament) and pushes it backward (toward the H-zone). Then the other foot comes down and imparts a second push. The belt soon moves smoothly while the runner (thin filament) remains stationary. And, like the legs of the runner, the myosin heads need a constant supply of energy to keep going. The power stoke repeats as long as ATP is available and the Ca^{2+} level near the thin filament is high.

This continual movement applies the force draws the Z-discs toward each other and the sarcomere shortens. The myofibrils thus contract and the whole muscle fiber shortens. During a maximal muscle contraction, the distance between Z-discs can decrease to half the resting length. However, the power stroke does not always result in shortening of the muscle fibers and the whole muscle. Isometric contraction or contraction without shortening occurs when the cross-bridges generate force but the filaments do not slide past one another.

Sustained small contractions give firmness to a relaxed skeletal muscle, known as *muscle tone*.

At any instant, a few muscle fibers are contracted while most are relaxed. This small amount of contraction firms up a muscle without producing movement and is essential for maintaining posture. Asynchronous firing of motor units allows muscle tone to be sustained continuously.

A single action potential in a motor neuron elicits a single contraction in all the muscle fibers of its motor unit. The contraction is said to be *all-or-none*, because individual muscle fibers will contract to their fullest extent. In other words, muscle fibers do not partially contract. The force of their contraction can vary only slightly, depending on local chemical conditions and whether or not a motor unit has just contracted previously.

In addition, other internal conditions in the muscle, such as temperature, pH, and viscosity change. A rise in temperature, for example, could provoke stronger contractions.

6.1.3 Electromyography

The electrical signal associated with the contraction of a muscle is called an *electromyogram* or EMG. *Electromyography*, which is the study of EMG, has revealed some basic information. Voluntary muscular activity results in an EMG that increases in magnitude with tension. However, other variables influencing the signal at any given time are velocity of shortening or lengthening of the muscle, rate of tension buildup, fatigue, and reflex activity.

Muscle tissue conducts electrical potentials somewhat similarly to axons of the nervous system. *Motor unit action potential* (m.u.a.p.) is an electrical signal generated in the muscle fibers because of the recruitment of fibers as the motor unit. Electrodes placed on the surface of a muscle or inside the muscle tissue will record the algebraic sum of all m.u.a.p.'s being transmitted along the muscle fibers at that point in time. Those motor units away from the electrode site will result in a smaller m.u.a.p. than those of similar size near the electrode.

For a given muscle there can be a variable number of motor units, each controlled by a motor neuron through special synaptic junctions called *motor end plates*. An action potential transmitted down the motor neuron arrives at the motor end plate and triggers a sequence of electrochemical events. A quantum of acetylcholine (ACh) is released. It then crosses the synaptic gap (200–500 Å wide) and causes a depolarization of the postsynaptic membrane. Such a depolarization can be recorded by a suitable microelectrode and is called an *end plate potential* (EPP). In normal circumstances, the EPP is large enough to reach a threshold level and an action potential is initiated in the adjacent muscle fiber membrane.

The beginning of the m.u.a.p. starts at the Z-disc of the contractile element by means of an inward spread of the stimulus along the transverse tubular system. This results in a release of Ca^{2+} in the SR. Ca^{2+} rapidly diffuses to the contractile filaments of actin and myosin where ATP is hydrolyzed to produce ADP plus heat plus mechanical energy (tension). The mechanical energy manifests itself as an impulsive force at the cross-bridges of the contractile element.

The depolarization of the transverse tubular system and the SR results in a depolarization wave along the direction of the muscle fibers. It is this depolarization wave front and the subsequent repolarization wave that are seen by the recording electrodes.

Two general types of EMG electrodes have been developed. Surface electrodes consist of disks of metal, usually silver/silver chloride, of about 1 cm in diameter. These electrodes detect the average activity of superficial muscles and give more reproducible results than do in-dwelling types. In-dwelling electrodes are required, however, for the assessment of fine movements or to record from deep muscles. A needle electrode is a fine hypodermic needle with an insulated conductor located inside and bared to the muscle tissue at the open end of the needle. The needle itself forms the other conductor.

In-dwelling electrodes are influenced by both waves that actually pass by their conducting surface and by waves that pass within a few millimeters of the bare conductor. The same is true for surface electrodes.

ATP is an important molecule for the life of living cells. It provides energy for various cellular activities such as muscular contraction, movement of chromosomes during cell division, movement of cytoplasm within cells, transporting substances across cell membranes, and putting together larger molecules from smaller ones during synthetic reactions. Structurally, ATP consist of three phosphate groups attached to an adenosine unit composed of adenine and five-carbon sugar ribose.

ATP is the energy reserve of living systems. When a reaction requires energy, ATP can transfer just the right amount, because it contains two high-energy phosphate bonds. When the terminal phosphate group P̲ is hydrolyzed by addition of a water molecule, the reaction releases energy. This energy is used by the cell to power its activities. The resulting molecule, after removal of the terminal phosphate groups, is ADP. This reaction may be represented as follows:

$$ATP \rightarrow ADP + \underline{P} + Energy. \qquad (6.1)$$

The energy supplied by the catabolism of ATP into ADP is constantly being used by the cell. Since the supply of ATP at any given time is limited, a mechanism exists to replenish it. A phosphate group is added to ADP to manufacture more ATP. The reaction may be represented as follows:

$$ADP + \underline{P} + Energy \rightarrow ATP. \qquad (6.2)$$

The energy required to attach phosphate groups to ADP to make ATP is provided by breakdown of glucose in the cellular respiration process, which has two phases:

1. *Anaerobic.* In the absence of oxygen, glucose is partially broken down by the glycolysis

process into pyruvic acid. Each glucose that is converted into a pyruvic acid molecule yields two molecules of ATP.

2. *Aerobic*. In the presence of oxygen, glucose is completely broken down into carbon dioxide and water. These reactions generate heat and ATP molecules from each glucose molecule.

A muscle fiber is about 100 μm in diameter and consists of fibrils about 1 μm in diameter. Fibrils in turn consist of filaments about 100 Å in diameter. These further are of smaller units of molecular chains called actin, myosin, and elastic elements. Electron micrographs of fibrils show the basic mechanical structure of the interacting actin and myosin filaments. The darker and wider myosin protein bands are interlaced with the lighter and smaller actin protein bands, as seen in electron micrographs. The space between them consists of a cross-bridge structure where the tension is created and elongation/contraction takes place. The term *contractile element* is used to describe the part of the muscle that generates the tension, and it is this part that shortens and lengthens as positive or negative work is done. The sarcomere, which is a basic length of the myofibril, is the distance between the Z-discs. It can vary from 1.5 μm at full shortening to 2.5 μm at resting length to about 4 μm at full lengthening.

The structure of the muscle is such that many filaments are in parallel and many sarcomere elements are in series to make up a single contractile element. Consider a motor unit of a cross-sectional area of 0.1 cm^2 and a resting length of 10 cm. The number of sarcomere contractile elements in series would be 10 cm/2.5 μm $= 40{,}000$ and the number of filaments (each with an area of 10^{-8} cm^2) in parallel would be $0.1/10^{-8} = 10^7$. Thus the number of contractile elements of sarcomere length packed into this motor unit would be 4×10^{11}.

The active contractile elements are contained within the fascia. These tissue sheaths enclose the muscles, separating them into layers and groups and ultimately connecting them to the tendons at either end. The mechanical characteristics of connective tissue are important in the overall biomechanics of the muscle. Some of the connective tissue is in series with the contractile element; some is in parallel. These tissues are modeled as springs and viscous dampers for modeling purposes.

Each muscle has a finite number of motor units (motor neuron plus muscle fibers it innervates), each of which is controlled individually by a separate nerve ending. Excitation of each unit is an all-or-none event. The electrical indication is a motor unit action potential with the mechanical result being a tension twitch. An increase in tension can be accomplished in two ways: by increasing the stimulation rate for the motor unit or by the excitation (recruitment) of an additional motor unit.

It is now generally accepted that the motor units are recruited according to the size principle, which states that the size of the newly recruited motor unit increases with the tension level at which it is recruited. This means that the smallest unit is recruited first and the largest unit last. In this manner, low-tension movements can be achieved in finely graded steps. Conversely, those movements requiring high forces but not needing fine control are accomplished by recruiting the larger motor units.

Successive recruitment can be described as follows: The smallest motor unit (MU-1) is recruited first, usually at an initial frequency ranging from about 5–13 Hz. Tension increases as MU-1 fires more rapidly until a certain tension is reached, at which MU-2 is recruited. Here MU-2 starts firing at its initial low rate, and further tension is achieved by the increased firing of both MU-1 and 2. At a certain tension, MU-1 reaches its maximum firing range (15–60 Hz) and therefore generates its maximum tension. This process of increasing tension reaching new thresholds and recruiting another larger motor unit continues until maximum voluntary contraction is reached.

At that point, all motor units will be firing at their maximum frequencies. For a detailed discussion of mammalian muscles, the reader is refered to Bobet and Stein [1] and Ding *et al.* [2, 3].

In the following section we present a brief review of electroactive polymers (EAP) as artificial muscles, in general.

6.1.4 Electroactive Polymers and Artificial/Synthetic Muscles

For a recent history on EAPs, the reader is referred to Bar-Cohen [4] and Shahinpoor *et al.* [19]. Some specific EAPs are as follows:

1. *Magnetically activated polymers.* Magnetically activated gels, also called *ferro-gels*, are chemically cross-linked polymer networks that change shape in the presence of a magnetic field [5].
2. *Electronic EAP/ferroelectric relaxer polymers.* Zhang *et al.* [6] have introduced defects into the crystalline structure of Poly(vinylidene fluoride) (also known as PVDF) using electron irradiation to dramatically reduce the dielectric loss in a PVDF Tri Fluoroethylene, or P(VDF-TrFE), copolymer. This copolymerization apparently permits AC switching with much less heat generated. It is the electric-field-induced change between nonpolar and polar regions that is responsible for the large electrostriction observed in this polymer. As large as 4% electrostrictive strains can be achieved, with low-frequency driving fields having amplitudes of about 150 V/μm.
3. *Electrets.* Electrets, which were discovered in 1925, are materials that retain their electric polarization after being subjected to a strong electric field [7]. Piezoelectric behavior in polymers also appears in electrets, which are essentially materials that consist of a geometrical combination of a hard and a soft phase [8]. The positive and negative charges within the material are permanently displaced along and against the direction of the field, respectively, making a polarized material with a net zero charge.
4. *Dielectric elastomer EAPs.* Polymers with low elastic stiffness modulus and high dielectric constant can be packaged with flexible and stretchable electrodes to generate large actuation strain by electrostatic attraction between the stretchable electrodes, like a parallel plate capacitor [9]. However, Roentgen [10] appears to have been the first to discover this effect as early as 1880 by observing the stretching of a rubber band that could change its shape by being charged or discharged electrostatically.
5. *Liquid crystal elastomer (LCE) materials.* Liquid crystal elastomers as artificial muscles were pioneered by Finkelmann and coworkers [11, 12]. These materials can be used to form an EAP actuator by inducing isotropic-nematic phase transition due to temperature increase via Joule heating.
6. *Ionic EAP/ionic polymer gels (IPG).* Polymer gels can be synthesized to produce strong actuators with the potential to match the force and energy density of biological muscles. These materials (e.g., polyacrylonitrile, PAN) are generally activated by a chemical reaction, changing from an acid to an alkaline environment and causing the gel to become dense or swollen, respectively. This reaction can be stimulated electrically, as was shown by Osada *et al.* [13], Osada and Ross–Murphy [14], Osada *et al.* [15], and Osada and Matsuda [16].
7. *Nonionic polymer gels/EAPs.* Nonionic polymer gels containing a dielectric solvent can be made to swell under a DC electric field with a significant strain. Hirai and coworkers at Shinshu University in Japan created bending and crawling nonionic EAPs using a poly(vinyl alcohol) gel with dimethyl sulfoxide [17, 18].

8. *Ionic polymer-metal composites (IPMCs).* Ionic polymer-metal composite (IPMC) is an EAP that bends in response to a small electrical field (5–10 V/mm) as a result of mobility of cations in the polymer network. Reference is made to Shahinpoor *et al.* for the introductory paper on IPMCs in 1998 [19]. Oguro *et al.* [20] and Shahinpoor [21, 22] should be consulted for some earlier similar development of electroactive inonic membrane gels.

9. *Conductive polymers (CP) or synthetic metals.* Conductive polymers operate under an electric field by the reversible counter-ion insertion and expulsion that occurs during redox cycling [23]. Oxidation and reduction occur at the electrodes, inducing a considerable volume change due mainly to the exchange of ions with an electrolyte. When a voltage is applied between the electrodes, oxidation occurs at the anode and reduction at the cathode. The presence of either a liquid eletrolyte containing conjugated ions or a solid polyelectrolyte medium in close proximity to conductive polymers, such as polypyrrole (pPy), is often necessary to cause charge migration into and out of the conductive polymer.

10. *Shape-memory polymers.* Shape-memory polymers (SMPs) are similar to shape-memory alloys (SMAs) in the sense that they remember their shape at a certain specific temperature and can recover their shape if they are heated to that temperature. For a very good coverage of this topic, see Behl *et al.* [24].

6.1.5 Ionic Polymer-Metal Composites

Ionic polymer-metal composites (IPMCs) are in fact nanocomposites and are biomimetic distributed nanosensors, nanoactuators, energy harvesters, and artificial muscles. For a good review of these materials, see Refs. 25–29.

Briefly, IPMCs are cationic capacitive actuators and sensors that operate dynamically due to ionic redistribution due to either an imposed electric field or an imposed deformation field. When we apply a voltage across their thickness in a membrane or strip form, they bend quickly (millisecond response) toward the anode because cations move away from the anode toward the cathode side and cause the cathode side to expand and the anode side to contract; thus, bending occurs on the anode side. If the signal is oscillatory, then the strip oscillates accordingly as the cations move back and forth across the membrane.

On the other hand, if they are mechanically bent by outside forces (mechanical deformation due to applied loads), they generate electricity across the two electrodes attached to them because ionic redistribution causes an electric field due to Poisson's effect. Thus they are actuators, sensors, and energy harvesters.

In Ref. 25, methods of fabrication of several electrically and chemically active ionic polymeric gel muscles, such as polyacrylonitrile (PAN), poly(2-acrylamido-2-methyl-1-propane sulfonic) acid (PAMPS), and polyacrylic-acid-bis-acrylamide (PAAMs) as well as a new class of electrically active composite muscle such as ionic polymeric conductor composites (IPCCs) or ionic polymer-metal composites (IPMCs) made with perfluorinated sulfonic or carboxylic ionic membranes, are introduced.

Mathematical theories and numerical simulations associated with ionic polymer nanocomposite electrodynamics and chemodynamics are also formulated for the manufactured materials.

In this chapter we concentrate on ionic biopolymers such as *chitosan* for biomimetic distributed nanoactuation, nanosensing, and energy harvesting as well as artificial muscle applications for medical and industrial needs.

Shahinpoor's group has been involved in performing research on combining the biopolymer chitosan with organic polymer electrolytes such as perfluorinated sulfonic or carboxylic ionic polymers for medical and implantation applications because of chitosan's amazing healing, medical, and diagnostic properties. Chitosan/ionic polymers containing equilibrated and conjugated ions within their molecular networks and capable of being chemically or electrolessly plated with a conductive phase such as carbon, metal, graphite, graphene, and carbon nanotube create a novel family of multifunctional materials with medical healing characteristics.

Shahinpoor and Schneider [30] have presented a larger family of multifunctional polymeric materials. Mac and Sun [31] have discussed the multifunctional characteristics of chitosan gels. On the IPMC side the reader is referred to Shahinpoor et al. [25] and five review articles by Shahinpoor and Kim [26–29] and Shahinpoor [32]. Furthermore, in Refs. 25 and 27, methods of fabrication of several electrically and chemically active ionic polymeric muscles have been introduced and investigated. Gel-based ionic polymer conductor composites have also been introduced and investigated [25, 33, 34].

As described in Ref. 30, several physical models have been developed to understand the mechanisms of ion transport in ionic polymers and membranes. Morphological features influence transport of ions in ionic polymers. These features have been studied using a host of experimental techniques, including small and wide-angle X-ray scattering, dielectric relaxation, and a number of microscopic and spectroscopic studies [35, 36].

The emerging picture of the morphology of ionic biopolymers is that of a two-phase system made up of a polar medium containing ion nanocluster networks surrounded by a hydrophobic medium. These nanoclusters, in the context of perfluorinated sulfonic membranes, have been conceptually described as containing an interfacial region of hydrated, sulfonate-terminated perfluoroether side chains surrounding a central region of polar fluids. Counterions such as Na^+ or Li^+ are to be found in the vicinity of the sulfonates. It must be noted that the length of the side chains has a direct bearing on the separation between ionic domains, where the majority of the polar fluids resides, and the nonpolar domains.

Perfluoroionomers show an unusual combination of a nonpolar, Teflon-like backbone with polar and ionic side branches under high-resolution NMR. Liu and Schmidt–Rohr [37] obtained high-resolution nuclear magnetic resonance (NMR) spectra of solid perfluorinated polymers by combining 28-kHz magic-angle spinning (MAS) with rotation-synchronized 19F pulses. Their NMR studies enable more detailed structural investigations of the nanometer-scale structure and dynamics of polytetrafluoroethylene (PTFE) or Teflon®-based ionomers. It has also been well established that anions are tethered to the polymer backbone and cations (H^+, Na^+, Li^+) are mobile and solvated by polar or ionic liquids within the nanoclusters of size 3–5 nm [25, 30]. For recent work on biopolymers/IPMC artificial muscles, see Shahinpoor [38] and Tiwari et al. [39].

A large class of ion-containing polymers exists and creates a rich source of ionic polymeric nanosensors and nanoactuators in nanocomposite form with conductive materials. Certain dopants in the form of charge-transfer agents can be used to generate positive or negative charges or pendant groups in an intrinsically conducting polymer by reduction/oxidation (redox) chemical operations. Ampholytic polymers (polyampholytes) that are composed of macromolecules containing both cationic and anionic groups are electoactive and generate the basis for biomimetic electroactive ionic biopolymer conductive composites such as chitosan, intercalated with ionic polyelectrolytes such as IPMCs.

It is worth noting that chitosan is a naturally occurring substance in shellfish such as shrimps, crabs, and lobsters and possesses many useful properties such as wound healing. Note that chitosan is a copolymer of glucosamine and N-acetyglucosamine units linked by 1–4 glucosidic bonds and can be obtained by N-deacetylation of chitin. *Chitin* is the second most abundant natural polymer on Earth. Chitosan is a polysaccharide derived from chitin, part of the shell structure of crustaceans and shellfish. The chitosan is produced commercially by deacetylation of chitin. Chitosan is also a cationic polyeletrolyte. The degree of deacetylation can be determined by NMR spectroscopy and can vary from 60% to 100%.

The structure of chitosan is similar to that of cellulose, with the presence of amino groups being the major difference (Figure 6.1). The fact that chitosan may be made electroactive with sensing and actuation capability is evidenced by the work of Cai and Kim [40] on electoactive papers based on cellulose, as well as the work of Mac and Sun [31] on chitosan gels. Chitosan is structurally related to cellulose, which consists of long chains of glucose molecules linked to each other.

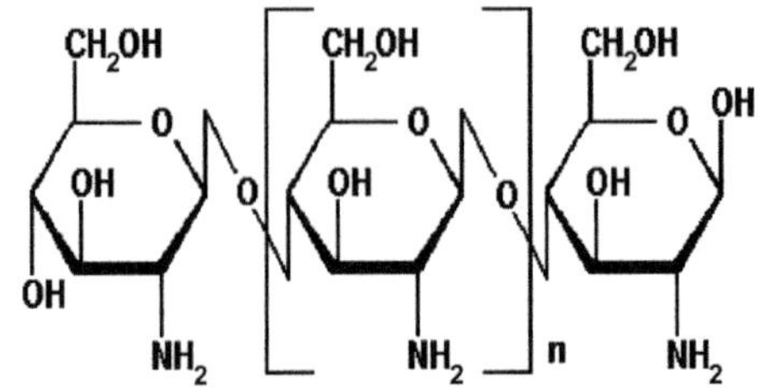

FIGURE 6.1   General structure of chitosan polyelectrolyte.

Chitosan comprises copolymers of N-acetylglucosamine and glucosamine and is a linear natural polysaccharide. Chitosan is prepared from chitin, which is closely related to both chitosan, a more water soluble derivative of chitin, and to cellulose, since it is a long unbranched chain of glucose derivatives, shown in Figure 6.2.

Note that protonated chitosan is cationic and is positively charged. These properties make chitosan ideal for use as a bio-adhesive, as it bonds to negatively charged surfaces such as mucosal membranes. Several studies have

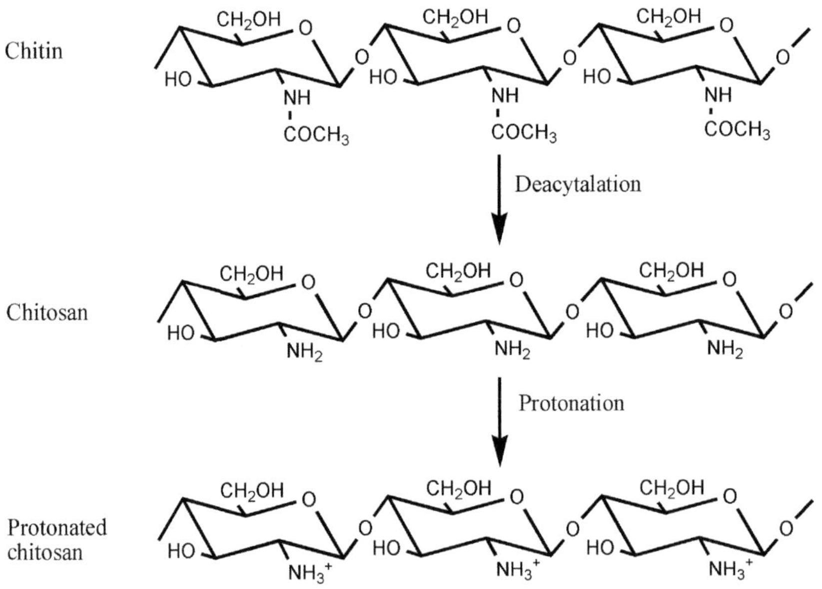

FIGURE 6.2   Manufacturing protonated chitosan from chitin by deacetylation in NaOH [30].

looked at chitosan gels and, incorporating different elements into the gel, to improve them for desired properties such as mechanical or binding properties. Chitosan polyethylene glycol forms a semi-interpenetration polymer network that increases the mechanical properties and pH-dependent swelling properties of the gel. For an excellent article on supramolecular interactions in chitosan gels, see Kato and Schneider [41].

The chitosan ionomers are useful in many different applications in a variety of fields. From cutting edge biomedical applications to agricultural applications, these gels are an essential asset for the future. One of the main advantages of these ionomers is their ability to contain and release various substances. They have also been recently developed as cationic membranes for fuel cell application to replace Nafion® [42–44]. Interestingly, Nafion® as a perfluorinated sulfoinc membrane is one of the basic materials used to manufacture IPMCs by a redox operation. Thus, it appears quite feasible to combine chitosan and Nafion® to manufacture chitosan/IPMC artificial muscles with healing and diagnostic capabilities.

6.2 THREE-DIMENSIONAL FABRICATION OF BIOPOLYMER NANOCOMPOSITES (IBMCs)

The fundamental procedure here is to manufacture chitosan membranes from chitin and then hydrolyze the chitosan membranes to give them ion-exchange capability, then boil them in an acid to protonate them for quick ion exchange with a noble metal such as platinum, gold, or palladium. The membrane form of these biopolymers has a typical thickness in the range of approximately 300–400 μm. Shahinpoor's group [25, 45] has devised a fabrication method that can scale up or down the IPMC artificial muscles in a strip size of micro-to-centimeter thickness, using a liquid form of perfluorinated ionic polymers. By meticulously

evaporating the solvent (isopropyl alcohol) out of the solution, recast ionic polymer can be obtained [45]. As a biopolymer is used, the IPMC is called an ionic biopolymer-metal nanocomposite (IBMC).

6.3 CHITOSAN/NAFION® COMPOSITE 3-D MANUFACTURING PROCEDURE

The general procedure in manufacturing cationic chitosan is to first obtain a chitosan from a vendor, say, Sigma–Aldrich, with a medium molecular weight for ease of acetylation. Acetic acid (CH_3COOH), hydrochloric acid (HCl), nitric acid (HNO_3), sodium hydroxide (NaOH), sodium tripolyphosphate ($Na_5P_3O_{10}$) and sodium sulfite (Na_2SO_3), 10% Nafion® solution and (DMSO) dimethyl sulfoxide (($CH_3)_2SO$), sodium or lithium borohydrides ($NaBH_4$, $LiBH_4$), tetra-amine platinum chlorides hydrate ($[Pt(NH_3)_4]Cl_2$ and $[Pt(NH_3)_6]Cl_4$), dichlorophenanthrolinegold (III) chloride (Au(phen) $Cl_2]Cl$), and ammonium tetrachloroaurate (III) hydrate ($NH_4AuCl_4 \cdot XH_2O$) in solution are also needed. All chemicals used should be of reagent grade.

The chitosan should be dissolved in 0.1 M acetic acid to prepare a 2% (by volume) chitosan solution. This solution should then be mixed thoroughly with a 10% Nafion® solution at room temperature. Subsequently, a solution of DMSO should be added to the mixture to act as a solvent. The resulting chitosan/Nafion® should be thoroughly mixed with acetic acid to make a Nafion®/chitosan composite that should then be sonicated for about 15 min in a bath sonicator to remove surface contaminants and impurities and then be vigorously stirred for 5 h. The chitosan/Nafion® mixture should then be poured into a glass petri dish to be cured for 13 h in an oven at 114 °C. The resulting membrane should be soaked in DI water at 85 °C for 2 h. The resulting chitosan/Nafion® membrane should further be hydrolyzed in 1M HCl for 2 h at 85 °C to protonate it.

The membrane will then be ready to be chemically plated by the following redox procedure:

1. The membrane material surfaces are roughened to increase the surface density and allow better molecular diffusion during oxidation and reduction processes to chemically plate the membrane with a metal to serve as an effective electrode. These steps include sandblasting, glass bead blasting, or sandpapering the surface of the biopolymer membrane to increase the surface-area density where platinum salt penetration and reduction occurs.

2. Ultrasonic cleaning and chemical cleaning by acid boiling (HCl or HNO_3-low concentrates) is carried out next.

3. The ion-exchanging process is incorporated using a metal complex solution such as tetra-amine platinum chlorides hydrate as an aqueous platinum salt such as ($[Pt(NH_3)_4]Cl_2$ and $[Pt(NH_3)_6]Cl_4$), or gold complexes such as dichlorophenanthrolinegold (III) chloride Au(phen) $Cl_2]Cl$, or ammonium tetrachloroaurate (III) hydrate $NH_4AuCl_4 \cdot XH_2O$ in solution. Although the equilibrium condition depends on the types of charge of the metal complex, such complexes have been found to provide good electrodes.

4. Here we perform the initial making of platinum ionic biopolymer nanocomposites, beginning with reducing the platinum complex cations to a metallic state in the form of nanoparticles using effective reducing agents such as an aqueous solution of sodium or lithium borohydride (5%) at favorable temperature (i.e., 60 °C). Platinum black-like layers deposit near the surface of the material.

5. The final step (surface electrode placement process) is intended to effectively grow platinum (or other novel metals of a few micron thickness) on top of the initial platinum surface to reduce the surface resistivity.

Although the equilibrium condition depends on the type of charge of the metal complex,

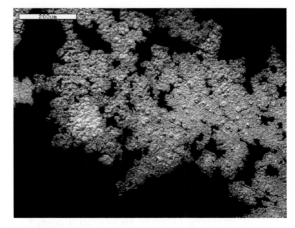

FIGURE 6.3 Fractal nature of reduced platinum within the chitosan/Nafion®.

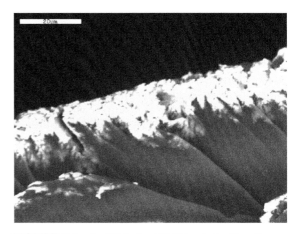

FIGURE 6.4 An SEM of an IBMC treated with a dispersing agent PVP.

such complexes were found to provide good electrodes. Figures 6.3 and 6.4 depict the fractal nature of platinum reduction within the macromolecular network. Note that platinum reduction within the macromolecular network nanoclusters branches out continuously to create fractal structures, as shown in Figure 6.3.

Figure 6.4 depicts a scanning electron micrograph (SEM) of a chitosan/Nafion® sample treated with a dispersing agent (polyvinylpyrolidone, or PVP).

The resulting chitosan IBMCs have been manufactured in various shapes and forms for a number of medical applications in connection with bodily fluid drainage problems such as in hydrocephalus, as shown in Figure 6.5.

A number of tests were performed on the new chitosan IBMCs in terms of actuation, sensing, and force exertion. Essentially the tests employed various samples of IBMCs, as shown in Figure 6.5, in a cantilever form between two electrodes at one end. A dynamic voltage generator was used to generate low voltage (4–6 V) and transient current (~100 mAs). For actuation purposes, the IBMCs were subjected to low voltage and current to study their bending deformations, as shown in Figures 6.6a–d.

FIGURE 6.5 Assortment of manufactured chitosan IBMC ionic biopolymer-metal composites of various complex shapes.

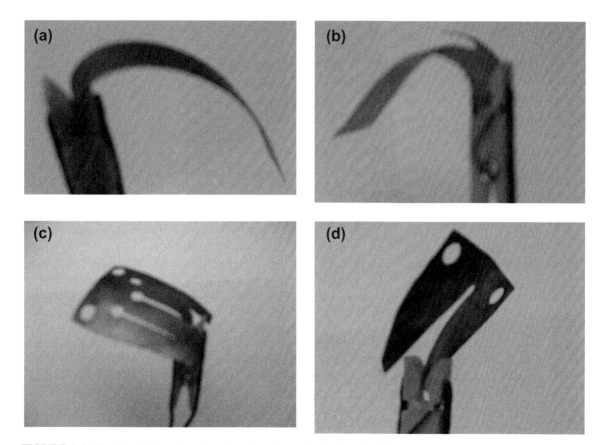

FIGURE 6.6 Bending deformation of various shaped samples (a, b, c, d) of chitosan IBMCs under a low voltage of 4–6 V.

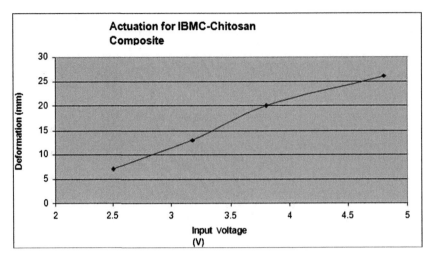

FIGURE 6.7   Displacement versus voltage for chitosan/IBMCs (size 3 mm × 40 mm × 0.24 mm).

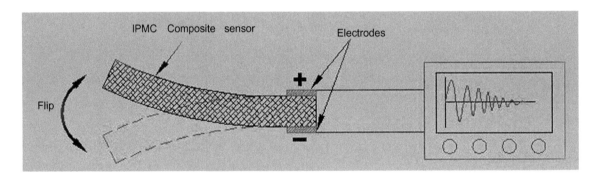

FIGURE 6.8   Sensing and energy-harvesting configuration for IBMCs.

For sensing purposes, the IBMC muscle strips were mechanically bent, either statically or dynamically, to record the voltage-generated output on an oscilloscope, to be observed and recorded.

Once an electric field is imposed on an IBMC strip in a cantilever configuration, the hydrated cations migrate to accommodate the local electric field. This creates a pressure gradient across the thickness of the beam, and thus the beam undergoes bending deformation (Figure 6.6) under small electric fields such as tens of volts per millimeter. Figure 6.7 depicts typical deflection characteristics of cantilever samples of IBMC artificial muscles. The resulting chitosan/ IBMC displayed good actuation and sensing properties, as shown in Figures 6.7–6.9.

The samples were then placed between two electrodes and the lead wires attached to an oscilloscope to observe the sensing and energy-harvesting capabilities of IBMCs, as shown in Figures 6.8 and 6.9. The samples generally showed very robust and fairly large output signals in the few millivolt range, which is

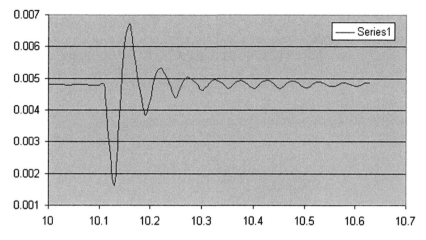

FIGURE 6.9 Voltage (volts, vertical axis) versus time (seconds, horizontal axis) for a chitosan/IBMC cantilever (sample size 3 mm × 40 mm × 0.24 mm) bent initially by 90° and then released to vibrationally damp out.

remarkable, as depicted in Figure 6.9, which depicts a typical sensing signal from a 3 mm × 40 mm × 240 μm IBMC strip.

To measure the force density of the IBMCs, high-resolution load cells (Mettler or Omega load cells with milligram resolution) are used as shown in Figure 6.10. The force exertion characteristics of chitosan IBMCs were quite good in the sense that the resulting force density was estimated to be about 35, which is close to IPMCs, force density in a cantilever form, which is about 45. *Force density* means the ratio of the blocking force for an IBMC cantilever strip at its tip to the weight of the cantilever strip itself. This will define an objective variable for actuator force generation and comparison. In this regard, a blocking force measurement set-up was used (Figures 6.10a and 6.11b) and a sample was loaded with a coin (a quarter, Figure 6.10c) to assess its blocking force. The force density of these IBMCs appears to be about 35 (normalized), which means in a cantilever form they can exert a blocking force more than 35 times the weight of the cantilever active part.

A chitosan blended with ionic polymers such as perfluorinated sulfonic or carboxylic membranes could lead to a new generation of chitosan/IBMCs that show bending towards the anode, as shown in Figure 6.10.

6.4 MODELING AND SIMULATION

De Gennes *et al.* [46] presented the first phenomenological modeling of sensing and actuation in IPMCs based on linear irreversible thermodynamics and equilibrium of forces and fluxes. Shahinpoor and Kim [47] further discussed the fundamental mechanisms of sensing and actuation in IPMCs based on osmotic diffusion and Nernst–Plank equations. Asaka *et al.* [48] presented an ion diffusion-based model for sensing and actuation of IPMCs. Nemat–Nasser and Zamani [49] also presented an electrochemomechanical modeling of the response of ionic polymer-metal composites with various solvents. Bonomo *et al.* [50] developed a software tool for simulation of actuation and sensing in IPMCs. Chen and Tan [51] presented a control-oriented and physics-based model for ionic polymer metal composite actuators. Del Bufalo *et al.* [52] discussed a mixture theory framework for modeling the mechanical actuation of ionic polymer-metal composites.

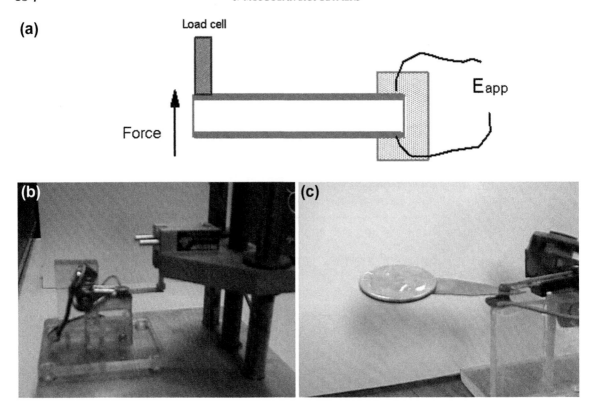

FIGURE 6.10   Cantilever and load-cell configuration for measuring the tip-blocking force of IBMC samples.

Porfiri [53] also discussed and modeled sensing and actuation of IPMCs based on charge dynamics. Wallmersperger *et al.* [54] discussed the thermodynamical modeling of the electromechanical behavior of ionic polymer-metal composites. It is beyond the scope of this chapter to elaborate on these models, but a brief discussion follows.

6.4.1   Linear Irreversible Thermodynamic Modeling of Forces and Fluxes

As recently as 2000, Gennes *et al.* [46] presented the first phenomenological theory for sensing and actuation in ionic polymer-metal composites. Note from Figure 6.11 that there are ionic fluxes and forces at work within the IBMCs.

Once an electric field is imposed on such a network, the conjugated and hydrated cations rearrange to accommodate the local electric field and thus the network deforms, which in the simplest of cases, such as in thin membrane sheets, spectacular bending is observed (Figure 6.6) under small electric fields such as tens of volts per millimeter.

Let us now summarize the underlying principle of the IBMC's actuation and sensing capabilities, which can be described by the standard Onsager formulation using linear irreversible thermodynamics. When *static conditions* are imposed, a simple description of *mechanoelectric effect* is possible based upon two forms of transport: *ion transport* (with a current density J, normal to the material) and *solvent transport* (with a flux Q, which we can assume is

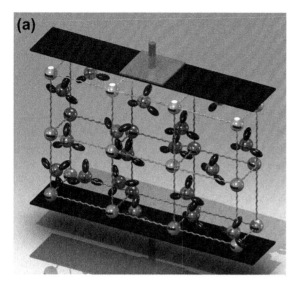

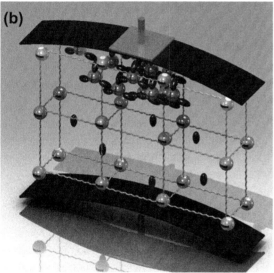

FIGURE 6.11 Hydrated cations migrate away from localized anode electrode toward the cathode electrode, causing the IBMC strip to bend toward the anode electrode.

water flux). The conjugate forces include the electric field E and the negative of the pressure gradient ∇p. The resulting equations are:

$$J(x,y,z,t) = \sigma E(x,y,z,t) - L_{12}\nabla p(x,y,z,t),$$
$$(6.3)$$

$$Q(x,y,z,t) = L_{21}E(x,y,z,t) - K\nabla p(x,y,z,t),$$
$$(6.4)$$

where σ and K are the material electric conductance and the Darcy permeability, respectively. The cross-coefficient is usually $L = L_{12} = L_{21}$. The simplicity of the preceding equations provides a compact view of the underlying principles of actuation, transduction, and sensing of the IBMCs, as also shown in Figures 6.7 and 6.8.

When the *direct* effect is measured (actuation mode, Figure 6.7), since ideally the electrodes are impermeable to ion species flux, it is observed that $Q = 0$. This gives:

$$\nabla p(x,y,z,t) = \frac{L}{K}E(x,y,z,t). \qquad (6.5)$$

This $\nabla p(x,y,z,t)$ will, in turn, induce a curvature κ proportional to $\nabla p(x,y,z,t)$. The relationships between κ and the pressure gradient $\nabla p(x,y,z,t)$ were fully derived and described by de Gennes *et al.* [46]. Let us just mention that $\kappa = M/YI$, where M is the local induced bending moment and is a function of the imposed electric field E, Y is the Young's modulus of the strip that is a function of the hydration H of the IPMC, and I is the moment of inertia of the strip. Note that locally M is related to the pressure gradient such that in a simplified format:

$$\nabla p(x,y,z,t) = M/I = \kappa_E. \qquad (6.6)$$

Now from Eq. (6.6) it is clear that the vectorial form of curvature κ_E is related to the imposed electric field E by:

$$\kappa_E = (L/KY)E. \qquad (6.7)$$

Based on this simplified model, the tip-bending deflection δ_{max} of an IPMC strip of length l_g should be almost linearly related to the imposed electric field due to the fact that

$$\kappa_E \cong \left[2\delta_{max} / \left(l_g^2 + \delta_{max}^2\right)\right] \cong 2\delta_{max}/l_g^2. \quad (6.8)$$

The experimental deformation characteristics depicted in Figure 6.6 are clearly consistent with the predictions obtained by the above linear irreversible thermodynamics formulation that is also consistent with Eqs. (6.7) and (6.8) in the steady-state conditions and has been used to estimate the value of the Onsager coefficient L to be of the order of 10^{-8} m^2/V s. Here, a low-frequency electric field is used in order to minimize the effect of back diffusion of loose water under a step voltage or a DC electric field. Other parameters have been experimentally measured to be $K \sim 10^{-18}$ m^2/CP and $\sigma \sim 1$ S/m [19]. The next section presents another approach to dynamic modeling of IBMCs in the context of charge dynamics and Poisson–Nernst–Planck equations.

6.4.2 Modeling Charge Dynamics and Actuation/Sensing Mechanisms of Biopolymers (IBMCs)

As an external voltage is applied at both sides of the chitosan/IBMC membrane, an electric field gradient across the membrane is induced. This is in accordance with the Nernst–Planck equations [47, 53, 55] such that

$$J = -D\left[\nabla C + \frac{ZF}{RT}C\nabla V\right], \qquad (6.9)$$

where J is the flux of ionic species in mol/(m^{2}s), C is the concentration of ionic species in mol/m^3, V is the electric potential field in volt, D is the diffusion coefficient in m^2/s, Z is the valence of ionic species, F is the Faraday constant, R is the universal gas constant, and T is the absolute temperature in Kelvin. Note that based on the Nernst–Planck equation (6.9), the second term on the right side acts as an external force that causes the movement of ions and results in changing ion concentrations across the membrane. The difference in ion concentrations results in lateral expansion and contraction of the biopolymer that consequently creates a mechanical pressure gradient due to ion diffusion from one side of the membrane to another side. This results in bending of the membrane. Now applying

a bending deformation induces ion diffusion across the membrane that results in a transient electric current in a short period of time (milliseconds) and also an electric potential across the electrodes plated on the IBMC. The exact mechanism that causes ion diffusion due to mechanical stimuli has been investigated by Porfiri [29] by considering the charge dynamics and micromechanics of ion diffusion in ion channels of a porous Nafion® membrane.

The electric potential between two electrodes lasts for a few seconds. Again, by using Eq. (6.3), the phenomenon can be described as follows: After moving cations to one side of membrane and generating an electric signal, a difference in ion concentration is created that causes ion diffusion at the reverse side, and ions tend to distribute evenly across the membrane to maintain a more stable condition. This causes the induced electric potential to disappear after a few seconds.

6.4.3 Poisson–Nernst–Planck Equation for Charge Dynamics

The most general governing equations for charge kinetic of ionic polymers are the Poisson–Nernst–Planck equations [53–56]. The equations are:

$$\frac{\partial c}{\partial t} = D\nabla \cdot \left[\nabla C + \frac{ZF}{RT}C\nabla V\right], \qquad (6.10)$$

$$\nabla^2 V + \frac{\rho}{\varepsilon} = 0, \qquad (6.11)$$

$$\rho = F(c^+ - c^-), \qquad (6.12)$$

where c is the general local charge concentration, c^+ is the local cation concentration, c^- is the local anion concentration, ρ is the local charge density, and ε is the permittivity of the medium.

As generally the thickness of the polymer membrane is significantly smaller than the other two lateral dimensions, it is a reasonable assumption that the ion diffusion is dominant over the thickness of the membrane with respect to two other dimensions; so the only spatial dependence is on x (thickness across the membrane). On the other hand, the ion species that contribute in transport phenomena are cations. Based on these assumptions, the one-dimensional form of the Nernst–Planck equation for cation transport will be:

$$\frac{\partial c^+}{\partial t} = D\frac{\partial}{\partial x}\left[\frac{\partial c^+}{\partial x} + \frac{F}{RT}c^+\frac{\partial V}{\partial x}\right]. \quad (6.13)$$

Because we are interested in the charge density dynamic of material to relate it to electric current, we rewrite Eq. (6.10) in terms of ρ using Eq. (6.13), which gives us the following equation:

$$\frac{\partial \rho}{\partial t} = D\frac{\partial}{\partial x}\left[\frac{\partial \rho}{\partial x} + \frac{F^2}{RT}\left(\frac{\rho}{F} + c^-\right)\frac{\partial V}{\partial x}\right]. \quad (6.14)$$

Since the anion concentration c^- is fixed across the membrane thickness, we can expand Eq. (6.14) into following form:

$$\frac{\partial \rho}{\partial t} = D\left[\frac{\partial^2 \rho}{\partial x^2} + \frac{F^2}{RT}\left(\frac{\rho}{F} + c^-\right)\frac{\partial^2 V}{\partial x^2} + \frac{F}{RT}\frac{\partial \rho}{\partial x}\frac{\partial V}{\partial x}\right]. \quad (6.15)$$

After assuming that the nonlinear terms $\frac{F}{RT}\frac{\partial \rho}{\partial x}\frac{\partial V}{\partial x}$ and $\frac{F}{RT}\rho\frac{\partial^2 V}{\partial x^2}$ are much smaller than the linear terms, Eq. (6.15) becomes:

$$\frac{\partial \rho}{\partial t} = D\left[\frac{\partial^2 \rho}{\partial x^2} + \frac{F^2}{RT}c^-\frac{\partial^2 V}{\partial x^2}\right]. \quad (6.16)$$

This equation is the linearized form of the Nernst–Planck equation for ionic polymers.

The Poisson equation is now given by:

$$\frac{\partial^2 V}{\partial x^2} + \frac{\rho}{\varepsilon} = 0. \quad (6.17)$$

Substituting the electric field term from Eq. (6.17) into Eq. (6.16), we have the following Poisson–Nernst–Planck governing partial differential equation for the kinetics of charge transport:

$$\frac{\partial \rho}{\partial t} = D\frac{\partial^2 \rho}{\partial x^2} - \alpha\rho, \quad 0 < x < h, \quad (6.18)$$

where $\alpha = \frac{DF^2}{RT\varepsilon}c^-$ and h is the thickness of the membrane.

This concludes the formulation of charge dynamics in chitosan/IPMC artificial muscles. This formulation indicates that the various dynamic phenomena in IBMCs are describable by well-known mathematical models.

6.5 CONCLUSIONS

The properties and characteristics of ionic biopolymer-metal nanocomposites as biomimetic multifunctional distributed nanoactuators, nanosensors, nanotransducers, and artificial muscles were discussed. Fundamental properties of biomimetic distributed nanosensing and nanoactuation of ionic polymer-metal composites and IBMCs were elaborated on, and some recent advances in the manufacturing techniques and 3-D fabrication of IBMCs were presented. Further, two modeling and simulation methodologies for actuation, sensing, and transduction of IBMcs were described. Procedures on how biopolymers such as chitosan and perfluorinated ionic polymers can be combined to make a new nanocomposite with actuation, energy harvesting, and sensing capabilities were also described. The fundamental properties and characteristics of biopolymeric muscles as artificial muscles were presented. Two ionic models based on linear irreversible thermodynamics as well as charge dynamics of the underlying sensing and actuation mechanisms were also presented.

Acknowledgment

This work was partially supported by Environmental Robots Inc.

References

[1] J. Bobet and R.B. Stein, A simple model of force generation by skeletal muscle during dynamic isometric contractions, *IEEE Trans Biomed Eng* **45** (1998), 1010–1016.

[2] J. Ding, A.S. Wexler, and S.A. Binder-Macleod, A mathematical model that predicts the force-frequency relationship of human skeletal muscle, *Muscle Nerve* **26** (2002), 477–485.

[3] J. Ding, A.S. Wexler, and S.A. Binder-Macleod, Development of a mathematical model that predicts optimal muscle activation patterns by using brief trains, *J Appl Physiol* **88** (2000), 917–925.

[4] Y. Bar-Cohen (ed.), *Electroactive polymer (EAP) actuators as artificial muscles: reality, potential and challenges*, SPIE Press, Bellingham, WA, USA (2001).

[5] M. Zrinyi, D. Szabo, and J. Feher, Comparative studies of electro- and magnetic field sensitive polymer gels, *Proc SPIE* **3669** (1999), 406–413.

[6] Q.M. Zhang, V. Bharti, and X. Zhao, Giant electrostriction and relaxor ferroelectric behavior in electron-irradiated poly(vinylidene fluoride-trifluorethylene) copolymer, *Science* **280** (1998), 2101–2104.

[7] M. Eguchi, Piezoelectric polymers, *Phil Mag* **49** (1925), 178–192.

[8] G.M. Sessler and J. Hillenbrand, Novel polymer electrets, *MRS Symp Proc* **600** (1999), 143–158.

[9] R. Pelrine, R. Kornbluh, Q. Pei, and J. Joseph, High speed electrically actuated elastomers with strain greater than 100%, *Science* **287** (2000), 836–839.

[10] W.C. Roentgen, About the changes in shape and volume of dielectrics caused by electricity, *Ann Phys Chem* **11** (1880), 771–786.

[11] H. Finkelmann and H.R. Brand, Liquid crystalline elastomers: a class of materials with novel properties, *Trends Polym Sci* **2** (1994), 222–226.

[12] H. Finkelmann and M. Shahinpoor, Electrically-controllable liquid crystal elastomer-graphite composites artificial muscles, *Proc SPIE* **4695** (2002), 459–464.

[13] Y. Osada, H. Okuzaki, and H. Hori, A polymer gel with electrically driven motility, *Nature* **355** (1992), 242–244.

[14] Y. Osada and S.B. Ross-Murphy, Intelligent gels, *Sci Am* **268** (5) (May 1993), 82–87.

[15] Y. Osada, H. Okuzaki, J.P. Gong, and T. Nitta, Electrodriven gel motility on the base of cooperative molecular assembly reaction, *Polym Sci* **36** (1994), 340–351.

[16] Y. Osada and A. Matsuda, Shape memory in hydrogels, *Nature* **376** (1995), 219.

[17] M. Hirai, T. Hirai, A. Sukumoda, H. Nemoto, Y. Amemiya, K. Kobayashi, and T. Ueki, Electrically induced reversible structural change of a highly swollen polymer gel network, *J Chem Soc Faraday Trans* **91** (1995), 473–477.

[18] T. Hirai, J. Zheng, and M. Watanabe, Solvent-drag bending motion of polymer gel induced by an electric field, *Proc SPIE* **3669** (1999), 209–217.

[19] M. Shahinpoor, Y. Bar-Cohen, J. Simpson, and J. Smith, Ionic polymer-metal composite (IPMCs) as biomimetic sensors, actuators and artificial muscles: a review, *Smart Mater Struct* **7** (1998), R15–R30.

[20] K. Oguro, K. Asaka, and H. Takenaka, Polymer film actuator driven by low voltage, *Proceedings of the 4th international symposium on micro machine and human science*, Nagoya, Japan (1993), 39–40.

[21] M. Shahinpoor, Conceptual design, kinematics and dynamics of swimming robotic structures using active polymer gels, *Proceedings of the ADPA/AIAA/ASME/SPIE conference on active materials & adaptive structures*, Alexandria, VA, USA (November 1991).

[22] M. Shahinpoor, Conceptual design, kinematics and dynamics of swimming robotic structures using ionic polymeric gel muscles, *Smart Mater Struct* **1** (1992), 91–94.

[23] T.F. Otero and J.M. Sansinena, Artificial muscles based on conducting polymers, *Bioelectrochem Bioenerg* **38** (1995), 411–414.

[24] M. Behl, R. Langer, and A. Lendlein, Intelligent materials: shape-memory polymers, in *Intelligent Materials* (M. Shahinpoor and H.-J. Schneider, eds.), Royal Society of Chemistry, Cambridge, UK (2008), 301–314.

[25] M. Shahinpoor, K.J. Kim, and M. Mojarrad, *Artificial muscles: applications of advanced polymeric nano composites*, CRC Press, Boca Raton, FL, USA (2007).

[26] M. Shahinpoor and K.J. Kim, Ionic polymer-metal composites – I. Fundamentals, *Smart Mater Struct* **10** (2001), 819–833.

[27] K.J. Kim and M. Shahinpoor, Ionic polymer-metal composites – II. Manufacturing techniques, *Smart Mater Struct* **12** (2003), 65–79.

[28] M. Shahinpoor and K.J. Kim, Ionic polymer-metal composites – III. Modeling and simulation as biomimetic sensors, actuators, transducers and artificial muscles, *Smart Mater Struct* **13** (2004), 1362–1388.

[29] M. Shahinpoor and K.J. Kim, Ionic polymer-metal composites – IV. Industrial and medical applications, *Smart Mater Struct* **14** (2005), 197–214.

[30] M. Shahinpoor and H.-J. Schneider, *Intelligent Materials*, Royal Society of Chemistry, Cambridge, UK (2008).

[31] A.F.T. Mak and S. Sun, Intelligent Chitosan-based hydrogels as multifunctional materials, in *Intelligent Materials* (M. Shahinpoor and H.-J. Schneider, eds.), Royal Society of Chemistry, Cambridge, UK (2008), 447–461.

[32] M. Shahinpoor, Ionic polymer-conductor composites as biomimetic sensors, robotic actuators and artificial muscles-a review, *Electrochim Acta* **48** (2003), 2343–2353.

[33] H.-J. Schneider, K. Kato, and R.M. Strongin, Chemome-chanical polymers as sensors and actuators for biologi-cal and medicinal applications, *Sensors* **7** (2007), 1578–1611.

[34] H.-J. Schneider and K. Kato, Chemomechanical poly-mers, in *Intelligent Materials* (M. Shahinpoor and H.-J. Schneider, eds.), Royal Society of Chemistry, Cam-bridge, UK (2008), 100–112.

[35] B. Stenger, *X-Ray diffraction and dielectric spectroscopy of poly(ethylene terephthalate) and PET/carbon nanotube nanocomposites*, Tufts University (May 2009).

[36] T.D. Gierke, G.E. Munn, and F.C. Wilson, The morphol-ogy in Nafion® perfluorinated membrane products, as determined by wide- and small-angle x-ray studies, *J Polym Sci Polym Phys Ed* **19** (1981), 1687–1704.

[37] S.-F. Liu and K. Schmidt-Rohr, High-resolution solid-state 13C NMR of fluoropolymers, *Macromolecules* **34** (2001), 8416–8418.

[38] M. Shahinpoor, Biopolymer/ionic polymer composite artificial muscles, in *Biotechnology in biopolymers* (A. Tiwari and R.B. Srivastava, eds.), Smithers-Rapra Pub-lishers, London, UK (2012), Chapter 10.

[39] A. Tiwari, R.B. Srivastava, R.K. Saini, A.K. Bajpai, L.H.I. Mei, S.B. Mishra, A. Tiwari, A. Kumar, M. Sha-hinpoor, G.B. Nando, S.C. Kundu, and A. Chadha, Biopolymers: an indispensable tool for biotechnology, in *Biotechnology in biopolymers* (A. Tiwari and R.B. Sriv-astava, eds.), Smithers-Rapra Publishers, London, UK (2012), Chapter 1.

[40] Z. Cai and J. Kim, Dry and durable electro-active paper actuator based on natural biodegradable polymer, *J Appl Polym Sci* **115** (2010), 2044–2049.

[41] K. Kato and H.J. Schneider, Supramolecular interac-tions in chitosan gels, *Eur J Org Chem* **2009** (2009), 1042–1047.

[42] P. Mukomaa, B.R. Jooste, and H.C.M. Vosloo, Synthesis and characterization of cross-linked chitosan mem-branes for application as alternative proton exchange membrane materials in fuel cells, *J Power Sources* **136** (2004), 16–23.

[43] Y. Wan, B. Peppley, K.A.M. Creber, V. Tam Bui, and E. Halliop, Preliminary evaluation of an alkaline chi-tosan-based membrane fuel cell, *J Power Sources* **162** (2006), 105–113.

[44] Y. Wan, B. Peppley, K.A.M. Creber, V. Tam Bu, and E. Halliop, Quaternized-chitosan membranes for possible

applications in alkaline fuel cells, *J Power Sources* **185** (2008), 183–187.

[45] K.J. Kim and M. Shahinpoor, A novel method of manu-facturing three-dimensional Ionic Polymer-Metal Com-posites (IPMCs) biomimetic sensors, actuators and artificial muscle, *Polymer* **43** (2002), 797–802.

[46] P.G. de Gennes, K. Okumura, M. Shahinpoor, and K.J. Kim, Mechanoelectric effects in ionic gels, *Europhys Lett* **50** (2000), 513–518.

[47] M. Shahinpoor and K.J. Kim, A solid-state soft actuator exhibiting large electromechanical effect, *Appl Phys Lett* **80** (2002), 3445–3447.

[48] K. Asaka, N. Mori, K. Hayashi, Y. Nakabo, T. Mukai, Z. Luo, Modeling of the electromechanical response of ionic polymer-metal composites (IPMC), *Proc SPIE* **5385** (2004), 172–181.

[49] S. Nemat-Nasser and S. Zamani, Modeling of electro-chemomechanical response of ionic polymer-metal composites with various solvents, *J Appl Phys* **100** (2006), 5253–5344.

[50] C. Bonomo, L. Fortuna, P. Giannone, S. Graziani, S. Strazerri, D.Harvey, G.S. Virk, A software tool for simulation of IPMCs as actuators and sensors, *Proceed-ings of the 11th Italian conference on sensors and microsys-tems*, Lecce, Italy, (February 2006), World Scientific, Singapore, 282–286.

[51] Z. Chen and X. Tan, A control-oriented and physics-based model for ionic polymer-metal composite actua-tors, *IEEE/ASME Trans Mechatron* **13** (2008), 519–529.

[52] G. Del Bufalo, L. Placidi, and M. Porfiri, A mixture theory framework for modeling the mechanical actua-tion of ionic polymer-metal composites, *Smart Mater Struct* **17** (2008), 045010.

[53] M. Porfiri, Charge dynamics in ionic polymer-metal composites, *J Appl Phys* **104** (2008), 104915.

[54] T. Wallmersperger, A. Horstmann, B. Kröplin, and D.J. Leo, Thermodynamical modeling of the electrome-chanical behavior of ionic polymer-metal composites, *J Intell Mater Syst Struct* **20** (2009), 741–750.

[55] Y. Bahramzadeh and M. Shahinpoor, Dynamic curva-ture sensing based on ionic polymer-metal composite (IPMC) sensors, *Smart Mater Struct* **20** (2011), 094011.

[56] M. Shahinpoor, Biomimetic robotic Venus flytrap (*Dionaea Muscipula Ellis*) made with ionic polymer-metal composites (IPMCs), *Bioinsp Biomim* **6** (2011), 046004.

ABOUT THE AUTHOR

Mohsen Shahinpoor is the Richard C. Hill Professor and Chair of the Mechanical Engineering Department at the University of Maine in Orono. He is also the director of Biomedical Engineering and Robotic Surgery Labs at the University of Maine. In the early 1990s he was one of the inventors of ionic polymer-metal composites (IPMCs), which are distributed nanoactuators, nanosensors, energy harvesters, and nanotransducers. He is also a pioneer in using IPMCs for left ventricular heart-assist systems (LVAS) as well as Ptosis (eyelid droop syndrome) and microcatheter stirrer for endovascular surgery and robotic surgery. He is currently involved with the applications and extensions of IPMCs to medicine and robotic surgery as well as biopolymers. He has also done a great deal lot of work on industrial applications of IPMCs, in particular a new family of smart materials called *robomorphs* that can be used for inspection in very tight and constrained spaces.

Bioscaffolds: Fabrication and Performance

Princeton Carter and Narayan Bhattarai

Department of Chemical and Bioengineering, North Carolina A&T State University, Greensboro, NC 27411, USA

Prospectus

The fabrication of three-dimensional (3D) scaffold architectures that closely approximate or effectively mimic native tissue extracellular matrix (ECM) is essential for regenerative success. In tissue engineering, native differentiable cells are incorporated into 3D scaffolds along with growth factors and other proteins. Materials used for the 3D scaffold construction must be biocompatible and bioresorbable to minimize adverse reactions during tissue regeneration. A 3D architecture is created by utilizing materials with specific surface properties, porosity, mechanical strength, etc., to improve desired cell activity and enhance tissue growth. Ideal 3D scaffolds should also not only have hierarchical macroporous structures comparable to those of living tissue, but they should also have surface features on the nanometer scale to improve cell adhesion and accelerate cell in growth.

Keywords

3D scaffolds, Bioresorbable materials, Composite materials, Electrospinning, Extracellular matrix, Hydrogels, Injectable scaffolds, Nanofibers, Natural polymers, Peptides, Porogen, Porosity, Resorbable biomaterials, Surface modification, Self-assembly, Tissue-derived scaffolds, Tissue engineering

7.1 INTRODUCTION

Since its inception centuries ago, scaffolding has remained an indispensable practice in the field of construction. Soaring around the perimeter of unfinished skyscrapers or encasing recently erected houses, scaffolding is a fixture in metropolises and suburban communities alike. These skeletal systems, typically composed of tubing affixed to wide planks, can be made from a variety of materials, though metallic components are often used. They function to critically support the safe and effective erection of edifices by allowing workers to immediately retrieve building materials and access various parts of the evolving structure. The scaffold must be strong, durable, and able to withstand its environment for the duration of the task. Most important, however, the supporting network must be removed with no consequence to the recently formed or restored structure.

The principles underlying frameworks in construction are analogous to those guiding scaffolding in tissue engineering and regenerative medicine. Support structures for regenerative purposes must be able to withstand the physiological stressors present *in situ,* allow cells to infiltrate the scaffold, and permit the exchange of nutrients and waste. The cells are comparable to workers, and the vasculature and extracellular matrix represent material stores enabling the cells to grow, proliferate, and produce tissue. Finally, as constructional scaffolds must be removed, biological skeletal aids must be resorbed by the body, leaving the newly formed tissue intact and functional [1,2].

The advent of engineered scaffolding systems for medical use was prompted by the desire of surgeon J. P. Vacanti to address the critical shortage of organ donors in the 1980s and 1990s. Regenerative strategies employing tailored scaffolds conceivably would remedy issues with grafts (*autologous*—from the same organism, *allogeneic*—from the same species, *xenogeneic*—from a different species). Although researchers have made significant gains in refining tissue engineering scaffolding, very few have been clinically implemented [3]. Consequently, suboptimal treatments as gold standards of care still persist.

7.1.1 Limitations of Current Practices in Tissue Regeneration

More than 3 million musculoskeletal procedures are done annually in the United States; about half of these involve bone grafting with either an autograft or an allograft [4]. An autograft is tissue taken from a person and implanted elsewhere in his body, whereas an allograft involves tissue borrowed from another person. As the gold standard of care for bone injuries, autografts account for the majority of the 500,000 bone-grafting procedures performed annually in the United States and the more than 2.2 million performed worldwide [5]. Bone autografts are ideal in their capacity to support new bone growth; however, grafts often lead to donor-site morbidity and are limited in supply [5]. Moreover, the

multiple surgeries involved with this procedure often lead to higher healthcare costs, prolonged recovery times, and an increased susceptibility to pathogens. To overcome these challenges, the development of tissue-engineered scaffolds is a primary objective for several researchers and medical practitioners around the globe.

Improved treatments are also needed for anterior cruciate ligament (ACL) ruptures or tears. The ACL, one of four ligaments that aids in stabilizing the knee, functions to support and strengthen the knee while preventing excessive anterior motion [6,7]. As a result of its position, there is a relatively high incidence of ACL injuries, with an estimated 200,000 patients per annum being diagnosed with disrupted ACLs. The relatively high frequency of these ligament injuries is a direct result of athletes participating in high-risk sports such as basketball, football, and soccer. Although the current standard of care for ACL reconstruction—the use of autografts—is often successful, there are still significant drawbacks to the procedure. The autograft tissue is typically taken from the patellar tendon, hamstring tendon, or quadriceps tendon. These tendons have limited availability and require multiple surgical procedures. Moreover, if allografts are employed, there is an increased likelihood of infection and transplant rejection [3].

Another prominent area of medicine in which inadequate care exists is in the treatment of coronary artery disease (CAD). It is estimated that CAD accounts for approximately 50% of the nearly 1 million deaths resulting from cardiovascular disease every year [8]. CAD involves the narrowing of arteries as a result of plaque, cholesterol, lipids, and other materials accumulating on the interior of arterial walls. As the luminal space progressively narrows, oxygen is effectively prevented from reaching the heart. When the severity of the blockage renders angioplasty ineffective, grafts are often used to circumnavigate the occluded region. The gold-standard therapy since 1968 has been the autotransplantation of the great saphenous vein, a vein that runs along the leg. The veinal graft, however, has short-term patency (condition of not being blocked or obstructed) success, with 12–27% of grafts becoming occluded the first year of transplantation and half of these occlusions occurring in the first month. Arterial grafts are preferred, given their long-term patency, but have limited use due to their insufficient availability, small diameter, and small length. Moreover, synthetic grafts, typically polymeric, are rarely used because they have had limited success as large-diameter vessels [9].

7.1.2 Utilizing 3D Scaffolds

Tissue engineering, as it is currently understood, was conceived in the late 1980s and early 1990s [10,11]. Since then, tissue-engineering strategies have expanded to include cells, genes, proteins, or other biomolecules. The bulk of a bioscaffold comprises biocompatible and bioresorbable materials that are comparable in 3D structure to the tissue area to be regenerated. Whereas progress has been made in the engineering of bioscaffolds, biomimetic scaffolds are a promising class of materials that helps to rectify the deficiencies of the preciously cited surgical treatments. Biomimetic scaffolds are those that closely approximate native tissue in chemical composition, architectural intricacies, and/or mechanical integrity.

The improved efficacy that biomimetic scaffolds exhibit over traditional scaffolds is a direct result of optimally presented physical and biochemical stimuli. These stimuli serve to increase cellular-matrix interaction and promote cell growth, proliferation, and subsequently tissue formation. The most mimicked physiological structure is the *extracellular matrix* (ECM), which is composed of a hydrated proteinaceous macromolecular network along with

other macromolecules such as carbohydrates. Several research groups over the past two decades have demonstrated the benefits of using biomimicry [1,12].

The ECM varies among tissue types but normally includes fibrillar collagens and elastins, with diameters ranging from 10 to several hundreds of nanometers. This network is covered with adhesive proteins such as laminin and fibronectin to provide sites for cell-adhesion molecules on the cell surface to interact with the ECM. The ECM also includes proteoglycans and glycoproteins—to fill the space between the fibers, act as a compression buffer against external stressors, and serve as a growth factor depot.

The benefits of harnessing biomimetic design in tissue engineering scaffolds and the mechanisms of its potential therapeutic effect are explored throughout the literature [13–17]. Most prominently detailed is the effect that nanoscale topography has on cellular behavior. To create an optimized cellular microenvironment conducive to the growth of tissues in natural form, biomimetic scaffolds with controllable physical, mechanical, and biological properties are required. Several traditional techniques of creating porous 3D structures, including salt leaching, free-form fabrication, and gas foaming, and fairly new techniques, including electrospinning and self-assembly, have been used to create biomimetic *3D scaffolds*. Older techniques result in macro- to micro-structured bioscaffolds, whereas later techniques result in micro- to nano-structured scaffolds.

The fabrication of hybrid 3D scaffolds by the merging of micro- and nanotechnologies is a promising alternative approach to designing 3D biomimetic scaffolds [14,18]. Nanofibrous scaffolds are effective substrates for cell cultivation and the assembly of thin tissues on the surface. The limitation of tissue formation only on the surface is a result of the scaffold inhibiting proper cellular infiltration to the core of the material. Hybrid scaffolds that consist of electrospun nanofibers and 3D microprinted pores provide the necessary topography and the macroporous structure needed for more efficient tissue growth. Such 3D scaffolds improve cell entrapment and proliferation, promote better cell differentiation, and ultimately enhance tissue regeneration.

7.1.3 3D Scaffold Requirements

Biomimetic 3D scaffold properties overwhelmingly influence the ability of the structure to promote functional tissue growth. When scaffolding is employed as part of the regenerative strategy, there are requirements for several characteristics. These characteristics include scaffold porosity, biocompatibility of scaffold constituents, appropriate time of degradation, suitable topographical features, ability to effectively encapsulate biomolecules and other therapeutics, and mechanical strength. If one of these characteristics is neglected in engineering a scaffold, there could be grave consequences for the material and the host.

The most important factor to consider is the material type that will comprise the bulk of the scaffold. Synthetic or natural polymeric materials, ceramics, metals, allotropic carbon, or any combination of these are the most commonly used material types for constructing a bioscaffold. The initial material selection affects all of the other physical properties of the scaffold and ultimately its behavior *in vivo*. Because the aforementioned materials have both significant benefits and limitations, composite or blended materials are created to obtain desired properties.

Polymers provide great diversity in chemistry and in the physical properties they impart to the engineered scaffold. Common biodegradable synthetic polymers are polyesters such as polylactide (PLA), polyglycolide (PGA) and polycaprolactone (PCL), polyanhydrides,

and polyorthoesters [19–22]. These synthetic materials can be converted into various geometries with predictable and controllable tensile and compressive mechanical strengths by using different processing techniques such as casting, extrusion, or drawing. In addition to precise control over mechanical strength, synthetic polymers are also easily manipulated to produce desired degradation rates. Optimal scaffold degradation allows the polymer support to subside as new tissue is formed, without comprising the stability of the healing site. Synthetic polymers typically degrade by hydrolytic cleavage of the polymer chain. Various rates of degradation can be achieved by altering the backbone of the polymer. Thus, placing more hydrophilic groups will produce faster rates, whereas increasing the hydrophobicity of the macromolecular chain will lead to slower rates.

These synthetic materials, however, typically lack sufficient biological cues found in the natural extracellular matrix, often resulting in unfavorable cell-material interactions. To overcome such biocompatibility problems, surface modification and biofunctionalization techniques can be utilized. Some of these have been discussed in Chapter 8 by Vogler. An example is the grafting of oligopeptides or oligosaccharides to the surface of the scaffold, thereby making the structure more amenable to the host. If greater mechanical strength is required, as with hard, calcified tissues, inorganic materials such as tricalcium phosphate or hydroxyapatite may be mixed with the polymer [23–26].

Natural polymer-based scaffolds have also garnered attention because of their proven biocompatibility and ability to enhance cell adhesion and proliferation [27–29]. Collagen, gelatin, hyaluronan, chitosan, silk, and alginate are among some of the most commonly used natural polymers in tissue engineering [30,31]. Utilizing these natural polymers, especially those derived from animal tissues, plants, and crustacean shells, may provide unlimited materials for making 3D scaffolds.

Despite the advances made with these materials, the majority of them cannot match the performance of autografts. Natural polymer-based 3D scaffolds have shown good tissue biocompatibility and produce safe degradation products but are mechanically weak, have fast and uncontrollable degradation rates, require immunosuppressive drugs, and retain batch-to-batch variation of physico-chemical properties. Synthetic materials typically have the advantage of being readily available, eliciting a limited immune response, and being produced with highly specific degradation rates. However, they lack cell-recognition signaling and release acidic by-products during degradation, leading to local inflammation [32]. Hybrid biomaterials that combine the favorable biological properties of natural polymers and mechanical properties of synthetic polymers represent a potential approach to enhance the properties of tissue engineering scaffolds [33–36].

Control of pore size and pore interconnectivity in a 3D scaffold are important considerations also. Although specific means to control porosity are mentioned in Section 7.2 for each fabrication technique, in general porosity is controlled through varying processing parameters. Pores allow effective transport of nutrients to regions throughout the bioscaffold as well as removal of waste from various regions of the bioscaffold. However, the porosity and mechanical strength of the material must be optimized to ensure that the bioscaffold is neither too brittle nor too dense. Optimal pore diameters are also needed to ensure that desired cellular behavior is achieved. Cells must maintain communication with one another and infiltrate the bioscaffold to its core, a necessary event for proper tissue formation.

The last issue to consider is the topographical features of the bioscaffold. Nanoscale features located uniformly on a bioscaffold will aid in proactively directing cellular arrangement [14,37]. Biomechanical cues can be transmitted to cells via micro- or nanoscale substrate topography. Morphological and functional changes have been observed for various types of cells, including mesenchymal stem cells (MSCs), when cultured on substrates with micro- or nanoscale topographical features such as fibers, pillars, grooves, or pits [38–42]. A variety of techniques can achieve nanoscale surface roughness, such as embedding nanofibers into a solid matrix, electrospraying onto a scaffold, and physically and chemically etching [43–45].

This chapter focuses on the fabrication and use of nanostructured materials as 3D biomimetic scaffolds for tissue regeneration. We also mention various techniques used to create micro/nanoporous structures for tissue regeneration.

7.2 FABRICATION OF 3D BIOSCAFFOLDS

In living tissues, cells reside in a 3D, cushioned ECM network and are protected from external mechanical stress [46,47]. The ECM also promotes cellular functions such as cell adhesion, migration, proliferation, and differentiation due to molecular interactions between specific cell membrane receptors and signaling cues from the ECM [14,40,48]. Therefore, as previously mentioned, the architecture of scaffolds should mimic the natural ECM and instruct cellular behavior while adequately housing the cells. Although enormous progress has been made in designing tissue engineering scaffolds using a variety of materials and various processing conditions, more research remains to be done. Table 7.1 summarizes some of the commonly used 3D scaffold fabrication techniques. In the following section, we discuss several representative

TABLE 7.1 Summary of fabrication techniques for porous 3D bioscaffolds.

Fabrication Techniques	Process Description	References
Conventional scaffolds		
Salt-leaching method	(i) Insoluble (porogen) particles are added to a polymer solution (ii) Solution is cast into a mold and the solvent allowed to evaporate (iii) Matrix is then placed in solvent to dissolve porogen particles	49–51
Gas-foaming method	(i) CO_2 and a polymer are placed in a chamber and the pressure increased until the two are miscible (ii) The pressure is lowered and the gas particles begin to cluster (iii) Pores form as gas evolves (iv) Often combine with particle-leaching step	52,53
Textile fiber-bonding method	(i) A fiber mesh is submerged into a polymer solution or placed in a mold and a polymer solution is poured into the mold (ii) The solvent is allowed to evaporate and the polymers are heated to the melting point of the fiber mesh to create welded points (iii) The non-fiber polymer is selectively dissolved and the bioscaffold is vacuum dried	54, 55
Solid free-form fabrication	(i) Computer-aided design (CAD) tools are used to produce a digital representation of the desired bioscaffold (ii) Polymeric material is then deposited in a layer-by-layer manner to produce the bioscaffold	50, 56, 57
Hydrogels	(i) Natural or synthetic hydrophilic polymers are placed in aqueous solution (ii) Chemical or physical crosslinks are used so the polymer does not dissolve	58, 59
Nanofibrous scaffolds		
Electrospinning	(i) A high voltage is applied to a polymer solution (ii) A stream forms if there is sufficient chain entanglement (iii) Fibers are deposited onto grounded collector of variable geometries	60–62
Thermally induced phase separation	(i) A polymer solution is frozen (ii) The polymer and solvent phases separate at lower temperatures (iii) The solvent phase is evaporated through lyophilization	12, 43
Self-assembly	(i) Amphiphilic molecules, typically peptides, are placed in aqueous solution (ii) Hydrogen bonding, ionic, electrostatic, hydrophobic, and van der Waals interactions can cause the molecules to form fibers	63, 65
Special scaffolds		
Native tissue derived	(i) Allogeneic or xenogeneic tissues are processed in proprietary ways to leave fibrous protein and carbohydrate ECM (ii) Other chemical and physical modifications are made to ensure bioscaffold's stability *in vivo*	66, 67
Injectable scaffolds	(i) Prepare hydrogel or ceramic solution (ii) Add cells and other material	68, 69
Inorganic ceramics/ composites	(i) Gas-foaming, free-form fabrication techniques	70–72

processes used to produce the various types of 3D scaffolds by traditional and recently developed fabrication techniques.

7.2.1 Conventional Scaffolds

Several conventional techniques used to produce porous scaffolds, including the salt-leaching and gas-foaming methods, have been used to introduce open pore structures and interconnected channels within bioscaffolds. Pores are important to increase the viability of seeded or injected cells within the scaffolds [73–76]. These traditional techniques for preparing 3D scaffolds are straightforward, cost-effective, and easy to scale up [77,78].

7.2.1.1 Salt-Leaching Method

Salt-leaching is a simple processing technique to produce 3D biomimetic scaffolds. This method involves making a polymer/organic solvent solution and incorporating porogen particles, which are insoluble in the organic solvent. The solution is then cast into a mold of the desired shape, and the solvent is evaporated away. After the solvent completely evaporates, the final step is to dissolve the porogen particles in an aqueous solution [49,50]. The resulting structure has significant porosity as a result of the void spaces left by the dissolved porogen particles. Typically, the porogen is a salt granule or particle.

As with all techniques, there are advantages and disadvantages in using this method. One of the main advantages is that porosity and pore size can be effectively controlled. Materials with porosity levels up to 90% and pore-size diameter ranging from 100 to 700 μm have been reported using the salt-leaching technique. The porosity is given by volume fraction occupied by leachable particles. The pore size and pore shape can be modified independently of the porosity by varying the leachable particles' characteristics (i.e., size and shape) [51,79,80]. Another advantage of using the salt-leaching

method is that it is one of the more convenient and straightforward methods for preparing porous 3D scaffolds. The technique involves a minimal number of steps and only requires basic laboratory methods.

One of the major drawbacks to salt leaching is that it only produces thin membranes with a dense surface skin. Moreover, the bioscaffold might contain residual salt particles used during the process. This could negatively impact cell behavior and ultimately tissue growth. In addition to residual salt or other particles remaining on the scaffold, there is often incomplete removal of the solvent during the drying process. Another disadvantage of this method (and other methods employing polymers) is that some of the polymers degrade into acidic by-products. These acidic degradation products could potentially have negative effects on cell adhesion and growth [50]. Finally, another potential drawback to the salt-leaching technique, especially for bioscaffolds needing lower porosity levels, is the lack of interconnectivity between pores. Salt particles that are not in contact with other particles lead to insufficient pore interconnectivity and often become trapped in the polymer scaffold [51].

To ameliorate the aforementioned problems, researchers have investigated means to make the process more benign. One method that has been employed is the use of melt polymer solutions for the solvent-casting stage, as opposed to using a polymer solution with harsh organic solvents. The melt-molding step consists of mixing the polymer powder with salt particles and melting the mixture [81]. The melting step eliminates the need for organic solvents, thereby preventing the possibility of the scaffold containing residual solvent and harming cells or tissue.

In an attempt to create better interconnectivity between pores and to increase the channel size between pores, a method where by salt particles are partially merged has been proposed. This method involves merging salt particles using moisture or using heat to merge sugar particles or paraffin spheres. By merging the porogen material, the probability of isolated particles remaining in the polymer mixture is significantly reduced. Moreover, cellular communication and the exchange of nutrients and waste should be improved with better interconnectivity between pores [81].

7.2.1.2 Gas-Foaming Method

The *gas-foaming method* has been demonstrated to produce effective bioscaffolds for tissue engineering applications [53,79,82,83]. This method takes advantage of a gas at high pressure to produce a porous structure instead of the harsh solvents often used in particle-leaching methods. The gas-foaming method begins by placing a material, typically polymeric, in a chamber with a gas such as CO_2 and increasing the pressure to the point where the gas becomes sufficiently soluble in the polymeric solid phase. This effectively saturates the polymer with the gas. The pressure is then lowered to the ambient pressure to induce thermodynamic instability of the gaseous phase. The gas begins to phase separate from the polymer, and in an effort to minimize free energy, the gas molecules begin to cluster and cause pore formation. The pores grow by the gas molecules diffusing to the pore nuclei. The resulting structure is highly porous, but is primarily a closed-pore structure because of the rapid depletion of the gas between pores [52,79]. The process is limited in its ability to produce consistently repeatable results. Pore formation does not occur in a predictable manner each time the technique is used.

To create better interconnectivity between pores in the gas-foaming method, this technique has been combined with the salt-leaching method. The hybrid technique first creates a composite of the polymer with a porogen. The composite is then placed with a gas in a high-pressure environment to allow the polymer and gas to mix. The pressure is then decreased to the ambient pressure, and the gas molecules create pores within the structure. Finally, the scaffold is submerged into deionized water (or some other solvent) to dissolve the porogen particle, creating more pores and enhancing pore interconnectivity [52].

Scaffolds produced using either the gas-foaming method or the hybrid method have been shown to support cellular functions critical for tissue regeneration [52]. These scaffolds have also been used for concomitant release of various biomolecules and small molecules in addition to structurally supporting tissue growth [52].

7.2.1.3 Textile Fiber Bonding

To produce fibrous 3D scaffolds with good mechanical properties, *textile fiber bonding* may be employed. Fiber-bonded structures have been shown to perform better than other materials in withstanding the *in-situ* stressors experienced during tissue growth [54].

Textile fiber was developed in 1993 by Mikos and colleagues [55]. The process was initially described in a series of steps that include the formation of a composite material of nonbonded fibers embedded in a polymer matrix, subsequent thermal treatment of the matrix, and finally, selective dissolution of the matrix. The nonbonded fibrous mesh is created by isolating fibers from a thicker multi-lamellar mat. This mat is then either submerged into another polymer solution, thereby ensuring the fibers are immiscible in the second polymer solution, or the fibers are placed into a mold and the other polymer solution is allowed to fill the remaining mold volume. After the solvent evaporates, the composite is heated to a temperature above the melting point of the polymer that comprises the fiber network to form welded points at the crosspoints of the fiber mesh. Finally, the non-fiber polymer is selectively dissolved using a solvent that is immiscible with the fiber network. The resulting fiber matrix is then vacuum-dried to completely remove solvents [55].

Two prominent drawbacks of the fiber-bonding technique include the inability to control pore size and the technique requires harsh solvents. As previously mentioned, harsh solvents and inadequate pore size could render the scaffold ineffectual for tissue engineering purposes.

7.2.1.4 Rapid Prototyping/Solid Free-Form Fabrication

Rapid prototyping is another technique to create porous 3D scaffolds. This method involves using computer-aided design tools to produce a digital representation of a bioscaffold. A depositor then layers polymeric or other material types to exactly replicate the desired shape. Layer-by-layer assembly, discussed in Chapters 3 and 11, allows for the exact control of morphological characteristics and ultimately the mechanical properties of the bioscaffold. The researcher easily controls pore location and size as well as surface characteristics of the scaffolds to ensure scaffold success as a regeneration vehicle. Moreover, the predictability of the rapid prototyping method allows for consistent reproduction of various scaffold types with complex shape, size, and other physical requirements [50, 84].

There are several variants of the rapid prototyping method; however, all variants are categorized by whether they use direct or indirect fabrication of the scaffold and whether they employ the melt-dissolution deposition (MDD) technique or the particle-bonding technique [56]. The direct fabrication methods that use the MDD technique are fused deposition method, 3D fiber-deposition technique, precision extrusion deposition, precise extrusion manufacturing, low temperature deposition, multi-nozzle deposition, pressure-assisted microsyringe, robocasting, 3D bioplotter, and rapid prototyping robotic dispensing system [56]. In general, MDD methods involve extruding a strand of material through a small opening in a lateral direction onto previous strands. The melted material eventually solidifies and remains attached to the preceding layer. By controlling the spacing between adjacent strands of polymeric material,

pore size is controlled in the horizontal plane. For manipulation in the vertical plane, strands are placed at varying angles to achieve specified pore dimensions [56].

A representative MDD method is the fused deposition method (FDM). It begins with feeding a strand of material into a liquefier. After melting, the material is extruded as a series of layers. The environment is controlled to maintain proper contact between layers. The subsequent scaffold has a honeycomb structure with channel diameters in the hundreds of micrometers. FDM has also produced scaffolds with polymeric and ceramic components for mechanical strength and scaffolds that support the growth and proliferation of various cell types [56].

A limitation of FDM is that the material being fed into the liquefier has to be of a specific diameter and possess certain material properties to physically fit through the rollers and nozzle. Most natural polymers cannot be used with the FDM because of the high operating temperatures required to melt them and produce strands. The inability to incorporate natural polymers limits the potential biomimicry of the scaffold. Finally, the ability to achieve sufficient microporosity is inhibited by the deposition of dense filaments.

To remedy these deficiencies, various modifications of the FDM have been employed. Newer methods that eliminate the need for a precursor filament produce scaffolds with sufficient porosity and allow for the incorporation of biopolymers and biomolecules through lower processing temperatures [56, 85]. Moreover, methods have been developed to significantly reduce the resolution of the scaffold, producing filaments that are in the tens of micrometers in diameter [56]. Another development with deposition methods is the ability to create hydrogel materials with well-defined pore structures. These hydrogels could provide the softer scaffolds needed for the regeneration of soft tissues, along with other scaffold properties [57].

The other class of rapid prototyping methods, particle-bonding techniques, imparts advantageous features to the resulting bioscaffold. In general, particle-bonding techniques involve the selective bonding of particles into a thin 2D layer. These 2D layers are then bonded one by one to produce the desired 3D structure. Various solvents have been used to bind the layers, including water with complex sugars [56] and organic solvents with synthetic polymers [56]. The technique gives rise to macroporous and microporous structures, but the extent of the micropores is limited by the size of polymer granules in the powder. The primary advantage of particle-bond methods over melt deposition methods is that the powder causes the scaffold to have a rough exterior, which has been shown to promote cell differentiation, growth, and proliferation [56]. Particle-bonding techniques also do not use heat, which allows for various material types to be used.

The final rapid prototyping methods are indirect methods. They take advantage of a mold to produce a 3D porous scaffold. Indirect methods involve casting material into a mold that will create desired external and internal structures. These methods are useful because they require less raw material and a variety of material blends can be accommodated. The original material properties are maintained throughout processing of the 3D scaffold. Adequate microporous and macroporous structures are produced [56,83,86].

7.2.1.5 Hydrogels

Hydrogels represent the final class of 3D bioscaffolds discussed in this survey. Hydrogels are ubiquitous in tissue engineering because they closely approximate the ECM. These bioscaffolds can be synthesized from a variety of materials, are highly biocompatible owing to their large water content (ranging between 70% and 90%), are processed under mild conditions, and can be augmented to achieve desired mechanical properties.

Hydrogel matrices are formed from a variety of synthetic and natural hydrophilic polymers as well as polymeric blends. The most common synthetic polymers used include polyvinyl alcohol (PVA), polyethylene glycol (PEG)/polyethylene oxide, and poly-2-hydroxyethyl methacrylate. Some natural polymers used to create hydrogels are chitosan, alginate, hyaluronic acid, and collagen [87–90].

The chosen material can significantly improve the function of the scaffold. To prevent dissolution of the hydrophilic polymer, crosslinking methods are needed. Some polymers used for hydrogels are charged, which serves to create a physical crosslink between oppositely charged macromolecules. Chemical methods involve covalently attaching a linker molecule to polymer chains and then reacting the chain-molecule complex with other macromolecular chains. Hydrogels can also be formed by polymerizing monomers in the presence of water [58,87–90].

The mechanical properties of hydrogels are enhanced through the use of ceramics and other blends. To make the scaffold more amenable to cell adhesion, given its smooth topography, a variety of peptide moieties may be attached to the polymer. The RGD (arginine, glutamic acid, aspartic acid) amino-acid sequence is among the oligopeptides used to mimic adhesion proteins that interact with cell surface receptors. Control over hydrogel porosity influences diffusion of material in and out of the scaffolds and dictates the ability of cells to infiltrate the scaffold [59]. For better control over hydrogel porosity, 3D printing techniques involving liquid-liquid bonding have been used [91,92].

7.2.2 Nanofibrous Scaffolds

Once used almost exclusively for textiles, *nanofibers*, designated by the National Science Foundation as fibers with diameters of 100 nm or less [93], are 3D materials with expanding interest for a variety of uses. Currently, these nanostructures are employed for applications ranging from fuel and energy storage to water filtration and electronic devices. This boom in nanofiber utilization over the past decade is directly related to Reneker's work [61, 94–96], which repopularized electrospinning, a common technique for fiber formation that is discussed in depth later.

The ubiquity of the fibers can also be attributed to their physical properties and facile adaptability. Specifically, nanofibers have a high surface-area-to-volume ratio, elasticity, and superior strength. In addition, nanofibers offer considerable flexibility in the materials that can be utilized for their fabrication. Nanofibers can be constructed from synthetic and natural polymers, allotropic carbon, and hybrids/composites made by combining two or more polymers or mixing with metallic or ceramic components.

Further physical modifications of fibers include axial alignment of fibers, fiber diameter manipulation, and tuning of the bioactivity of the structures [97]. Typical chemical manipulations involve the reaction of a functional group on the polymer backbone with another molecule to produce the desired functionality of the fibers. Furthermore, a variety of manufacturing processes can be used to create nanofibers. These include electrospinning, self-assembly, phase separation, templating, extraction, and polymerization techniques [61, 94–100].

Given nanofibers' pliancy in processing and achievable conformations, these nanostructures are prime candidates for tissue engineering use. The nanofibers provide multifunctional 3D substrates needed to heal and restore tissue function in a timely manner. The three prevalent fabrication techniques for nanoscale fibers in tissue engineering and regenerative medicine are electrospinning, phase separation, and self-assembly [12].

7.2.2.1 Electrospinning Technique

Electrospinning is a widely used technique to produce fibers in the nanoscale. In 1934, Formhals was the first to publish a work concerning the electrospinning process. He described it as a method to produce thin polymer filaments by exploiting electrostatic repulsions between surface charges. He also gave a rigorous description of the apparatus he used to create the polymer filaments. As previously mentioned, however, the increased popularity in the use of electrospinning for producing nanofibers for tissue engineering and other applications was sparked by Reneker in the early 1990s [61, 62]. Prior to Reneker's work, the effect of fiber diameter on the performance and manufacturing of more complex fibrous structures was known, but the practical generation and implementation of nanoscale fibers were limited [101]. During the early 1990s, the technique would also assume its current appellation, becoming a portmanteau of the words *electrostatic* and *spinning* [60]. This technique remains viable primarily because of its simplicity.

The electrospinning process involves applying a voltage, typically 1–30 kV, to charge a polymer solution (or melt) loaded into a syringe (Figure 7.1). This high applied voltage causes the polymer solution to be sufficiently charged, and the induced charge distributes evenly through the surface. At this point, the solution experiences electrostatic repulsion forces from the surface charges and Coulombic forces exerted by the external electric field [60]. When the electrostatic repulsion forces combined with the Coulombic force are sufficient to overcome the surface tension of the solution, a stream erupts from the deformed droplet, known as a *Taylor cone*, at the end of the nozzle. If the cohesion is sufficient, then the stream is elongated and eventually deposited onto a grounded collector plate. If cohesion or chain entanglement is not sufficient, then electrospraying or droplet formation usually occurs. Jet elongation happens during

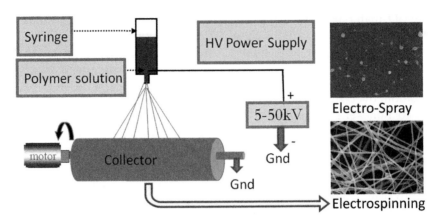

FIGURE 7.1 Schematic of an electrospinning setup.

the stream's travel toward the collector. During its flight, the jet undergoes a stretching and whipping process to draw the fiber into an ultrafine long filament. The solvent simultaneously evaporates and the fibers are deposited on the grounded collector, thereby creating a nonwoven, randomly aligned fibrous mat [60].

The parameters chosen during electrospinning greatly influence the collected fibers. These parameters are typically divided into three categories: polymer parameters, polymer solution parameters, and parameters of the apparatus. The type of polymer used and its physical properties greatly affect the nanofibers. These properties include the molecular weight, the molecular weight distribution, and the branching of the polymer [12]. Solution properties found to have an integral role in fiber formation include viscosity, polymer concentration, conductivity, and surface tension. Important apparatus parameters are applied flow rate, voltage, distance from syringe needle tip to collector, type of collector and whether it is static or dynamic, the type of needle used, and the ambient conditions during electrospinning [61,94–96].

One of the most studied dependent properties of the fibers is the fiber diameter. Several researchers have attempted to sum up the effects of the many independent variables of electrospinning

on fiber diameter into a succinct mathematical model. Rutledge *et al.* [102] developed a mathematical model that related surface tension γ, static relative permittivity ϵ, flow rate Q, current carried by the fiber I, and the ratio of initial jet length to the nozzle diameter χ to fiber diameter d as follows:

$$d = \left[\gamma \epsilon \frac{Q^2}{I^2} \frac{2}{\pi(2\ln\chi - 3)} \right]^{1/3}. \tag{7.1}$$

This equation was derived by fitting an exponential model to empirical data. According to this equation, increasing the current-carrying capability of fibers by adding more conductive materials to the polymer solution will significantly reduce fiber diameter [62]. It is also possible to reduce fiber diameter by manipulating other independent variables such as a reduction in either the flow rate Q or the nozzle diameter γ.

Another model relates fiber diameter to the molecular weight of the polymer and the concentration of the polymer in the spinning solution. It also uses the dimensionless parameter called the Berry number B. The Berry number is the product of the intrinsic viscosity η and polymer concentration C, i.e.,

$$B = \eta C. \tag{7.2}$$

The Berry number has four distinct domains. According to the value assigned to B of a polymer, a researcher can determine the likelihood the polymer will produce nanofibers [62]. Region I, where $B < 1$, is representative of a very dilute polymer solution with limited chain entanglement. This results in only polymer droplets being formed. In Region II, $1 < B < 3$, the fiber diameter increases within the range of 100–500 nm as B increases. This region is indicative of molecular entanglement that is just sufficient for fiber formation. Although fiber formation is observed, there is still some droplet formation as a result of polymer relaxation and surface tension. In Region III, $3 < B < 4$, the fiber diameter increases rapidly with B and is in the range of 1,700–2,800 nm. The rapid increase in fiber diameter is attributed to the intensive molecular entanglement resulting in an increase in polymer viscosity. The consequence of increased polymer viscosity also means that a higher electric field is required to produce fibers. Finally, in Region IV, where $B > 4$ and there is significant chain entanglement within and among chains fiber diameter is more dependent on the applied voltage/electric field and other process parameters than it is on B [62].

The two quantitative models for calculating fiber diameter are a sampling of what researchers have produced thus far, with more complex examples existing. They serve as a suitable starting point for researchers, suggesting parameter boundaries to achieve desired fibers. There are also numerous other variations that can be made to electrospinning parameters to engineer favorable fibers.

7.2.2.1.1 Electrospinning for uniaxially aligned nanofibers

Aligning nanofibrous arrays is useful for a variety of purposes. When fibers are used to impart additional mechanical integrity as a composite component, control of the alignment dictates the degree of structural support the fibers provide. In the areas of tissue engineering and regenerative medicine, aligned fibers can serve as a physical guide for cellular growth, influence cell adhesion, and modulate cellular patterns found in native tissue [103]. They are particularly useful when employed to regenerate tissues that require directional recruitment and assembly of cells, for example, in neural tissue engineering. Cooper et al. [103] demonstrated the benefits of aligned chitosan-PCL fibers over films and randomly oriented fibers of the same materials in promoting the attachment and proliferation of Schwann cells, which are important cells of the peripheral nervous system. The aligned fibers induced cellular responses as a result of topographical and chemical cues for the modulation of neurite outgrowth [104].

One way to fabricate uniaxially aligned nanofibers or parallel arrays of fibers is to use a rapidly rotating drum or cylinder as the collector. The rotating collector forces fibers to align in a perpendicular orientation to the axis of rotation of the drum. This method, however, only produces partially aligned fibers [62]. To improve alignment, researchers modified the drum by adding a sharp edge.

Another method to fabricate aligned fibers is to use a pair of split electrodes [60]. Two conductive strips separated by a gap of up to several centimeters allow for the synthesis of aligned nanofibers in the gap. Researchers believe that the insulating gap alters the electrostatic forces acting on the fibers in the gap. As a result, electrostatic forces act in opposing directions and fibers are stretched, aligning themselves perpendicular to the edge of the gap. The electrostatic repulsion between deposited fibers can further enhance the alignment of the collected fibers. This technique can also be used to readily stack the aligned fibers into films or mats for practical applications [105].

The overarching concept governing the previously described methods to fabricate aligned fibers is the ability to control the electric field. Some researchers have modeled the dominant role of the electric field in controlling the trajectory of the fiber and ultimately its collection. Manipulation of the field, either through the applied voltage or by the type of collector used, provides the proper fiber geometry for the application.

7.2.2.1.2 Blend and composite nanofibers

As with other fabrication techniques, composites and blend systems can be created to enhance the biocompatibility and structural properties of nanofibrous scaffolds. Blend nanofibers incorporate two or more polymers into a single solution that can be electrospun. Figure 7.2 shows one example, developed by Cooper *et al.* [106], of polyblend nanofibers of PCL-chitosan for the application of skeletal muscle tissue reconstruction. The combinations of polymeric materials provide increased structural support, enhanced biocompatibility, and desired degradative properties. The material choice is critically important in engineering fiber scaffolds with suitable physicochemical properties. Synthetic polymers are widely used because of their uniformity (physical

and chemical properties) and because they are consistently reproduced. Some synthetic polymers commonly used in tissue engineering applications are PCL and poly-lactide-co-glycolide acid (PLGA). For nanofibrous scaffold production, these polymers are particularly suited because they easily form nanostructures from electrospinning, and like many other polymers, they degrade *via* hydrolysis of their ester linkage. However, this degradation often occurs at an unfavorably slow rate. To rectify this issue, polymers such as PCL, with good mechanical integrity but inherent hydrophobicity, can be altered with natural or more hydrophilic synthetic polymers to produce structurally sound and biocompatible scaffolds. Several research reports have been published showing the ability of PCL and a naturally occurring polymer to form blended copolymeric nanofibrous structures [35,106–110].

The resulting fiber morphology appeared to be smooth, and fibers had a mean diameter of approximately 100 nm, a result researchers attribute to the solvent system used. Contact angle

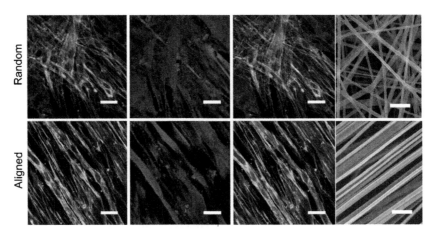

FIGURE 7.2 Electrospun composite nanofibers for a skeletal muscle tissue reconstruction application. Confocal microscopy images showing immunocytochemistry analysis of actin (left column, green) and myosin heavy chain (MHC) (middle column, red) expressed by muscle cells grown on chitosan-PCL randomly oriented and aligned nanofibrous scaffolds after culture in fusion media for five days. The merged images with nuclei stained with DAPI (blue) are shown in the right column. SEM images showing the morphology of chitosan-PCL nanofibers. Scale bars represent 40 μm for cell/nanofiber structures and 20 μm for SEM images. Reprinted with permission from Ref. 106; copyright 2012 Royal Society of Chemistry. (For interpretation of the references to color in this figure legend, the reader is referred to the Web version of this book.)

measurements were used to determine the relative hydrophilicity of the scaffolds. The blend scaffold of chitosan and PCL was shown to exhibit a significant decrease in hydrophobicity and degradation rates compared to the pure PCL fiber scaffold. The blend fibers were also optimally mixed to produce the most bioactive structures, resulting in a larger percentage of chitosan being used to sustain cells more favorably.

Studies have also been conducted demonstrating that optimizing biocompatibility does not cause severe losses of mechanical integrity [111]. The natural polymer chitosan was again combined with the synthetic polymer PCL. The ultimate goal was to produce a synergistic blend of the two polymers that had the mechanical integrity of PCL and the biocompatibility and bioactivity of chitosan.

Composite nanofibers are a product of ceramics or metals added to polymers to increase fiber mechanical properties and enhance bioactivity. Hydroxyapatite or some other form of calcium phosphate or bioglass is a commonly used ceramic in bioscaffolds because of its chemical similarity to natural mineral components. The increased osteoconductivity, the ability to support bone formation, obtained by adding ceramics aids in the formation of natural bone. Although it is simple to add the mineral component to the polymer solution to be electrospun, this typically results in the mineral being embedded or encapsulated in the fibers, rendering it ineffective. To remedy this problem, several researchers have deposited the mineral on the fibers as a post-fabrication treatment. Incubation in simulated body fluid allows the pores and fiber surfaces to get sufficiently covered in minerals; this process can take several hours to weeks. To shorten the time it takes to mineralize a scaffold, electrodeposition technology was used to coat the fibers, accomplishing mineralization in less than an hour. The resulting fibers had the same morphology compared to simulated body fluid (SBF) incubated fibers as well as similar mechanical strength [43].

7.2.2.1.3 Centrifugal electrospinning

Centrifugal electrospinning is another modified electrospinning technique that produces highly aligned nanofibers. This process involves loading a spinneret and nozzle onto a circular disk that is attached to a rotating axle. A metallic cylindrical shell is then placed around the disk and grounded to serve as the collector. Centrifugal action on the polymer solution provides a uniform distribution of stress, which stretches the polymer into a long fiber if the solution viscosity is ample. This process also allows lower-molecular-weight polymer solutions to be electrospun.

Polymer solution viscosity is a result of the friction between polymer chains. The frictional forces are dependent on the speed of the centrifuge: as the speed increases, the frictional force increases and consequently the solution viscosity increases. The electrocentrifugal technique produces fibers that are better aligned compared to electrospinning, due to its ability to reduce the bending stability of the polymer jet. These fibers, however, exhibit only marginal improvements to mechanical properties [112].

7.2.2.1.4 Coaxial electrospinning

Coaxial or core-shell electrospinning is another common modification to the traditional electrospinning technique to obtain multifunctional nanofibrous scaffolds. The fibers can be spun from many different polymers and polymeric combinations, and a variety of material (synthetic or natural) can be placed in the core, all aimed at efficient tissue formation. Most often, biomolecules, which easily lose their bioactivity in the harsh solvents used to dissolve polymers, are encapsulated as the core. The polymer shell surrounding the biomolecule core has the property of tunable degradability, depending on its composition, which allows these biochemical agents to be released over a favorable time period. The ability to optimize the temporal release of the molecules promotes more efficacious behavior of the biomolecules [113,114].

The most common *core* molecules are sensitive biomolecules like the growth factors bone morphogenic protein (BMP) and fiberblast growth factor (FGF), which elicit favorable responses from cells. The favorable responses of cells in the presence of growth factors include enhanced growth, cell proliferation, and cell differentiation, all of which contribute to tissue growth. It is necessary that molecules like growth factors be introduced into the wound or defective area in a sustained manner to ensure favorable cellular function throughout the wound-healing or tissue-formation process [113,114].

The polymer shell, in addition to protecting the encapsulated biomolecules and moderating biomolecular release, can function to make cells and the implant interact favorably. This behavior is normally achieved through simple chemical modifications [115] to the polymer shell's surface. Finally, the core-shell method is applicable for encasing fibers that are mechanically strong or have some other benefit but are not benign to the cells at the site of implantation [115].

In addition to encasing biomolecules and other polymers, the core-shell method can be used to alter the physical properties of electrospun nanofibers in an optimal way [114]. The dynamic evaporation of solvent from the electrospun fluid jets presents an opportunity for manipulation of the resulting collected nanofibers. As previously explained, electrospinning has an instability region where the fluid polymer jets are whipped and bent, stretched and elongated. This is a result of the interactions of the applied electrostatic force, the solution's intrinsic viscoelasticity, and its surface tension. As the fluid jets become increasingly sticky due to solvent molecule transfer and evaporation, the electrical forces gradually lose their influence on the fluid jets because electrons can only easily interact with fluids. As a result, the electrical drawing process stops when the entire jet, or sometimes the surface of the jet, solidifies. The use of surfactant solutions as the shell or sheath fluid helps remedy this problem. As a result of using surfactant solutions, the collected fibers have reduced diameters and are smoother compared to fibers spun with no sheath solution [114].

To achieve the core-shell geometry, the outer solution consisting of a solution or a solvent is loaded into a syringe and placed into a pump apparatus. The core solution is then loaded into another syringe and placed into a pump. A metallic capillary that has an inner and outer diameter is attached to the syringes using some type of polymer piping. The piping from the inner diameter is connected to the core solution, and the piping to the outer diameter is connected to the sheath solution. Like traditional electrospinning, a voltage is applied to the metallic capillary, then a jet forms, and eventually fibers are deposited on a grounded collector [114].

7.2.2.1.5 Melt electrospinning

Although electrospun nanofibers from polymer solutions have been successful in producing fibers in the submicron region, the potential clinical use of these nanofibers could be limited due to the use of harsh solvents and incomplete drying [116]. Therefore, a need for solvent-free processing exists. Zhmayev *et al.* showed that by using gas-assisted polymer melt electrospinning, it is possible to achieve fibers with submicrometer diameters [116]. Air drag in the electrospinning setup, as well as heating provided by the air stream, aid in thinning the fiber [116]. This presents a viable alternative to solutions of synthetic polymers such as polylactic acid (PLA). However, this technique would not suffice for some natural polymers because heating could cause structural damage [117].

7.2.2.2 *Thermally Induced Phase Separation*

Thermally induced phase separation (TIPS) is another technique that can be used to produce nanofibers. TIPS takes advantage of the thermal instability of polymer solutions [43] and can readily produce a 3D nanofibrous scaffold with a five-step process.

The five steps of the TIPS method are polymer dissolution, phase separation and gelation, solvent extraction, freezing, and lyophilization. Polymer dissolution simply involves mixing the polymer, synthetic or natural, with a suitable solvent. The temperature of the solution is subsequently lowered by freezing the polymer solution. At these lower temperatures, the solution becomes thermodynamically unstable, leading to a spontaneous separation of phases. The mostly solvent phase is evaporated through lyophilization (freeze drying) and, if optimal process parameters are used, the mostly polymeric phase forms a nanofibrous matrix with fibers ranging in diameter between 50 and 500 nm [43]. The ability to produce fiber distributions in this range make this technique particularly utile in tissue engineering, as the Type I collagen fibers of the ECM have diameters in this range [43].

The resulting nanofibrous shape is controlled by fabrication parameters such as gelation temperature and polymer concentration of the solution [12]. In addition, this process can also be combined with a particle-leaching step to easily create macropores in the construct.

There are, however, limitations to the TIPS method. The sensitivity of this technique to process variations makes it difficult to reproduce scaffolds. Moreover, the solvent used for phase separation is often difficult to completely remove from the scaffold. As a result, the scaffolds could have grave unintended consequences as a tissue engineering product [118].

7.2.2.3 Self-Assembly

Self-assembly involves the autonomous aggregation of molecules into thermodynamically favored nanostructures through non-covalent interactions. Peptidic molecules are most commonly used for self-assembled nanofibers as a result of the multiple arrays of amino acids that can be generated. Due to substantive diversity in peptide creation, self-assembly is a suitable technique to produce fibers with physical and chemical properties suitable for a host of biomedical applications. In addition, the amphipathic (having hydrophilic and hydrophobic qualities) molecules can be tailored to elicit multiple biological responses after implantation, a property central to optimizing regenerative strategies.

The chemical mechanisms of peptide self-assembly are highly specific in that molecular recognition mediates hydrogen bonding, ionic, electrostatic, hydrophobic, and van der Waals interactions [63]. Therefore, amino acids can be engineered so that certain secondary structures (β-sheets, β-hairpins, and α-helices) can optimally interact and give rise to nanofibrils and 3D nanofibrous networks. The self-assembly process has been shown to produce nanofibers with diameters well below 1,000 nm, with a distribution of 5–25 nm [64].

One common method of self-assembly is to first synthesize an amino-acid sequence and then chemically attach an alkyl chain to the oligopeptide, a peptide chain of 2–20 amino acids in length. The oligopeptide can serve a dual function in that it is hydrophilic, and the oligopeptide can be synthesized to mimic part of a native protein of interest. The hydrophilic oligopeptide covalently linked to an alkyl chain causes the molecules to assemble into nanofibrous geometry when the proper conditions are applied. The specific amino-acid sequences are often derived from common ECM proteins such as laminin, mimicking the peptide sequences that interact with cell receptors of progenitor cells and promote their differentiation to the desired cell [65]. Attaching an oligopeptide domain is preferred over incorporating the entire peptidic chain due to the ease in synthesizing oligopeptides, stability of the shorter sequences, and their demonstrated bioactivity [65].

When the amphiphile, a surfactant molecule, is placed under aqueous conditions, the alkyl chain packs into the center, and the peptide domain is exposed to the aqueous environment. It has been well documented in the literature that a 16-carbon alkyl chain length is optimal for nanofiber formation [64]. These micellar structures are typically cylindrical in shape due to the assembly of the amphiphiles that have an overall conical shape. Once the cylindrical micelles aggregate, they form nanofibrous structures [64].

Synthetic polymers can also be used to fabricate self-assembled nanofibers. The process begins with synthesizing a block or a block copolymer of more than two polymers in order to introduce different regions of functionality into the backbone. These polymers often include positively and negatively charged domains, which, once introduced into an aqueous solution, interact to form nanofibers.

7.2.3 Special Scaffolds

7.2.3.1 Native-Tissue-Derived Scaffolds

Biological scaffolds, particularly extracellular matrices derived from various animal tissues, are an increasingly attractive option for tissue engineering applications. The native extracellular matrix affords the researcher an intact 3D bioscaffold that can effectively modulate the necessary events for regeneration. In addition, there is some degree of tailorability to the researcher in constructing the bioscaffolds. The tissue from which the matrix is derived can be excised from an organism at various time points of development. This dictates the chemical constituents and, consequently, the mechanical properties of the bioscaffold. There are also inherent differences in the extracellular matrix owing to ECM variability between tissue types and species. Together these differences contribute to variations in the microstructure, quantities of non-collagenous proteins, glycosaminoglycans (GAGs), and other factors, mechanical properties, and ratios of collagen fibers present. For instance, fetal and neonatal bovine dermises have 3–5 times more Type III collagen than that of adult dermis, while as the calf develops, the thickness of the dermis increases as well as the diameter of the fibers [66].

Currently, there are a variety of commercially available extracellular matrix products that aid in the regeneration of human tissue. These extracellular matrix scaffolds come from human autograft, human autogenous, and allogeneic sources and include tendon, fascia, ligament, and dermis (e.g., Alloderm™). The xenogeneic sources of extracellular matrix used to manufacture scaffolds include porcine, equine, and bovine tissues. The specific tissues processed from the xenogeneic sources include heart valves, dermis, pericardium, and parts of the intestine. Depending on the processing conditions, the host biological response can be tailored. A collagen-based bioscaffold from the small intestine submucosa (SIS) has been shown to induce site-specific regeneration in numerous tissues, including blood vessels, tendons, hernias, ligaments, skin, urinary bladder, musculoskeletal repair, and dural substitute [119–123].

The human response to extracellular matrix scaffolds can range from the foreign body encapsulation observed with permanent implants to degradation and resorption, as well as being populated with fibroblasts and vascularized to support new tissue growth. Chemical crosslinking is the most prominent way to influence the human body's response to extracellular matrix-based bioscaffolds (e.g., Contigen™). Some researchers have used hexamethylene diisocyanate (HMDI) and carbodiimide (CDI) to chemically crosslink the scaffold to prevent rapid degradation and also mitigate the immunogenicity of the xenogeneic tissue derived ECM products. This technique, however, results in the chemical crosslinker remaining in the final scaffold product–a potential detriment to a host. To remedy this, 1-ethyl-3-(3-dimethylaminopropyl) carbodiimide (EDC) was used, and the EDC allowed the natural scaffold to be crosslinked without the retention of the crosslinker molecule. There have been negative effects associated with unnatural crosslinking of scaffolds, including the decreased binding affinity of growth factors toward the modified matrix as well as decreased cell attachment and proliferation. As a result, current methodologies allow for the ECM to be processed in a way in which crosslinking is unnecessary and the scaffolds have clinically viable degradability properties [66].

These natural scaffolds represent an interesting class of bioscaffolds that will become more useful as processing methods are further refined, fostering a new generation of 3D bioscaffolds that could potentially aid in more efficient regeneration of human tissue.

7.2.3.2 Injectable Scaffolds

Injectable scaffolds are attractive options for tissue engineering applications because they are delivered in a minimally invasive manner, have the potential to set *in situ*, and can potentially conform to complex, intricate tissue defects. These scaffolds avoid the invasive surgeries required to implant grafts and preformed scaffolds. In addition, injectable scaffolds afford the ability to seed and deliver cells more efficiently, avoiding the inability to seed cells deep into preformed scaffolds. Moreover, injectable scaffolds shorten operating times, minimize post-operative pain and scar tissue, and reduce cost [68, 124–131]. Figure 7.3 shows a schematic of *in situ* cross-linked poly(ethylene glycol) (PEG)-hyaluronic acid (HA) injectable hydrogel. A blend of PEG, HA, the photoinitiator, and eosin Y are injected transdermally and then photocrosslinked using a light-emitting diode (LED) [131].

These bioscaffolds can be formed from a variety of materials, depending on the tissue being regenerated. Calcium-phosphorous ceramic injectable bioscaffolds have been created for bone tissue engineering applications. Once these scaffolds were mixed with cells encapsulated in alginate hydrogel beads and fibers, they exhibited suitable biological and mechanical properties [68]. For soft-tissue defects, several injectable hydrogels have been used. Natural polymer systems, which use collagen and

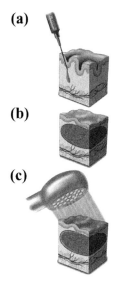

FIGURE 7.3 Schematic showing a transdermally injected, photoactivated soft-tissue bioscaffold. An *in situ* crosslinked injectable hydrogel was obtained by combining PEG, hyaluronic acid (HA) and photoinitiator, eosin Y. A mixture of these components was injected transdermally (a) and could be formed into a desired shape (b). The material was then crosslinked by exposing the bioscaffold to an array of LEDs emitting light (c). This light was shown to penetrate tissue depths of up to 4 mm with a 2 min. exposure time, sufficient to activate the eosin Y and photocrosslink the hydrogel scaffold. Reprinted with permission from Ref. 131; copyright 2012 American Association for the Advancement of Sciences.

glycosaminoglycans (GAGs) reinforced with synthetic polymers for mechanical integrity and decreased gelation time, have shown promise to aid in soft-tissue regeneration [69].

Although numerous engineering approaches have been developed to create injectable scaffolds, new classes of peptide-based *in situ* gelling scaffold materials have garnered significant interest [124–127]. Rationally designed peptides are rapidly assembled into hydrogel scaffolds in physiological condition by physical or chemical cross-linking mechanisms providing encapsulated cells with an artificial ECM environment. Peptides with environmental stimuli responsiveness are physically assembled by controlling the temperature, pH, or ionic strength while peptides with reactive functional groups are *in situ* chemically crosslinked by using multifunctional crosslinkers, enzymes, photopolymerization, and gamma irradiation [124–130]. Using these physical or chemical means, a liquid-like precursor solution can be easily mixed with target cells, administered into the potential tissue defect site, and then gelled within several minutes to create hydrogel scaffolds. In particular, ECM-derived polypeptides, including elastin-based peptides and collagen-like proteins as well as fibrins and silk proteins have been widely used to synthesize injectable hydrogel scaffolds. They show good biocompatibility, minimal immune responses, and controllable degradation rates *in vivo* [124–131].

7.2.3.3 Inorganic Ceramics/Composites

Ceramic materials are central components in bioscaffolds manufactured for hard-tissue engineering. This is primarily because bone, the predominant hard mineralized tissue of the human body, has a large mineral component of a low crystalline form of the ceramic hydroxyapatite (HA) ($Ca_{10}(PO_4)_6(OH)_2$). As a result, there have been attempts to create ceramics and composites of ceramics that elicit the same biological response and have the same physical properties of the native inorganic matter. Some absorbable ceramic materials that have been investigated for bone tissue engineering include $CaCO_3$ (argonite), $CaSO_4$-$2H_2O$ (Plaster of Paris), and $Ca_3(PO_4)_2$ (beta-whitlockite, a form of tri-calcium phosphate, and TCP) [70]. The most commonly studied CaP ceramics, however, are TCP, HA, and tetracalcium phosphate. Calcium-phosphorous scaffolds are advantageous because they are biocompatible, elicit a minimal immunologic response, and can be processed to avoid systemic toxicity [71]. They are also osteoconductive and integrate well with natural bone [71].

To make 3D porous ceramic scaffolds, a variety of techniques may be employed. These methods include the gas-foaming method, template

casting, phase mixing, and free-form fabrication methods [72,132–134]. Depending on the processing method and the parameters chosen, the pore size and mechanical integrity of the scaffold can be tailored. Advanced manufacturing techniques such as free-form fabrication are particularly attractive in that they offer improved control over scaffold architecture [72]. This allows for the formation of almost any shape as well as defined pores and topography.

7.3 SURFACE MODIFICATION OF SCAFFOLDS

The surface of a tissue engineering scaffold is important in tissue engineering because it can directly affect cellular functions such as cell adhesion, growth, migration, and differentiation and ultimately the tissue regeneration [135–137]. Although many synthetic polymers have been used to create 3D scaffolds for directing the repair and regeneration of damaged tissues, active control of cell adhesion and downstream cellular events is still challenging on these scaffolds due to the lack of biological recognition sites that can instruct cells [136,138,139]. Therefore, efforts have been made to incorporate bioactive ECM molecules such as collagen-I, III, IV, laminin, fibronectin, RGD peptide, etc., and growth factors onto the surfaces of scaffolds using various modification techniques, including non-covalent and covalent binding [135,140].

ECM and growth factor molecules can be non-covalently adsorbed on synthetic scaffolds by secondary bonding such as electrostatic interaction and van der Waals forces. Surface-modified synthetic porous scaffolds such as PCL and PLGA have the ability to preferentially differentiate mesenchymal stem cells (MSCs) into osteogenic cells [140–142]. However, non-covalent adsorption of ECM molecules is a relatively weak force and may, therefore, not be appropriate for tissue engineering applications for which prolonged signaling is required.

Covalently conjugated bioactive molecules have been shown to be stable under physiological conditions and maintain biological activity after prolonged implantation in the localized microenvironment. For covalent binding, surfaces of preformed scaffolds are converted into reactive functional groups such as OH, COOH, and NH_2. These functional groups can undergo further reactions with ECM proteins, peptides, and growth factors on the surfaces of various synthetic scaffolds, including PLLA, PLGA, and PCL. These surface-modified scaffolds have significantly enhanced biological performance after covalent modifications [143–146]. Surface modification can be performed after a porous scaffold has been fabricated. It therefore does not usually affect the scaffold structure and mechanical properties significantly. However, density and distribution of the bioactive molecules on the scaffold surface should be optimized for appropriate control of cell behavior [143, 147].

7.4 BIOACTIVE MOLECULE DELIVERY WITH SCAFFOLDS

In addition to a scaffold's microstructural, topographical, and other physical properties, biological signaling is also a key component for cell function and tissue regeneration. Endogenous signaling molecules such as growth factors are often not sufficient for the repair of large size defects, necessitating the addition of exogenous signaling molecules. These biomolecules usually have short half-lives, and their concentration gradients play a prominent role in cellular responses. For effective tissue regeneration, the delivery system must be sustained and should overcome the short half-life of these biomolecules *in vivo*.

Several promising approaches are available to couple growth factors to the biomaterial surface so they are readily bioavailable. Controlled release systems using micro- and nanospheres have been shown to be effective in retaining the bioactivities of various therapeutic agents [148–151]. There also have been attempts to modify a scaffold with biological factors using the layer-by-layer process [152]. A multilayered heparin-based polyelectrolyte delivery system is constructed on the surface of a PLGA layer that facilitates the loading of basic fibroblast growth factor (bFGF) and increases growth factor stability. When growth factor molecules are released in a sustained manner and not instantaneously, greater cell proliferation was observed *in vitro*. Biological factors have been incorporated in porous scaffolds for a sustained release system by adding them into a polymer solution or emulsion and further processing the polymer solution [153, 154]. The gas-foaming process has been shown to entrap PLGA microspheres containing biomolecules in a porous scaffold for sustained delivery at a tissue defect site [155]. This technique allows for the release of single or multiple biological factors in a spatially and temporally controlled manner.

7.5 CONCLUSIONS AND PERSPECTIVES

In reviewing various approaches for the design and engineering of 3D scaffolds that closely approximate human tissue ECM, it is clear that the scaffold's geometry, topography, porosity, density, and other physicochemical properties regulate many cellular processes. New materials and combinations of materials, as well as improved scaffold designs based on novel processing techniques, are continuously advancing the tissue engineering field, particularly in recent years. However, the biological performance of scaffolds for specific applications *in vitro* and *in vivo* is not fully understood. When the biological performance of scaffolds is clear to researchers, a rationale for the design of multifunctional scaffolds can be implemented.

Recent studies have found that matrix stiffness is an instructive signal for cell differentiation. For example, mesenchymal stem cells preferentially differentiate into neuronal, myogenic, and osteogenic cells when cultured on hydrogels with the lowest (0.1 kPa), intermediate (11 kPa), or highest (34 kPa) moduli, respectively. In addition, micro/nanoscaled topography created on 3D scaffold surfaces appears to have a profound impact on cell shape, organization of the cytoskeletal structure, and intercellular signaling [156]. It is also worth noting that cells release many soluble biomolecules that dynamically remodel ECM structures by changing the local modulus and chemical composition of the tissue ECM [157]. Therefore, identifying the 3D architecture of the ECM and the interactions between ECM components in each tissue type is another key step toward designing more effective bioscaffolds.

The generation of vascularized tissues also remains a key challenge in tissue engineering. Numerous approaches have been conceived in the last decade to overcome this problem. Today, only *in vitro* engineered tissues like skin, cartilage, and cornea are used in clinics. This limited success is due to the fact that cells of these tissues can be functional, with nutrients and oxygen via diffusion from blood vessel systems that are further away. Currently, there are two main strategies: prevascularization of synthetic scaffolds or the use of natural tissue-based 3D scaffolds. Furthering the success of tissue engineering in the clinic and realizing the ultimate goal of whole-organ generation hinge on the advancement of these strategies and others by the continued collaboration of researchers and technologists across the spectrum of academia and industry, utilizing each person's expertise.

References

[1] S. Liao, B. Li, Z. Ma, H. Wei, C. Chan, and S. Ramakrishna, Biomimetic electrospun nanofibers for tissue regeneration, *Biomed Mater* 1 (2006), R45–R53.

[2] J.R. Fuchs, B.A. Nasseri, and J.P. Vacanti, Tissue engineering: a 21st century solution to surgical reconstruction, *Ann Thorac Surg* 72 (2001), 577–591.

[3] C.T. Laurencin and J.W. Freeman, Ligament tissue engineering: an evolutionary materials science approach, *Biomaterials* 26 (2005), 7530–7536.

[4] T. Neighbour, *The global orthobiologics market: players, products and technologies driving change*, Espicom Business Intelligence, Chicester, UK (2008).

[5] C. Laurencin, Y. Khan, and S.F. El-Amin, Bone graft substitutes, *Expert Rev Med Dev* 3 (2006), 49–57.

[6] J.W. Freeman, M.D. Woods, D.A. Cromer, E.C. Ekwueme, T. Andric, E.A. Atiemo, C.H. Bijoux, and C.T. Laurencin, Evaluation of a hydrogel-fiber composite for ACL tissue engineering, *J Biomech* 44 (2011), 694–699.

[7] P.F. O'Loughlin, B.E. Heyworth, and J.G. Kennedy, Current concepts in the diagnosis and treatment of osteochondral lesions of the ankle, *Am J Sports Med* 38 (2010), 392–404.

[8] V.L. Roger, A.S. Go, D.M. Lloyd-Jones, R.J. Adams, J.D. Berry, T.M. Brown, M.R. Carnethon, S. Dai, G. de Simone, E.S. Ford, C.S. Fox, H.J. Fullerton, C. Gillespie, K.J. Greenlund, S.M. Hailpern, J.A. Heit, P.M. Ho, V.J. Howard, B.M. Kissela, S.J. Kittner, D.T. Lackland, J.H. Lichtman, L.D. Lisabeth, D.M. Makuc, G.M. Marcus, A. Marelli, D.B. Matchar, M.M. McDermott, J.B. Meigs, C.S. Moy, D. Mozaffarian, M.E. Mussolino, G. Nichol, N.P. Paynter, W.D. Rosamond, P.D. Sorlie, R.S. Stafford, T.N. Turan, M.B. Turner, N.D. Wong, and J. Wylie-Rosett, Heart disease and stroke statistics—2011 update: a report from the American Heart Association, *Circulation* 123 (2011), e18–e209.

[9] R.A. Vorp, T. Maul, and A. Nieponice, Molecular aspects of vascular tissue engineering, *Front Biosci* 10 (2005), 768–789.

[10] R. Langer and J.P. Vacanti, Tissue engineering, *Science* 260 (1993), 920–926.

[11] S. Cohen, M.C. Bano, L.G. Cima, H.R. Allcock, J.P. Vacanti, C.A. Vacanti, and R. Langer, Design of synthetic polymeric structures for cell transplantation and tissue engineering, *Clin Mater* 13 (1993), 3–10.

[12] V. Beachley and X. Wen, Polymer nanofibrous structures: Fabrication, biofunctionalization, and cell interactions, *Prog Polym Sci* 35 (2010), 868–892.

[13] B.M. Willie, A. Petersen, K. Schmidt-Bleek, A. Cipitria, M. Mehta, P. Strube, J. Lienau, B. Wildemann, P. Fratzl, and G. Duda, Designing biomimetic scaffolds for bone regeneration: why aim for a copy of mature tissue properties if nature uses a different approach? *Soft Matter* 6 (2010), 4976–4987.

[14] T. Dvir, B.P. Timko, D.S. Kohane, and R. Langer, Nanotechnological strategies for engineering complex tissues, *Nat Nanotechnol* 6 (2011), 13–22.

[15] S.C. Owen and M.S. Shoichet, Design of three-dimensional biomimetic scaffolds, *J Biomed Mater Res A* **94** (2010), 1321–1331.

[16] J.J. Mercuri, S.S. Gill, and D.T. Simionescu, Novel tissue-derived biomimetic scaffold for regenerating the human nucleus pulposus, *J Biomed Mater Res A* **96** (2011), 422–435.

[17] G. Kumar, C.K. Tison, K. Chatterjee, P.S. Pine, J.H. McDaniel, M.L. Salit, M.F. Young, and C.G. Simon, Jr , The determination of stem cell fate by 3D scaffold structures through the control of cell shape, *Biomaterials* **32** (2011), 9188–9196.

[18] S. Panseri, C. Cunha, J. Lowery, U. Del Carro, F. Taraballi, S. Amadio, A. Vescovi, and F. Gelain, Electrospun micro- and nanofiber tubes for functional nervous regeneration in sciatic nerve transections, *BMC Biotechnol* **8** (2008), 39.

[19] P. Gunatillake, R. Mayadunne, and R. Adhikari, Recent developments in biodegradable synthetic polymers, *Biotechnol Annu Rev* **12** (2006), 301–347.

[20] X. Liu, J.M. Holzwarth, and P.X. Ma, Functionalized synthetic biodegradable polymer scaffolds for tissue engineering, *Macromol Biosci* **12** (2012), 911–919.

[21] J.P. Bruggeman, B.J. de Bruin, C.J. Bettinger, and R. Langer, Biodegradable poly(polyol sebacate) polymers, *Biomaterials* **29** (2008), 4726–4735.

[22] C.A. Vacanti, J.P. Vacanti, and R. Langer, Tissue engineering using synthetic biodegradable polymers, *Polym Biol Biomed Sig* **540** (1994), 16–34.

[23] K. Rezwan, Q.Z. Chen, J.J. Blaker, and A.R. Boccaccini, Biodegradable and bioactive porous polymer/inorganic composite scaffolds for bone tissue engineering, *Biomaterials* **27** (2006), 3413–3431.

[24] V. Guarino, F. Causa, and L. Ambrosio, Bioactive scaffolds for bone and ligament tissue, *Expert Rev Med Dev* **4** (2007), 405–418.

[25] Q. Lv, L. Nair, and C.T. Laurencin, Fabrication, characterization, and in vitro evaluation of poly(lactic acid glycolic acid)/nano-hydroxyapatite composite microsphere-based scaffolds for bone tissue engineering in rotating bioreactors, *J Biomed Mater Res A* **91** (2009), 679–691.

[26] I.O. Smith, X.H. Liu, L.A. Smith, and P.X. Ma, Nanostructured polymer scaffolds for tissue engineering and regenerative medicine, *Wiley Interdisc Rev Nanomed Nanobiotechnol* **1** (2009), 226–236.

[27] J.K. Suh and H.W. Matthew, Application of chitosan-based polysaccharide biomaterials in cartilage tissue engineering: a review, *Biomaterials* **21** (2000), 2589–2598.

[28] X. Zhang, M.R. Reagan, and D.L. Kaplan, Electrospun silk biomaterial scaffolds for regenerative medicine, *Adv Drug Deliv Rev* **61** (2009), 988–1006.

[29] S.H. Hsu, S.W. Whu, S.C. Hsieh, C.L. Tsai, D.C. Chen, and T.S. Tan, Evaluation of chitosan-alginate-hyaluronate complexes modified by an RGD-containing protein as tissue-engineering scaffolds for cartilage regeneration, *Artif Organs* **28** (2004), 693–703.

[30] R. Langer and D.A. Tirrell, Designing materials for biology and medicine, *Nature* **428** (2004), 487–492.

[31] M.N. Kumar, R.A. Muzzarelli, C. Muzzarelli, H. Sashiwa, and A.J. Domb, Chitosan chemistry and pharmaceutical perspectives, *Chem Rev* **104** (2004), 6017–6084.

[32] B.S. Kim and D.J. Mooney, Development of biocompatible synthetic extracellular matrices for tissue engineering, *Trends Biotechnol* **16** (1998), 224–230.

[33] E.S. Place, J.H. George, C.K. Williams, and M.M. Stevens, Synthetic polymer scaffolds for tissue engineering, *Chem Soc Rev* **38** (2009), 1139–1151.

[34] C.Y. Wang, K.H. Zhang, C.Y. Fan, X.M. Mo, H.J. Ruan, and F.F. Li, Aligned natural-synthetic polyblend nanofibers for peripheral nerve regeneration, *Acta Biomater* **7** (2011), 634–643.

[35] N. Bhattarai, Z. Li, J. Gunn, M. Leung, A. Cooper, D. Edmondson, O. Veiseh, M.-H. Chen, Y. Zhang, R.G. Ellenbogen, and M. Zhang, Natural-synthetic polyblend nanofibers for biomedical applications, *Adv Mater* **21** (2009), 2792–2797.

[36] G. Chan and D.J. Mooney, New materials for tissue engineering: towards greater control over the biological response, *Trends Biotechnol* **26** (2008), 382–392.

[37] R. Vasita and D.S. Katti, Nanofibers and their applications in tissue engineering, *Int J Nanomed* **1** (2006), 15–30.

[38] E.K. Yim and K.W. Leong, Significance of synthetic nanostructures in dictating cellular response, *Nanomedicine* **1** (2005), 10–21.

[39] D.M. Le, K. Kulangara, A.F. Adler, K.W. Leong, and V.S. Ashby, Dynamic topographical control of mesenchymal stem cells by culture on responsive poly(epsilon-caprolactone) surfaces, *Adv Mater* **23** (2011), 3278–3283.

[40] S.H. Lim and H.Q. Mao, Electrospun scaffolds for stem cell engineering, *Adv Drug Deliv Rev* **61** (2009), 1084–1096.

[41] R.G. Flemming, C.J. Murphy, G.A. Abrams, S.L. Goodman, and P.F. Nealey, Effects of synthetic micro- and nano-structured surfaces on cell behavior, *Biomaterials* **20** (1999), 573–588.

[42] L. Nivison-Smith and A.S. Weiss, Alignment of human vascular smooth muscle cells on parallel electrospun synthetic elastin fibers, *J Biomed Mater Res A* **100** (2012), 155–161.

[43] J.M. Holzwarth and P.X. Ma, Biomimetic nanofibrous scaffolds for bone tissue engineering, *Biomaterials* **32** (2011), 9622–9629.

[44] N. Tran and T.J. Webster, Nanotechnology for bone materials, *Wiley Interdiscip Rev Nanomed Nanobiotechnol* **1** (2009), 336–351.

[45] M.A. Pattison, S. Wurster, T.J. Webster, and K.M. Haberstroh, Three-dimensional nano-structured PLGA scaffolds for bladder tissue replacement applications, *Biomaterials* **26** (2005), 2491–2500.

[46] M. Kjaer, Role of extracellular matrix in adaptation of tendon and skeletal muscle to mechanical loading, *Physiol Rev* **84** (2004), 649–698.

[47] J.P. Arokoski, J.S. Jurvelin, U. Vaatainen, and H.J. Helminen, Normal and pathological adaptations of articular cartilage to joint loading, *Scand J Med Sci Sports* **10** (2000), 186–198.

[48] S. Mathews, R. Bhonde, P.K. Gupta, and S. Totey, Extracellular matrix protein mediated regulation of the osteoblast differentiation of bone marrow derived human mesenchymal stem cells, *Differentiation* **84** (2012), 185–192.

[49] W.L. Murphy, R.G. Dennis, J.L. Kileny, and D.J. Mooney, Salt fusion: An approach to improve pore interconnectivity within tissue engineering scaffolds, *Tissue Eng* **8** (2002), 43–52.

[50] H.J. Chung and T.G. Park, Surface engineered and drug releasing pre-fabricated scaffolds for tissue engineering, *Adv Drug Deliv Rev* **59** (2007), 249–262.

[51] J. Reignier and M. Huneault, Preparation of interconnected poly(ε-caprolactone) porous scaffolds by a combination of polymer and salt particulate leaching, *Polymer* **47** (2006), 4703–4717.

[52] Y. Huang and D.J. Mooney, Gas foaming to fabricate polymer scaffolds in tissue engineering, in *Scaffolding in Tissue Engineering* (P.X. Ma and J. Elisseeff, eds.), CRC Press, Boca Raton, FL, USA (2005), 155–165.

[53] L.D. Harris, B.S. Kim, and D.J. Mooney, Open pore biodegradable matrices formed with gas foaming, *J Biomed Mater Res* **42** (1998), 396–402.

[54] B.S. Kim and D.J. Mooney, Engineering smooth muscle tissue with a predefined structure, *J Biomed Mater Res* **41** (1998), 322–332.

[55] A.G. Mikos, Y. Bao, L.G. Cima, D.E. Ingber, J.P. Vacanti, and R. Langer, Preparation of poly(glycolic acid) bonded fiber structures for cell attachment and transplantation, *J Biomed Mater Res* **27** (1993), 183–189.

[56] W.Y. Yeong, C.K. Chua, K.F. Leong, and M. Chandrasekaran, Rapid prototyping in tissue engineering: challenges and potential, *Trends Biotechnol* **22** (2004), 643–652.

[57] S.M. Peltola, F.P. Melchels, D.W. Grijpma, and M. Kellomaki, A review of rapid prototyping techniques for tissue engineering purposes, *Ann Med* **40** (2008), 268–280.

[58] B.V. Slaughter, S.S. Khurshid, O.Z. Fisher, A. Khademhosseini, and N.A. Peppas, Hydrogels in regenerative medicine, *Adv Mater* **21** (2009), 3307–3329.

[59] J.L. Drury and D.J. Mooney, Hydrogels for tissue engineering: scaffold design variables and applications, *Biomaterials* **24** (2003), 4337–4351.

[60] D. Li and Y.N. Xia, Electrospinning of nanofibers: Reinventing the wheel? *Adv Mater* **16** (2004), 1151–1170.

[61] D.H. Reneker and I. Chun, Nanometre diameter fibres of polymer, produced by electrospinning, *Nanotechnology* **7** (1996), 216–223.

[62] F. Ko, Nanofiber technology, in *Nanomaterials Handbook* (Y. Gogotsi, ed.), CRC Press, Boca Raton, FL, USA (2006), 553–565.

[63] Y. Loo, S. Zhang, and C.A. Hauser, From short peptides to nanofibers to macromolecular assemblies in biomedicine, *Biotechnol Adv* **30** (2012), 593–603.

[64] J.Y. Lee, J.E. Choo, Y.S. Choi, J.S. Suh, S.J. Lee, C.P. Chung, and Y.J. Park, Osteoblastic differentiation of human bone marrow stromal cells in self-assembled BMP-2 receptor-binding peptide-amphiphiles, *Biomaterials* **30** (2009), 3532–3541.

[65] J.B. Matson, R.H. Zha, and S.I. Stupp, Peptide self-assembly for crafting functional biological materials, *Curr Opin Solid State Mater Sci* **15** (2011), 225–235.

[66] K.G. Cornwell, A. Landsman, and K.S. James, Extracellular matrix biomaterials for soft tissue repair, *Clin Podiatr Med Surg* **26** (2009), 507–523.

[67] S. Badylak, S. Arnoczky, P. Plouhar, R. Haut, V. Mendenhall, R. Clarke, and C. Horvath, Naturally occurring extracellular matrix as a scaffold for musculoskeletal repair, *Clin Orthop Relat Res* (1999), S333–S343.

[68] L. Zhao, M.D. Weir, and H.H. Xu, An injectable calcium phosphate-alginate hydrogel-umbilical cord mesenchymal stem cell paste for bone tissue engineering, *Biomaterials* **31** (2010), 6502–6510.

[69] R. Hartwell, V. Leung, C. Chavez-Munoz, L. Nabai, H. Yang, F. Ko, and A. Ghahary, A novel hydrogel-collagen composite improves functionality of an injectable extracellular matrix, *Acta Biomater* **7** (2011), 3060–3069.

[70] J.P. Levine, J. Bradley, A.E. Turk, J.L. Ricci, J.J. Benedict, G. Steiner, M.T. Longaker, and J.G. McCarthy, Bone morphogenetic protein promotes vascularization and osteoinduction in preformed hydroxyapatite in the rabbit, *Ann Plast Surg* **39** (1997), 158–168.

[71] C.H. Hammerle, A.J. Olah, J. Schmid, L. Fluckiger, S. Gogolewski, J.R. Winkler, and N.P. Lang, The biological effect of natural bone mineral on bone neoformation on the rabbit skull, *Clin Oral Implants Res* **8** (1997), 198–207.

[72] T.M. Chu, D.G. Orton, S.J. Hollister, S.E. Feinberg, and J.W. Halloran, Mechanical and in vivo performance of hydroxyapatite implants with controlled architectures, *Biomaterials* **23** (2002), 1283–1293.

[73] H.R. Ramay, Z. Li, E. Shum, and M. Zhang, Chitosan-alginate porous scaffolds reinforced by hydroxyapatite nano- and micro-particles: structural mechanical

and biological properties, *J Biomed Nanotechnol* **1** (2005), 151–160.

[74] S.J. Hollister, Porous scaffold design for tissue engineering, *Nat Mater* **4** (2005), 518–524.

[75] L.R. Madden, D.J. Mortisen, E.M. Sussman, S.K. Dupras, J.A. Fugate, J.L. Cuy, K.D. Hauch, M.A. Laflamme, C.E. Murry, and B.D. Ratner, Proangiogenic scaffolds as functional templates for cardiac tissue engineering, *Proc Natl Acad Sci* **107** (2010), 15211–15216.

[76] T. Osathanon, M.L. Linnes, R.M. Rajachar, B.D. Ratner, M.J. Somerman, and C.M. Giachelli, Microporous nanofibrous fibrin-based scaffolds for bone tissue engineering, *Biomaterials* **29** (2008), 4091–4099.

[77] J. Chen, J. Xu, A. Wang, and M. Zheng, Scaffolds for tendon and ligament repair: review of the efficacy of commercial products, *Exp Rev Med Dev* **6** (2009), 61–73.

[78] U.G. Longo, A. Lamberti, N. Maffulli, and V. Denaro, Tendon augmentation grafts: a systematic review, *Brit Med Bull* **94** (2010), 165–188.

[79] F. Dehghani and N. Annabi, Engineering porous scaffolds using gas-based techniques, *Curr Opin Biotechnol* **22** (2011), 661–666.

[80] X. Liu, Y. Won, and P.X. Ma, Porogen-induced surface modification of nano-fibrous poly(L-lactic acid) scaffolds for tissue engineering, *Biomaterials* **27** (2006), 3980–3987.

[81] J. Reignier and M.A. Huneault, Preparation of interconnected poly(epsilon-caprolactone) porous scaffolds by a combination of polymer and salt particulate leaching, *Polymer* **47** (2006), 4703–4717.

[82] K. Chatterjee, A.M. Kraigsley, D. Bolikal, J. Kohn, and C.G. Simon, Jr., Gas-foamed scaffold gradients for combinatorial screening in 3D, *J Funct Biomater* **3** (2012), 173–182.

[83] E. Sachlos and J.T. Czernuszka, Making tissue engineering scaffolds work. Review: the application of solid freeform fabrication technology to the production of tissue engineering scaffolds, *Eur Cell Mater* **5** (2003), 29–39 discussion 39–40

[84] M.E. Hoque, Y.L. Chuan, and I. Pashby, Extrusion based rapid prototyping technique: an advanced platform for tissue engineering scaffold fabrication, *Biopolymers* **97** (2012), 83–93.

[85] A. Gloria, T. Russo, R. De Santis, and L. Ambrosio, 3D fiber deposition technique to make multifunctional and tailor-made scaffolds for tissue engineering applications, *J Appl Biomater Biomech* **7** (2009), 141–152.

[86] J.W. Lee, P.X. Lan, B. Kim, G. Lim, and D.W. Cho, Fabrication and characteristic analysis of a poly(propylene fumarate) scaffold using micro-stereolithography technology, *J Biomed Mater Res B* **87** (2008), 1–9.

[87] N. Bhattarai, J. Gunn, and M. Zhang, Chitosan-based hydrogels for controlled, localized drug delivery, *Adv Drug Deliv Rev* **62** (2010), 83–99.

[88] D.W. Pack, A.S. Hoffman, S. Pun, and P.S. Stayton, Design and development of polymers for gene delivery, *Nat Rev Drug Discov* **4** (2005), 581–593.

[89] K.Y. Lee and D.J. Mooney, Hydrogels for tissue engineering, *Chem Rev* **101** (2001), 1869–1879.

[90] N.A. Peppas, J.Z. Hilt, A. Khademhosseini, and R. Langer, Hydrogels in biology and medicine: From molecular principles to bionanotechnology, *Adv Mater* **18** (2006), 1345–1360.

[91] R.A. Barry, R.F. Shepherd, J.N. Hanson, R.G. Nuzzo, P. Wiltzius, and J.A. Lewis, Direct-write assembly of 3D hydrogel scaffolds for guided cell growth, *Adv Mater* **21** (2009), 2407–2410.

[92] N.E. Fedorovich, J. Alblas, J.R. de Wijn, W.E. Hennink, A.J. Verbout, and W.J. Dhert, Hydrogels as extracellular matrices for skeletal tissue engineering: state-of-the-art and novel application in organ printing, *Tissue Eng* **13** (2007), 1905–1925.

[93] M.C. Roco, C.A. Mirkin, and M.C. Hersam (eds.), *Nanotechnology research directions for societal needs in 2020*, Springer-Verlag, New York, NY, USA (2011).

[94] D.H. Reneker, A.L. Yarin, H. Fong, and S. Koombhongse, Bending instability of electrically charged liquid jets of polymer solutions in electrospinning, *J Appl Phys* **87** (2000), 4531–4547.

[95] A.F. Spivak, Y.A. Dzenis, and D.H. Reneker, A model of steady state jet in the electrospinning process, *Mech Res Commun* **27** (2000), 37–42.

[96] G. Srinivasan and D.H. Reneker, Structure and morphology of small-diameter electrospun aramid fibers, *Polym Int* **36** (1995), 195–201.

[97] A. Greiner and J.H. Wendorff, Electrospinning: A fascinating method for the preparation of ultrathin fibres, *Angew Chem Int Ed* **46** (2007), 5670–5703.

[98] K. Jayaraman, M. Kotaki, Y. Zhang, X. Mo, and S. Ramakrishna, Recent advances in polymer nanofibers, *J Nanosci Nanotechnol* **4** (2004), 52–65.

[99] Z.M. Huang, Y.Z. Zhang, M. Kotaki, and S. Ramakrishna, A review on polymer nanofibers by electrospinning and their applications in nanocomposites, *Compos Sci Technol* **63** (2003), 2223–2253.

[100] V. Leung and F. Ko, Biomedical applications of nanofibers, *Polym Adv Technol* **22** (2011), 350–365.

[101] Z. Ma, M. Kotaki, R. Inai, and S. Ramakrishna, Potential of Nanofiber Matrix as Tissue-Engineering Scaffolds, *Tissue Eng* **11** (2005), 101–109.

[102] G.C. Rutledge and S.V. Fridrikh, Formation of fibers by electrospinning, *Adv Drug Deliv Rev* **59** (2007), 1384–1391.

[103] A. Cooper, N. Bhattarai, and M. Zhang, Fabrication and cellular compatibility of aligned chitosan–PCL fibers for nerve tissue regeneration, *Carbohydr Polym* **85** (2011), 149–156.

[104] C. Cunha, S. Panseri, and S. Antonini, Emerging nanotechnology approaches in tissue engineering for peripheral nerve regeneration, *Nanomedicine* **7** (2011), 50–59.

[105] V. Chaurey, P.C. Chiang, C. Polanco, Y.H. Su, C.F. Chou, and N.S. Swami, Interplay of electrical forces for alignment of sub-100 nm electrospun nanofibers on insulator gap collectors, *Langmuir* **26** (2010), 19022–19026.

[106] A. Cooper, S. Jana, N. Bhattarai, and M. Zhang, Aligned chitosan-based nanofibers for enhanced myogenesis, *J Mater Chem* **20** (2010), 8904–8911.

[107] K.T. Shalumon, K.H. Anulekha, K.P. Chennazhi, H. Tamura, S.V. Nair, and R. Jayakumar, Fabrication of chitosan/poly(caprolactone) nanofibrous scaffold for bone and skin tissue engineering, *Int J Biol Macromol* **48** (2011), 571–576.

[108] Y.Z. Zhang, J. Venugopal, Z.M. Huang, C.T. Lim, and S. Ramakrishna, Characterization of the surface biocompatibility of the electrospun PCL-collagen nanofibers using fibroblasts, *Biomacromolecules* **6** (2005), 2583–2589.

[109] Y. Zhang, H. Ouyang, C.T. Lim, S. Ramakrishna, and Z.M. Huang, Electrospinning of gelatin fibers and gelatin/PCL composite fibrous scaffolds, *J Biomed Mater Res B* **72** (2005), 156–165.

[110] Y.Z. Zhang, Z.M. Huang, X. Xu, C.T. Lim, and S. Ramakrishna, Preparation of core-shell structured PCL-r-gelatin Bi-component nanofibers by coaxial electrospinnin, *Chem Mater* **16** (2004), 3406–3409.

[111] J. Gunn and M. Zhang, Polyblend nanofibers for biomedical applications: perspectives and challenges, *Trends Biotechnol* **28** (2010), 189–197.

[112] D. Li, C. Yang, B. Zhao, M. Zhou, M. Qi, and J. Zhang, Investigation on centrifugal impeller in an axial-radial combined compressor with inlet distortion, *J Therm Sci* **20** (2011), 486–494.

[113] T.T.T. Nguyen, O.H. Chung, and J.S. Park, Coaxial electrospun poly(lactic acid)/chitosan (core/shell) composite nanofibers and their antibacterial activity, *Carbohydr Polym* **86** (2011), 1799–1806.

[114] D.G. Yu, G.R. Williams, L.D. Gao, S.W. Bligh, J.H. Yang, and X. Wang, Coaxial electrospinning with sodium dodecylbenzene sulfonate solution for high quality polyacrylonitrile nanofibers, *Colloid Interf A* **396** (2012), 161–168.

[115] J. Zhang, Y. Duan, D. Wei, L. Wang, H. Wang, Z. Gu, and D. Kong, Co-electrospun fibrous scaffold-adsorbed DNA for substrate-mediated gene delivery, *J Biomed Mater Res A* **96** (2011), 212–220.

[116] E. Zhmayev, D. Cho, and Y.L. Joo, Nanofibers from gas-assisted polymer melt electrospinning, *Polymer* **51** (2010), 4140–4144.

[117] X. Li, H. Liu, J. Wang, and C. Li, Preparation and characterization of poly(ε-caprolactone) nonwoven mats via melt electrospinning, *Polymer* **53** (2012), 248–253.

[118] J. Reignier, J. Tatibouet, and R. Gendron, Batch foaming of poly(epsilon-caprolactone) using carbon dioxide: Impact of crystallization on cell nucleation as probed by ultrasonic measurements, *Polymer* **47** (2006), 5012–5024.

[119] S. Badylak, S. Arnoczky, P. Plouhar, R. Haut, V. Mendenhall, R. Clarke, and C. Horvath, Naturally occurring extracellular matrix as a scaffold for musculoskeletal repair, *Clinical Orthop Relat Res* **367** (1999), S333–S343.

[120] M. Franklin and K. Russek, Use of porcine small intestine submucosa as a prosthetic material for laparoscopic hernia repair in infected and potentially contaminated fields: long-term follow-up assessment, *Surg Endosc* **25** (2011), 1693–1694.

[121] S. Hong and G. Kim, Electrospun micro/nanofibrous conduits composed of poly(epsilon-caprolactone) and small intestine submucosa powder for nerve tissue regeneration, *J Biomed Mater Res B* **94** (2010), 421–428.

[122] G. Feil, M. Christ-Adler, S. Maurer, S. Corvin, H.O. Rennekampff, J. Krug, J. Hennenlotter, U. Kuehs, A. Stenzl, and K.D. Sievert, Investigations of urothelial cells seeded on commercially available small intestine submucosa, *Eur Urol* **50** (2006), 1330–1337.

[123] J.P. Iannotti, M.J. Codsi, Y.W. Kwon, K. Derwin, J. Ciccone, and J.J. Brems, Porcine small intestine submucosa augmentation of surgical repair of chronic two-tendon rotator cuff tears. A randomized, controlled trial, *J Bone Joint Surg Am* **88** (2006), 1238–1244.

[124] D.L. Nettles, A. Chilkoti, and L.A. Setton, Applications of elastin-like polypeptides in tissue engineering, *Adv Drug Deliv Rev* **62** (2010), 1479–1485.

[125] W. Kim and E.L. Chaikof, Recombinant elastin-mimetic biomaterials: Emerging applications in medicine, *Adv Drug Deliv Rev* **62** (2010), 1468–1478.

[126] R.E. Sallach, W. Cui, J. Wen, A. Martinez, V.P. Conticello, and E.L. Chaikof, Elastin-mimetic protein polymers capable of physical and chemical crosslinking, *Biomaterials* **30** (2009), 409–422.

[127] R.G. Ellis-Behnke, Y.X. Liang, S.W. You, D.K. Tay, S. Zhang, K.F. So, and G.E. Schneider, Nano neuro knitting: peptide nanofiber scaffold for brain repair and axon regeneration with functional return of vision, *Proc Natl Acad Sci* **103** (2006), 5054–5059.

[128] G.M. Peretti, J.W. Xu, L.J. Bonassar, C.H. Kirchhoff, M.J. Yaremchuk, and M.A. Randolph, Review of injectable cartilage engineering using fibrin gel in mice and swine models, *Tissue Eng* **12** (2006), 1151–1168.

[129] S.R. MacEwan and A. Chilkoti, Elastin-like polypeptides: biomedical applications of tunable biopolymers, *Biopolymers* **94** (2010), 60–77.

[130] J.A. Kluge, O. Rabotyagova, G.G. Leisk, and D.L. Kaplan, Spider silks and their applications, *Trends Biotechnol* **26** (2008), 244–251.

[131] A.T. Hillel, S. Unterman, Z. Nahas, B. Reid, J.M. Coburn, J. Axelman, J.J. Chae, Q. Guo, R. Trow, A. Thomas, Z. Hou, S. Lichtsteiner, D. Sutton, C. Matheson, P. Walker, N. David, S. Mori, J.M. Taube, and J.H. Elisseeff, Photoactivated composite biomaterial for soft tissue restoration in rodents and in humans, *Science Transl Med* **93** (2011), 93ra67.

[132] J.M. Bouler, M. Trecant, J. Delecrin, J. Royer, N. Passuti, and G. Daculsi, Macroporous biphasic calcium phosphate ceramics: influence of five synthesis parameters on compressive strength, *J Biomed Mater Res* **32** (1996), 603–609.

[133] D.M. Liu, Fabrication of hydroxyapatite ceramic with controlled porosity, *J Mater Sci Mater Med* **8** (1997), 227–232.

[134] S.H. Li, J.R. De Wijn, P. Layrolle, and K. de Groot, Synthesis of macroporous hydroxyapatite scaffolds for bone tissue engineering, *J Biomed Mater Res* **61** (2002), 109–120.

[135] X. Liu and P.X. Ma, Polymeric scaffolds for bone tissue engineering, *Ann Biomed Eng* **32** (2004), 477–486.

[136] P.X. Ma, Biomimetic materials for tissue engineering, *Adv Drug Deliv Rev* **60** (2008), 184–198.

[137] T. Dvir, B.P. Timko, D.S. Kohane, and R. Langer, Nanotechnological strategies for engineering complex tissues, *Nat Nanotechnol* **6** (2011), 13–22.

[138] H. Shin, S. Jo, and A.G. Mikos, Biomimetic materials for tissue engineering, *Biomaterials* **24** (2003), 4353–4364.

[139] H.L. Holtorf, J.A. Jansen, and A.G. Mikos, Modulation of cell differentiation in bone tissue engineering constructs cultured in a bioreactor, *Adv Exp Med Biol* **585** (2006), 225–241.

[140] Y. Hu, S.R. Winn, I. Krajbich, and J.O. Hollinger, Porous polymer scaffolds surface-modified with arginine-glycine-aspartic acid enhance bone cell attachment and differentiation in vitro, *J Biomed Mater Res A* **64** (2003), 583–590.

[141] J.M. Curran, Z. Tang, and J.A. Hunt, PLGA doping of PCL affects the plastic potential of human mesenchymal stem cells, both in the presence and absence of biological stimuli, *J Biomed Mater Res A* **89** (2009), 1–12.

[142] S.R. Chastain, A.K. Kundu, S. Dhar, J.W. Calvert, and A.J. Putnam, Adhesion of mesenchymal stem cells to polymer scaffolds occurs via distinct ECM ligands and controls their osteogenic differentiation, *J Biomed Mater Res A* **78** (2006), 73–85.

[143] R.A. Marklein and J.A. Burdick, Controlling stem cell fate with material design, *Adv Mater* **22** (2010), 175–189.

[144] Y.P. Jiao and F.Z. Cui, Surface modification of polyester biomaterials for tissue engineering, *Biomed Mater* **2** (2007), R24–R37.

[145] R.S. Bhati, D.P. Mukherjee, K.J. McCarthy, S.H. Rogers, D.F. Smith, and S.W. Shalaby, The growth of chondrocytes into a fibronectin-coated biodegradable scaffold, *J Biomed Mater Res* **56** (2001), 74–82.

[146] J.E. Leslie-Barbick, J.E. Saik, D.J. Gould, M.E. Dickinson, and J.L. West, The promotion of microvasculature formation in poly(ethylene glycol) diacrylate hydrogels by an immobilized VEGF-mimetic peptide, *Biomaterials* **32** (2011), 5782–5789.

[147] U. Hersel, C. Dahmen, and H. Kessler, RGD modified polymers: biomaterials for stimulated cell adhesion and beyond, *Biomaterials* **24** (2003), 4385–4415.

[148] R. Langer, Drug delivery and targeting, *Nature* **392** (1998), 5–10.

[149] X. Wang, E. Wenk, X. Zhang, L. Meinel, G. Vunjak-Novakovic, and D.L. Kaplan, Growth factor gradients via microsphere delivery in biopolymer scaffolds for osteochondral tissue engineering, *J Control Release* **134** (2009), 81–90.

[150] K. Ladewig, Drug delivery in soft tissue engineering. *Expert Opin Drug Deliv* **8** (2011), 1175–1188.

[151] G. Wei, Q. Jin, W.V. Giannobile, and P.X. Ma, The enhancement of osteogenesis by nano-fibrous scaffolds incorporating rhBMP-7 nanospheres, *Biomaterials* **28** (2007), 2087–2096.

[152] D.P. Go, S.L. Gras, D. Mitra, T.H. Nguyen, G.W. Stevens, J.J. Cooper-White, and A.J. O'Connor, Multi-layered microspheres for the controlled release of growth factors in tissue engineering, *Biomacromolecules* **12** (2011), 1494–1503.

[153] X. Shi, Y. Wang, R.R. Varshney, L. Ren, Y. Gong, and D.A. Wang, Microsphere-based drug releasing scaffolds for inducing osteogenesis of human mesenchymal stem cells in vitro, *Eur J Pharm Sci* **39** (2010), 59–67.

[154] J. Bonadio, E. Smiley, P. Patil, and S. Goldstein, Localized direct plasmid gene delivery in vivo: prolonged therapy results in reproducible tissue regeneration, *Nat Med* **5** (1999), 753–759.

[155] D.D. Hile, M.L. Amirpour, A. Akgerman, M.V. Pishko, and Active growth factor delivery from poly(D, L-lactide-*co*-glycolide) foams prepared in supercritical CO(2), *J Control Release* **66** (2000), 177–185.

[156] C.J. Bettinger, R. Langer, and J.T. Borenstein, Engineering substrate topography at the micro- and nanoscale to control cell function, *Angew Chem Int Ed* **48** (2009), 5406–5415.

[157] S.C. Owen and M.S. Shoichet, Design of three-dimensional biomimetic scaffolds, *J Biomed Mater Res A* **94** (2010), 1321–1331.

ABOUT THE AUTHORS

Princeton Carter is a Master's candidate in the chemical, biological, and bioengineering department at North Carolina Agricultural and Technical State University. He holds a BS in biomedical engineering from the University of Virginia, Charlottesville, VA. His research interests include the fabrication and characterization of bioscaffolds for various tissue engineering applications. He intends to pursue a degree in medicine upon graduation.

Narayan Bhattarai serves as an assistant professor of bioengineering at North Carolina A&T State University (NCAT), Greensboro, NC. He graduated from Tribhuvan University of Nepal with an MSc degree in chemistry. He obtained his PhD in materials engineering from Chonbuk National University, South Korea. His graduate research was on the synthesis and characterization of biodegradable polymers for biomedical applications. Before joining NCAT, he was a postdoctoral researcher at the University of Washington, Seattle, where his research projects included developing tissue engineering scaffolds and diagnostic nanoparticles for cancer imaging and treatments. Currently, his research and teaching interests are in biomaterials, polymer science, nanomedicine, and tissue engineering.

Surface Modification for Biocompatibility

Erwin A. Vogler

Department of Materials Science and Engineering,
Pennsylvania State University, University Park, PA 16802, USA

Prospectus

The principal motivation behind surface engineering and modification for improved biocompatibility of a biomaterial is to control interactions of the biomaterial with components of living systems or subsets thereof in a manner that mimics the normal physiological state or produces a desired change in biological state. This pursuit of biomimicry is discussed in this chapter within the context of the core mechanisms of the biological response to materials. A tutorial on surfaces, interfaces, and interphases leads to the identification of specific targets for surface engineering and modification. These targets include water wettability (surface energy), surface chemistry, surface chemical patterns and surface textures, and surface presentation of biomimetic motifs. The chapter concludes with a discussion of the essential conceptual tools required for building a biomaterials surface science laboratory, illustrated with an example of modifying surfaces for improved cardiovascular biomaterials.

Keywords

Biocompatibility, Biological response, Biomaterials, Surface chemistry, Surface energy, Surface engineering, Surface modification, Surface texture, Wettability

8.1 INTRODUCTION

8.1.1 The Healthcare Pyramid

Biomaterials are materials used in the fabrication of medical devices and various tools of modern biotechnology. Examples include sterile disposables such as syringes, specimen collection tubes/plates, and a wide variety of containers used in animal cell culture and microbiology; temporary implants such as access ports, catheters, and sutures; and long-term implants generally intended to service the recipient for a lifetime, such as dental abutments, heart valves, or orthopedic plates and screws. Collectively, these biomedical devices comprise a *healthcare pyramid* when arranged according to volume in use and technical sophistication, as shown in Figure 8.1.

Sterile-disposable medical-procedure devices are used by the billions worldwide and form the base of the healthcare pyramid. These devices are critical components of modern medical practice and biotechnology. Indeed, sterile disposables and the biomaterials from which these are fabricated

Engineered Biomimicry

http://dx.doi.org/10.1016/B978-0-12-415995-2.00008-8

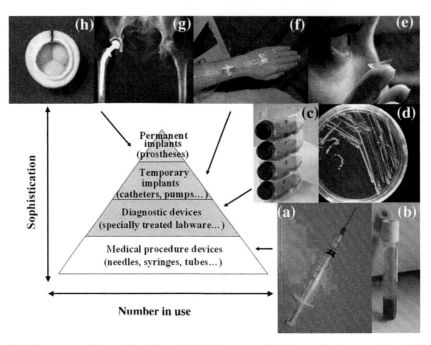

FIGURE 8.1 A *healthcare pyramid* organizing biomedical devices according to number in use and sophistication. The base of the healthcare pyramid comprises the ubiquitous sterile-disposable medical devices such as syringes (a) and specimen collection tubes (b) that underlie modern medical practice. These medical devices are used in the billion annually and represent the greatest volume application of biomaterials. At a slightly higher level of technical sophistication are diagnostic devices that have surface treatments and/or additives such as tissue-culture flasks (c) and microbiological plates (d). Temporary implants such as contact lenses (e) and peripheral catheters (f) lie yet higher in the pyramid, with commensurate technical sophistication and lower-volume use of biomaterials. The highest level of sophistication at the apex of the pyramid is occupied by permanent implants such as artificial hips (g) and heart valves (h). Boundaries between strata of the healthcare pyramid are not actually sharp as diagrammed but rather overlap to a significant degree, depending on actual end-use application. Images taken from Wikimedia Commons with permission to copy, distribute, or modify.

represent the quintessential component of human and veterinary healthcare without which medical practice would effectively collapse. One only needs to reflect on dire situations created by natural disasters, poverty, or war that limit access to sterile disposables to see how critical these components of the healthcare pyramid really are. The technical sophistication of sterile disposables may be rudimentary compared to that employed in higher strata of the healthcare pyramid, but the absolute technological sophistication should not be underestimated, especially with regard to high-volume manufacturing. The underlying technology is not simple, merely taken for granted.

Diagnostic devices are slightly more sophisticated than medical-procedure devices because biomaterials used in fabrication typically require special surface treatments or packaged culture media of some kind. Examples shown in Figure 8.1 are tissue-culture labware for the growth of animal cells in the laboratory and microbiological plates for detection of bacterial pathogens. As with the medical-procedure devices, diagnostic devices are used in very high volume and are core to routine delivery of modern healthcare. Diagnostic devices and medical-procedure devices together represent the largest volume usage of biomaterials.

The technical sophistication of implants, whether temporary or permanent, is higher than that of sterile disposables and diagnostic devices simply because contact with biology (a patient) is longer in duration. Also, implanted biomaterials can be, and typically are, in simultaneous intimate contact with several tissue types. For example, catheters and access ports providing semi-routine access to the vascular system contact epidermis, dermis, muscle, vascular tissue, and blood. Devices such as peripheral catheters shown in Figure 8.1 might be in use for a few days to many months. Very common temporary implants of a kind are contact lenses, used daily by millions of people worldwide.

Technical issues associated with long-term implants that repair dysfunctional parts of the body to restore health and mobility are yet more complex, confounded by issues of wear in service. Failure of orthopedic implants (arthroplasty) or artificial heart valves, shown in Figure 8.1 as examples, can require inconvenient and expensive replacement surgeries or can be life threatening for the recipient. Advanced medical devices of the future, such as a total-replacement artificial heart, have extremely stringent service requirements because failure is catastrophic [1], but they will become available as soon as biomaterials can be designed that are well suited for these ultra-sophisticated applications.

8.1.2 Biomaterials and Biocompatibility Defined

It is apparent from the preceding discussion that biomaterials must meet performance criteria that vary considerably with the medical or biotechnical application. Stringency requires very careful definition of terms, engineering requirements, and measures of performance against these requirements. As a consequence, biomaterial scientists take great effort to carefully define all terms in use. A biomaterial is currently defined as "a substance that has been engineered to take a form which, alone or as part of a complex system, is used to direct, by control of interactions with components of living systems, the course of any therapeutic or diagnostic procedure" [2]. Key phrases in this definition are "engineered to take a form" and "control of interactions with components of living systems." It is evident from these key words that a biomaterial is designed for a particular application with the intent to control interactions with cells, fluids, and tissues.

If a biomaterial meets these carefully defined engineering requirements, the material is said to be *biocompatible*. It is important to stress that biocompatibility is not a material property but rather a measure of how successfully the biomaterial meets different engineering requirements for very different biomedical applications. That is to say, there is no single material that is biocompatible in all applications. Biocompatibility is defined as "the ability of a material to perform with an appropriate host response in a specific application" [3]. Here, the term *host* broadly refers to biological cells, tissues, and fluids. Perhaps the cells are cultured in the laboratory (*in vitro*) or are cells that comprise certain tissues of a patient's body (*in vivo*). Therein lays the primary technical challenge of biomaterials science: how to create materials that are biocompatible.

Effectively, engineering that leads to "control of interactions with components of living system" is a bit of biomimicry. Ideally, the biomaterial integrates with and functions in the biomedical context in a manner that mimics the biological circumstance. As a relatively simple example (but not at all trivial), consider the just-mentioned act of growing cells taken from an animal tissue in the laboratory. This is typically performed in various kinds of sterile-disposable flasks or dishes for which an engineering requirement frequently is that the cells attach and proliferate on the inside surface of these containers. Here, the interaction between the cell and the biomaterial *in vitro* should in some way mimic the *in vivo* interaction among cells and supporting surfaces occurring in the tissue from which cells were extracted. A more complex example is an orthopedic implant, such as a hip replacement. Here the objective is to have the implant integrate with bone tissue, forming a strong bond between the implant surface and bone.

8.1.3 Scope

The foregoing examples serve to broadly introduce the scope and topic of this chapter: how to modify the surface of biomaterials to achieve biocompatibility. First on the agenda is to identify exactly what a surface is and why it is important to biomaterial performance. Second is a discussion of how surfaces influence the "biological response" to artificial materials that controls biocompatibility, to the extent this is currently understood. Third, examples of modifying surfaces to achieve biocompatibility are discussed within the context of the basic mechanisms of biocompatibility. Importantly, this chapter does not provide a review of the substantial literature on biomaterial surface-modification techniques. That task has been admirably carried out in many good books [4–8] and review articles [9,10] appearing elsewhere. The agenda concludes with a discussion of the general technology that enables a researcher to engage in biomaterial surface modification and a specific example using ideas and technology discussed in this chapter.

8.2 THE SURFACE REGION

8.2.1 Ordinary and Scientific Definitions

The ordinary dictionary definition of a surface given in Box 8.1 is effectively the same as that used in surface science. A surface is the topmost boundary of a material that has thickness no less than the atoms or molecules that occupy this boundary region. The mathematical definition of a surface as a planar region with no thickness

---

> ### BOX 8.1
>
> ## A DICTIONARY DEFINITION OF A SURFACE
>
> sur·face (sûr"fəs) *n. Abbr.* **sur. 1. a.** The outer or the topmost boundary of an object. **b.** A material layer constituting such a boundary. **2.** *Mathematics.* **a.** The boundary of a three-dimensional figure. **b.** The two-dimensional locus of points located in three-dimensional space. **c.** A portion of space having length and breadth but no thickness. (*Merriam-Webster*, www.merriam-webster.com)

> ### BOX 8.2
>
> ## A SCIENTIFIC DEFINITION OF A SURFACE
>
> The outermost region of a material that is chemically and/or energetically unique by virtue of being located at a boundary.

---

is a simplified concept of the surface region. In reality, of course, the surface region is three-dimensional because atoms and molecules are three-dimensional.

An important question asks just how thick the surface region is. A useful answer to this question is given in Box 8.2. Accordingly, a surface is not defined by an arbitrary thickness such as a nanometer or micrometer but rather as a chemical or energetic distinction from bulk-phase properties. As long as the surface region is a small portion of the total system, the beginnings and ends of the surface region can be determined, at least in principle, by comparing properties of the bulk phase to any test plane. Start in the middle of the material and proceed outward in any direction, comparing properties until at last chemical or energetic differences are found. However, when the surface region is a substantial part of the system because the bulk phase has been significantly reduced in volume (a large surface-area-to-volume ratio), the definition in Box 8.2 becomes ambiguous. Such a circumstance arises in very thin films when an upper bounding surface comes within close proximity to the lower bounding surface and there is little *bulk* in between these bounds. Otherwise, for most macroscopic applications of materials, the conceptual definition in Box 8.2 works well. In most practical applications of surface science it is not necessary to know the surface-region thickness in absolute terms, but it is important to realize that the surface region is very different from anywhere in the bulk phase of a material.

8.2.2 Surface Energy

Surfaces are in a unique energetic predicament because atoms or molecules at the boundary are deprived of nearest-neighbor interactions otherwise enjoyed in the bulk phase. Irving Langmuir, widely regarded as the father of modern surface science and namesake of the American Chemical Society journal of surface science *Langmuir*, realized this by stating in his landmark papers of 1916 [11, 12]: "Since energy must be expended in breaking apart a solid, the surfaces of solids must contain more potential energy than do the corresponding number of atoms in the interior. Since this potential energy is probably electromagnetic energy in the field between atoms, the inter-atomic forces are more intense on the surface than in the interior." Even though at this early time in the history of science the six fundamental intermolecular interactions

---

> ### BOX 8.3
>
> ## SURFACE ENERGY
>
> An intensive thermodynamic property of a material that arises from the loss of nearest-neighbor interactions among atoms or molecules at the boundary. This excess energy most prominently manifests itself in adhesion and adsorption reactions at the surface.

---

[13] had not yet been clearly identified, Langmuir understood that the energy expended (work done) to separate atoms or molecules along a plane (by cleavage, for example) left the two surfaces so created in a state of excess energy compared to the energetic state of identical atoms or molecules in an equivalent plane within the bulk phase. Thus it is apparent that surface energy is the excess energy per unit area of boundary plane (ergs/cm^2 or mJ/m^2) and is an intensive thermodynamic property of materials (Box 8.3).

Surface energetics are quite large and are not to be ignored in many important technical applications (such as adhesives, biomaterials, colloids, paints, etc.), as well as in nearly any circumstance when surfaces come in contact with liquids, solids, or vapors.

On thinking of atoms as ball-bearing-like sphere lying on a plane (Figure 8.2a), it becomes quickly evident that a particular ball bearing can have six nearest neighbors when close-packed into a group that comprises a small portion of a hypothetical condensed-phase material. This close-packed arrangement of ball bearings is a conceptual model of atoms at a planar surface. That is to say, an atom or molecule at the surface of the hypothetical material has six nearest neighbors in the plane of the surface. But these surface atoms would also be in contact with the

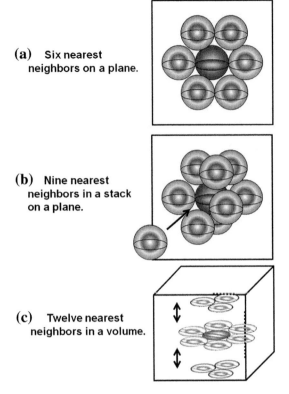

(a) Six nearest
neighbors on a plane.

(b) Nine nearest
neighbors in a stack
on a plane.

(c) Twelve nearest
neighbors in a volume.

FIGURE 8.2 Nearest-neighbor interactions among spherical atoms comprising a hypothetical material. Figure 8.2a shows that, on a planar surface, a central atom (darker color) can be surrounded by 6 nearest neighbors (lighter color) in a close-packed arrangement. An adjacent layer brings 3 additional nearest neighbors into contact with the central atom (Figure 8.2b, arrow points to the location of a third atom displaced to reveal the central atom). Figure 8.2c illustrates the packing arrangement relative to the central atom immersed in the bulk of the hypothetical material. Accounting leads to the conclusion that a surface atom (Figure 8.2b) interacts with 9 nearest neighbors, whereas a bulk atom interacts with 12 nearest neighbors, causing the surface atom to be deprived of energy-stabilizing interactions. This leads to excess energy at the surface and is the origin of surface energy.

atoms of the second layer of the material, which would be in contact with the third layer and so forth, propagating further into the *bulk phase*. This effect can be modeled by adding a second

layer of ball bearings in contact with the surface atoms comprising the six close-packed planar arrangements (Figure 8.2b). Inspection reveals that the central ball bearing in Figure 8.2b would have three additional ball bearings lying in the interstices between contacting neighbors in the plane.

Thus, the central surface atom has six nearest neighbors in the plane and three additional nearest neighbors in a contacting second layer, or a total of nine nearest neighbors (Figure 8.2b). The bulk phase can be modeled by placing a central ball bearing between two such layers of contacting ball bearings (Figure 8.2c). Inspection of this arrangement reveals that the central ball bearing and its neighbors within the plane are sandwiched between three upper and three lower neighbors, yielding a total of 12 nearest neighbors (three on the bottom, three on the top, and six in the plane). So then, in summary, a surface ball bearing has nine nearest-neighbor interactions, whereas a bulk-phase ball bearing enjoys 12 nearest-neighbor interactions. The surface ball bearing is thus bound by $9/12 = 3/4$ of the energy of a bulk ball bearing. In other words, the surface ball bearing has $1/4 = 25\%$ excess energy compared to a ball bearing in the bulk phase [14].

Although this ball-bearing model has a very limited range of applicability to real-world materials, it does give a sense of scale of surface energetics. Surface energy is about 25% of the cohesive energy of materials that must be overcome to create a surface. Thus, most or all of the energy required to cleave a material ends up as surface energy. It definitely takes work to create a surface. Surface energy falls within a broad range depending on material characteristics, between approximately 10–1,000 mJ/m². When surfaces are a large fraction of the system (large surface area, as occurs with particles or thin films) or when surface reactions are important, as in biomaterials, then surface energetics are necessary to take into careful consideration (Box 8.4).

BOX 8.4

PROTEINS

Biological macromolecules comprising amino acids that serve as antibodies, enzymes, hormones, and structural components of cells and tissues. The *proteome* is the entire set of proteins made by an organism that subsumes thousands of different kinds of proteins. Proteins are Mother Nature's tools for accomplishing nearly all physiological functions; proteins are Mother Nature's agents of change.

BOX 8.5

THE BIOMATERIAL INTERFACE

A pseudo two-dimensional region separating the physical surface of a biomaterial from a contacting aqueous biological milieu that instantly arises because of the interaction with water that can cause the redistribution of ions near the surface.

8.2.3 Why Surfaces are Important in Biomaterials

Water is the biological solvent system and the majority molecule in all biological environments [15]. Biology is dissolved or suspended in an aqueous solution of various ions (such as Na^+, Mg^{+2}, and Cl^-) and many different biological macromolecules, such as proteins and sugars. For example, human blood is approximately 45% by volume water and salts and about 10% by volume proteins [16]. There are more than 1,000 different types of proteins in blood, varying some 10 decades in concentration [17, 18]. Such mixtures are not typically chemically defined in a precise way and are frequently referred to as a *biological milieu* or simply a *milieu*. This milieu may also contain cells of various kinds. Human blood is about 45% by volume "formed elements" that include platelets, red and white blood cells (Box 8.5).

8.2.3.1 The Interface

An immediate consequence of bringing a biomaterial into contact with a biological milieu is interaction with water. Water is very small, only about the size of atomic oxygen, and hydrates surfaces to varying degrees depending on surface chemistry/energy [19]. These hydration reactions

create a thin pseudo two-dimensional zone of water directly adjacent to the surface, referred to as *interfacial* or sometimes *vicinal* (near to) water [20]. Vicinal-water chemistry is different from that of water within the bulk solution due to the physicochemical interactions with the surface that constitute hydration reactions. The exact thickness of the vicinal-water region no doubt depends on surface chemistry of the biomaterial. The structure/reactivity of vicinal water is both an important and a controversial subject in materials science [21, 22]. But the evidence seems to be that the vicinal-water region is probably no thicker than a few layers of water molecules.

The reason that vicinal-water chemistry is thought to be influential in the biological response to biomaterials is that water solvent properties correlate with the extent of hydrogen bonding [21, 22]. Water is a relatively poor solvent at low temperatures near the density maximum (at 3.98 °C) because nearly all hydrogen bonds are used up in a self-associated network of water molecules. At 0 °C, water ice is entirely self-associated by hydrogen bonds that form the familiar ice crystal structure. By contrast, water steam (100 °C) is quite corrosive because nearly all hydrogen bonds are available to participate in solvation reactions. Thus it may be anticipated that changes in hydrogen bonding induced by contact with surfaces at ambient temperatures

---

BOX 8.6

THE BIOMATERIAL INTERPHASE

A three-dimensional region separating the physical surface of a biomaterial from a contacting aqueous biological milieu wherein important physicochemical reactions that catalyze, mediate, or moderate the biological response to the biomaterial occur.

BOX 8.7

PROTEIN ADSORPTION

Adsorption can concentrate Mother Nature's agents of change (Box 8.4) within the biomaterial interphase, conferring biological activity to a synthetic material that affects biocompatibility.

---

will have a significant effect on vicinal-water solvent properties [19] that, in turn, will influence the distribution of ions near the water-contacting surface [21–23] and possibly affect pH within the vicinal-water region. A biological entity such as a protein or a cell entering the vicinal-water region can encounter significantly different water chemistry than experienced in bulk solution, depending on the extent to which self-association of vicinal water has been affected by the presence of the surface (Box 8.6).

8.2.3.2 *The Dynamic Interphase*

Subsequent to the initial hydration reactions mentioned previously, macromolecules such as proteins might adsorb to the hydrated surface, creating a complex and truly three-dimensional region referred to as an *interphase* [20, 22]. The term *interphase*, rather than *interface*, is used to emphasize that this region can be significantly thicker than the vicinal-water region (interface) created by surface hydration. Adsorbed proteins can form multilayers depending on biomaterial surface properties [16], and with molecular diameters in the 5–10 nm range, multiple layers of proteins might constitute an interphase that is tens of nm thick [24] (Box 8.7).

Protein concentration within the interphase can be very high, much higher than bulk solution and, in fact, even higher than protein solubility

limits [16, 21]! Water concentration within the interphase is commensurately lower than bulk solution since two objects, water and protein molecules, cannot occupy the same place at the same time. The viscosity of the interphase is also quite different from that of bulk solution [25]. Interphase chemistry is very different from bulk solution chemistry in nearly every way.

Not all biomaterial surfaces adsorb protein, however, at least in the early phase of material contact with a biological milieu. As hydrophilicity increases, protein adsorption decreases to vanishing quantities near water wetting, characterized by a water contact angle $\theta \rightarrow 65°$ for most types of materials. (See Ref. 20 for a discussion of water wetting relevant to biomaterial applications.) Minerals and other surfaces with ion-exchange properties are an exception to this general rule [16].

Water molecules are so strongly bound to hydrophilic materials ($\theta < 65°$) that protein cannot displace water from the interphase and enter the adsorbed state. The biological response to hydrophilic materials is observed to be quite different than that to hydrophobic materials ($\theta > 65°$), presumably because of the influence of adsorbed proteins [16, 21, 26, 27]. Examples of this hydrophilic/hydrophobic contrast in the biological response to materials [21, 22, 27] include contact activation of blood coagulation [28] and mammalian cell adhesion [26]. Long-term contact of a material with a biological milieu might invite considerable biological processing, as in, for example, attachment of cells that excrete and deposit proteins directly onto a surface through a process quite different from adsorption from solution.

Presumably, then, at least in the immediate or acute phase of interaction of a biomaterial with a biological milieu, interphase thickness depends on biomaterial water-wetting characteristics, with a thick proteinaceous interphase near hydrophobic materials and a thinner interphase (interface) near hydrophilic materials. Regardless of biomaterial hydrophilicity, we can anticipate that the interphase is a dynamic region with chemistry quite different from that of bulk solution. The term *dynamic* has been applied here because proteins might both adsorb and desorb, causing a flux of water, proteins, and associated ions into, and out of, the interphase. The overall biological response to materials is really a response to interphase chemistry.

8.2.4 The Biological Response to the Dynamic Interphase

The observed macroscopic biological response to biomaterials can, and most usually does, involve many complex biological and biochemical reactions, even in the relatively simple application of sterile disposables that are not implanted into a body. These reactions can include triggering of linked enzyme reactions that can amplify the biological response to a biomaterial, such as occurs in blood coagulation [29]. Cells can participate in biological response to material by becoming activated or by adhering to the surface. The initial surface hydration and adsorption reactions occurring when the biomaterial surface is first immersed into a biological milieu are thus obscured in whole or part by these secondary reactions.

In general, the biological environment is quite corrosive and the biomaterial can degrade with time. Also, in general, biological and biochemical reactions propagate outward from the surface of the biomaterial and can cause systemic effects distant from the surface. An example is formation

of a blood clot on the surface of a catheter (an embolus) that detaches from the catheter due to shear stress under blood flow, travels in the bloodstream (embolization), and becomes lodged in a distal part of the circulatory system (an arm or a leg), thereby causing swelling and pain (thrombophlebitis) or sometimes death if the clot lodges in a critical organ such as the brain (stroke).

8.2.4.1 Williams' Four Components of Biocompatibility

The foregoing pattern of events is summarized by *Williams' Four Components of Biocompatibility*, diagrammed in Figure 8.3 to represent the essence of David Williams' original description [30], here modified to emphasize a cascade of causes and effects that spans both time and space. The time coordinate ranges from the nanoseconds involved in biomaterial surface hydration reactions, through the milliseconds involved in protein adsorption (*acute* or short term), to full service life of the biomaterial lasting hours, days, or years (*chronic*, or long term). The spatial coordinate ranges from a few tenths

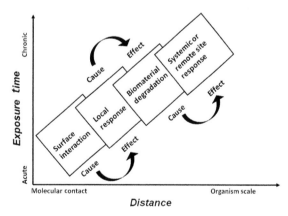

FIGURE 8.3 Diagrammatic representation of Williams' Four Components of Biocompatibility concept. A cascade of cause-and-effect reactions is caused by bringing a biomaterial into contact with a biological milieu, propagating along both spatial and temporal coordinates. Both coordinates span several orders of magnitude, making the biological response to materials a very complex biophysical phenomenon. See also Figure 8.4.

of nanometers associated with water molecules (molecular contact) to the scale of the biomedical device or patient into whom an implant might be placed (organism scale). An example is thrombophlebitis, mentioned in Section 8.2.4. The initial molecular interactions occurring at the catheter surface lead to embolus formation that dislodges from the catheter surface, traveling to a distal site. Clearly, then, the biological response to materials spans a diversity of time and length scales and, as such, is a very complex biophysical phenomenon.

Sometimes the overall biological response can be favorable, or at least not entirely unfavorable. But in most situations the linked cause-and-effect reactions of Figure 8.3 lead to poor biocompatibility. Generally speaking, the higher in the healthcare pyramid of Figure 8.1 the biomaterial application falls, the more difficult it is to achieve biocompatibility. As examples, sterile-disposable cell-culture containers are widely and successfully used in biotechnology and would thus be considered biocompatible. But there are no materials that do not induce blood coagulation to some degree when used in a total-replacement artificial heart, and this is a major

impediment to developing such advanced cardiovascular medical devices. Most materials eventually become encapsulated in fibrous scar tissue when implanted into soft tissues, greatly complicating functionality of implanted electronic sensors, such as a glucose sensor to be used in treatment of diabetes. Williams' Four Components of Biocompatibility make it very clear that biocompatibility must be engineered at the first step of Figure 8.3. Otherwise, the cause-and-effect cascade leading to the systemic biological response will lead to poor biocompatibility. The following section examines this critical first surface interaction step in somewhat more detail.

8.2.4.2 Target of Surface Modification

Figure 8.4 is a kind of descriptive chemical reaction that provides more detail to the surface interaction step of Figure 8.3. According to the preceding discussion, the cause of the biological response to materials is the formation of a dynamic interphase. Whereas all of the various steps leading to this interphase might be amenable to comprehension in detail someday, none is subject to any proactive manipulation to

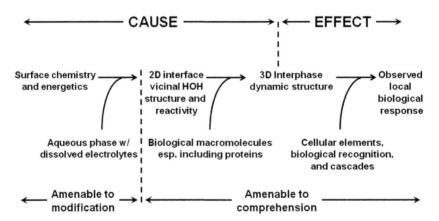

FIGURE 8.4 Descriptive chemical reaction expanding the surface interaction step of Williams' Four Components of Biocompatibility (Figure 8.3), identifying formation of a dynamic interphase region as the primary cause of the local biological response. Whereas the steps leading to the local biological response to a biomaterial may someday be amenable to full comprehension, only the surface chemistry that interacts with the aqueous phase in establishing vicinal-water structure and reactivity can be modified to control biocompatibility.

BOX 8.8

THE BIOLOGICAL
RESPONSE TO
MATERIALS

Surface hydration is the first step in the interaction of a biomaterial with a biological milieu that controls protein adsorption and all other subsequent steps in the biological response to materials.

improve biocompatibility because it is the properties of water at the surface resulting from the initial hydration reaction that control all downstream events. The only step that is subject to manipulation is the first one—the interaction of water with the biomaterial surface (Box 8.8).

This then is the target of engineering biocompatibility—learning to manipulate surface chemistry so that the dynamic interphase region created upon immersion of a biomaterial into a biological milieu leads to "control of interactions with components of living systems," as discussed in the Introduction to this chapter. This engineering must, of course, be specific to the nature of the milieu in which it is immersed. That is to say, engineering biocompatibility has to be tailored to the biomedical application, as discussed in Section 8.1.2.

Both Figures 8.3 and 8.4 may be broadly descriptive of the biological response to materials, but neither provides the detail required to prospectively design surface-engineered solutions to biocompatibility issues. At this stage in the development of biomaterials science, these details remain well out of reach and are part of cutting-edge research in biomaterials surface science [16]. Stated in the language of materials science, structure-property relationships linking material properties to utility in biomedical applications are effectively unknown. Nevertheless,

in spite of this lack of knowledge and because of the great socioeconomic value of biomaterials, design-directed and trial-and-error approaches have been applied to engineer biocompatibility. The remainder of this chapter presents applications of surface modification of biomaterials, pointing out, to the extent currently possible, how surface modification affects biocompatibility in reference to the concepts presented in Figures 8.3 and 8.4.

8.2.5 Limitations of Trial and Error

Before discussing specific applications of surface modification, it is of interest to contemplate why design-directed and trial-and-error approaches to engineering biocompatibility are not entirely adequate for biomaterials development. That is to ask, why is it necessary to understand the fundamentals summarized by the descriptive chemical reaction of Figure 8.4? After all, design-directed engineering with trial-and-error experimentation was how all of the biomaterials used in the medical devices represented in the healthcare pyramid of Figure 8.1 were developed. What value is a fundamental understanding of biocompatibility?

One answer to this probing question is that more of the same approach can be expected to invent more of the same sort of biomaterials. But we need much better biocompatibility for advanced medical devices; more of the same is simply not adequate to meet evolving medical needs. There is, of course, the possibility of a happenstance breakthrough discovery by trial and error, and this would be highly desirable. But other than functionality for the particular purpose under investigation, how would we know how to leverage this breakthrough in a general way? At issue here is that discovery without understanding the science underlying discovery provides no basis for prospective optimization or generalization. If a material just happens to be biocompatible for a specific application, there is no learning that teaches how to make that miracle happen again in another application. Conversely, and much more frequently, a particular material is found not be biocompatible, and even failure does not teach how or why.

A second problem with purely design-directed and trial-and-error approaches to engineered biocompatibility is that invention alone does not readily allow prediction of liability in end use. This is extremely important to medical-device manufacturers who need good estimates of service lifetimes and the ability to predict modes of failure. A case in point is the extensive litigation associated with silicone breast implants [31–33]. Patients claimed that connective-tissue disease, autoimmune responses, and pain were among many other health issues associated with silicone breast implants. A lengthy litigation took place involving a number of companies. Among the many serious outcomes of this litigation, Dow Corning was forced into bankruptcy, negatively affecting the many stakeholders in this company. Finally, after nearly a decade of legal and scientific wrangling, the Institute of Medicine of the National Academy of Science concluded that there was no direct relationship between silicone implants and major diseases [34]. Although there are many facets to this long and complex story, the need to define and measure liability clearly stands out.

A third reason why prospective engineering is much needed is that it seems that most of the low-hanging fruit has been picked. Most of the materials used in medical devices, especially the sterile disposable devices, have been in existence for about 50 years [35]. Few new materials are on the horizon. Methods of tissue engineering promise a new approach to implanted medical devices, but it seems unlikely that tissue engineering can be used to make medical devices in high volume, such as the temporary implants or sterile disposables of Figure 8.1. And even so, tissue engineering relies on materials as scaffolds and rigid supports. Development of materials at the nanoscopic level is a very active area

of research at this writing, and many promising pathways toward engineered biocompatibility have been uncovered. But especially with these new material forms, liability in use is important to thoroughly understand.

Finally on this topic, it is to be admitted that full reliance on trial-and-error development of biomaterials bankrupts the future of biomaterials because, in the words of David Williams, " it is difficult to optimize biomaterial characteristics through purely physiology-based assay techniques because all that can be observed are macroscopic medical outcomes—not molecular interactions with biomaterials that lead to these outcomes" [30]. Without understanding core mechanisms of biocompatibility, biomaterials development is condemned to an uncertain future controlled by serendipity.

8.2.5.1 Medical Mediocrity

Yet another downside to purely discovery-based materials development is that there is little economic motivation to improve on what happens to work or to find out why the material works well in the first place. As a consequence, biomaterials comprising commodity medical devices generally meet minimal medical requirements, and that is all. Healthcare works around the remaining biocompatibility issues that cause mediocre medical-device performance.

A primary reason why few new biomaterials enter the medical field is that it is difficult to economically justify continuous improvement in medical-device performance unless these improvements substantially reduce time and expense of medical care. Patients and medical doctors might want continuous medical-device improvement, but the fact of the matter is that patients and doctors are seldom the customer who can demand such improvements. Much more frequently, the choices of medical devices are made by hospital purchasing departments or insurance companies that focus on medical costs and return on investment. Introduction of new biomaterials into healthcare most usually accompanies a new medical device with high value in end use that can substantiate the high cost of research and development, testing, and building of new manufacturing lines.

An example is furnished by polyurethanes widely used in catheter manufacturing. It was discovered that certain polyurethane formulations just happen to work well as catheters. This is not to say that all polyurethanes are biocompatible or that existing polyurethane catheters are perfect [36]. Certainly, academic and some industrial researchers have invested considerable effort in improving polyurethane biocompatibility and to understand why these materials happen to exhibit these excellent properties [37]. But because the original aromatic polyurethane formulations *Pellethane*, *Biomer*, and *Tecoflex,* introduced more than 20 years ago by the Upjohn, Ethicon, and Thermetics corporations, respectively [38], worked fairly well, these are the formulations still in widespread use by catheter manufacturers today. Companies and technologies have been bought and sold, but aromatic polyurethane formulations remain basically the same. Effectively, what has happened is that trial and error led to polyurethane formulations that meet basic medical and manufacturing requirements. The success of these imperfect materials effectively blocks continuous improvement, and catheter development is frozen in time.

8.3 SURFACE MODIFICATION OF BIOMATERIALS

The primary motivation behind surface modification is that materials with desirable physical properties, such as strength or flexibility, are frequently not biocompatible due to adverse surface-mediated reactions with the biological milieu. Modifying the surface allows medical-device designers to retain desirable physical properties while improving biocompatibility to a useful level. There are many methods of surface modification ranging from chemical modification to coatings, sometimes used in combination. Surface modification is also used as a tool to study the biological response to materials. A range of surface chemistries or energies incrementally sampling the available range can be prepared, allowing researchers to explore how the biological response to materials changes in response to these different surface properties [19–22]. Surface modification is a very important tool in the medical device designer's kit.

As mentioned in Section 8.1.3, there are several books [4–8] and review articles [9, 10] that supplement a substantial literature detailing specific biomaterial surface modification techniques and methods. The following subsections neither repeat these summaries nor review recent literature in detail. Rather, the scope of this section is to discuss broad classes of biomaterial surface modification methods in the context of the mechanism of biocompatibility diagrammed in Figures 8.3 and 8.4.

8.3.1 Wettability

One of the early discoveries in the development of biomaterials was that water *wettability* (surface energy) had a great effect on the biological response. It is unclear just when and for what purpose the first discovery of the effect of water wettability was made, but certainly changing the surface chemistry of materials used in culturing animal cells in the laboratory, called tissue culture, discussed further in Section 8.3.1.1, must have been among the earliest [26].

Altering water wettability of biomaterial surfaces influences adsorption and adhesion—the two most important manifestations of surface chemistry/energy (Box 8.3). Interpreted in terms of Figure 8.4, changing water wettability affects the chemistry of vicinal water that guides formation of the dynamic interphase upon contact with a biological milieu. Exactly how this occurs is not known. Understanding the various molecular interactions involved and how these molecular interactions control the

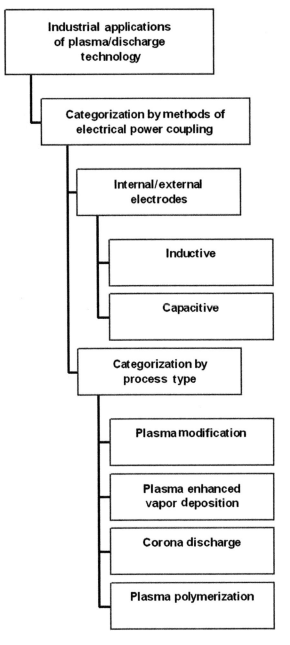

systemic biological response is among the most important fundamental research problems in biomaterials surface science [16]. Whatever the mechanism, altering biomaterial hydrophilicity works like magic in many applications.

The hydrophilic/hydrophobic terminology used to describe water wetting has caused considerable confusion in biomaterials because these are relative terms with no universally accepted reference scale [19, 39]. A sharp bifurcation in biomaterial surface properties near a water wettability characterized by a 65° contact angle serves as a convenient dividing line that distinguishes hydrophilic from hydrophobic for biomaterials applications, as mentioned in Section 8.2.3.2.

8.3.1.1 Reactive Gas-Discharge Surface Treatment

Soon after Rappaport's pioneering studies of mammalian cell adhesion [40–42] using chemical methods to vary surface chemistry, a variety of surface synthesis strategies were explored, ranging from use of liquid-phase chemical oxidants [43, 44] to the application of gas-discharge surface treatments [45]. These latter developments set the precedent for widespread application of modern gas-discharge (plasma) technology [46, 47] in biomedicine [48, 49] and biotechnology [50, 51]. Various kinds of gas-discharge technologies coarsely categorized in Figure 8.5 have become an essential tool in surface modification of biomaterials. For example, sterile-disposable polystyrene tissue-culture dishes and flasks, shown in Figure 8.1, are almost universally oxidized using oxygen plasma technology at high manufacturing speeds [20]. Certain kinds of soft contact lenses are treated similarly to prevent lenses from sticking to

FIGURE 8.5 A coarse categorization of industrial applications of plasma/discharge technology in biomaterials modification. Two levels of organization, categories of electrical power coupling and process type, are shown. Internal and external electrodes can be used in both inductive and capacitively coupled modalities. Among the discharge types, plasma modification and corona discharge are widely used to affect biomaterial wettability. Plasma-enhanced vapor deposition and plasma polymerization are related methods that deposit smooth, conformal coatings on biomaterial surfaces.

the eye [52, 53]. Reactive gas plasmas can also be used to deposit thin, conformal coatings on medical devices by a process referred to as *plasma-enhanced vapor deposition and plasma polymerization* [54]. Some applications of plasma polymerization are followed by oxidation via an oxygen discharge to create hydrophilic, thin conformal coatings.

As an example, polystyrene used in the manufacture of tissue-culture labware has excellent physical properties such as high clarity and can be molded at high speed, but it is inherently hydrophobic. This is a big problem because, generally speaking, mammalian cells do not prosper on hydrophobic surfaces [26, 55, 56]. Plasma oxidation renders this surface hydrophilic and conducive to the adhesion of cells. Exactly why hydrophilicity is essential to mammalian cell adhesion is not known with clarity [26,57–59], but the outcome is nevertheless pronounced. Therefore, plasma-treated, sterile disposables have become essential tools in modern biotechnology.

8.3.1.2 The Biological Response to Hydrophilicity

Hydrophilicity does not improve biocompatibility of all medical devices, as might be anticipated from the fact that biocompatibility is different for different biomedical or biotechnical applications. For example, hydrophilic surfaces are potent activators of blood plasma coagulation [28] and therefore not likely biomaterial candidates for cardiovascular devices. On the other hand, hydrophobic surfaces adsorb proteins such as fibrinogen that can activate platelets that cause blood coagulation.

This contrast in the behavior of biology in contact with surfaces leads to the speculation that there are two basic kinds of biological response to changing water wettability: a Type I response mediated by adsorbed protein, and a Type II response in which adsorbed protein does not play a significant role [16, 21, 22]. At the dividing line between hydrophilic and hydrophobic, the biological response to materials appears to be muted, perhaps because surfaces near this dividing line do not perturb the structure and reactivity of vicinal water [27]. It is to be emphasized that whereas this speculation is possibly true for the acute biological response, chronic exposure of a surface to a biological milieu is very likely to be much more complex and subject to *processing,* represented by biomaterial degradation in Figure 8.3. Also, this trend in water wetting (surface energy) substantially ignores the role of surface chemistry, contemplating only the effects of surface energy but not how this surface energy is created by surface chemistry. This brings us to the subject of chemically defined surfaces.

8.3.2 Well-defined Surface Chemistry

Surface engineering is a term that came into popular use in the mid- 1980s, usually in reference to creating well-defined surface chemistries using self-assembled monolayers (SAMs). Various organic compounds, such as silanes on silicon oxides (glass) [60] or thiols/sulfides on metal-coated substrata [61], can be used to create dense monolayers presenting synthetically defined surface functionalities, such as carboxyl or hydroxyl groups, at the surface. In this way, surfaces can be engineered with a high degree of chemical precision. These developments are a cornerstone of modern nanotechnology [62–64]. But the 1952 work of Shafrin and Zisman with alcohols and amines on platinum [65, 66] significantly predates SAM methods. And before Shafrin and Zisman, Blodgett and Langmuir were preparing organized layers of fatty acid salts using what would come to be called the *Langmuir-Blodgett film balance* [67, 68]. And well before Blodgett and Langmuir, Pockels was spreading monolayers of oil on the surface of water [69–71]. But the introduction of thiol-on-gold SAMs by Allara *et al.* greatly expanded the range of chemistries that could be explored and opened use of SAMs to the general scientific community because preparation methods are as simple as dip coating.

8.3.2.1 The Biological Response to Well-Defined Surface Chemistry

Well-defined surface chemistry was of great interest to biomaterial surface science because there was a widespread opinion within the biomaterials community that poor biocompatibility was related to heterogeneous surface chemistry [72]. It was thought that well-defined, homogeneous surface chemistries might provide a route to improved biocompatibility. Although studies of the biological response to SAMs did not lead to a general solution to the biocompatibility problem, these studies conclusively showed that the outermost surface functional groups that influence water wetting are directly responsible for the nature and intensity of the biological response to material surfaces. By using SAM chemistry, all other aspects of the material and surface could be held constant while varying only the terminal functional group; Refs. 4, 7, and 8 provide good summaries of what now is a broad field of endeavor.

But exactly how, or if, terminal-functional group chemistry influences the observed biological response to a SAM surface, over and above the commensurate variation in water wettability, is not at all clear. That is to ask, does the chemical specificity of a particular functional group shine through the interaction with water and influence the biological response more than, say, oxygen-plasma treatment to an equivalent water contact angle? To the extent that Figure 8.4 is true, one might not expect to observe such chemical specificity unless the chemical functionality had an extraordinary affect on local pH or ionic strength beyond the capacity of a simple water contact angle to measure.

There are relatively few studies that measure a biological response to surfaces with well-defined chemistry, incrementally sampling a full range of surface energy that allows chemical specificity to be isolated from surface energy effects. A primary reason is that making a number of well-defined surfaces with incrementally changing water wettability is experimentally

quite difficult. Early literature provided some evidence that a particular surface functional group—hydroxyl or carboxyl, for example—was stimulatory to cell adhesion and proliferation [59, 73–76] over other functional groups. However, it has proven difficult in subsequent research to clearly separate cause and effect in the cell adhesion/proliferation process, especially in the ubiquitous presence of proteins, and by doing so unambiguously separate surface chemistry from all other influences (such as surface energy/water wettability) [55, 77].

The most general rule connecting material properties with cell-substratum compatibility emerging from decades of focused research is that anchorage-dependent mammalian cells strongly favor hydrophilic surfaces [20–22, 57, 58, 78–80], as mentioned in Section 8.3.2. The surface chemistry that attains this wettability seems secondary or possibly unimportant.

Work from my laboratory has compared the catalytic potential to induce blood plasma coagulation by contacting plasma with various SAM surfaces, finding that a basic trend in surface energy was followed with little or no evidence of chemical specificity. The only exception appeared to be SAM surfaces with terminal carboxyl functionalities that were more activating than anticipated based on a purely water-wettability trend [81]. This latter effect may well be related to low surface pH related to $-CO_2H$ ionization [82, 83]. Surfaces bearing strong Lewis acid/base functionalities are found to exhibit extraordinary protein-adsorption capacity [84] and unique blood-contact behavior [85], but this effect is related to exceptional ion-exchange properties of functionalities such as sulfopropyl ($-CH_2 - SO_3^-$), carboxymethyl (CH_2COOH), iminodiacetic acid ($HN(CH_2CO_2H)_2$), quaternary ammonium (NR_4^+), or dimethyl aminoethyl ($(CH_3)_2N$ (CH_2CH_2)) [86].

Thus it would appear from this limited survey that chemical specificity of surface functional groups in the biological response to materials can be observed only in those cases when that chemistry is highly unusual (such as ion-exchange properties), or perhaps when ordinary chemistry is densely packed on the surface (such as carboxyl-terminated SAMs). There is no doubt that surface chemistry and wettability are inextricably convolved properties, because it is the hydrogen bonding of water to surface functional groups that most profoundly influences wettability; see, as examples, Refs. 19–21. Wetting and surface chemistry are not separate factors, as is sometimes asserted in the biomaterials literature [87]. Indeed, surface chemistry is responsible for wetting properties. But it appears that sometimes surface chemistry does shine through the generic effect on water wettability.

Returning to Figure 8.4, any surface chemistry, whether homogeneous or heterogeneous, interacts with water in a manner particular to that surface chemistry, and a dynamic interphase will be formed by contact with a biological milieu. Unless that interphase happens to properly direct subsequent interactions with the biological milieu, a homogeneous surface chemistry will not necessarily lead to improved biocompatibility. Although SAM chemistries yet may prove to be a necessary part of an effective surface engineering strategy, well-defined surfaces are not likely to be sufficient in and of themselves. A biomimetic component seems necessary to carry out the function "control of interactions with components of living systems."

8.3.3 Surface Chemical Patterns and Textures

Not long after it was discovered that surface chemistry can influence the biological response to materials, the idea of creating patterns of chemistry at various size scales on a surface was pursued (see Refs. 88–90 as examples), leading to a rapidly inflating literature on the subject for many different applications [91]. The idea here is that an alternating presentation of a particular surface chemistry mixed with a second particular chemistry, such as a checkerboard pattern,

for example, would influence the biological response to a material in a different way than observed using a macroscopic surface of either surface-chemical scheme. Likewise, topographical features (surface textures) have been found to influence the behavior of biology at textured surfaces [92]. Biomaterials quickly embraced this approach to engineering biocompatibility and functionality [93–95].

The length scale of patterns can range from centimeter to nanometer [93], depending on application and purpose. For example, surface chemistry or texture patterns can be used to create islands onto which cells specifically adhere in a sea of non-adherent surface [96]. When surface patterns approach the size scale of cells or biological molecules, the biological response can be much more than the sum or average of effects obtained on macroscopic surfaces consisting of a single surface chemistry or texture type. As examples, it is found that microscale chemical patterns on a surface can control cell shape and phenotypic behavior [90]. Nano-patterned surface chemistry is found to greatly affect blood plasma coagulation induced by contact with these surfaces compared to activation of coagulation by equal surface area of macroscopic uniform chemistries comprising the pattern [97]. Likewise, it is found that activation of blood plasma coagulation induced by surfaces with molecularly dispersed chemical functionalities is different than a physical mixture of uniform surface chemistry with the same net hydrophilicity [98].

8.3.3.1 The Biological Response to Surface Chemical Patterns and Textures

The motivation behind deliberately imposed surface heterogeneity in the form of surface patterns and textures is almost the opposite of that discussed in Section 8.3.2 for SAM surfaces. The motivation for well-defined surfaces was that chemical heterogeneity was responsible for poor biocompatibility. The motivation behind ordered patterns of heterogeneity is that heterogeneity will somehow direct favorable surface interactions that ultimately lead to improved biocompatibility, as these interactions propagate through the chain of cause and effect outlined in Williams' Four Components of Biocompatibility, diagrammed in Figure 8.3.

Using Figure 8.4 as a guide, we can surmise that the size scale of the pattern relative to the size of interacting constituents will be very important. If the size of the pattern is large compared to the scale of these constituents, then it seems likely that the net effect will be more like a weighted average of biological responses to pure, macroscopic-pattern constituents. This seems to be evident in cell responses to surfaces. Cell shape and phenotypic response seem most pronounced when the scale of the surface feature is some fraction of cell size [90, 99–101]. The effects of patterning at the size scale of proteins are much less clear. Not only are nanoscale patterns technically challenging to make and characterize [93], but also understanding this surface interaction phase of the biological response remains well outside the grasp of modern biomaterial surface science.

As pattern size decreases, the relative contribution of edges between domains necessarily increases. The transition in chemistry between edges seems likely to be important in the orientation of proteins that differentially adsorb to surfaces with different surface chemistry/energy [102]. If molecular simulations are a guide, it can also be expected that the structure and reactivity of water at molecular edges will be quite different than within bulk solution [103]. It seems reasonable to speculate that macroscopic biological responses to nano-patterned surfaces such as blood plasma coagulation, mentioned in the preceding section, might be traced to such difficult-to-characterize phenomena.

8.3.4 Biomimetic Surface Engineering

Biomimetic surfaces take design cues from the biophysical and/or biochemical properties observed in nature in an attempt to accomplish what nature does through a purely synthetic strategy. Examples include engineering super-hydrophobic properties of lotus leaves onto surfaces [104–110] and immobilizing biological molecules with specific directed function, or even immobilizing living cells [111–113]. Biomimetic molecules include glycoproteins, peptides, phospholipids and proteins (enzymes), and saccharides. Many different strategies can be employed to achieve biomimicry, including nonspecific adsorption of macromolecules and covalent bonding to a surface, frequently using chemical grafting reactions [114–116] or through application of plasma technology, briefly mentioned in Section 8.3.2.1 [116–118]. Many applications of biomimetic surfaces are not explicitly designed for improved biocompatibility of biomaterials, as defined in Section 8.1.2, but a level of compatibility with biology in general is required to retain biological activity.

8.3.4.1 Biomimetic Biomaterials

As applied to biomaterials, biomimetic surfaces are intended to guide the biological response by recruiting proteins and cells, typically to influence integration and healing of implants. In those cases when the activity of a particular biological molecule is known to direct a particular desirable biological response, these specific molecules can be immobilized on a surface. An example is immobilization of bone morphogenic protein (BMP) onto orthopedic and dental implant materials to induce bone in-growth and healing of the implant into surrounding bone. BMPs constitute a family of growth factors that induce formation of bone and cartilage. There are many similar biological macromolecules that are known to exhibit specific stimulatory effects.

An alternative to immobilizing biologically active molecules onto a surface is to create a surface that specifically recognizes and binds selected proteins from the biological milieu into which the biomimetic biomaterial is placed [117, 119, 120]. Here chemical or plasma-based reactions are used to template a nanocavity with an interior surface that exactly matches a particular protein used to create the nanocavity. These *template chemistry* techniques offer a number of advantages over biomolecule immobilization including preparation of a dry surface that does not require the special storage or handling that may be required for immobilized biological molecules.

8.3.4.2 The Biological Response to Biomimetic Surfaces

Biomimetic surfaces do not subvert the sequence of steps outlined in Figure 8.4, but these surfaces can amplify or initiate biological responses that might not occur to ordinary materials. If the biomimetic surface bears immobilized biomolecules, it most likely is already hydrated because desiccation can cause loss in bioactivity. In this case, the initial hydration reactions have already occurred before the biomimetic surface is immersed in the biological milieu. Nevertheless, it can be expected that there will be an exchange of ions between the water of hydration and the biological milieu, leading to vicinal-water region that is different than the water of hydration. Thereafter, the biomimetic surface will be subject to the same physicochemical rules that control adsorption of proteins as apply to any other surface. The dynamic interphase, which includes immobilized biological molecules, interacts with the biological milieu through adsorbed protein if protein adsorbs to these surfaces. And, in a manner paralleling the discussion of water interaction with a surface functional group on a SAM surface in Section 8.3.3.1, the immobilized biological molecule must shine through water and adsorbed protein to interact with constituents of the biological milieu. The fact that biological cells can specifically interact with surfaces bearing immobilized RGD amino acid sequences (arginine-glycine-aspartic acid or similar) or bind to various so-called *adhesion molecules* (proteins such as cadherins, neural cell adhesion molecules or N-CAMs,

integrins, and selectins) [99, 121, 122] suggests that immobilized biological molecules are indeed active in this regard. Another example is that immobilized heparin can control blood coagulation [123].

The interaction of template surfaces with the biological milieu is more like considerations already outlined in Sections 8.3.2 and 8.3.3. The dry template undergoes the instantaneous hydration reactions that establish a vicinal-water layer that interacts with the proteins of the biological milieu. In this case, however, the nanocavity created by templating can recognize target proteins through shape selectivity and hydrogen-bonding reactions through a *recognition-of-the-fittest* effect [117] that leads to adsorption selectivity of the target protein(s) that may subsequently direct a desirable biological response through "control of interactions with components of living systems."

The full power of the biomimetic surfaces will be unleashed if and when the sequence of steps of the biological response labeled *amenable to comprehension* in Figure 8.4 can be understood in a way that identifies the biological molecules most important in obtaining a desired biological response. After all, one cannot mimic what one cannot observe. Template chemistry has particular potential in rational design of biomaterials because it promises to deliver biological selectivity without the technical difficulties imposed by fragile biological molecules, attributes that greatly facilitate manufacturing and distribution of medical devices lying low in the healthcare pyramid of Figure 8.1.

8.4 A BIOMATERIALS SURFACE SCIENCE LAB OF YOUR OWN

It seems appropriate to close this chapter with a brief discussion of the tools required to successfully engineer surfaces or to modify surfaces of existing materials for improved biocompatibility. The intent of doing so is not to create a list of specific equipment required or to design laboratory space but rather to discuss the concepts underlying surface engineering of biocompatibility. Experience shows that there are three primary capabilities that a biomaterials surface science laboratory must have: (1) the ability to prepare surfaces, incrementally varying the characteristic to be modified, (2) surface characterization tools that quantify the incrementally varying characteristic, and (3) a method of quantifying a biological response of interest to this incrementally varying surface characteristic.

The essential measurement relates the dose (surface characteristic or independent variable) to the response (biological response or dependent parameter) in a way that quantitatively explores the available response surface. Ideally, these data are amenable to mathematical modeling. The resulting mathematical model enables statistical fitting and interpolative optimization and possibly is a basis for the formulation of hypotheses stating how the linked cascade of causes and effects shown in Figures 8.3 and 8.4 work. Without this quantitative information, surface engineering or modification is effectively unguided and little more than grasping at straws from a very large pile.

8.4.1 Incrementally Varying Surface Characteristics

The surface characteristic to be varied depends on the kind of surface engineering or modification to be explored, as outlined in Section 8.3. For examples, if water wettability (surface energy) is the primary surface characteristic to be modified (Section 8.3.1), then methods that incrementally vary water contact angle over the observable 0°–120° range, or at least some substantial portion of this range, are essential. Polymers can frequently be oxidized with short bursts of an oxygen or air plasma [20] (Section 8.3.1.1) to increase wetting from the (typically) native hydrophobic state. Glass in a native hydrophilic state can be silanized with silanes bearing different terminal functional groups that incrementally sample the water-wetting range. Similarly, SAMs can be deposited on gold-coated substratum (Section 8.3.2), possibly prepared by mixing two or more different thiols with different terminal functionalities (e.g., OH mixed with CH_3 [124, 125]. A multiplicity of wetting properties can sometimes be created using chemical gradients that vary with position on a gradient panel (such as along the length of a microscope slide) [126, 127], sometimes combined with a surface-texture gradient [128] (Section 8.3.3).

Methods of creating surface patterns and textures cited in Section 8.3.3 can be used to vary the areal density of a pattern. Returning to the checkerboard analogy mentioned in Section 8.3.3, surface coverage of the black squares can be systematically varied relative to the red squares, creating a surface that ranges from all black to all red in stages. Similarly, textures can, in principle, range from entirely smooth to entirely textured at different degrees, or the texture can assume different rugosity in depth, sometimes prepared using plasma-processing methods mentioned in Section 8.3.1.1. Incremental variation in biomimetic surface engineering, discussed in Section 8.3.4, can follow the same basic plan for surface patterns and textures by varying the areal density of the immobilized biomimetic molecule or, in the case of templating, the surface coverage of the molecular motif(s) templates.

8.4.2 Quantitative Surface Characterization

Surface characterization of biomaterials is technically challenging. The ideal characterization method for biomaterials offers three attributes: surface sensitivity, biomedical interpretability, and biomedical relevance. As shown in the Venn diagram of Figure 8.6, the intersection of

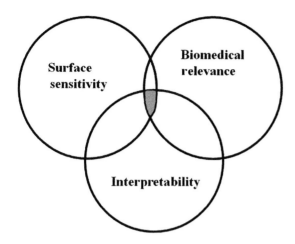

FIGURE 8.6 Venn diagram expressing the overlap of surface analytical techniques simultaneously offering surface sensitivity, biomedical relevance, and interpretability of the analytical information in the context of biomedical applications. The overlap is small, perhaps vanishingly small, creating considerable analytical difficulties in biomaterials surface science.

analytical techniques where these ideal characteristics converge is very small, perhaps vanishingly small. This is to say, that there are numerous methods that, on an individual basis, may offer any one, or possibly two, of these ideal attributes, but there are few that offer all three. A restatement of the old project-management adage "quick, cheap, good…choose any two" relevant to biomaterials surface characterization would be "sensitive, interpretable, relevant… choose any two (or maybe one)."

Surface-sensitivity requirements are at the upper nanometer or so of the surface, where the terminal functional groups reside that influence the biological response to materials, as discussed in Sections 8.2 and 8.3.2.1. Surface sensitivity at this level is all but limited to high-vacuum spectroscopies [129, 130] and a few relatively new and sophisticated optical spectroscopies, such as secondary harmonic generation (SHG) and sum frequency generation (SFG) techniques [131–134].

Biomedical interpretability requires that the data gathered by the surface analytical method comes in a form that can be interpreted in terms of the biological response to material surfaces. It is not immediately evident that chemical composition detected by spectroscopies, or morphological evidence obtained by microscopy, or the surface energetics detected by tensiometry are immediately applicable to understanding the biological response. After all, knowing chemical composition, even in exquisite detail, does not provide the mechanistic information captured in Figure 8.4 that permits formulation of structure-property relationships [77]. The same is true for surface energetic and morphological lines of evidence.

Perhaps the most stringent attribute of the three ideal analytical criteria stated in Section 8.4.2 is that of biomedical relevance, as can be appreciated by considering Figure 8.4 in this regard. Biomedical relevant characterization obtains analytical results from hydrated surfaces, is sensitive to the energetics that drive adsorption and adhesion to hydrated surfaces, and is applicable to a chemically undefined milieu. That is to say, it would be beneficial to somehow characterize the hydrated biomaterial surface in contact with the biological milieu rather than in the vacuum of an electron microscope or spectrometer [77].

8.4.3 A Quantitative Measure of the Biological Response

Biomedical devices are used in specialized biotechnical applications or come into contact with specific physiological compartments. For example, sterile disposables used in the culture of animal cells in the laboratory come into contact with cells and protein-containing media, ophthalmic materials contact the ocular environment, and blood contacts vascular grafts and artificial heart valves. From this end-use environment, meaningful measures of the biological response, hopefully predictive of biocompatibility in use, can be deduced and ways of quantifying this response be devised. Related examples include measurement of cell attachment and proliferation kinetics [56, 135]; measurement of protein adsorption from artificial tear fluid onto contact lens materials [136–141]; and the potential of materials to activate blood plasma coagulation can be assessed using plasma coagulation methods [28, 81].

8.4.4 Interpreting the Data

Because surface-analytical methods are seldom ideal in the sense of the intersection of Figure 8.6, a means of utilizing analytical information that may only be indirectly related to the biological response to materials must be found. This situation is hardly unique in science because, in general, reality is so complex that we can measure only certain aspects of the phenomenon under study. Relationships among the intensity of the phenomenon (the dependent parameter or response) and variables under experimental control (the independent parameters or dose) are inferred by plotting dose-response on Cartesian coordinates. A cause-and-effect relationship between dose and response is then inferred from a curvilinear interpolation between authentic data points or lack of cause and effect is suggested if no such interpolation is apparent.

Effectively, these inferred dose-response relationships are a kind of calibration curve. Figure 8.7 diagrams two complementary levels of calibration that relate surface characteristics (Section 8.4.2) to a quantitative measure of the biological response (Section 8.4.3). For example, a measure of cell adhesion might be plotted against water/buffer wettability of the substratum surface to which cells attach [26, 55].

Substratum surface energy can be incrementally varied by methods suggested in Section 8.4.1 and the number of cells attached to this surface measured according to some protocol [56, 135]. As long as the substratum exhibits a particular

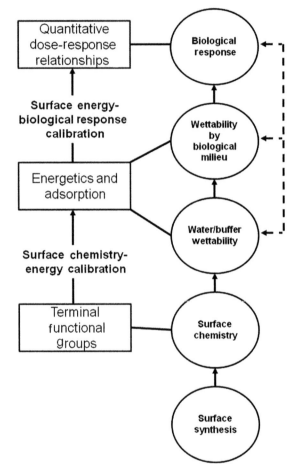

FIGURE 8.7 A data interpretation scheme that relates the biological response to materials to surface chemistry and energy. Surface synthesis (engineering and modification) creates a surface chemistry that can be quantified by one of a number of surface analytical methods measuring terminal functional group chemistry. Surface composition is not directly connected to the biological response but rather influences surface energetics that are responsible for the adhesion and adsorption events underlying the biological response to materials (see Figures 8.3 and 8.4). A surface chemistry-energy calibration quantifies these connections, which, in turn, can be related to quantified measures of the biological response to materials from which dose-response (cause and effect) can be inferred.

level of water wetting, it can be expected that cells will adhere to the extent indicated by a calibration curve plotting cell adhesion against substratum wettability (dashed arrows connecting water/buffer wetting to the biological response in Figure 8.7). No doubt, this kind of calibration curve is used by manufacturers of sterile-disposable tissue-culture flasks and dishes as a rapid quality-control tool. According to Figure 8.4, this calibration scheme works because water wetting is predictive of interphase characteristics that lead to bioadhesive outcomes, at least over the short time interval of the acute biological response to materials. A similar strategy has been applied to blood plasma coagulation induced by contact with materials to infer a relationship between surface energy and the propensity of blood plasma to coagulate (i.e., clot) [28, 98].

A limitation of this calibration approach to biomaterials surface modification is that it does not reveal how the observed correlation between water wetting and the biological response occurs. A full appreciation of the part of Figure 8.4 labeled amenable to comprehension would minimally require an inventory of proteins adsorbed to the surface, if any. Since this remains out of our collective grasp [16], it seems that our understanding will remain at the level of calibration curve. Nevertheless, it is generally possible to measure surface wetting by the biological milieu understudy (plasma, serum, cell-culture fluids, etc.) that includes the effect of protein adsorption. Compared to wetting in pure water or buffer (dashed arrows on the right side of Figure 8.7), the effect of adsorption on interfacial energetics is obtained in a phenomenological way that does not require an inventory of adsorbed proteins. A surface energy-biological response calibration made at this level is that much closer to the cause and effects suggested by Figure 8.4, and a reproducible correlation provides some clues about the nature of the dynamic interphase that guides the observed response. For example, a comparison of relationships between mammalian cell adhesion and water wettability in the presence and absence of proteins provides evidence that initial stages of

bioadhesion are controlled by surface energetics and that adsorbed proteins affects bioadhesion by altering surface energetics of the adhesive process [79, 80, 135]. The role of protein adsorption in controlling surface-contact-induced coagulation of blood plasma has been clarified using a similar approach [28, 98].

Surface-analytical methods measuring chemical composition are at least one step removed from the cause-and-effect approach suggested by Figure 8.4, because surface chemistry is not in and of itself predictive of water/buffer wettability or the surface energetics that drive interfacial phenomena of adsorption and adhesion. A two-stage calibration scheme suggested by Figure 8.7 can be used in this instance [77]. Surface engineering or modification (Section 8.3) creates surface chemistry that can be characterized by any one of many surface analytical tools, such as electron spectroscopy (ESCA) or secondary ion mass spectroscopy (SIMS); see, for examples, Refs. 95 and 130. Tensiometric measurement of water wettability of these surfaces enables a primary calibration that relates surface chemistry to surface energy (for example, oxygen composition measured by ESCA to water contact angles [20]). Measurement of surface wetting by the authentic biological milieu under study expands this primary calibration by including the effects of protein adsorption as mentioned earlier.

When surfaces bear immobilized biomimetic functional groups, such as functional biological molecules or templates of biological motifs, data interpretation almost assuredly will be at the level of a calibration. In this situation, dose-response relationships are vastly more complex because of the inclusion of biological activity. The surface does not acquire biological activity only through processes outlined in Figure 8.4 but rather is biologically active before imposition into the biological milieu. The reactivity of the surface is thus not directly related to physics and chemistry but involves a level of biological activity that physics and chemistry do not necessarily predict. In this case, the calibration might relate the biological response to the number of immobilized biomimetic groups per unit area, perhaps as measured by radiometry, for example. Alternatively, net enzymic activity of the surface might be correlated with the biological response.

8.4.5 An Example

The following is an example relating wettability of engineered surfaces to the adsorption of blood factor XII and the related activation of blood plasma coagulation. This example is based on methods and results reported in Refs. 81, 142, and 143 and illustrates some of the principles discussed in the preceding section, showing how surface modification can be related to a complex biological response to these surfaces. Some brief background about blood coagulation and protein adsorption is helpful as an introduction.

8.4.5.1 Contact Activation of Blood Coagulation

Blood coagulates in contact with all biomaterials [28]. The exact reasons for this are not entirely known. Thus, the contact activation of blood coagulation problem has been, and continues to be, an active area of research in biomaterials surface science. Understanding the molecular details of contact activation is crucial to development of advanced cardiovascular biomaterials that occupy the upper strata of the healthcare pyramid diagrammed in Figure 8.1.

An important discovery made in the late 1960s (see Refs. 144–148 and citations therein for historical reviews) was that a particular protein, blood factor XII, sometimes referred to as *Hageman factor* or *FXII*, became activated by contact with hydrophilic surfaces, producing an enzyme (or possibly enzymes) that potentiate a cascade of biochemical reactions that ultimately causes blood plasma to coagulate or clot [28]. It was postulated that FXII adsorbs or assembles onto

activating surfaces together with other proteins, constituting a reaction complex that ultimately produces the principle activated enzyme FXIIa from FXII. It is thus essential to understand the adsorption properties of FXII and FXIIa toward designing surfaces that do not induce blood coagulation. Likewise, understanding protein adsorption from the plasma milieu is crucial to understanding how FXII adsorbs to surfaces in competition with more than a thousand other proteins that constitute the blood plasma proteome [18].

8.4.5.2 Adsorption Mapping

There are many methods of measuring protein adsorption from purified solutions of a single protein, but few are designed to measure protein adsorption from binary solutions, let alone a mixture of more than a thousand proteins [16]. A modification of standard surface thermodynamics was invented that correlates protein adsorption to the water wettability of an adsorbent surface, which is equally applicable to purified protein solutions and biological milieu such as blood plasma [142]. This graphical method was termed *adsorption mapping* and has been applied to the blood coagulation problem [20–22, 81, 143].

Briefly, an adsorption map is the plot of the difference in adhesion tension $[\tau' - \tau^o]$ against τ^o, where $\tau' = \gamma'_{lv} \cos \theta'$ and $\tau^o = \gamma^o_{lv} \cos \theta^o$. Adhesion tension measures the strength of adhesion of a droplet of fluid to a test surface in mJ/m^2 exhibiting an advancing or receding contact angle θ on that surface. The liquid-vapor (lv) interfacial tensions γ_{lv} are measured separately from θ. The prime superscript denotes a solution containing a protein or many proteins comprising a milieu of interest whereas the "o" superscript denotes pure buffer solution containing no proteins. Thus, the difference $[\tau' - \tau^o]$ is an adsorption index that measures the effect of protein adsorption on surface wetting, with positive values corresponding to protein-adsorbent surfaces and zero or negative values corresponding to surfaces that do not adsorb protein (see Ref. 142 and citations therein for more details). The adsorption map is a kind of calibration curve relating protein adsorption to the surface energy (water wettability) of the adsorbent.

Plotting $[\tau' - \tau^o]$ against τ^o is a way of accomplishing the correlations indicated by the dashed arrows on the upper portion of Figure 8.7 that readily identify the surface energy (water wettability) at which protein adsorption does not occur. Figure 8.8 is an adsorption map for purified FXII and FXIIa corresponding to each purified protein at 300 μg/mL, nearly 10X the nominal physiological concentration of FXII [149, 150]. Thermodynamic boundary conditions enclose all observable data for a particular protein or surfactant system in what is termed an *adsorption triangle* (shaded portion of Figure 8.8). According to the theory of adsorption maps, $[\tau' - \tau^o] > 0$ is characteristic of surfaces that support adsorption, whereas surfaces that do not support adsorption are characterized by $[\tau' - \tau^o] \leq 0 \text{ mJ/m}^2$. Inspection of Figure 8.8

leads to the immediate observation that there is no measurable difference between FXII and FXIIa in adsorption to solid surfaces. Furthermore, it can be concluded from these data that *neither FXII nor FXIIa adsorb at surfaces exhibiting* $\tau^o > 40 \text{ mJ/m}^2$ ($\theta \sim 55^o$) because the trend line through the data passes through $[\tau' - \tau^o] = 0$ near this point.

Figure 8.9a is an adsorption map corresponding to 10% plasma from which it is also noted that plasma proteins, taken as a whole, do not

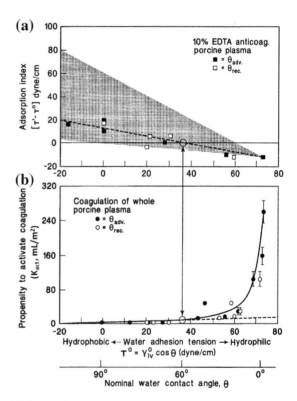

FIGURE 8.9   An adsorption map for 10% EDTA anticoagulated porcine plasma correlated with a quantitative measure of the catalytic potential of a material to activate porcine plasma coagulation. Panel A setup is the same as Figure 8.8 and shows that blood plasma proteins do not adsorb hydrophilic surfaces (see vertical arrow), in a manner parallel to that discovered for human FXII and FXIIa. Comparison to Panel B shows that hydrophilic surfaces that do not adsorb blood proteins are the most efficient activators of plasma coagulation.

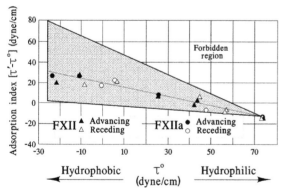

FIGURE 8.8   An adsorption map for blood factor XII and activated enzyme form FXIIa. All physically possible wetting data falls inside the adsorption triangle (shaded). Data corresponding to FXII (triangles) cannot be distinguished from FXIIa (circles). Open and closed symbols refer to adhesion tensions calculated from advancing and receding contact angles (see Section 8.4.4 for further discussion). Data falls along a linear trend line passing through $[\tau' - \tau^o] = 0$ near $\tau^o = 40 \text{ mJ/m}^2$ ($\theta \sim 55^\circ$), showing that neither protein adsorbs to more hydrophilic surfaces ($\theta < 55^\circ$).

adsorb to surfaces exhibiting $\tau^0 > 40 \, \text{mJ/m}^2$. Comparison of Figures 8.8 and 8.9a reveals that purified FXII and FXIIa and plasma proteins adsorb similarly to surfaces incrementally sampling the observable water-wetting range. These studies are confirmed by an extensive survey of blood protein adsorption using both tensiometry and solution-depletion methods [16].

8.4.5.3 Quantification of Blood Plasma Coagulation

The biological response of interest in this example is coagulation of blood plasma. A means of quantifying coagulation was developed and a mathematical model of coagulation used to interpret results [81, 151, 152]. The outcome was a single parameter K_{act} that measured the catalytic potential of a test surface to activate plasma coagulation. In turn, K_{act} was scaled against the wettability of the test activator surface, τ^0, as shown in Figure 8.9b. Figure 8.9b is a kind of calibration curve relating blood plasma coagulation to the water wettability of the activating surface in the manner contemplated by Figure 8.7. It is evident from Figure 8.9 that hydrophilic surfaces are the most activating surfaces with the highest catalytic potential. This outcome is completely consistent with early findings of the late 1960s mentioned in Section 8.4.5.1.

8.4.5.4 Interpretation of Results

Comparison of Figures 8.9a to 8.9b leads immediately to the conclusion that surfaces that do not adsorb plasma proteins are surfaces that efficiently activate blood plasma coagulation with large K_{act}. Conversely, surfaces that adsorb plasma proteins are the least activating with K_{act} near zero. Comparison of Figures 8.9a and 8.9b to Figure 8.8 further shows that surfaces that adsorb FXII and FXIIa are the least activating, whereas surfaces that do not adsorb FXII and FXIIa are surfaces that efficiently activate blood plasma coagulation. All taken together, it is apparent that the traditional mechanism of plasma coagulation involving FXII adsorption or assembly onto activating surfaces is inconsistent

with experimental results. Continued studies have shown that the reason protein-adsorbent hydrophobic surfaces do not contact activate FXII in plasma is because adsorption of a plethora of higher-concentration blood proteins effectively blocks FXII contact with a surface [98].

8.4.5.5 Summary

Quantification of the catalytic potential of a material to induce plasma coagulation and development of a mathematical model thereof revealed a striking relationship between plasma coagulation and material surface energy. The adsorption map is a calibration tool equally applicable to purified proteins and plasma that revealed a relationship to adsorbent surface energy. Although neither of these tools afforded direct, molecular insights into the steps of Figure 8.4 labeled amenable to comprehension, the combination of perspectives revealed heretofore unknown mechanistic information about contact activation of blood coagulation that can help direct rational design of cardiovascular biomaterials.

8.5 CONCLUSION

There are many methods of modifying biomaterial surfaces to influence biocompatibility. The full power of the methods will be unleashed when mechanisms detailing how cells and proteins involved in the biological response to materials are revealed. With this information in hand, rational surface engineering can be applied to direct biocompatibility by control of interactions with components of living systems. Without this information, surface engineering and modification will continue to be a product of design-directed, trial-and-error, or phenomenological approaches. In this latter pursuit, methods relating quantitative measures of the biological response to variables such as surface energy or areal density of immobilized functional proteins are invaluable interpolative tools that can yield insights into the long-sought mechanisms of biocompatibility.

References

[1] M. Lavine, M. Roberts, and O. Smith, Bodybuilding: the bionic human, *Science* **295** (2002), 995–1032.

[2] D.F. Williams, On the mechanisms of biocompatibility, *Biomaterials* **29** (2008), 2941–2953.

[3] D.F. Williams, Definitions in biomaterials, *Proceedings of a consensus conference of the European society for biomaterials*, Elsevier, New York, NY, USA (1987).

[4] B.D. Ratner and A.S. Hoffman, Physicochemical surface modification of materials used in medicine, in *Biomaterials science* (B.D. Ratner, A.S. Hoffman, F.J. Schoen, and J.E. Lemons, eds.), Elsevier Academic Press, San Diego, CA, USA (2004).

[5] R. Williams, *Surface modification of biomaterials. Methods analysis and applications*, Woodhead Publishing, Oxford, UK (2011).

[6] B.D. Ratner and D.G. Castner, Surface modification of polymeric biomaterials, *International symposium on surface modification of polymeric biomaterials*, Plenum Press, Anaheim, CA, USA (1997).

[7] J.A. Burdick and R.L. Mauck, *Biomaterials for tissue engineering applications: a review of the past and future trends*, Springer, Vienna, Austria (2010).

[8] D.F. Williams, *The biomaterials silver jubilee compendium: the best papers published in biomaterials, 1980–2004*, Elsevier, Amsterdam, The Netherlands (2006).

[9] Y. Ikada, Surface modification of polymers for medical applications, *Biomaterials* **15** (1994), 725–736.

[10] B.D. Ratner, Surface modification of polymers: chemical, biological and surface analytical challenges, *Biosens Bioelectron* **10** (1995), 797–804.

[11] I. Langmuir, The constitution of liquids with especial reference to surface tension phenomena, *Metall Chem Eng* **15** (1916), 468–470.

[12] I. Langmuir, The constitution and fundamental properties of solids and liquids. Part I. Solids, *J Am Chem Soc* **38** (1916), 2221–2295.

[13] S.N. Vinogradov and R.H. Linnell, *Hydrogen bonding*, Van Nostrand Reinhold, New York, NY, USA (1971).

[14] G.W. Castellan, *Physical chemistry*, Addison-Wesley, Reading, MA, USA (1964).

[15] E.A. Vogler, Biological properties of water, in *Water in biomaterials surface science* (M. Morra, ed.), Wiley, New York, NY, USA (2001), 4–24.

[16] E.A. Vogler, Protein adsorption in three dimensions, *Biomaterials* **33** (2012), 1201–1207.

[17] F.W. Putnam, Alpha, beta, gamma, omega—the roster of the plasma proteins, in *The plasma proteins: structure, function, and genetic control* (F.W. Putnam, ed.), Academic Press, New York, NY, USA (1975), 58–131.

[18] N.L. Anderson and N.G. Anderson, The human plasma proteome: history, character, and diagnostic prospects, *Mol Cell Proteom* **1** (2002), 845–867.

[19] E.A. Vogler, How water wets biomaterials, in *Water in biomaterials surface science* (M. Morra, ed.), Wiley, New York, NY, USA (2001), 269–290.

[20] E.A. Vogler, Interfacial chemistry in biomaterials science, in *Wettability* (J. Berg, ed.), Marcel Dekker, New York, NY, USA (1993), 184–250.

[21] E.A. Vogler, Structure and reactivity of water at biomaterial surfaces, *Adv Colloid Interface Sci* **74** (1998), 69–117.

[22] E.A. Vogler, Water and the acute biological response to surfaces, *J Biomat Sci Polym Edu* **10** (1999), 1015–1045.

[23] K.D. Collins, Sticky ions in biological systems, *Proc Natl Acad Sci* **92** (1995), 5553–5557.

[24] P. Kao, P. Parhi, A. Krishnan, H. Noh, W. Haider, S. Tadigadapa, D.L. Allara, and E.A. Vogler, Volumetric interpretation of protein adsorption: interfacial packing of protein adsorbed to hydrophobic surfaces from surface-saturating solution concentrations, *Biomaterials* **32** (2010), 969–978.

[25] F. Ariola, A. Krishnan, and E.A. Vogler, Interfacial rheology of blood proteins adsorbed to the aqueous-buffer/air interface, *Biomaterials* **27** (2006), 3404–3412.

[26] P. Parhi, A. Golas, and E.A. Vogler, Role of water and proteins in the attachment of mammalian cells to surfaces: a review, *J Adhes Sci Tech* **24** (2010), 853–888.

[27] E.A. Vogler, The goldilocks surface, *Biomaterials* **32** (2011), 6670–6675.

[28] E.A. Vogler and C.A. Siedlecki, Contact activation of blood plasma coagulation, *Biomaterials* **30** (2009), 1857–1869.

[29] E.A. Vogler, J.C. Graper, G.R. Harper, H.W. Sugg, L.M. Lander, and W.J. Brittain, Contact activation of the plasma coagulation cascade I. Procoagulant surface chemistry and energy, *J Biomed Mater Res* **29** (1995), 1005–1016.

[30] D.F. Williams, General concepts of biocompatibility, *Handbook of biomaterial properties* (J. Black and G. Hastings, eds.), Chapman and Hall, London, UK (1998), 481–488.

[31] M.L. Vanderford and D.H. Smith, *The silicone breast implant story: communication and uncertainty*, L. Erlbaum Associates, Mahwah, NJ, USA (1996).

[32] D.E. Bernstein, The breast implant fiasco, *Calif Law Rev* **87** (1999), 457–510.

[33] M. Angell, *Science on trial: the clash of medical evidence and the law in the breast implant case*, W.W. Norton, London, UK (1996).

[34] S. Bondurant, V.L. Ernster, and R. Herdman, *Safety of silicone breast implants*, Institute of Medicine, Washington, DC, USA (2000).

[35] B.D. Ratner, A.S. Hoffman, F.J. Schoen, and J.E. Lemmons, *Biomaterials science: an introduction to materials in medicine*, Elsevier Academic Press, New York, NY, USA (2004).

[36] V. Tagalakis, S.R. Kahn, M. Libman, and M. Blostein, The epidemiology of peripheral vein infusion thrombophlebitis: a critical review, *Am J Med* **113** (2002), 146–151.

[37] N.M.K. Lamba, K.A. Woodhouse, S.L. Cooper, and M.D. Lelah, *Polyurethanes in biomedical applications*, CRC Press, Boca Raton, FL, USA (1998).

[38] H.E. Kambic, S. Murabayashi, and Y. Nose', Biomaterials in artificial organs, *C&E News* **64** (15) (April 1986), 31–48.

[39] E.A. Vogler, On the origins of water wetting terminology, in *Water in biomaterials surface science* (M. Morra, ed.), Wiley, New York, NY, USA (2001), 150–182.

[40] C. Rappaport, Studies on properties of surface required for growth of mammalian cells in synthetic medium: II. The monkey kidney cell, *Exp Cell Res* **20** (1960), 470–494.

[41] C. Rappaport, Studies on properties of surface required for growth of mammalian cells in synthetic medium: III. The L cell strain 929, *Exp Cell Res* **20** (1960), 495–510.

[42] C. Rappaport, J.P. Poole, and H.P. Rappaport, Studies on properties of surfaces required for growth of mammalian cells in synthetic medium: I. The HeLa cell, *Exp Cell Res* **20** (1960), 465–479.

[43] T. Matsuda and M.H. Litt, Modification and characterization of polystyrene surfaces used in cell culture, *J Polym Sci* **12** (1974), 489–497.

[44] H.G. Klemperer and P. Knox, Attachment and growth of BHK cells and liver cells on polystyrene: effect of surface groups introduced by treatment with chromic acid, *Lab Pract* **26** (1977), 179–180.

[45] R.W. Benedict and M.C. Williams, Bonding erythrocytes to plastic substrates by glow-discharge activation, *Biomat Med Art Org* **7** (1979), 477–493.

[46] S. Eliezer and Y. Eliezer, *The fourth state of matter: an introduction to the physics of plasma*, Institute of Physics Publishing, Bristol, UK (2001).

[47] R. D'Agostino, P. Favia, and F. Fracassi, *Plasma processing of polymers*, Kluwer, Dordrecht, The Netherlands (1997).

[48] T. Desmet, R. Morent, N.D. Geyter, C. Leys, E. Schacht, and P. Dubruel, Nonthermal plasma technology as a versatile strategy for polymeric biomaterials surface modification: a review, *Biomacromolecules* **10** (2009), 2351–2378.

[49] B.O. Aronsson, J. Lausmaa, and B. Kasemo, Glow discharge plasma treatment for surface cleaning and modification of metallic biomaterials, *J Biomed Mater Res* **35** (1997), 49–73.

[50] P.K. Chu, J.Y. Chen, L.P. Wang, and N. Huang, Plasma-surface modification of biomaterials, *Mater Sci Eng R* **36** (2002), 143–206.

[51] P. Favia and R. Agostino, Plasma treatments and plasma deposition of polymers for biomedical applications, *Surf Coat Tech* **98** (1998), 1102–1106.

[52] N.J. Bailey, Soft silicone lenses: the beginning, Contact lens, *Spectrum* (1987) 17–21.

[53] N.J. Bailey, Bausch & Lomb pitches in for Dr. Simon, Contact lens, *Spectrum* (1987), 53–60.

[54] H. Yasuda, *Plasma polymerization*, Academic Press, Orlando, FL, USA (1985).

[55] X. Liu, J.Y. Lim, H.J. Donahue, R. Dhurati, M.M. Andrea, and E.A. Vogler, Influence of substratum surface hydrophilicity/hydrophobicity on osteoblast adhesion, morphology, and focal adhesion assembly, *Biomaterials* **28** (2007), 4535–4550.

[56] E.A. Vogler and R.W. Bussian, Short-term cell-attachment rates: a surface-sensitive test of cell-substrate compatibility, *J Biomed Mater Res* **21** (1987), 1197–1211.

[57] T.A. Horbett and L.A. Klumb, Cell culturing: surface aspects and considerations, in *Interfacial phenomena and bioproducts* (J.L. Brash and P.W. Wojciechowski, eds.), Marcel Dekker, New York, NY, USA (1996), 351–445.

[58] D. Barngrover, Substrata for anchorage-dependent cells, in *Mammalian cell technology* (W.G. Thilly, ed.), Butterworths, Boston, MA, USA (1986), 131–149.

[59] W.S. Ramsey, W. Hertl, E.D. Nowlan, and N.J. Binkowski, Surface treatments and cell attachment, *In Vitro* **20** (1984), 802–808.

[60] J. Sagiv, J. Gun, R. Maoz, and L. Netzer, Self-assembling monolayers: a study of their formation, composition, and structure, in *Surfactants in solution* (K.L. Mittal and P. Bothorel, eds.), Plenum Press, New York, NY, USA (1986), 965–978.

[61] J.D. Swalen, D.L. Allara, J.D. Andrade, E.A. Chandross, S. Garoff, J. Israelachivili, T.J. McCarthy, R. Murray, R.F. Pease, J.F. Rabolt, K.J. Wynne, and H. Yu, Molecular monolayers and films, *Langmuir* **3** (1987), 932–950.

[62] M. Jacoby, New tools for tiny jobs, *C&E News* **78** (16) (October 2000), 33–35.

[63] R. Dagani, Building from the bottom up, *C&E News* **78** (16) (October 2000), 27–32.

[64] W. Schulz, Crafting a national nanotechnology effort, *C&E News* **78** (42) (October 2000), 39–42.

[65] E.G. Shafrin and W.A. Zisman, The spreading of liquids on low-energy surfaces. IV. Monolayer coatings on platinum, *J Colloid Sci* **7** (1952), 166–177.

[66] E.G. Shafrin and W.A. Zisman, The adsorption on platinum and wettability of monolayers of terminally fluorinated octadecyl derivatives, *J Phys Chem* **61** (1957), 1046–1053.

[67] K.B. Blodgett, Films built by depositing successive monomolecular layers on a solid surface, *J Am Chem Soc* **57** (1935), 1007–1022.

[68] K.B. Blodgett and I. Langmuir, Built-up films of barium stearate and their optical properties, *Phys Rev* **51** (1937), 964–982.

[69] A. Pockels, On the spreading of oil upon water, *Nature* **50** (1894), 223–224.

[70] A. Pockels, On the relative contamination of the water surface by equal quantities of different substances, *Nature* **46** (1892), 418–419.

[71] N.T. Southall, K.A. Dill, and A.D.J. Haymet, A View of the hydrophobic effect, *J Chem Phys B* **106** (2002), 521–533.

[72] M. Jacoby, Custom-made biomaterials, *C&E News* **79** (6) (February 2001), 30–35.

[73] A.S.G. Curtis, J.V. Forrester, C. McInnes, and F. Lawrie, Adhesion of cells to polystyrene surfaces, *J Cell Biol* **97** (1983), 1500–1506.

[74] A. Curtis and C. Wilkinson, Ambiguities in the evidence about cell adhesion problems with activation events and with the structure of the cell-contact, *Studia biophysica* **127** (1988), 75–82.

[75] N.F. Owens, D. Gingell, and A. Trommler, Cell adhesion to hydroxyl groups of a monolayer film, *J Cell Sci* **91** (1988), 269–279.

[76] S. Margel, E.A. Vogler, L. Firment, T. Watt, S. Haynie, and D.Y. Sogah, Peptide, protein, and cellular interactions with self-assembled monolayer model surfaces, *J Biomed Mater Res* **27** (1993), 1463–1476.

[77] E.A. Vogler, On the biomedical relevance of surface spectroscopy, *J Electron Spectrosc Rel Phenom* **81** (1996), 237–247.

[78] E.A. Vogler and R.W. Bussian, Short-term cell-attachment rates: a surface sensitive test of cell-substrate compatibility, *J Biomed Mater Res* **21** (1987), 1197–1211.

[79] E.A. Vogler, Thermodynamics of short-term cell adhesion *in vitro*, *Biophys J* **53** (1988), 759–769.

[80] E.A. Vogler, A thermodynamic model of short-term cell adhesion *in vitro*, *Colloid Surface* **42** (1989), 233–254.

[81] E.A. Vogler, J.C. Graper, G.R. Harper, L.M. Lander, and W.J. Brittain, Contact activation of the plasma coagulation cascade. 1. Procoagulant surface energy and chemistry, *J Biomed Mater Res* **29** (1995), 1005–1016.

[82] S.R. Holmes-Farley, R.H. Reamey, T.J. McCarthy, J. Deutch, and G.M. Whitesides, Acid-base behavior of carboxylic acid groups covalently attached at the surface of polyethylene: the usefulness of contact angle in following the ionization of surface functionality, *Langmuir* **1** (1985), 725–740.

[83] C.D. Bain and G.M. Whitesides, A study by contact angle of the acid-base behavior of monolayers containing w-Mercaptocarboxylic acids adsorbed on gold: an example of reactive spreading, *Langmuir* **5** (1989), 1370–1378.

[84] H. Noh and E.A. Vogler, Volumetric interpretation of protein adsorption: ion-exchange adsorbent capacity, protein pI, and interaction energetics, *Biomaterials* **29** (2008), 2033–2048.

[85] C.-H.J. Yeh, Z.O. Dimachkie, A. Golas, A. Cheng, P. Parhi, and E.A. Vogler, Contact activation of blood plasma and factor XII by ion-exchange resins, *Biomaterials* **33** (2011), 9–19.

[86] X. Xu and A.M. Lenhoff, A Predictive Approach to Correlating Protein Adsorption Isotherms on Ion-Exchange Media, *J Phys Chem B* **112** (2008), 1028–1040.

[87] Y. Arima and H. Iwata, Effect of wettability and surface functional groups on protein adsorption and cell adhesion using well-defined mixed self-assembled monolayers, *Biomaterials* **28** (2007), 3074–3082.

[88] M. Thery, V. Racine, A. Pepin, M. Piel, Y. Chen, J.-B. Sibarita, and M. Bornens, The extracellular matrix guides the orientation of the cell division axis, *Nat Cell Biol* **7** (2005), 947–953.

[89] M. Thacry, V. Racine, M. Piel, A. Pacpin, A. Dimitrov, Y. Chen, J.-B. Sibarita, and M. Bornens, Anisotropy of cell adhesive microenvironment governs cell internal organization and orientation of polarity, *Proc Natl Acad Sci* **103** (2006), 19771–19776.

[90] R. Singhvi, A. Kumar, G.P. Lopez, G.N. Stephanopoulos, D.I.C. Wang, G.M. Whitesides, and D.E. Ingber, Engineering cell shape and function, *Science* **264** (1994), 696–698.

[91] Z. Nie and E. Kumacheva, Patterning surfaces with functional polymers, *Nat Mater* **7** (2008), 277–290.

[92] B. Chehroudi and D.M. Brunette, Effects of surface topography on cell behavior, in *Encyclopedic handbook of biomaterials and bioengineering Part A: materials* (D.L. Wise, D.J. Trantolo, D.E. Altobelli, M.J. Yaszemski, J.D. Gresser, and R. Schwartz, eds.), Marcel Dekker, New York, NY, USA (1995), 813–842.

[93] J.L. Charest and W.P. King, Engineering biomaterial interfaces through micro and nano-patterning BioNanoFluidic MEMS, in *BioNanoFluidic MEMS* (P.J. Hesketh and G.W. Wodruff, eds.), Springer, New York, NY, USA (2008), 251–277.

[94] S. Mitragotri and J. Lahann, Physical approaches to biomaterial design, *Nat Mater* **8** (2009), 15–23.

[95] P. Kingshott, G. Andersson, S.L. McArthur, and H.J. Griesser, Surface modification and chemical surface analysis of biomaterials, *Curr Opin Chem Biol* **15** (2011), 667–676.

[96] A. Welle and E. Gottwald, UV-based patterning of polymeric substrates for cell culture applications, *Biomed Microdev* **4** (2002), 33–41.

[97] R. Miller, Z. Guo, E.A. Vogler, and C.A. Siedlecki, Plasma coagulation response to surfaces with nanoscale heterogeneity, *Biomaterials* **27** (2006), 208–215.

[98] A. Golas, P. Parhi, Z.O. Dimachkie, C.A. Siedlecki, and E.A. Vogler, Surface-energy dependent contact activation of blood factor XII, *Biomaterials* **31** (2010), 1068–1079.

[99] C. Roberts, C.S. Chen, M. Mrksich, V. Martichonok, D.E. Ingber, and G.M. Whitesides, Using mixed self-assembled monolayers presenting RGD and (EG)3OH groups to characterize long-term attachment of bovine

capillary endothelial cells to surfaces, *J Am Chem Soc* **120** (1998), 6548–6555.

[100] A. Brock, E. Chang, C.C. Ho, P. LeDuc, X.Y. Jiang, G.M. Whitesides, and D.E. Ingber, Geometric determinants of directional cell motility revealed using microcontact printing, *Langmuir* **19** (2003), 1611–1617.

[101] G.M. Whitesides, E. Ostuni, S. Takayama, X. Jiang, and D.E. Ingber, Soft lithography in biology and biochemistry, *Annu Rev Biomed Eng* **3** (2001), 335–373.

[102] M.S. Lord, M. Foss, and F. Besenbacher, Influence of nanoscale surface topography on protein adsorption and cellular response, *Nano Today* **5** (1) (February. 2010), 66–78.

[103] N. Giovambattista, P.G. Debenedetti, and P.J. Rossky, Hydration behavior under confinement by nanoscale surfaces with patterned hydrophobicity and hydrophilicity, *J Phys Chem C* **111** (2007), 1323–1332.

[104] H.C.V. Baeyer, The lotus effect, *The Sciences* **40** (2000), 12–15.

[105] J. Zhang and Y. Han, A topography/chemical composition gradient polystyrene surface: toward the investigation of the relationship between surface wettability and surface structure and chemical composition, *Langmuir* **24** (2007), 796–801.

[106] N. Zhao, X. Lu, X. Zhang, H. Liu, S. Tan, and J. Xu, Progress in superhydrophobic surfaces, *Prog Chem* **19** (2007), 860–871.

[107] A. Marmur, Super-hydrophobicity fundamentals: implications to biofouling prevention, *Biofouling* **22** (2006), 107–115.

[108] D.M. Spori, T. Drobek, S. Rcher, M. Ochsner, C. Sprecher, A. Hlebach, and N.D. Spencer, Beyond the lotus effect: roughness influences on wetting over a wide surface-energy range, *Langmuir* **20** (2008), 5411–5417.

[109] Y. Su, B. Ji, K. Zhang, H. Gao, Y. Huang, and K. Hwang, Nano to micro structural hierarchy is crucial for stable superhydrophobic and water-repellent surfaces, *Langmuir* **26** (2010), 4984–4989.

[110] L. Gao and T.J. McCarthy, Wetting 101, *Langmuir* **25** (2009), 14105–14115.

[111] T. Cass and F.S. Ligler, *Immobilized biomolecules in analysis: a practical approach*, Oxford University Press, New York, NY, USA (1998).

[112] Z.-H. Xing, Y. Chang, and I.-K. Kang, Immobilization of biomolecules on the surface of inorganic nanoparticles for biomedical applications, *Sci Tech Adv Mat* **11** (2010), 014101.

[113] D. Samanta and A. Sarkar, Immobilization of biomacromolecules on self-assembled monolayers: methods and sensor applications, *Chem Soc Rev* **40** (2011), 2567–2592.

[114] P. Baumann, P. Tanner, O. Onaca, and C.G. Palivan, Biodecorated polymer membranes: a new approach in diagnostics and therapeutics, *Polymers* **3** (2010), 173–192.

[115] M. Nosonovsky and B. Bhushan, Other biomimetic surfaces, in *Multiscale dissipative mechanisms and hierarchical surfaces* (M. Nosonovsky and B. Bhushan, eds.), Springer-Verlag, Berlin, Germany (2008), 243–250.

[116] K.S. Siow, L. Britcher, S. Kumar, and H.J. Griesser, Plasma methods for the generation of chemically reactive surfaces for biomolecule immobilization and cell colonization—a review, *Plasma Process Polym* **3** (2006), 392–418.

[117] H. Shi, W.-B. Tsai, M.D. Garrison, S. Ferrari, and B.D. Ratner, Template imprinted nanostructured surfaces for protein recognition, *Nature* **398** (1999), 593–597.

[118] B.D. Ratner, Plasma deposition for biomedical applications: a brief review, *J Biomat Sci Polym E* **4** (1993), 3–11.

[119] H. Shi and B.D. Ratner, Template recognition of protein-imprinted polymer surfaces, *J Biomed Mater Res* **49** (2000), 1–11.

[120] B.D. Ratner, The engineering of biomaterials exhibiting recognition and specificity, *J Mol Recogn* **9** (1996), 617–625.

[121] Y. Hirano, Y. Kando, T. Hayashi, K. Goto, and A. Nakajima, Synthesis and cell attachment activity of bioactive oligopeptides: RGD, RGDS, RGDV and RGDT, *J Biomed Mater Res* **25** (1991), 1523–1534.

[122] Y. Hirano, Y. Kando, K. Goto, and A. Nakajima, Synthesis and cell attachment activity of oligopeptides: RGD, RGDS, RGDV, and RGDT, *J Biomed Mater Res* **25** (1991), 1523–1534.

[123] J.A. Neff, K.D. Caldwell, and P.A. Tresco, A novel method for surface modification to promote cell attachment to hydrophobic substrates, *J Biomed Mater Res* **40** (1998), 511–519.

[124] A.E. Aksoy, V. Hasirci, and N. Hasirci, Surface modification of polyurethanes with covalent immobilization of heparin, *Macromol Symp* **269** (2008), 145–153.

[125] A. Ulman, Wetting studies of molecularly engineered surfaces, *Thin Solid Films* **273** (1996), 48–53.

[126] A. Ulman, S.D. Evans, Y. Shnidman, R. Sharma, J.E. Eilers, and J.C. Chang, Concentration-driven surface transition in the wetting of mixed alkanethiol monolayers on gold, *J Am Chem Soc* **113** (1991), 1499–1506.

[127] H. Elwing, S. Welin, A. Askendal, U. Nillson, and I. Lundstrom, A wettability gradient method for studies of macromolecular interactions at the liquid/solid interface, *J Colloid Interf Sci* **119** (1987), 203–210.

[128] T. Ueda-Yukoshi and T. Matsuda, Cellular responses on a wettability gradient surface with continuous variations in surface compositions of carbonate and hydroxyl groups, *Langmuir* **11** (1995), 4135–4140.

[129] L. Wei, E.A. Vogler, T.M. Ritty, and A. Lakhtakia, A 2D surface morphology-composition gradient panel for protein-binding assays, *Mater Sci Eng C* **31** (2011), 1861–1866.

[130] D.G. Castner and B.D. Ratner, Biomedical surface science: foundations to frontiers, *Surf Sci* **500** (2002), 28–60.

[131] J.C. Vickerman and I.S. Gilmore, *Surface analysis: the principal techniques*, 2nd ed., Wiley, Chichester, UK (2009).

[132] J.M. Perry, A.J. Moad, N.J. Begue, R.D. Wampler, and G.J. Simpson, Electronic and vibrational second-order nonlinear optical properties of protein secondary structural motifs, *J Phys Chem B* **109** (2005), 20009–20026.

[133] J. Wang, M.L. Clarke, X. Chen, M.A. Even, W.C. Johnson, and Z. Chen, Molecular studies on protein conformations at polymer/liquid interfaces using sum frequency generation vibrational spectroscopy, *Surf Sci* **587** (2005), 1–11.

[134] X. Chen, *Investigating biointerfaces using sum frequency generation vibrational spectroscopy*, PhD Dissertation, University of Michigan (2007).

[135] S.V. Le Clair, K. Nguyen, and Z. Chen, Sum frequency generation studies on bioadhesion: elucidating the molecular structure of proteins at interfaces, *J. Adhesion* **85** (2009), 484–511.

[136] P. Parhi, A. Golas, and E.A. Vogler, Role of proteins and water in the initial attachment of mammalian cells to biomedical surfaces: a review, in *Surface and interfacial aspects of cell adhesion* (A. Carré and K.L. Mittal, eds.), Koninklijke Brill NV, Leiden, The Netherlands (2010), 103–138.

[137] Q. Garrett, B. Laycock, and R.W. Garrett, Hydrogel lens monomer constituents modulate protein sorption, *Invest Ophthalmol Vis Sci* **41** (2000), 1687–1695.

[138] J.L. Bohnert, T.A. Horbett, B.D. Ratner, and F.H. Royce, Adsorption of proteins from artificial tear solutions to contact lens materials, *Invest Ophthalmol Vis Sci* **29** (1988), 362–373.

[139] E. Castillo, J. Koenig, and J. Anderson, Characterization of protein adsorption on soft contact lenses I. Conformational changes of adsorbed human serum albumin, *Biomaterials* **5** (1984), 319–325.

[140] E. Castillo, J. Koenig, and J. Anderson, Protein adsorption on hydrogels II. Reversible and irreversible interactions between lysozyme and soft contact lens surfaces, *Biomaterials* **6** (1985), 338–345.

[141] E. Castillo, J. Koenig, J. Anderson, and N. Jentoft, Protein adsorption on soft contact lenses. III. Mucin, *Biomaterials* **7** (1986), 9–16.

[142] E.J. Castillo, J.L. Koenig, and J.M. Anderson, Characterization of protein adsorption on soft contact lenses IV. Comparison of *in vivo* spoilage with the *in vitro* adsorption of tear proteins, *Biomaterials* **7** (1986), 89–96.

[143] E.A. Vogler, D.A. Martin, D.B. Montgomery, J.C. Graper, and H.W. Sugg, A graphical method for predicting protein and surfactant adsorption properties, *Langmuir* **9** (1993), 497–507.

[144] E.A. Vogler, J.C. Graper, H.W. Sugg, L.M. Lander, and W.J. Brittain, Contact activation of the plasma coagulation cascade. 2. Protein adsorption on procoagulant surfaces, *J Biomed Mater Res* **29** (1995), 1017–1028.

[145] H.R. Roberts, Oscar Ratnoff: his contributions to the golden era of coagulation research, *Brit Haematol* **122** (2003), 180–192.

[146] A.H. Schmaier, The elusive physiologic role of factor XII, *J Clin Invest* **118** (2008), 3006–3009.

[147] R.W. Colman, C.F. Scott, A.H. Schmaier, T.T. Wachtfogel, R.A. Pixley, and L.H. Edmunds, Initiation of blood coagulation at artificial surfaces, *Ann NY Acad Sci* **516** (1987), 253–267.

[148] I.M. Sainz, R.A. Pixley, and R.W. Colman, Fifty years of research on the plasma kallikrein-kinin system: from protein structure and function to cell biology and in-vivo pathophysiology, *Thromb Haemost* **98** (2007), 77–83.

[149] O.D. Ratnoff, Hemostasis and blood coagulation, in *Physiology* (R.M. Berne and M.N. Levy, eds.), Mosby Year Book, St. Louis, MO, USA (1993).

[150] G.G.M. Fuhrer, W. Heller, and H.E. Hofmeister, F XII, *Blut* **61** (1990), 258–266.

[151] K. Fujikawa and E.W. Davie, Human factor XII (Hageman factor), *Methods Enzymol* **80** (1981), 198–211.

[152] R. Zhuo, R. Miller, K.M. Bussard, C.A. Siedlecki, and E.A. Vogler, Procoagulant stimulus processing by the intrinsic pathway of blood plasma coagulation, *Biomaterials* **26** (2005), 2965–2973.

[153] Z. Guo, K. Bussard, E.A. Vogler, and C.A. Siedlecki, Mathematical modeling of material-induced blood plasma coagulation, *Biomaterials* **27** (2006), 796–806.

ABOUT THE AUTHOR

Erwin A. Vogler is a Professor of Materials Science and Engineering and Bioengineering at the Pennsylvania State University where he is a member of the Materials Research Institute and Huck Institutes of Life Science. Prior to joining Penn State in 1999, he was employed as a Research Fellow at the Becton Dickinson Research Center, Research Triangle Park, North Carolina from 1988 and as a Research Scientist at the Du Pont Experimental Station Wilmington, Delaware from 1980. He received a PhD in Chemistry at Indiana University in 1979 where he also held a National Aeronautics and Space Administration Post-Doctoral Fellowship through 1980.

Flight Control Using Biomimetic Optical Sensors

Javaan Chahl[a] *and Akiko Mizutani*[b]

[a]School of Engineering, University of South Australia, Adelaide,
SA 5001, Australia
[b]Odonatrix Pty. Ltd., One Tree Hill, SA 5114, Australia

Prospectus

Insects are dependent on the spatial, spectral, and temporal distributions of light in the environment for flight control and navigation. This chapter reports on flight trials of implementations of insect-inspired behaviors on unmanned aerial vehicles. Optical-flow methods for maintaining a constant height above ground and a constant course have been demonstrated to provide navigational capabilities that are impossible using conventional avionics sensors. Precision control of height above ground and ground course were achieved over long distances. Other demonstrated vision-based techniques include a biomimetic stabilization sensor that uses the ultraviolet and green bands of the spectrum and a sky polarization compass. Both of these sensors were tested over long trajectories in different directions, in each case showing performance similar to low-cost inertial heading and attitude systems.

Keywords

Autopilot, Dragonfly, Insect, Locust, Navigation, Optical flow, Polarization, Sensors, Spectral opponency, Sun detector, Unmanned aerial vehicle

9.1 INTRODUCTION

The key principles of insect vision have been deduced over decades of biological research [1, 2]. A common theme that has emerged is the reliance by insects on the spatial and temporal distribution of light in the environment for controlling flight. Computational foundations and control laws derived from insects [3] have become increasingly well formulated. The specialization of an insect's sensors to the environment in which it operates is an important principle in reverse engineering insect sensory systems.

Unmanned aerial vehicles (UAVs) have emerged as a revolution in air power. UAVs are not as autonomous as we would like. They are dependent on artificial external signals for position information and do not operate close to obstacles. They are also typified by a strong dependence on a small number of reliable and accurate sensors that are each critical for operation. As we attempt to miniaturize UAVs, scaling problems emerge

Engineered Biomimicry

http://dx.doi.org/10.1016/B978-0-12-415995-2.00009-X

beyond the low fidelity of small sensors. For example, the vibration of flapping wings and propellors can degrade inertial attitude sensing, and flight close to the ground results in distorted measurements of the Earth's magnetic field.

Insects have overcome these problems using a highly integrated visuomotor system. In addition to compound eyes, insects typically have three simple eyes (a retina behind a single lens), called the *ocelli*. The ocelli in dragonflies and locusts provide a highly evolved optical stabilization function by using the horizon as a visual reference. Polarization sensitivity in the dorsal rim area of the eyes of many insects provides a measure of heading with respect to the sun. The small size of insect sensors makes biomimetic solutions based on insect eyes desirable. Many of these reflexes will only function robustly as part of an ensemble of reflexes, analogous to the biologically inspired subsumption architecture proposed by Brooks [4].

The characterization of systems deep within the nervous system of insects is difficult due to the complex multimodel nature of insect sensorimotor neurons [5, 6]. Direct measurement of the responses of neurons requires glass tubes with a tip diameter of <1 μm filled with electrolyte to be placed and held either inside (intracellular) or outside (extracellular) the cell wall of the neuron. Ideally, many thousands of neuronal responses would be measured simultaneously while the insect is in flight in a natural environment. Such an experiment is beyond the state of the art in neurophysiology and quite impractical.

In the absence of direct measurement, another approach is to implement artificial versions of the sensors involved, in an attempt to learn what problems an integrated system must solve. The remainder of this chapter demonstrates four biomimetic optical technologies that have been implemented and flight tested on UAVs. The resulting systems reveal many challenges faced by the insect brain and provide cues as to what neurophysiology experiments should be done to fully understand the system.

The head of a dragonfly (*Hemianax papuensis*) is shown in Figure 9.1. The dorsal rim area of

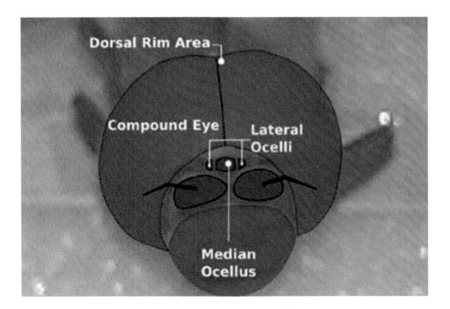

FIGURE 9.1 Detail of the head of a dragonfly, *Hemianax papuensis*. The dorsal rim area along the seam between the left and right eyes of many insects is polarization sensitive. The median ocellus and lateral ocelli receive light from the forward direction and sideways, respectively.

many insects is adapted to measure polarization from sky light. The ocelli have wider fields of view than any of the ommatidia (popularly known as *facets* or *eyelets*) of the compound eye of the dragonfly. From the figure it is also clear that ocelli have an aperture at least an order of magnitude greater than those of the ommatidia of the compound eye, suggesting an adaptation to low-light conditions.

Ocellar stabilization and polarization compassing are mediated by aspects of the environment that are invisible to the human eye. Ocelli accept input from the ultraviolet region of the spectrum. The polarization of light is usually imperceptible to the unaided human eye. The geometry of insect vision is also outside the human experience since the compound eye provides a near-spherical view of the world.

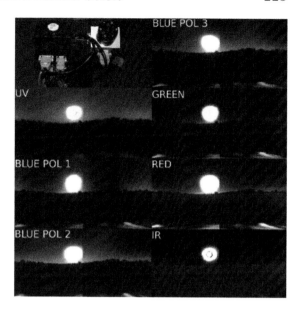

FIGURE 9.2 The panoramic scanning device (top left) produces a polar-coordinate image in multiple spectra and three polarization angles. Images are shown from each of the spectral channels sampled by the device. The spectrum of each channel was centered on the following wavelengths: ultraviolet (340 nm), blue (400 nm), green (500 nm), red (600 nm), and infrared (1,000 nm). Each channel was resolved to 16 bits resolution and each image required 60 s to generate.

9.2 STRUCTURE OF THE VISUAL WORLD OF INSECTS

It is not possible to view the multispectral and multipolarization world of insects. Understanding the sensors and their limitations as well as development of flying biomimetic sensors requires simulators and control system design. Instrumentation and models of the insect visual system have been developed to inform the process of biomimetic engineering of insect sensory systems.

To understand and reverse engineer the insects' view of the static world, we built a device that approximates the geometry and spectral response of an insect eye. Figure 9.2 shows the multispectral, multipolar, and panoramic scanning device and the output of the seven channels in polar coordinates. The scanning device produced a polar-coordinate-system image from 400×200 discrete samples over a capture period of one minute. The polarization pattern distributed across the sky is revealed in the differences between the three blue channels. The photodiodes in the device and all implementations

described in this chapter were interfaced in *current mode* [7] in order to achieve a linear voltage representation of light intensity.

Images of the environment were scanned under different conditions. From these images, observations about the optical structure of the environment and quantitative measurements of the performance of postulated insect-sensor algorithms could be made.

The polarization pattern of the sky on a clear day is shown in Figure 9.3a. The saturation of colors indicates the degree of polarization. In Figure 9.3b, the ground plane is devoid of pseudo color apart from one area, which is a body of water. The water beetle *Notonecta glauca* has been shown to use the polarized reflections off bodies of water on the ground plane to navigate [8],

FIGURE 9.3 Polarization images from the scanning device. The images through the three blue-filter alignments are composited into a single image, mapped to red, green, and blue. The left image (a) was captured on a fine, clear morning. The right image (b) was captured on a cloudy afternoon beside a lake.

indicating other promising applications of biomimetic sensing for autonomous systems. A partly cloudy sky still provides a strong polarization signal in clear patches, even though the location of the sun is not apparent in the image.

9.3 AIRBORNE COMPASS BASED ON SKY POLARIZATION

The sun is a clearly defined landmark in the sky that can be used to determine bearings. When the sky is partially occluded by cloud, foliage, terrain, or dirt on the optical sensor, the sun may be obscured. In conventional solar compasses, significant effort is expended to ensure that the sun is the landmark detected [9] rather than a cloud.

The sky polarization pattern is the result of *Rayleigh scattering* of sunlight off molecules in the atmosphere [10], discussed in detail by Coulson [11]. The magnitude of the pattern is modified by atmospheric effects; however, the direction of polarization vectors is reliable [12]. Degree of polarization of reflected sunlight has been used in the past as a remote measure of atmospheric density on the planets. Measurements taken by Dollfus [13] and others [14] in the 1950s provided an estimate of the density of the Martian atmosphere and later an indication of the value of polarization compassing for autonomous systems operating on Mars [15].

The sky-polarization pattern relative to the location of the sun, represented in Figure 9.4,

is predicted by the Rayleigh atmosphere model [16]. To a first-order approximation, the sun is orthogonal to the direction of polarization of any patch of sky. A *polarization sensor* measures power and is directional but unsigned. In contrast, magnetometers provide an unambiguous measurement of the magnetic field vector that is also effectively constant for any given location [17].

Insects use the polarization patterns cast by the sun and moon as an orientation reference in preference to the actual angular position of the celestial bodies on the eye [18, 19]. The advantage of the polarization pattern is that it is distributed across the sky. A measurement in any direction gives a measurement of the direction of the sun from that point [12]. Many insects use polarization in the blue region of the spectrum, sensed with the dorsal rim area of the compound eye [20], shown in Figure 9.1. In this region the ommatidia are arranged to respond maximally to a series of polarization directions. The radius of curvature of the dorsal rim is significantly lower than for other parts of the eye, ensuring that there is little variation in the direction of view of dorsal rim ommatidia [20]. The population of ommatidia provides the information required to code the direction of polarization.

Polarization compasses have been used for human navigation, possibly by the Vikings a millennium ago, through the use of naturally occurring dichroic crystals [21]. Scandinavian airlines revived this form of navigation for commercial

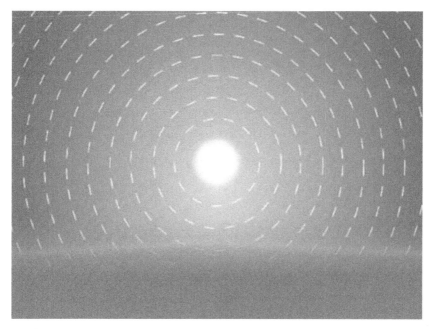

FIGURE 9.4 The sky polarization pattern cast by the sun is approximately tangential to the sun at every point.

flights over the Arctic in the early 1950s using a polarization-sensitive astrolabe device [22]. This form of navigation was driven by the arctic environment. At high latitudes the sun is either at low elevations or below the horizon. The magnetic field is oriented almost vertically and varies significantly over small distances. Gyrocompasses do not perform well, because navigation is over the axis of rotation of the Earth [23]. Prior to the advent of the global positioning system (GPS), polarization provided a self-sufficient measurement of bearing advent of the in an environment unforgiving of technical faults.

Implementations of polarization compassing have also been developed for a ground robot by Lambrinos *et al.* [24]. Their implementation used a polarization sensor to guide dead-reckoning navigation. The approach was demonstrated to be effective on the ground but is not suitable for a dynamic airborne implementation due to the need to rotate in position. Artificial polarization compasses

have been postulated by others since then [12, 25]. The most promising applications of polarization compass are in the unmanned aerospace field.

9.3.1 Sensor Signal Processing

Polarization sensing requires a series of elements. First, the optics capture light from a region of the sky, which is then analyzed in at least three polarization angles. In our work, measurement was achieved using three photodiodes, each with its own optics and polarization filters. A version of the sensor is shown in Figure 9.5. The outputs of the photodiode amplifiers were digitized and processed on a microcontroller.

Any light-sensitive element will be imperfect, with different units responding to the same signal with different output voltages. A simple model of the response, which is effective if the supporting electronics are designed well, is to assume that the output of each sensor is linear,

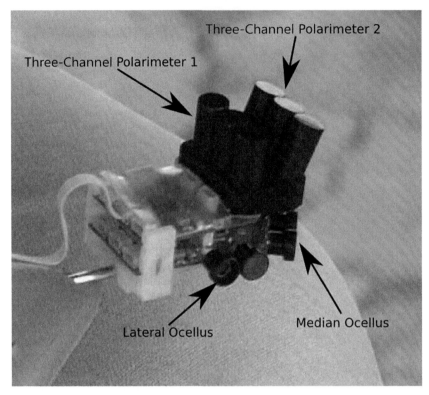

FIGURE 9.5 Integration of the ocelli and polarization sensors is designed to provide improved compass measurements and precise control of roll angle by stabilizing the sensor in the roll direction.

with individual variations of *bias* above 0 volts when responding to darkness and individually varying scale factors or *gains*. Not only do the light-sensitive elements differ from each other, but the supporting electronics will have the same variables. To simplify the discussion, we consider how direction is computed from three individual assemblies of a photodiode, a polarization filter, and an electronic amplifier. We assume for now that the polarization filters on each assembly are aligned at known angles with respect to an external reference.

We represent the response of each photodetector to incident light in terms of voltage

$$v = b + \vec{P} \cdot \vec{F} + q, \qquad (9.1)$$

which includes the scalar response q to unpolarized light, which is a result of optical input, and a scalar bias term b that is an electrical signal. The response of the sensor to the polarized component of the light is given by $\vec{P} \cdot \vec{F}$, where $\vec{F}$ is a two-dimensional Cartesian vector representing the orientation of the polarization axis of the filter (as the direction of the vector given by $\angle \vec{F} = \tan^{-1} F_x/F_y$) and its attenuation as $||\vec{F}||$. Similarly, $\vec{P}$ is a vector representing the direction and magnitude of the incident polarized light. The maximum response occurs when $\vec{F}$ and $\vec{P}$ are parallel and the minimum when the two vectors are orthogonal.

To eliminate electronic and optical biases, b and q, the difference between the responses of the three samples v_1 to v_3 is taken; thus,

$$v_1 - v_2 = \vec{P} \cdot \vec{F}_1 - \vec{P} \cdot \vec{F}_2, \qquad (9.2)$$

$$v_1 - v_3 = \vec{P} \cdot \vec{F}_1 - \vec{P} \cdot \vec{F}_3. \qquad (9.3)$$

Therefore,

$$P^T = \begin{bmatrix} \vec{F}_1 - \vec{F}_2 \\ \vec{F}_1 - \vec{F}_3 \end{bmatrix}^{-1} \begin{bmatrix} v_1 - v_2 \\ v_1 - v_3 \end{bmatrix}. \qquad (9.4)$$

The direction of polarization is given by $\angle \vec{P} = \tan^{-1} P_x / P_y$. The solution in Eq. (9.4) assumes a calibrated system for which $\vec{F}, q$, and b are known. The amplifier electronics and photodiode must be calibrated such that the three channels of the polarimeter are matched.

Considering differences between the photodiodes, our sensor model incorporated an unknown scale factor for the conversion of light into voltage, k_i for each of i diode channels; hence,

$$v_i = b_i + k_i (\vec{P} \cdot \vec{F}_i + q). \qquad (9.5)$$

We used the sinusoidal response of all three polarimeter channels while they were directed vertically upward and rotated about the vertical axis (azimuth). The turntable moved discretely at 200 steps per 180°. The three polarization filters were oriented at approximately 0°, 60°, and 120° to avoid parallel polarization-sensitivity axes that would prevent the solution in Eq. (9.4) from being well formed. Despite apparently accurate manual alignment, the exact filter polarization angles were not known well enough to obtain a solution for $\vec{P}$ that would be useful for aircraft navigation.

The relative gains between the channels were established by removing the mean measurement, $\bar{v}_i$, over a 180° rotation of each sensor, which left a full-wavelength sinusoidal curve of mean 0 for each of the i channels, with phase of the sinusoid dependent on $\vec{P}$ and i^{th} channel-specific filter properties represented in $\vec{F}_i$. The integral of the absolute values of each curve provided a distributed measure of the relative amplitudes of the response from the two

sensors. The solution only required relative measures, so k_i was normalized relative to the amplitude of k_1, the gain of the first channel. Then, recalling that the machine performed 180° rotation in 200 steps, we get

$$\frac{k_1}{k_i} = \frac{\sum_{n=0}^{200} |v_1 - \bar{v}_1|}{\sum_{n=0}^{200} |v_i - \bar{v}_i|}. \qquad (9.6)$$

The factor k_1 / k_i was used to correct the difference in b_i between the channels, since q was common to all sensors and not dependent on sensor parameters; hence,

$$\left(b_i \frac{k_1}{k_i} - b_1 \right) = \bar{v}_i \frac{k_1}{k_i} - \bar{v}_1. \qquad (9.7)$$

The calibrated response c_i of each sensor was then given by

$$c_i = \frac{k_1}{k_i} v_i - \left(b_i \frac{k_1}{k_i} - b_1 \right), \qquad (9.8)$$

without needing to determine the actual value of any of the bias terms, yet ensuring that the responses of all channels were matched to the first channel.

The orientation of the polarization filters was determined for each channel by computing the phase of the sinusoidal response

$$\vec{F}_i = \left[\sum_{n=0}^{200} \frac{(v_i)_n - \bar{v}_i}{200} \sin \frac{2\pi n}{200} \right. $$
$$\left. \times \sum_{n=0}^{200} \frac{(v_i)_n - \bar{v}_i}{200} \cos \frac{2\pi n}{200} \right], \qquad (9.9)$$

where $0.5 \angle \vec{F}_i$ is the orientation of the ith channel polarization filter.

So values of $\vec{F}_i, k_1 / k_i$, and $\left(b_i \frac{k_1}{k_i} - b_1 \right)$ were calculated as above, and Eq. (9.4) was used to solve for polarization direction. Figure 9.6a shows the signal from the sensors over a half-rotation of the turntable.

The geometrical calibration step, in which $\vec{F}$ is computed for each channel, is critical for

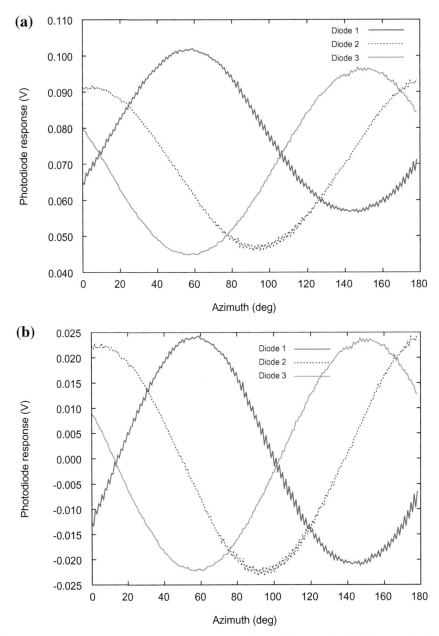

FIGURE 9.6 (a) The response of individual photodiodes oriented vertically to angular displacement in the yaw direction. (b) The calibrated output of the photodiodes after bias and scale factors were matched between the channels.

calculating accurate bearings. Consider a case where the polarization filters on each channel were aligned as well as possible by hand and then assumed to be mutually separated by 60°. Equation (9.4) was applied to normalized but geometrically uncalibrated sensors. Sensor responses

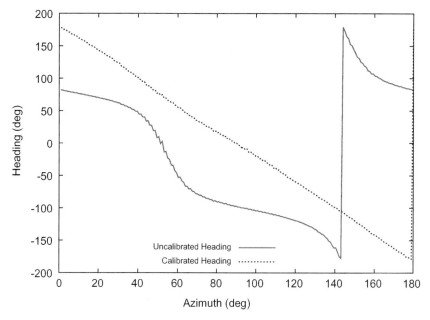

FIGURE 9.7 The dotted line shows the response of the calibrated sensor to angular displacement. The solid line is the output of an uncalibrated sensor in which polarization-filter angles were assumed to be separated by 60°.

are shown in Figure 9.6b. Heading errors in excess of 20°, shown in Figure 9.7 (solid line), occur. After calibration of the filter geometry, the problem is resolved and the signal is usable for navigation, as shown in Figure 9.7 (dotted line). Determination of $\vec{F}_i$ revealed that polarization filters were separated by 55° and 62° with respect to the first sensor channel filter orientation.

9.3.2 Flight Test

The remotely piloted aircraft shown in Figure 9.8 was instrumented with the calibrated polarimeter being tested and an inertial/magnetic attitude and heading reference system that provided an independent measure of heading. The flight was run early in the morning when the sun was low to minimize the probability that direct sunlight might saturate the sensor.

The output of the polarization heading solution and the compass bearing is shown in Figure 9.9. As predicted, the polarization-compass output produced a discontinuity several times during the flight as the solution passed through 180°. There were minor differences between the two measures. The absence of any correction for attitude probably contributed to the most deviation between the magnetic and north-aligned polarization heading.

9.4 OCELLI ATTITUDE REFERENCE

Dragonflies are an ancient airborne predator with clearly recognizable ancestors found in fossil records from the carboniferous era 300 million years ago [26]. Since that time, their basic structure has changed little [27]. As hovering animals, they are extremely reliant on vision for flight control. Although dragonflies have compound eyes with high visual acuity relative to other species [28], it appears that stabilization of

FIGURE 9.8   The small UAV used for polarization-compassing research.

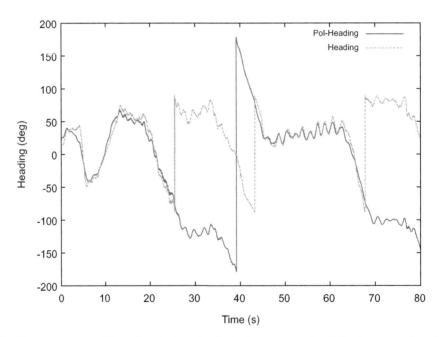

FIGURE 9.9   The output of a calibrated and aligned polarization compass (solid line) compared to the output of a calibrated magnetic compass (dotted line) during flight on a UAV.

their flight is significantly influenced by an auxiliary visual system known as the *ocelli*.

The ocelli are simple eyes mounted on the front of the insect's head. The fields of view of the ocelli in the species *H. papuensis* were mapped by

Stange [29] using the observed *eye-shine*, which term describes the tapetal retro-reflections off the intact ocellar retinas. A graphical representation of the fields of view is shown in Figure 9.10. Each of the side-looking or lateral ocelli has a field of

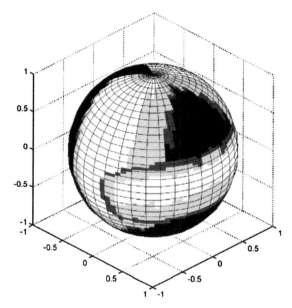

FIGURE 9.10 The field of view of the dragonfly ocellar system: left lateral, right lateral, and median ocelli, mapped onto the view sphere. The measurements were made by observing the points that showed tapetal retro-reflections in each ocellus from an external light source. The hot-scale of color from white (maximum) to black (minimum) indicates the intensity of the retro-reflection.

view 120° wide by 150° high. The forward-looking median ocellus has a horizontal field of view of approximately 160° and a vertical field of view of 40° [29]. Behind each ocellus is a retina containing green- and ultraviolet-sensitive photoreceptors [30].

The dynamic and static properties of the ocelli have been known for some time due to work by Stange and others [30–32]. Individual L-neurons on the retina of median ocelli have been shown by Berry *et al.* [33] to have receptive fields of as little as 12°, allowing spatial information to be preserved for possible higher-order processing in the brain. For the purposes of biomimcry, we have considered the well-known intensity-based responses of dragonfly ocelli originally mapped by Stange [30]. The full functionality of the ocellar visual system of the dragonfly has not yet been determined.

Gradients or steps of intensity between the sky and ground can be used to stabilize an aircraft or ground vehicle. The sky is much brighter than the ground in the blue through ultraviolet wavelengths [34] under almost all daylight conditions in almost all environments. Notable exceptions are environments covered in snow. In the simple case, consider ultraviolet- or blue- sensors looking laterally outward on the right and left sides of a vehicle. Simply balancing the left and right intensities will ensure that both sensors are looking at the horizon. A representative implementation of this concept is shown in Figure 9.11 (right), with a photodiode looking into each quadrant of the horizon. In the vertical plane, the sensors are arranged to look outward, with the center of the field of view aligned with the horizon. A variation of this system has been proposed by Stange in studies of dragonflies [30] and by Taylor in studies of locusts [35, 36], where the arrangement is equivalent to that of Figure 9.11 (left) with no rear-looking ocellus and a wide-field-of-view median ocellus.

The proposed control law was that the aircraft (or insect) would tilt its head or body away from the darker side. In an ideal environment with a level horizon, the consequences of this control law are that the head and ultimately the wings would be level. Consider an aircraft flying toward the observer. When the aircraft is rolled left, as in Figure 9.12 (top), the right sensor sees more sky and the left sees less. In the case of a sensor that sees a bright sky and dark ground, the right sensor will have a higher output than the left. When the aircraft is rolled right as in Figure 9.12 (middle), the imbalance between left and right sensors reverses. When the craft is flying with wings level, as in Figure 9.12 (bottom), the sensors register the same output signal.

The same rule may be applied on the pitch axis for a device that looks forward and backward; the resulting equilibrium point is with the fuselage oriented horizontally. Clearly, the combination of

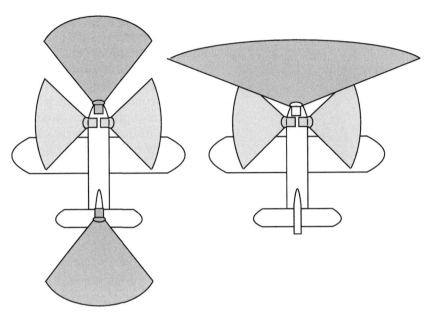

FIGURE 9.11 Arrangements of sensors for intensity-based horizon stabilization. The configuration on the left is ideal but impractical in most aircraft due to the separation between nose and tail. The configuration on the right is convenient, at the cost of less definitive control over pitch.

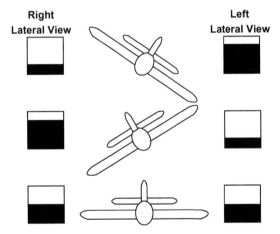

FIGURE 9.12 The opposed sensors detect an imbalance in light levels. To correct the attitude error, the aircraft-control surfaces must rotate the craft away from the dark side. The equilibrium position resulting from following this simple rule is level wings.

lateral- and longitudinal-looking sensors would allow full stabilization of attitude in pitch and roll (in an ideal environment).

Since insects do not have rear-looking ocelli, it can be assumed that the pitch reflexes caused by the median ocellus are referenced to the light-level information sensed by the two lateral sensors or the compound eyes. The solution of using overall light levels as a reference in the pitch axis is less reliable than the differential response on the roll axis. It appears that dragon-flies have evolved higher-order processing in the median ocelli, including a selective response to the rate of pitch axis rotation [32], analogous to a rate gyroscope.

9.4.1 Sensor Signal Processing

Embedded hardware was constructed to test the concept developed in theory and simulation. The resulting device is shown in Figure 9.13. The ultraviolet channel used silicon-carbide diodes with integral ultraviolet-pass filter. The green channel used silicon diodes with an interference filter of bandwidth 100 nm centered at 550 nm.

FIGURE 9.13 The device developed to mimic the function of ocelli consisted of eight photodiodes composed of four ultraviolet/green pairs. Two pairs were used to emulate the broad frontal field of view of the median ocellus.

For the sake of simplicity, we did not use optics to shape the fields of view of the ocelli. We thus depended on the intrinsic geometrical properties of the diodes. Fields of view of the each photodiodes were tested by rotationally sweeping the field of view of each photodiode over a small, bright target (the sun). Their responses are plotted in Figure 9.14. The ultraviolet photodiode had a circular field of view (full width at half maximum) of 100°, compared to approximately 120° for the green photodiode.

From the data in Figure 9.2, it is evident that the sky is bright in all visible wavelengths and ultraviolet. The ground is relatively bereft of shorter wavelengths while remaining almost as bright as the sky in the green and red wavelengths. The sun, on the other hand, is very bright in all wavelengths captured by the scanning device. Our main concern was to limit the effect of the sun and clouds on horizon stabilization.

The ultraviolet band showed a ratio of sky intensity to ground intensity of 15:1 or more, whereas the green and red bands had an overall ratio of 1.1:1 or less.

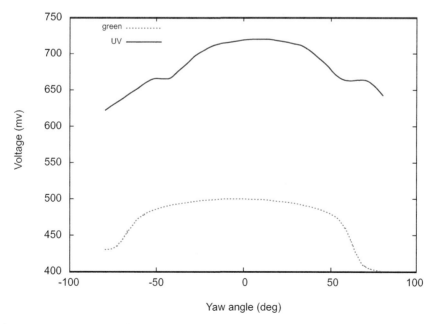

FIGURE 9.14 Responses of the green (dotted line) and ultraviolet (solid line) photodiodes to being rotated over the sun on a rotary table. The field of regard of the green *photodiode* is slightly larger than that of the ultraviolet photodiodes due to the shorter *metal can* packaging of the component. We determined that the fields of view were representative enough of the dragonfly anatomy to form a lateral ocelli, or one-half of a median ocellus.

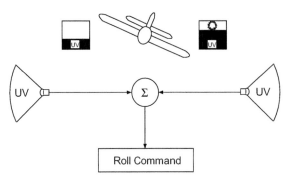

FIGURE 9.15 A simple implementation of the ocelli using only ultraviolet light will fail under some circumstances.

Reliance on the green band for stabilization would fail due to lack of reliable contrast. Reliance on the ultraviolet band alone would be satisfactory under some circumstances but would fail in others. For example, in Figure 9.15, where the sun is within the field of view of one ocellus, the aircraft will bank to turn toward the sun.

For the purposes of horizon stabilization in fair weather, the green band can be seen as a *sun detector*, contributing little to the distinction between sky and ground but removing biases caused by the sun. The effect is shown graphically in Figure 9.16.

Considering the signals received by the two calibrated point-light sensors on one side of the aircraft, we get

$$O_l(t) = U_l(t) - G_l(t), \qquad (9.10)$$

where $O_l(t)$ is the left-side output signal, whereas $U_l(t)$ and $G_l(t)$ are the ultraviolet and green signals

from the photodiodes on the left side. If the sun is not within the field of view of the pair of sensors, then $G_l(t)$ can be treated as constant, i.e.,

$$O_l(t) = U_l(t) - G, \qquad (9.11)$$

and changes in O_l are driven only by changes in U_l. Should the sun be visible to the sensor pair, the output becomes

$$O_l(t) = [U_l(t) + S] - (G + S) = U_l(t) - G, \qquad (9.12)$$

where S is the change in intensity caused by the sun, a constant and almost equal in the green and ultraviolet bands for our purposes. There are other bright objects in the sky, such as clouds, which diffuse sunlight, and these also can be considered as a common-mode signal S.

We refer to this method as *spectral opponency* due to the subtraction of the two spectral bands to extract the feature of interest, the horizon.

The signal that drives the actuator on the axis orthogonal to the view directions of the two sets of photodiodes is produced from the difference of the signals from either side; thus,

$$A(t) = O_l(t) - O_r(t), \qquad (9.13)$$

where $A(t)$ is the processed actuator signal. This processing eliminates the effects of green light intensity on the output signal yet also varies in amplitude with the absolute level of ultraviolet light in the environment (which may complicate control system design). The basic function of the

FIGURE 9.16 The environment measured by a green-sensitive sensor is illustrated on the left; in the center is the same measurement by the ultraviolet sensor. With the correct weighted sum of the two sensors, the effect of the sun can be removed.

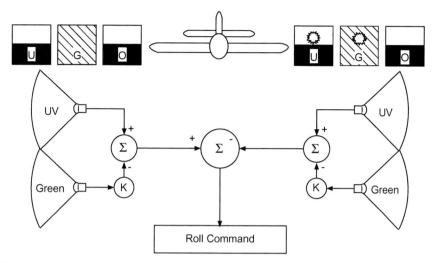

FIGURE 9.17   The environment measured by a green-sensitive sensor is illustrated on the left; in the center is the same measurement by the ultraviolet sensor. With the correct weighted sum of the two sensors, the effect of the sun can be removed.

control system shown in Figure 9.17 will cause the behavior shown in Figure 9.12.

9.4.2 Static Test

The ocelli were simulated using data from the scanning device carrying the ocelli implementation. Intrinsic to the simulation was the ability to roll, pitch, and yaw in order to determine the response of the elements of the ocelli to movements of the horizon.

When the simulation was run with lateral ocelli only, the advantage of spectral opponency became clear. The simulated vehicle was subjected to roll, pitch, and yaw motions within a virtual viewsphere created with the simulated data.

On the roll axis, the difference between the signal from each side in corresponding wavelengths was always non-zero at zero roll angle. Following the simple light-balancing rule, this approach would lead to a significant bank angle. When the signal considered was the spectrally opponent signal from each side, it was apparent that the bias caused by the sun was much

smaller, with a crossover point very close to zero (Figure 9.18). Rotation about the yaw axis simply showed that variations in skylight levels in the ultraviolet and green wavelengths had minimal effect on the difference between the opposed spectrally opponent signals (Figure 9.19). In contrast, the individual photodiode voltages showed significant variations between right and left throughout the rotation.

The simulation revealed that the spectrally opponent signal-processing technique is a significant improvement over the ultraviolet intensity-balancing approach proposed after the first investigation of the dragonfly ocelli [30] while remaining biologically plausible. In earlier work [37], we attempted to use the originally described function of the ocelli for stabilization on both pitch and roll axes. Adequate stabilization was observed in roll, with inconsistent results in pitch. In these flight-test experiments, we chose to use a longitudenal axis control system based on barometric pressure that indirectly maintained a stable pitch attitude. Advanced biomimetic ocelli with designs inspired by recent advances in biology [32] might control the pitch axis more robustly.

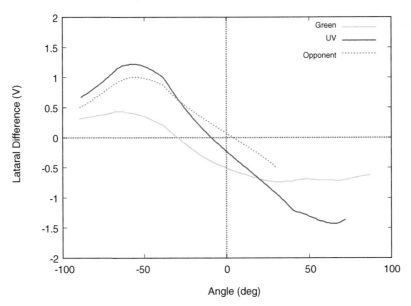

FIGURE 9.18 The signals from the photodiode configuration when the simulated body holding the ocelli was rolled. Ideally, each curve, representing the difference between left and right signals, would pass through the origin. Only the spectrally, opponent signal crossed the origin when wings were level. Biases away from the origin at zero roll angle were caused by the position of the sun.

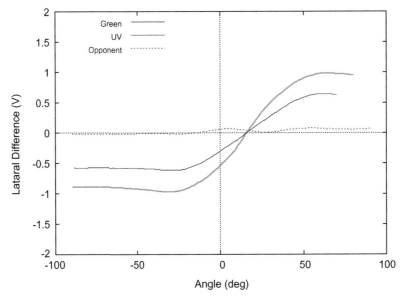

FIGURE 9.19 The signals from the photodiode configuration when the simulated body holding the ocelli was yawed. As the fields of view of the ocelli pan around the horizon, the difference between left and right light levels should be zero or close to zero. Due to the asymmetrical light distribution caused by the sun, the ultraviolet and green diode signals do not remain equal on the left and right sides. The sensor channel using the difference between left and right spectrally opponent signals did not deviate substantially from zero throughout the yaw motion.

9.4.3 Flight Test

The ocelli-based autopilot was given control of the aircraft for periods of approximately 10 s, and the inputs and outputs were logged. The control system commanded aileron deflection proportionally to the sensed differences between the spectrally opponent signal from the photodiodes. A transient occurred on all photodiode responses at approximately 0.25 s at the transition between the pilot and the autopilot.

The raw unscaled data from the photodiodes in Figure 9.20 show stable green-channel outputs and less stable ultraviolet-channel outputs. This is due to the already observed lack of contrast in the green band between sky and ground on clear days (Figure 9.2). The control system held the difference between left and right opponent signals in Figure 9.21 at low levels.

Figure 9.22 shows a series of GPS trajectories over the horizontal plane from the aircraft under control of the ocelli. The estimated radius of curvature indicates a roll angle of less than 2° in each case.

9.5 OPTICAL-FLOW CONTROL OF HEIGHT AND COURSE

Navigation requires more than holding a heading and maintaining an upright attitude. Cross-wind must be compensated by adjusting heading, and the ground must be avoided for most of the flight.

9.5.1 Optical Flow

Apart from the spectral and polarization aspects of insect vision, there is also optical flow. Humans sense this aspect of vision as well, but flying insects are unusually dependent on it. *Optical flow* is the angular motion of the visual field as perceived by an imaging device. A moving platform observes the relative motion of the

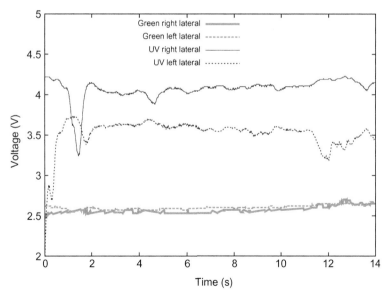

FIGURE 9.20 Raw data from the four photodiodes of interest while the autopilot commanded the control system. Noise levels from the amplifier were of the order of 2 bits in 16 bits; thus observed level changes are predominantly driven by motion of the aircraft and changes in the environment.

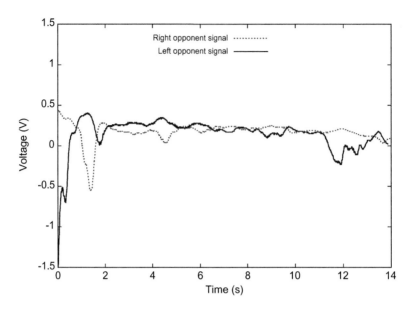

FIGURE 9.21 Opponent signals computed using the scaled difference between the ultraviolet and green photodiodes on respective sides of the aircraft. The difference between the two signals is small after the transient during manual-to-automatic handover at 1 s and before the transient during automatic-to-manual takeover at 11 s.

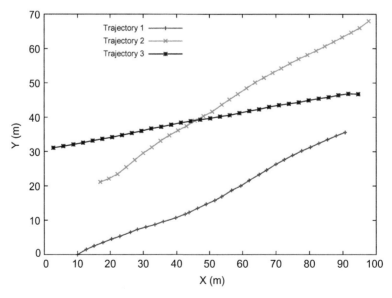

FIGURE 9.22 The GPS trajectories of the aircraft during three periods in which the ocelli controlled roll and pitch for up to 12 s.

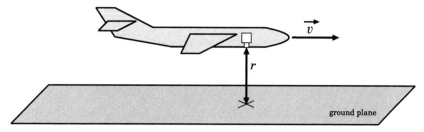

FIGURE 9.23 The simplest case of an optical-flow measurement from an aircraft: a downward-looking camera, a forward-moving aircraft, over a flat surface.

world around it as patterns of optical flow. For the simplest case of a downward-looking perspective-projection camera mounted on a forward-moving aircraft flying over flat ground, as shown in Figure 9.23, the optical-flow vector $\vec{f}$ is given by

$$\vec{f} = -k\frac{\vec{v}}{r}, \qquad (9.14)$$

where $\vec{v}$ is the velocity, r is range to the ground, and k is the camera constant that scales world coordinates to image coordinates. Note that $\vec{f}$ is a 2-dimensional vector on the perspective-projection camera's imaging plane. From this simple equation, it is also clear that if k, $\vec{v}$, and $\vec{f}$ are known, then r can be computed. There is a substantial literature debating the exact means used by insects to compute optical flow [2, 39]; however, it is well established that insects use optically detected relative movement of the world extensively for flight control.

Rotation in pitch and roll in the case of a flying insect or vehicle will also appear as optical flow. If $\vec{f}$ is optical flow in pixels per second, then

$$\vec{f} = -k\vec{\omega}, \qquad (9.15)$$

where $\vec{\omega}$ is a rotation orthogonal to the direction of view. The effect of rotation is not significant for insects if they stabilize their heads (in dragonflies, using the ocelli), so rotation angles do not always enter the visual system. If not,

at any significant height, optical flow induced by rotation of the head will be the primary measurement produced by the optical-flow system. In a *strap-down* system [38], rotation can be canceled by measuring the rotation using gyroscopes and subtracting the effect from the optical-flow field. Unlike the effect of translation, rotation induces an optical Flow pattern that does not vary with distance to the surface viewed.

In all experiments and equations that follow, rotation was removed by software compensation of the optical-flow vector using gyro measurements from the attitude reference in the aircraft.

9.5.2 Height

The difficulty faced by aircraft navigation systems is that there is no absolute instantaneous measure of inertial-frame velocity. Airspeed is a measure of dynamic pressure relative to the airmass in which the aircraft is flying. True airspeed (v_{tas}) is the airspeed measure compensated for the barometric pressure and temperature. Velocity is the resultant of the v_{tas} along the well-calibrated and compensated magnetic heading unit vector ($\vec{m}$) and the wind ($\vec{w}$), as shown in Figure 9.24; thus,

$$\vec{v} = v_{tas}\frac{\vec{m}}{|\vec{m}|} + \vec{w}. \qquad (9.16)$$

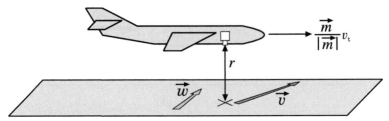

FIGURE 9.24 If navigation is done purely by heading, wind causes the direction of travel to be the resultant of the true airspeed along the heading direction and the wind vector.

Wind may not be a significant issue for a fast platform; however, the forward velocity of a micro UAV might approach 0. A simple feedback loop attempting to reduce altitude to increase optical flow to some expected value might reduce altitude until contact with the ground occurs.

The difficulties of computing height from a single optical-flow sensor at an instant are intractable. Obvious solutions, such as modulating speed and observing changes in direction of flow or magnitude of flow, will succeed under some conditions but fail under others. In the experiments described in the remainder of this section, when altitude was being computed, $|\vec{v}|$ was the speed output of a GPS. A fast or robust system could ignore the GPS velocity and rely entirely on v_{tas} with a potentially insignificant reduction in the accuracy of altitude flown.

9.5.3 Course

Correction of the course over the ground is possible using optical flow and a heading reference. Substituting from Eq. (9.14), we get

$$\vec{f} = -\frac{v_{tas}\frac{\vec{m}}{|\vec{m}|} + \vec{w}}{kr};$$

(9.17)

hence, any changes in r have no effect on the angle $\angle\vec{f}$, but only on the magnitude $|\vec{f}|$. Thus measurements of optical-flow direction and magnetic compass (or gyro compass) heading

provide the direction of flight. The direction of flight is calculated from the sum of the magnetic heading vector angle and the optical-flow vector angle; thus,

$$\angle\vec{v} = \angle\vec{m} - \angle\vec{f}.$$

(9.18)

Obviously, the solution to following a desired path is to fly the desired heading on the bearing reference $\angle\vec{m}$, measure $\angle\vec{f}$, and compensate the desired heading. This could be considered as a single step, repeated at discrete intervals of minutes or seconds, as is appropriate compared to a human pilot. The alternative is to run the system as a closed-loop regulator, with the difference between the actual heading angle and the desired heading direction as the error to be nulled. The regulator shown in Figure 9.25 was implemented in the flight test. A simple proportional/integral/differential (PID) regulator was used to manage the bank angle that causes change of heading.

9.5.4 Flight Test

The small UAV used in the experiment is shown in Figure 9.26. Optical flow was implemented using a single sensor from an optical computer mouse. The sensor provides a reliable optical-flow signal under most daylight conditions. It was designed for high optical-flow rates (as would be experienced within 1 mm of a desktop). The optics we fitted had a narrow 5° field

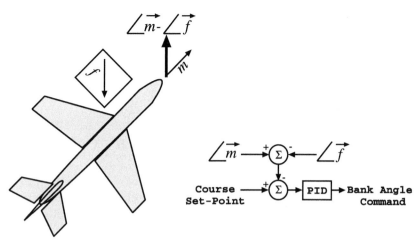

FIGURE 9.25 A simple PID regulator using compass heading and direction of optical flow removes wind drift in flight.

FIGURE 9.26 The UAV used for the flight trials was an electric-powered model aircraft with a 2m wingpsan. It was fitted with an autopilot capable of conventional GPS/inertial flight and customized autonomous modes of flight. The optical-flow sensor was configured to gaze directly downward.

of view to provide enough optical-flow signal when in flight.

The flight-control system was configured to control height using optical flow and GPS velocity, and heading using the direction of optical flow, combined with a magnetic compass. The environment was a flat expanse of desert. The ground track was maintained to

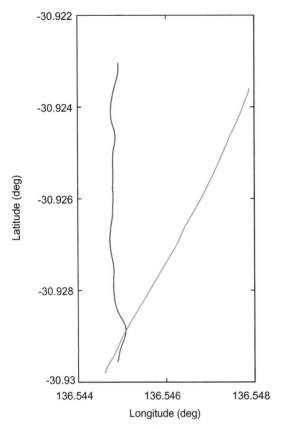

FIGURE 9.27 The red/left trace shows a course that was corrected for lateral drift using optical flow; the green/right trace is a flight immediately afterward in which the optical-flow system was turned off. In both cases, the UAV was commanded to fly grid North.

within 1° over an 800 m trajectory, shown in Figure 9.27.

The altitude computed using optical flow was offset using the known height of the ground. This allowed the flight-control system to switch between GPS altitude and optical-flow altitude without large steps. Optical flow depends on visual texture on the ground and is thus

somewhat dependent on the environment. As Figure 9.28 shows, there is close agreement between the optical-flow computation and GPS over the 800 m flight.

9.6 CONCLUSION

Integration of sensors and controls into a robust system is a challenge in biomimetic flying systems. Implementations of each of the sensors function as part of a flying system, but the system does not yet operate as an integrated whole. Each sensor has weaknesses. Optical flow requires texture, ocelli require a distinct horizon, and polarization requires visibility of the sunlit or moonlit sky. There are layers of behavior above the simple reflexes we have implemented that would allow systems to continue to operate in adverse conditions. What we have shown is a series of reflexes that are capable of almost complete control of a UAV using only light from the environment, which is radically different from conventional navigation avionics.

We have demonstrated substantial autonomy with an array of simple sensors in a simple environment. Insects use optical flow over the entire visual field for flight control; they have a distributed view of the polarization pattern above and below the horizon. The polarization-sensitive array in the dorsal rim extends over a large arc of the sky. Ocelli may have spatial resolution required to detect motion and resolve the horizon. A comprehensive sensor suite emulating the insect optical and neural system would provide the means to implement these behaviors robustly in complex environments.

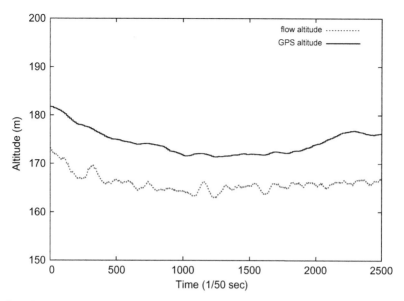

FIGURE 9.28 The solid trace shows the height reported by GPS, the dotted trace shows the height reported by optical flow. The flight was 800 m in length.

References

[1] K.V. Frisch, *Tanzsprache und Orientierung der Bienen*, Springer-Verlag, Heidelberg, Germany (1965).

[2] B. Hassenstein and W. Reichardt, System theoretische Analyse der Zeit-Reihenfolgen und Vorzeichenauswertung bei der Bewegungsperzeption des Russelkafers Chlorophanus, *Z Naturforsch* **116** (1956), 513–524.

[3] M.V. Srinivasan, S.W. Zhang, M. Lehrer, and T.S. Collett, Honeybee navigation en route to the goal: visual flight control and odometry, *J Exp Biol* **199** (1996), 237–244.

[4] R.A. Brooks, Elephants don't play chess, *Robot Auton Syst* **6** (1990), 3–15.

[5] J. Wessnitzer and B. Webb, Multimodal sensory integration in insects–towards insect brain control architectures, *Bioinsp Biomim* **1** (2006), 63–75.

[6] U. Homberg, *Methods in insect sensory neuroscience*, CRC Press, Boca Raton, FL, USA (2005).

[7] T.P Pearsall, *Photonic essentials*, McGraw-Hill, New York, NY, USA (2003).

[8] R. Schwind, Polarization vision in water insects and insects living on a moist substrate, *J Comp Physiol A* **169** (1991), 531–540.

[9] R. Doraiswami and R.S. Price, A robust position estimation scheme using sun sensor, *IEEE Trans Instrum Meas* **47** (1998), 595–603.

[10] J.W. Strutt, On the transmission of light through an atmosphere containing small particles in suspension, and on the origin of the blue sky, *Phil Mag* **47** (1899), 375–384.

[11] K.L. Coulson, *Polarization and intensity of light in the atmosphere*, Deepak Publishing, Hampton, VA, USA (1989).

[12] G.P Können, *Polarized light in nature*, Cambridge University Press, Cambridge, UK (1985).

[13] A. Dollfus, Étude photométrique des contrées sombres de la surface de la planéte Mars, *Compt Rend Acad Sci* **244** (1957), 1458–1460.

[14] G. de Vaucouleurs, *Physics of the planet Mars*, Faber and Faber, London, UK (1954).

[15] S. Thakoor, J.M. Morookian, J. Chahl, B. Hine, and S. Zornetzer, BEES: Exploring Mars with bioinspired technologies, *IEEE Comput Mag* **37** (2004), 38–47.

[16] J.W. Strutt (Lord Rayleigh), On the light from the sky, its polarisation and colour, *Phil Mag* **41** (1871), 107–120, 274–279.

[17] M. Kayton, *Avionics navigation systems*, Wiley, New York, NY, USA (1997).

[18] T. Labhart and E.P. Meyer, Neural mechanisms in insect navigation: polarization compass and odometer, *Curr Opin Neurobiol* **12** (2002), 707–714.

[19] T.W. Cronin, E.J. Warrant, and B. Greiner, Celestial polarization patterns during twilight, *Appl Opt* **45** (2006), 5582–5589.

[20] G. Horváth and D. Varjú, *Polarized light in animal vision: polarization patterns in nature*, Springer-Verlag, Heidelberg, Germany (2004).

[21] D. Pye, *Polarized light in science and nature*, Institute of Physics Publishing, Bristol, UK (2001).

[22] E.S. Pedersen, Airline navigation in polar areas, *J Navigation* **11** (1958), 356–360.

[23] E.W. Anderson, I-Navigation in polar regions, *J Navigation* **10** (1957), 156–161.

[24] D. Lambrinos, H. Kobayashi, R. Pfeifer, M. Maris, T. Labhart, and R. Wehner, An autonomous agent navigation with a polarized light compass, *Adap Behav* **6** (1997), 131–161.

[25] K. Zhao, J. Chu, and T.W.Q. Zhnag, A novel angle algorithm of polarization sensor for navigation, *IEEE Trans Instrum Meas* **58** (2009), 2791–2796.

[26] D. Grimaldi and M. Engel, *Evolution of the insects*, Cambridge University Press, Cambridge, UK (2005).

[27] J. Silsby, *Dragonflies of the world*, CSIRO Publishing, Collingwood, Victoria, Australia (2001).

[28] G.A. Horridge, The compound eye of insects, *Sci Am* **237** (7) (July 1977), 108–120.

[29] G.F. Stange, S. Stowe, J.S. Chahl, and A. Massaro, Anisotropic imaging in the dragonfly median ocellus: a matched filter for horizon detection, *J Comp Physiol* **188** (2002), 455–467.

[30] G. Stange and J. Howard, An ocellar dorsal light response in a dragonfly, *J Exp Biol* **83** (1979), 351–355.

[31] R. Berry, J. van Kleef, and G. Stange, The mapping of visual space by dragonfly lateral ocelli, *J Comp Physiol* **193** (2007), 495–513.

[32] J. von Kleef, R. Berry, and G. Stange, Directional selectivity in the simple eye of an insect, *J Neurosci* **12** (29) (2008), 2845–2855.

[33] R. Berry, G. Stange, R. Olberg, and J. van Kleef, The mapping of visual pace by identified large second-order neurons in the dragonfly mediam ocellus, *J Comp Physiol* **192** (2006), 1105–1123.

[34] M. Wilson, The functional organisation of locust ocelli, *J Comp Physiol* **124** (1978), 297–316.

[35] C.P. Taylor, Contribution of compound eyes and ocelli to steering of locusts in flight: I. Behavioural analysis, *J Exp Biol* **93** (1981), 1–18.

[36] C.P. Taylor, Contribution of compound eyes and ocelli to steering of locusts in flight: II. Timing changes in flight motor units, *J Exp Biol* **93** (1981), 19–31.

[37] J.S. Chahl, S. Thakkor, N. Le Bouant, G. Stange, M.V. Srinivasan, B. Hine, and S. Zoornetzer, Bioinspired engineering of exploration systems: A horizontal sensor/attitude reference system based on the dragony ocelli for Mars exploration applications, *J Robotic Sys* **20** (2009), 35–42.

[38] M.V. Srinivasan, Visual motor computations in insects, *Annu Rev Neurosci* **27** (2004), 679–696.

[39] D.H. Titterton and J.L. Weston, *Strapdown inertial navigation technology*, Peter Peregrinus, Stevenage, Hertsfordshire, UK (1997).

ABOUT THE AUTHORS

Javaan Chahl obtained his bachelor's degree in computer engineering from the University of Newcastle, Australia, in 1991. He completed a graduate diploma of neuroscience at the Australian National University (ANU) in 1992 and was awarded his doctorate in 1997 for biologically inspired machine vision systems. He joined the Defence Science and Technology Organisation (DSTO) in 1999 and continued at the ANU on attachment until 2005, when he relocated his laboratories to DSTO. In 2011, Javaan joined RMIT University as the first Professor of Unmanned Aerial Vehicles. In July 2012, Javaan joined the University of South Australia as Chair of Sensor Systems.

Akiko Mizutani completed her doctorate in biology from the Kyushu University in Japan in 1996. She obtained a computer science graduate degree in 1999 from Monash University, Australia. Since 2004 she has been the CEO of Odonatrix Pty. Ltd., a technology company specializing in research on integrated aerospace systems.

CHAPTER

10

Biomimetic Textiles

Michael S. Ellison

School of Materials Science and Engineering, Clemson University,
Clemson, SC 29634-0971, USA

Prospectus

In a sense, the archetype of bioinspiration for materials design and use is textiles. The field of biomimesis has spawned many new materials and continues to be a fruitful field of investigation. This chapter presents the current state of bioinspiration in textiles, how this has resulted in improved fibrous materials, how it may inform our continued progress. Because I have found many preconceived notions about the field that need addressing before the application of biomimetics to textiles can be truly appreciated, I begin with an introduction to textiles. Next, naturally enough, the discipline of biomimesis is introduced and then fleshed out in terms of its textile engineering importance. Following this, some details on fiber and textile science and engineering and biological concepts germane to our topic are presented. In the last step in this journey, the marriage of biomimesis and textiles is performed and some consequences revealed. Finally, I offer some prognostications on the topic.

Keywords

Biomimetic, Cellulose, Clothing, Cotton, Fabric, Fiber, Hagfish slime threads, Inspiration, Mussel byssus, Natural fiber, Polypeptide, Resilience, Spider silk, Strength, Sustainable, Technical textiles, Textiles, Toughness, Wool, Yarn.

10.1 TEXTILES

Clothing is quintessentially human. Although there are, of course, reports of other species using natural materials for shelter, there are no reports, for example, of monkeys wearing clothing of their own making, the organ-grinder's friend notwithstanding. It is a common belief that humans used animal skins and even vegetation (e.g., fig leaves) as coverings for protection from the elements. Body covering may also have been concocted for purposes such as decoration or prestige. Whatever their inception, textiles and clothing have been present in human history since the earliest records and reflect both the raw materials available to people and the technologies that they developed.

Textiles, which here includes the fibers that are then made into yarn and thence into fabrics (as we shall see, there are many intermediate and transitional steps in the general process), apparently appeared in the Middle East during the late Stone Age. In fact, humans' use of biological fibers is a venerable art: We have been spinning natural fibers into yarns and forming yarns into fabrics for over 8,000 years [1].

Engineered Biomimicry

http://dx.doi.org/10.1016/B978-0-12-415995-2.00010-6

From those early times to the present, the continued development of methods of textile production has broadened the available choices of textiles and thus influenced how people clothe themselves and protect themselves from their surroundings. The study of the DNA of the body louse has allowed us to infer the length of time during our evolution that humans have been wearing clothing. The common louse particular to humans diverged from its ancestors possibly as long as 650,000 years ago [2]. Scientific research, based on the study of the lice that infest modern humans estimates that humans have been wearing clothing for at least 107,000 years [3]. Thus, it is clear that the manufacture of clothing and related textiles is a venerable and integral part of human culture. In fact, the ubiquitous nature of textile materials renders them essentially invisible; nonetheless, as it was in ancient times, textiles remain a critical component of our modern infrastructure.

The first actual textile, instead of skins simply sewn together, was probably felt [4]. In modern incarnations, felts can be made from essentially any fiber through the nonwoven fabric manufacturing process, which is introduced in Section 10.3.1.3; originally, however, felts were made only from animal hair fibers. The surface of a hair fiber (keratin) such as wool has scales (Figure 10.1). When a bundle of such fibers, randomly oriented, is subjected to mechanical motion, especially in the presence of moisture, the scales interlock in a ratcheting fashion, causing a relatively permanent fabric structure, i.e., felt, to be formed. Animal hair felt is the earliest form of a nonwoven fabric. Other fabric types require first that the fibers be formed into yarns.

It is not clear when humans first discovered that fibers, such as from wool or cotton, could be twisted together into the larger and more robust structure of yarns, but it is clear that this has been done from prehistoric times. Nonetheless, the engineering aspects of the processes for mechanized yarn formation were not well developed until the Industrial Revolution. Clothing

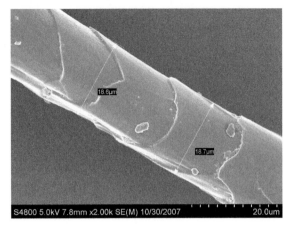

FIGURE 10.1   SEM micrograph of a Merino wool fiber. (Photo: Dr. Mevlut Tascan)

production continued to be accomplished by hand until sewing machines became common in the 19th century, thereby facilitating production of apparel.

In essence, to produce a yarn, the randomly oriented fibers must first be aligned and rarefied and then twisted together. *Carding*, the process of combing the fibers into alignment, was originally done by hand using *card flats*, flat or curved boards with wire teeth. The carding machine was developed to increase the speed of production by using rotating drums with the card wire on them. The output of the card, the *card web*, is a thin mat of oriented fibers. This mat is pulled together to form the *card sliver* (the *i* is pronounced *eye*). Although there are ancillary steps in the process, we will not review them here; the sliver is then drawn out to a smaller diameter and fed into the spinning machine. The familiar spinning wheel was a human-powered machine developed to spin the fibers into yarns. An earlier system, the *spinning bobbin*, was gravity powered.

At the yarn-spinning machine, the most obvious level of hierarchical structure is introduced by making the yarn (having a high aspect ratio) from fibers (also with high aspect ratio, albeit of smaller diameter). However, the fibers, which are thin structures with a high aspect ratio, are made of *polymers*, which are long chain-like molecules. These molecules are, of course, very thin, to say the least, but the transverse dimension of fibers is somewhat larger, ranging from as large as 100 μm down to the nanometer scale. *Yarns*, whose diameter is again increased in relation to that of fibers, are composed of these long, thin fibers. This type of hierarchy of form is a dominant feature in textiles, as it is in nature. Because of this development of structure, yarns and fibers have many of the same properties, such as high aspect ratio and good flexibility, but the yarn is stronger and may be much longer. In fact, the length of natural fibers ranges from about 1 in. for cotton to up to 6 in. for wool, so spinning makes the natural material more useful.

Fibers and yarns are, in a sense, one-dimensional. To provide the function of protection mentioned earlier, the dimensionality must be increased. Fabric-forming processes provide higher dimensions, from the two-dimensional clothing fabric to robust three-dimensional structures. Fabrics that are composed of yarns have the most broad application field, and the three basic classes of yarn-based fabrics are woven, knitted, and braided. These are briefly covered in Section 10.3.1.3.

It is generally agreed that there are two essential methods of fabric formation from yarns: *interlacing* and *interlooping*. Knits basically are formed by connecting loops of yarn. There are several types of knits, and each has a specific set of properties, but the basic idea is represented by the simple jersey knit. This fabric structure exhibits a great deal of dimensional fluidity and flexibility. Some other knit structures offer increased control over the dimensional stability and also impart a higher flexural modulus [5].

In contrast with knits, woven fabrics and braids are formed by the interlacing of yarns. The quintessential braiding process is that of a maypole celebration, where two or more groups of people dance in opposite-directed twirling and interlacing circles around a pole, each person

holding the end of a ribbon of a (woven) fabric, the other end of which is tied to the top of the pole. In manufacturing a braid, yarns or monofilaments are used. There are multiple rotating yarn carriers, determining the desired complexity of the braid [5]. Renderings of two braids are shown in Figures 10.2 and 10.3, one with 6 yarns (Figure 10.2) and one with 16 (Figure 10.3).

The braid structure, sometimes referred to as a *plait*, has a resulting dominant property best represented by the monkey trap. When the braid is pulled (lengthened), the inner diameter of the tube reduces and tightens onto whatever is

FIGURE 10.2 Rendering of a braid comprising six yarns. (Used with permission: Kevin John, www.kevinjohn3d.com.)

FIGURE 10.3 Rendering of a braid comprising 16 yarns. (Used with permission: Kevin John, www.kevinjohn3d.com.)

inside. One very common use of a braid is to enclose a collection of elastic strips to make bungee cord. When the braid is extended to the point where the yarns are pressed against one another, the braid cannot be extended further and is quite strong and rigid.

Simple plaiting results in one component lying over and under its nearest neighbors, with the number of interlacings depending on the complexity of the braid. It is thus structurally analogous to a woven fabric, with the two major differences being that the plain weave is a flat structure (although it is possible to weave complex 3D shapes), whereas the braid is intrinsically a tube and there is a skew angle of the structure relative to the major axis. In a plain weave, nearest neighbors cross over one another, whereas the yarn interlacing may occur with some complexity in braids. The mechanical properties are, of course, significantly different between the two. Tension at 45° applied to a plain weave results in a skew deformation (square to diamond). Tension (compression) applied axially to a braided tube, which is analogous to the skew load on a plain weave because the plaiting lies at 45° to the tube axis, results in an extension (shortening) with associated narrowing (expansion) of the tube. These actions have been exploited in making artificial pneumatic muscles. These artificial muscles have an elastic membrane (hose) inside the braid, which, when inflated with air, expands, causing the braid to shorten [6]. Although perhaps not a pertinent example of biomimetics *per se*, this is nonetheless an application of a textile material to mimic a biological function. Some plant structures do resemble braids [7].

Woven fabrics, perhaps the most ubiquitous of all fabrics, are produced by interlacing yarns on a weaving loom. The most basic weave is a plain weave, shown in Figure 10.4. There are two sets of yarns (at the most essential level), called the *warp* and the *weft*. The warp runs the length of the fabric, and the weft is in the cross-wise direction. In the rendering shown in Figure 10.4, the

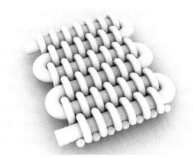

FIGURE 10.4   A visual model rendering of a plain-weave fabric. (J. Manganelli, personal communication. Used with permission.)

larger-diameter element depicts the weft yarn; in real fabrics, the warp and weft yarns are more nearly matched in size. This is a representation of a woven structure as it would be produced on a shuttle loom. The weft yarns are continuous, so the edges of the fabric (the *selvedge*) are neat, without loose ends. Fabric formed using shuttleless weaving machines—there are several types of these looms and they are faster—have loose selvedges.

The word *textile* is often assumed to be synonymous with cloth used for apparel, sheets, towels, and similar items. Admittedly, the predominant and most venerable use of fibers and yarns is for apparel fabric and bed sheeting. Nonetheless, there is a broad range of common textiles that sometimes goes unnoticed, ranging from floor coverings and upholstery to tenting material. Beyond these textile materials, we find fibrous materials in industrial and architectural applications, some with structures and functions very different from everyday textiles. The commonly used fibers and some of the specialty, high-performance fibers are discussed in Section 10.3.1.1.

Development in the field of modern textiles parallels that of synthetic fibers. The cost and limited survivability of natural fibers, with the possible exception of mineral fibers, limit their use in demanding environments. Extremely high loads and rates of loading, high temperatures or chemically corrosive situations, prolonged exposure to radiation, or attack and degradation by biological entities (e.g., microbes) are severe limiting factors for engineering of structures using natural fibers. Synthetic fibers made from polymers such as polyester, nylon, and polypropylene provide the textile engineer with a broad palette of reasonably inexpensive fibrous materials, which are capable of displaying very good physical and chemical properties. If we add to this list high-performance fibers, such as aromatic polyamides or aramids (e.g., Kevlar®), polybenzimidazole (PBI), carbon, and others, extremely high temperatures or direct exposure to flame are about the only factors limiting the engineer. In fact, though, PBI and the aramids (such as Nomex®) are highly flame resistant. It is important to have knowledge of all the properties of the fiber in use, since there are strong trade-offs between properties. For example, PBI does not easily ignite and burn, but it is not stable to strong acids; Kevlar is very strong, but it is not very resistant to abrasion; and carbon fiber is strong but extremely brittle [5]. Engineers must realize the limitations of each fiber type when designing a material in terms of the stresses of the production techniques needed to form fibrous materials.

The 20th century saw the invention of synthetic polymers, in particular nylons and polyesters. Given the low prices of the petroleum feedstock at that time, volumes of these synthetic fibers, eventually surpassed the natural fibers, and today synthetics make up two-thirds of the textiles produced worldwide. Synthetic fibers have benefits over natural fibers for several reasons. For example, synthetic fiber diameters can range from the nanoscale to monofilaments suitable for fishing, with material properties consistent from batch to batch [5].

Synthetic polymers are attractive in terms of their properties; nevertheless, their use is coming under close scrutiny owing to the fact that the feedstock for synthetic fibers, petroleum, is finite. Furthermore, global environmental awareness has tarnished some of the luster of synthetic polymers.

Given this situation, the synthetic fiber industry as it currently exists will change and be replaced by an industry based on renewable feedstock.

One type of relatively modern fabric is the class of nonwoven fabrics, in which the composing fibers (almost exclusively synthetic) are combined directly into the fabric structure. There are two broad classes of nonwovens, with several members in each class. In one class, the synthetic fibers are extruded directly into a fabric structure; in the other, the fibers are first formed and then the fabric is made from them, with no intermediate step such as yarn formation. In the first class, the direct polymer-to-fiber lay-down process, there are a few subclasses, but in essence the fibers as they leave the spinneret are laid down on a moving belt to form a fabric, which is generally relatively thin. Melt blowing and spun bonding are the names of two such processes. These materials are not considered further in this chapter.

The second class of nonwoven fabrics requires the fibers be first made by a fiber production process, packed and shipped, and then made into the fabric. Applications range from paper (made from processed wood pulp, generally) to automobile headliners (made from synthetic fibers). In papermaking, slurry containing the fibers is cast onto a porous belt and vacuum pressed and dried. It is an example of a *wet-lay* process. Consolidation is accomplished by hydrogen bonding between the cellulose molecules in the pulp as it is dried. The dominant nonwoven process not using a slurry (so, *dry-lay*) uses needlepunching, i.e., entangling of the fibers, generally synthetic, by multiple penetration of a web of the fibers by barbed needles. The fabrics produced in this process can vary from relatively thin to quite thick and heavy. Even for the thin, lightweight materials, nonwoven fabrics do not have the drape properties required for comfortable apparel. Nonetheless, they are relatively inexpensive to manufacture and find many disparate uses, mostly in industrial settings and technical textiles.

10.2 BIOMIMICRY

10.2.1 Biology

We have argued the importance for having an understanding of textiles to developing an appreciation of biomimetic textiles, and so it is with biology. Since my background is in polymer and textile physics, my presentation of the foundations of biology is consequently limited.

In trying to make sense out of what we see in the world about us, we first classify things into ordered categories so that we can have rational discussions about what we see. In biology, this exercise has reached a pinnacle of development, and rightly so: The living world is intensely complex, and one must have a name for whatever is being discussed. However, biological classification schemes are only a part of scientific *taxonomy*. A common example of a general taxonomy is of the processes in learning [8]. In fact, the term taxonomy can simply refer to systematic knowledge organization. One use reflects the biologist's penchant for the classification of living organisms and is a hierarchical classification of things or concepts in what may be considered a tree structure. A more recent usage of the term is to refer to any controlled vocabulary of terms for a subject area domain or a specific purpose. The terms may or may not be arranged in a hierarchy, and they may or may not have even more complex relationships between each other. Thus, *taxonomy* has taken on a broader meaning that encompasses subject-specific glossaries and thesauri, controlled vocabularies, and ontologies [9]. Biological classification is different from some other classification systems because the similarity between organisms placed in the same class is not arbitrary but is instead a result of shared descent from common ancestry; that is, evolution determines the class.

With the advent of molecular biology, which was heralded by the elucidation of the structure and function of DNA, the field of biology was given a firm footing in the science of biochemistry, and from that, chemistry, and by extension, physics. Modern biology thus has become more accessible to practitioners in the physical sciences and engineering. It is because of this interaction that the field of biomimetics has entered its adolescent growth spurt, as it were. If we consider the work of Leonardo da Vinci in developing his flying machines in the late 1400s, we witness inspiration from nature. Within the arcane knowledge base that is biological science, there is a veritable trove of materials and processes that is only now beginning to be tapped.

Finally, as regards biology, the natural world has several intricate examples of systems engineering if we look at it from the appropriate perspective. Germane to the present chapter, the quintessential example of systems engineering in biology is a spider's orb web. The system aspect of these webs is apparent when one takes into account the different properties of the various silk fibers in the web. The mooring lines and the radial framework are composed of the very tough dragline silk; the flagelliform capture silk spiral linking the radial members is more extensible and still tough. This member allows the web to absorb the kinetic energy of the flying insect without breaking, while relying on the dragline silk for support [10].

10.2.2 Key Principles of Biomimicry

The key principle of biomimesis is to use the natural biological world as source of inspiration and as a guide in the development of new materials. Detailed study of systems and organisms within natural systems, which may be used as models in science and engineering of new materials, is required. However, one must recognize that such a study must focus on elucidation of the relationships between structure and function in natural systems in order to apply the fruit of such study in engineering.

Biomimicry has experienced a rapid development since Janine Benyus published her seminal work in 1997. In that work, she defined biomimicry as having three primary components [11]:

- *Nature as model.* New solutions to human problems.
- *Nature as measure.* Ecology/evolution as the standard of what works.
- *Nature as mentor.* Learn from nature, not take from nature.

The view Benyus presented is that of natural systems as a source for inspiration, which is predicated on an active study of what nature presents and focused on what solutions nature employs. Nature is a mentor, not a resource to exploit; thus, we are directed to nature to assess the viability of our solution. This is sustainability, and the standard is high.

The evolution of biomimicry has spawned several terms around which to form discussions. One of the earlier terms for this field is *bionics*, and *biomimetics* was coined in the early 1950s. *Bionic* became a much more popularized term, thanks in part to the entertainment industry (*The Bionic Woman* television show, for example), while biomimetics and biomimicry remained the purview of the scientific community. Other terms are also in use. The NRC report *Inspired by Biology* provided the following definitions [12]:

- *Biomimicry.* Learning the mechanistic principle of a natural function and then trying to achieve that function in a synthetic material.
- *Bioinspiration.* From observing a particular task performed by a natural system, gain inspiration for a synthetic system to effect the same task.
- *Bioderivation.* Hybrid between natural and synthetic materials and functions.

As we explore how we may take inspiration from nature to design fibrous material structures—textiles—that mimic natural systems in process, form, and function, it is important to keep in mind that fibrous structures are inherently hierarchical: From the long-chain polymer comes high-aspect-ratio fibers; the fibers are combined to make yarns and threads, which are then compiled into fabrics, which increase the dimensionality of the hierarchy. Thus the materials we consider include the polymers that comprise the fibers, which are often synthetic, but there is a broad palette of natural fibers from which to choose. In fact, the available fiber materials have been bolstered by molecular biology, the extremely interesting and fruitful study of recombinant DNA, which has provided researchers with new tools with which to study nature's fiber materials in order to mimic them and to make new ones [13]. In terms of higher-order constructions, plants and animals have both provided insight into textile structures. One thing we find is that there is still much to learn about using nature's models of sustainable manufacturing in our own manufacturing of mimicked materials.

10.3 BIOMIMESIS IN TEXTILE-MATERIALS ENGINEERING

10.3.1 Textile-Materials Engineering

One of the themes in this chapter has been recognizing that, in order to appreciate how the science and engineering of textiles and fibers have been informed by nature, it is helpful to have at least a basic understanding of the field. This section lays the foundation for later exposition of some bio-inspired textiles by further developing the reader's fiber and textile base.

10.3.1.1 Fiber

Fiber is a dietary necessity. Fibers are a clothing necessity. It is this latter use of the term *fiber* that is relevant here. Interestingly, dietary fiber at the microscopic level resembles a cellular structure [14] reminiscent of plastic insulating foams, and, of course, synthetic fibers are made from plastic.

Organic fibers, whether natural or synthetic, are made up of linear-chain polymers, with the most common natural fibers being cotton, wool, and (silk worm) silk, and common synthetics being polyesters, polyolefins, and polyamides. Other commercial polymers used in synthetic fibrous materials include the aromatic polyamides (such as Kevlar®) and other high-performance polymers. The common traits among these materials derive from their being composed of linear chains that have been oriented and locally crystallized during fiber production, as discussed later in this section. In contrast to fiber production in natural systems, synthetic fiber production generally requires the use of high temperatures and pressures and/or unfriendly solvents. Consequently, there has been much exploration into the natural systems (silkworms and spiders being the most common) in order to devise more sustainable methods of fiber production.

Natural fibers are essentially exclusively composed of either cellulose (e.g., cotton and flax) or protein (e.g., wool and silk). As in the case of the synthetic fibers, these natural polymers also comprise (mostly) linear polymers with orientation and crystallinity. Because the cellulose molecule is consistent between plants, cotton, flax (linen), ramie, and hemp all have the same basic polymer structure. There are other constituents, such as lignin, which vary in amount, however, depending on the source. In addition, the degree of crystallinity may also be different, which contributes to the different properties of the fibers. The cross-sectional shape of the fiber also contributes to its properties. Cotton is bean-shaped in cross-section, whereas ramie tends to be more angular [5]; hence, the flexural rigidity, being strongly influenced by the moment of inertia, is much different for each of these fibers. Cotton is a seed hair, whereas the other fibers derive from the plant itself, such as the stem. This difference in function helps us understand why there are marked differences in structure: Form follows function. Common to all plant-derived fibers is cellulose. An organic compound with the chemical formula $(C_6H_{10}O_5)_n$, cellulose is, by this formula, a polysaccharide. Unlike other saccharidic materials, such as starch, the cellulose chain

consists of up to 10,000 $\beta(1\rightarrow4)$ linked D-glucose units.

Wool is the hair of sheep. As such it is a protein, a polypeptide composed of a specific set of amino acids. Silk (from the silkworm) is the cocoon material housing the pupae, and, like wool, silk is also a polypeptide. The different amounts of the various amino acids in wool vis-à-vis silk account for the differences in the properties of the two fiber types. In addition, owing to their different functions, the physical structure of the fibers is different. Wool fibers have scales (see Figure 10.1), whereas the surface of silk fibers is relatively smooth. The pupae are delicate and so need a smooth cradle in which to grow. It is generally agreed [5] that the function of scales on hair, which imparts a roughness to the fiber, is to help keep the skin of the animal clean. The protruding edge of the scale points away from the skin, making more difficult the progression of detritus from the tip of the hair to the skin. Also, dead epidermal cells will be preferentially moved from the surface of the skin toward the tip of the fiber, i.e., away from the skin, as the hair grows.

It seems reasonable to presume that the natural world was the inspiration for the earliest engineering forays into fiber manufacturing, or fiber spinning. The process by which the natural fibers, notably the silk materials but also hair, are produced is essentially that of extrusion. Not all fibrous materials in nature are formed through extrusion, however. The growth of seed hairs and plant stems does not follow an extrusion model. Nor does the byssus thread of the mussel. Nonetheless, the resulting fibers all have the same microstructural characteristics. Of course, natural extrusion is conducted under ambient conditions, which is not the case for the synthetics, but nonetheless the similarities are remarkable.

Irrespective of whether the process is natural or synthetic, the polymer, in a fluid state, is pumped through a small orifice or, in the case of synthetics, simultaneously through many small orifices. The orifice is referred to as a *spigot* in spider silk spinning. The plate containing a collection of spinning orifices for synthetic fiber manufacturing is called a *spinneret*. The spinneret resembles a showerhead. After the polymer fluid passes out of the spigot or spinneret, the natural and synthetic processes begin to diverge. In the natural process, the filament is immediately put to work; it is in its final, useable form, having solidified just before exiting the spigot. In synthetic fiber manufacturing, the filament remains fluid close to the face of the spinneret. The filament is stretched and solidified (by cooling or solvent removal) prior to being wound up on a package. There are more details to this process, but that is the process at its most basic.

Orientation of the chains is achieved by the applied longitudinal stress and the extensional flow field during the fluid or semifluid phase of fiber production. Crystallization is a consequence of annealing (aided by applied stress, in some instances). Both of these material structural attributes are critical to the final fiber having the desired properties, and they must be carefully controlled [15].

The interesting science and engineering from the standpoint of structure development occur at distinctly different points in the fiber production process, natural vs. synthetic. In natural systems, the development of structure occurs within the host. In synthetic extrusion, with apologies to polymer extruder engineers, very little of interest occurs prior to the face of the spinneret. It is in this region that, under the influence of mechanical and thermal stresses, the fiber structure is developed.

Since the next step, yarn production, was developed using natural fibers, which have restricted lengths and some degree of roughness to their surface, and since neither property occurs automatically in synthetic fiber production, the synthetics must be made to have more natural features: They are cut to short lengths to yield staple fibers, and they are crimped to be not smooth or straight. Then they can be processed into spun yarns. It is important to note that filament yarns of silk and synthetic fibers are useful

in their own right. Fabrics made from filament yarns have excellent drape characteristics. Filament yarns are also useful as the reinforcement phase in composites.

10.3.1.2 Yarn

The next step up in the hierarchy of textiles is the ordered collection of fibers into a yarn, which is achieved by twisting, called *spinning* by practitioners, of the staple fibers into a yarn. The feedstock for yarn spinning is a several hundred pound bale of randomly oriented fibers, either natural or synthetic, which must be opened and individual fibers separated and aligned. Hence, at the most fundamental, all the yarn-forming processes have the goal of entropy reduction and the result of improvement in strength. The fibers are twisted so as to increase the strength of the interfiber cohesion. The most common of the spinning processes currently in use is *ring spinning*, so named because of the ring that carries the nascent yarn in a circle as it is being twisted. The degree of fiber-to-fiber overlap, the surface characteristics of the fibers, and the degree of twist and the tightness of the twist, in addition to the fiber strength, all contribute to the yarn strength and flexibility. There have been and continue to be many developments in yarn spinning. I commend the interested reader to Goswami *et al.* [16].

10.3.1.3 Fabric

Fabric as a concept has many facets. The common thread running through the manifestations of the concept is that the elemental component is a fiber. The fiber is paramount, but the intermediate structures, if any, are critical in their own right. The properties of the individual fibers do have an impact, albeit limited, on the ultimate mechanical properties of the fabric and strongly influence its chemical and thermal properties. The properties of a fabric that influence comfort are strongly influenced by both the fiber type and the fabric structure.

The earliest type of loom, and still the most commonly used in the production of fine fabrics, is the *shuttle loom*, so named because the weft yarns are woven into the warp yarn sheet by being carried across the width of the fabric on a shuttle. Other means of weft insertion include the airjet, projectile, and rapier, each of which has its own benefits and liabilities. For example, weaving speeds can be very high on a loom with airjet insertion in comparison with a shuttle, and so the airjet is most often used for commodity flat goods such as bedsheets and pillowcases.

For our purposes, let us consider only the shuttle loom. For the weft to interlace with the warp, the relative position of neighboring warp yarns must be changed; the term for this is that a *shed* must be formed in the warp through which the weft yarn is inserted. The most elementary manner of accomplishing this is through a *harness-and-heddle* system. The *heddle* is a wire-like structure with a hole in it through which a warp yarn is passed. There is one heddle for each warp yarn (in a plain weave). The heddles are collected in a set of *harnesses*, the number ranging from two or four up to eight. The harnesses are raised and lowered programmatically by a *cam*; the machinery limitations of space preclude a loom from having more than eight harnesses when using a cam mechanism. Other shed-forming methods do exist [5,17]. Two harnesses are used to weave the plain weave, with nearest-neighbor warp yarns passing through every other harness. The harnesses are raised and lowered in an alternating fashion, forming a simple alternating shed. With each pass of the shuttle the shed is reversed. The final step after weft insertion is *beating up* by the reed, during which the weft is pressed against the prior weft, forming the *fell line* of the cloth. The force with which the reed is pressed to the fell line determines of the tightness of the weave.

With the use of more than two harnesses, other weaves may be produced. Each type of weave has its own set of properties. For example, the *satin weave* (Figure 10.5) has the very characteristic feel of satin-smoothness because of

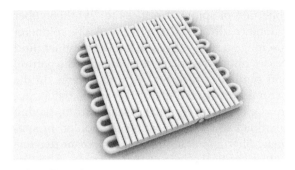

FIGURE 10.5   A visual model rendering of a satin-weave fabric. (J. Manganelli, personal communication. Used with permission.)

the long floats, i.e., yarns that are not interlaced with each neighbor, in the warp direction. (The satin weave identifies the long floats as being in the warp direction.) Of course, this structure is prone to snagging, also as a consequence of the long float. This weave has been used as the reinforcement in composites because there is not so much bending stress on the warp yarns. Many other woven structures are possible [17].

10.3.2 Bioinspiration as it Informs Engineering

There are many popular examples of nature-inspired engineering. The most common example is Velcro®, modeled as it is after the hook shape of burrs. The most recent entry is the gecko, with a multitude of small hair-like appendages on its feet [18]. The question remains, however, as to the order of, or the route to, the inspiration.

Consider bamboo as a composite. It seems clear, at least in terms of the time sequence of events, that the field of fiber-reinforced composites research was not predicated on a morphological study of bamboo; if anything, the language of composites has been used to describe the observed structure of bamboo. Or, take synthetic fiber spinning, discussed in Section 10.3.1.1. Again, this technology was developed through scientific and engineering research, not through study of silkworms and spiders. As with composites, the language of fiber spinning developed a lexicon with which to discuss natural fiber spinning: *spinneret*, *extensional flow*, and *liquid crystalline phases* were not terms used by the early entomologists to describe their observations.

Those comments notwithstanding, what is currently happening is that biology is informing engineering practice, in particular with respect to sustainability, as well as novel structures. Again, take the gecko, for example. The mystery of the ability of the gecko to traverse smooth surfaces at virtually any angle has been resolved as a consequence of the nanoscaled hair-like structures on its feet. Contemporaneously, humans have learned how to make nanofibers, so we now have gecko-like material [18].

It is important to note that, without the required maturity of fiber science, this development would not have been possible and that the development of nanofibers was not in response to discovering gecko feet. It was instead the curiosity of physical scientists and engineers, striving to make ever-smaller fibers for their own sake. As my good friend Julian Vincent would say, however, "Physics deals with nature at such an adaptable level that physicists think of their own science as underpinning everything else … but it is worth considering that there are so many mechanisms waiting to be discovered in biology that perhaps the study of living organisms is the basic science, and physics is just a special case" [19]. I may choose to say physics has the source of its power as a reductionist science, whereas biology is proud to explore complexity.

10.3.3 The Marriage: Bioinspiration with Fibrous (Textile) Materials Engineering

I have reviewed a broad set of the work on biomimetics in textiles *per se* in a recent book chapter [20]. My current area of interest encompasses the study of biomimetic fibers more than general textiles, and so I conclude this chapter with a discussion of biomimetics in the fiber area of textiles. Readers interested in a more broad treatment of textiles in biomimesis are commended to the above reference. However, I will intersperse a few points of textile interest here.

10.3.3.1 Apparel

One of the emerging new methods of coloration of textile materials is through the interference phenomenon that is the basis for coloration of butterfly wings. In general, the butterfly wing consists of two or more layers of small scales resident on a membrane, which allow diffraction to occur. See the accompanying chapter on structural colors.

Researchers in the Advanced Fiber-Based Materials (AFBM) Center of Economic Excellence at Clemson University demonstrated some fibers that mimic the coloration process used by the natural world, from butterfly wings to beetle backs. Those materials display color by the interference of white light reflected from several layers within each fiber, resulting in a fiber that changes color with viewing angle, without the use of dyes [21]. This is a different approach to some of the work done by commercial fiber manufacturer Teijin, which produced a fiber called Morphotex® that mimics the microstructure of *Morpho* butterfly wings to produce structural color. The fiber, made of either polyester or nylon, has more than 60 laminated layers of nanometer dimension [18].

In addition to the fashion aspect, protective equipment is an important area of textiles and apparel. One such application is reviewed in Section 10.3.3.4 in the discussion of byssus thread sheathing. It is relevant to note here that recent applications of shear thickening fluids, although perhaps not directly biomimetic, have gained some traction. In this instance, incorporation of a viscous material with a viscosity that increases with shear rate has been shown to improve ballistic protection. An armor composite material that contains a ballistic fabric impregnated with shear thickening fluid has been invented [22].

10.3.3.2 Biomedical Materials

As discussed by Sinclair [23], the alignment of cells in the direction of the grooves on an etched surface, i.e., contact guidance, governs the growth behavior of many cell types, including epithelial cells, oligodendrocytes, astrocytes, and fibroblasts. Capillary-Channeled Polymer™ (C-CP™) fibers are fabricated with micrometer-scale surface channels aligned parallel to the fiber axis by melt-extrusion through Clemson University proprietary spinnerets at Specialty Custom Fibers (www.specialtycustomfibers.com). The microscale surface topography and comparable groove dimensions of the C-CP fibers provide the surface topography necessary to align cells along the axis of each fiber. The specific target for this work was the anterior cruciate ligament (ACL), which is composed primarily of fibroblasts and extracellular matrix (ECM) organized in parallel structural alignment consistent with their biomechanical function in resisting tensile loading. Sinclair *et al.* concluded that the deep channels of the C-CP fibers appeared to be potential candidates to serve as scaffolding for a tissue engineering approach to ligament regeneration.

10.3.3.3 Technical Materials

ACTUATORS

The pneumatic muscle, which incorporates a braid and a bladder, as mentioned in Section 10.1, is one type of actuator. However, actuators modeled after squid tentacles provide us with an excellent example of biomimicry [24]. Whereas the braid/bladder muscle is an example of mimicking rudimentary muscle action using the textile structure of a braid, those workers report using the more complex structure of the squid tentacle as a versatile mobile arm. The

other natural actuator that has been studied is that of the octopus [25]. In these instances, it was the structure of these types of appendages, as known to biologists, that led the engineering researchers to design an arm that would be as mobile.

SELF-CLEANING MATERIALS

Researchers at Clemson University developed a super-hydrophobic fabric based on the "lotus" effect. To create the lotus effect on a fabric surface, they employed the deposition of both polystyrene (PS) and the triblock copolymer polystyrene-β-(ethylene-co-butylene)-β-styrene (SEBS) simultaneously on a model substrate. The selective dissolution of PS by ethyl acetate (EA), which acts as a solvent for PS and as a non-solvent for SEBS, created a porous hydrophobic rough surface on the substrate. This method of surface modification was applied to a polyester fabric and resulted in a practically nonwettable textile material [26, 27].

RESPONSIVE MATERIALS

Observation of the spontaneous unfolding of tree leaves led Kornev of Clemson University to develop deployable wet-responsive fibrous materials [28]. This is an application of super-absorbency in which the physicochemical energy of wetting is converted into mechanical energy to effect fabric bending. Electrospun nanofibers of alginate, a natural biopolymer extracted from seaweed and that a gels upon cross-linking with calcium chloride, were incorporated between fabric layers. When wetted, the alginate layer swells to orders of magnitude. One result of this work is a scroll that unrolls itself upon wetting.

Natural mechanical deformation sensors use filiform hairs (cilia) as the transducer. One common example is the inner ear of humans. The cilia are responsible for acoustic perception as well as for balance. Other examples include the detection of flow or inertial forces in other vertebrates and insects. Optimized through natural selection, these transducers surpass any current technological embodiment in terms of sensitivity. Nonetheless, there are devices based on micro electro mechanical systems (MEMS) that make use of such things as piezoelectricity. Most commercially available vibration and acceleration sensors function by spring-mass systems and electronic detectors. A cilia-mimetic system utilizing conductive fibers has been suggested [29]. Nanofibers of polypyrrole, tagged with a magnetic material, were used as detectors on giant magneto resistive multilayer sensors. As the artificial cilia are deformed, the magnetic tip interacts with the underlying sensor.

10.3.3.4 Strong and Tough Fibers (Spiders, Hagfish, and Mussels)

The basic components of animal fibers are mainly structural fibrous proteins as found in hair, tendon, cartilage, skin, arteries, and muscles of mammals or in cuticles and silks of arthropods, such as insects and arachnids. The individual proteins making up these fibers are of a specific amino-acid sequence, often sharing specific amino-acid motifs from one type of material to another. In addition, the individual fibrous proteins have the ability to assemble into a supramolecular network, such as macroscopic fibers. The resulting network structure, generally insoluble in water, is maintained by a combination of labile cross-linking, hydrophobic interactions, hydrogen bonding, and electrostatic interactions that are created between distinct amino-acid molecules in the protein [30].

The molecular architecture (sequence and composition) of the individual proteins forming the network, together with the amount and type of intermolecular interactions involved, determine fiber mechanical and physical properties. Some of the fibers are composed of rubber-like or elastomeric proteins such as elastin or resilin that confer a high degree of elasticity to the fiber while retaining relative strength. Embedded within the matrix are crystalline domains, imparting the elasticity needed for specific function.

Engineering new customized protein fibers that have designed mechanical properties presents a challenge. To achieve this goal, the molecular architecture of the molecules comprising these fibers, as well as their assembly process, needs to be fully understood in order to control assembly in any manufacturing process. Protein fibers are more accessible to needed engineering analysis, and tremendous efforts are being made in this field to understand the structure/function relationship of protein polymers. Because of the availability of technologies allowing the manipulation of genes encoding for these proteins, the perspective of designing and manufacturing new protein-based polymers is promising [31].

SPIDER SILK

Like insects, spiders also manufacture silk; however, the silk gland location and associated spinning systems are somewhat different in that spiders do not use modified salivary glands as silk glands. Spiders possess one or more silk glands located in their abdomen that are each linked to specialized external structures called *spinnerets* located on the ventral part of their abdomen. A duct of a particular length and shape, depending on the spider and silk gland considered, links each type of gland to a certain type of spinneret. The soluble silk made by specialized cells in the gland wall is collected in the lumen. The soluble silk in a liquid crystalline phase [32] is slowly processed into a fiber as the molecules move through the long duct and are subjected to chemical and physical stresses and modifications. The silk fiber that the spider pulls out with its legs is insoluble and exits through specialized hair-like structures called *spigots* covering the spinnerets [33].

The most characterized spider silks are the ones produced by spiders belonging to the genera *Nephila* and *Araneus*. A picture of *Nephila calvipes* from my laboratory is shown in Figure 10.6. These spiders, considered highly evolved spiders,

FIGURE 10.6   Female *Nephila clavipes* on a female *Homo sapiens* digit. (Photo: Janci Despain.)

are all orb-web weaver spiders whose survival relies on catching insects in an elaborate web structure. These types of spiders possess seven types of glands and are able to produce seven types silks of very different mechanical properties and compositions that each have a determined function and use (i.e., web-building silks, cocoon silk, dragline, and swathing silk). More primitive spiders (i.e., *Mygalomorphae*) only have one type of silk gland producing a silk used for cocoon swathing or, in some cases, lining the borrow of certain underground trapdoor spiders such as *Antroaediatus unicolor*.

The best-characterized silks are the dragline silk (major ampullate silk) used for construction of the web frame or as a safety line when the orb weaver spider drops from elevations, and the viscid silk (or flagelliform silk) forming the capture spiral of the web. These silks differ in their repeat structures, resulting in different tensile strengths and elasticities.

Amino-acid analysis of the spider silk used for orb-web building showed that these silks are characterized by the presence, in very high proportion, of glycine and alanine residues for dragline silks and glycine and proline residues for flagelliform silks [34]. This finding has strong implications for silk properties in that glycine and alanine are very simple molecules, comprising hydrogen and a methyl group, respectively, leading to linear polymers.

X-ray diffraction and nuclear magnetic resonance spectroscopy studies demonstrated that the dragline silk fiber protein contains glycine-rich repeats alternating with alanine rich-repeats; crystalline regions made of linked polyalanine regions are organized in anti-parallel β-sheets alternating with amorphous regions [35]. The polyalanine domains can form noncovalent cross-links or hydrogen bonds between the individual proteins and alternate either with a glycine-rich proline containing pentapeptide, supposed to be responsible for the fiber's elasticity [36], or with a glycine-rich tripeptide, which forms a small 3_{10} helix (3 turns in 10 amino-acid units), supposed to link highly crystalline and more amorphous regions [37]. The structure of the pentapeptide motif is modeled after that of the elastin pentapeptide and is thus expected to form β-turns, which in tandem would take the shape of an elastic β-spiral and act as a spring [37]. Spider dragline silks are very tough fibers due to a combination of high tensile strength (400 kpsi) and high extensibility (35%). Flagelliform silks are also very tough fibers, far more elastic than dragline silks (up to 200%) as a result of the presence of these long regions containing repeats of several elastic motifs [31].

HAGFISH SLIME THREADS

In recent years, spider silks have been presumed to be a model material for the design of protein fibers that possess substantial physical properties and are ultimate sustainability; unfortunately, attempts at making artificial spider silks

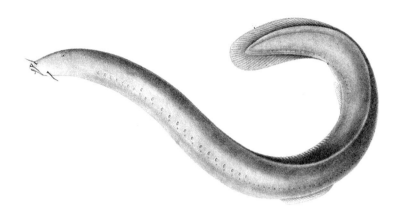

FIGURE 10.7   Drawing of *Eptatretus polytrema* (hagfish). Source: Wikimedia Commons.

through recombinant DNA approaches have not met this promise.

The hagfish, which resembles an eel in appearance (Figure 10.7), has a skeleton entirely made of cartilage and no scales. Its body covering is a soft skin containing many glands, which can be seen in Figure 10.7, that produce very large amounts of slime as a defense mechanism. This slime contains long microfibers, thought to entrain water, which accounts for the gel-like property [38].

Some work on the mechanics of protein fibers from the slime of hagfish suggests that these fibers might provide an additional biomimetic model. The proteins within these slime threads adopt conformations that are similar to those in spider silks when they are stretched. Draw processing of slime threads were found to yield fibers comparable to spider dragline silk in strength and toughness. From a biomimetic textile vantage point, the slime itself is a soft, fiber-reinforced composite with very interesting properties [39].

MUSSELS

Mussels, which survive in ocean wave-swept tidal zones, owe their success in these extreme conditions to a unique fiber system that provides robust attachment to any surface, including rocks, piers, and bridges [12]. This attachment system, the *mussel byssus*, is deployed as an array of threads composed essentially of collagen. Distinct from the extrusion process of silk, the byssus is assembled in the foot of the mussel [40]; consequently, and what makes the byssus threads unique, the threads have the unusual mechanical property of being stiff at one end and 160% extensible at the other end. The extensive literature on the distinct structure and molecular composition of the two regions of the byssus has been reviewed by Waite *et al.* [41]. The proximal region contains loosely packed coiled fibrils, providing the extensibility; the distal region contains dense bundles of filaments that account for the stiffness. In addition, thread stiffness and damping increase with increasing strain rate [42].

The mussel employs the strategy, relatively novel among most natural fiber producers but not unlike the scales present on wool, of coating the threads with a thin (2–4 mm) cuticle to protect the core. To match the compliance of the byssus, the cuticle is infused with nanoscale granules, which increase the extensibility of the hard coating to 70%, making it more compliant than many synthetic polymer coatings [43]. Although toughening of polymers (mostly epoxies) by incorporating rubber particles is not new, this example still may provide us with some new mechanisms to explore. The mussel cuticle may thus inspire new thin composite coatings that are hard yet extensible. Such a system can have benefits as a bioinspired material—for example, as a protective sheath on elastomeric fibers.

10.4 CONCLUDING REMARKS

The extent of discovery of structure and function in biology apropos fibers and fibrous materials is merely beginning. The more that materials scientists and engineers engage with biological scientists and (genetic) engineers, the richer the possibilities become in both domains. Biologists can discover new motivations to drive their explorations, and the materials community will be taught new ways of making new materials.

As just one example, consider hagfish slime, discussed in Section 10.3.3.4. It is, after all is said and done, essentially a fiber-reinforced composite, albeit one whose function is not structural in the sense we normally associate with composites. Slime, as the name implies, is soft and malleable, not firm. It may be possible to effect a slight modification of the matrix (the slime), which is largely composed of mucins, which are in the class of glycoproteins. Such a modification might be simply encouraging the formation of the disulfide bridges from the cysteine amino acids present in the protein [44] and have the goal of making the slime lightly less compliant. If the slime were more rubbery (elastic), it could serve a function, such as biodegradable packaging, and with the intrinsic fiber reinforcement, the packing would be fairly robust.

The worldwide effort to devise a method for production of spider silk mimetic fibers from synthetic proteins will eventually come to fruition. There have been some marked successes in this regard, and I suggest the reader follow the work of Florence Teulé in Randy Lewis' group and Thomas Scheibel at Bayreuth University to see the remaining issues likely to be resolved under such intense effort. Nonetheless, the fiber is still a protein and so will have the associated limitations on function; but then, all materials have limitations.

In summary, I hope I have piqued the curiosity of the reader to explore beyond my tantalizingly brief review. My aim was to give the reader a better appreciation of the truly remarkable world of textile materials, together with a glimpse into the audacity present at the interface between this venerable human endeavor and the ancient wisdom of biology. Just as textile materials underpinned the Industrial Revolution, exploring this interface with purpose and an engineering mindset will produce a materials evolution that we currently can barely appreciate.

References

[1] D. Jenkins (ed.), *The Cambridge history of western textiles*, Cambridge University Press, Cambridge, UK (2003).

[2] D.L. Reed, J.E. Light, J.M. Allen, and J.J. Kirchman, Pair of lice lost or parasites regained: the evolutionary history of Anthropoid primate lice, *BMC Biol* **5** (2007), 7.

[3] R. Kittler, M. Kayser, and M. Stoneking, Molecular evolution of pediculus humanus and the origin of clothing, *Curr Biol* **13** (2003), 1414–1417.

[4] C.M. Baldia and K.A. Jakes, Photographic methods to detect colourants in archaeological textiles, *J Archaeol Sci* **34** (2007), 519–525.

[5] M.L. Joseph, P.B. Hudson, A.C. Clapp, and D. Kness, *Joseph's introductory textile science*, Harcourt Brace Jovanovich, Fort Worth, TX, USA (1992).

[6] F. Daerden and D. Lefeber, Pneumatic artificial muscles: actuators for robotics and automation, *Eur J Mech Environ Eng* **47** (2002), 11–21.

[7] A. Abbott and M.S. Ellison (eds.), *Biologically inspired textiles*, Woodhead Publishing, Cambridge, UK (2008).

[8] B.S. Bloom, M.D. Engelhart, E.J. Furst, W.H. Hill, and D.R. Krathwohl, *Taxonomy of educational objectives. The classification of educational goals* (B.S. Bloom, ed.), Handbook I: The Cognitive Domain, David McKay, New York, NY, USA (1956).

[9] J. Abbas, *Structures for organizing knowledge: exploring taxonomies, ontologies, and other schemas*, Neal-Schuman, New York, NY, USA (2010).

[10] L. Romer and T. Scheibel, The elaborate structure of spider silk: structure and function of a natural high performance fiber, *Prion* **2** (2008), 154–161.

[11] J.M. Benyus, *Biomimicry*, Harper-Collins, New York, NY, USA (2002).

[12] National Research Council (US), *Inspired by biology: from molecules to materials to machines*, National Academies Press, Washington, DC, USA (2008).

[13] A. Lazaris, S. Arcidiacono, Y. Huang, J.-F. Zhou, F. Duguay, N. Chretien, E.A. Welsh, J.W. Soares, and C.N. Karatzas, Spider silk fibers spun from soluble recombinant silk produced in mammalian cells, *Science* **295** (2002), 472–476.

[14] S. Aoe, M. Nakaoka, K. Ido, Y. Tamai, F. Ohta, and Y. Ayano, Availability of dietary fiber in extruded wheat bran and apparent digestibility in rats of coexisting nutrients, *Cereal Chem* **66** (1989), 252–256.

[15] D.R. Salem (ed.), *Structure formation in polymeric fibers*, Hanser, Munich, Germany (2001).

[16] B.C. Goswami, J.G. Martindale, and F.L. Scardino, *Textile yarns: technology, structure, and applications*, Wiley, New York, NY, USA (1977).

[17] K. Hatch, *Textile science*, West, St. Paul, MN, USA (1993).

[18] L. Eadie and T.K. Ghosh, Biomimicry in textiles: past, present and potential. An overview, *J R Soc Interf* **8** (2011), 761–775.

[19] J.F.V. Vincent, Stealing ideas from nature, in *Deployable structures* (S. Pellegrino, ed.), Springer-Verlag, Vienna, Austria (2001), 51–58.

[20] M.S. Ellison, Biomimetics and textile materials, in *Handbook of natural fibres volume 2: processing and applications* (R. Kozlowski, ed.), Woodhead Publishing, Cambridge, UK (2012).

[21] M. Mignanelli, K. Wani, J. Ballato, S. Foulger, and P. Brown, Polymer microstructured fibers by one-step extrusion, *Opt Express* **15** (2007), 6183–6189.

[22] N.J. Wagner and E.D. Wetzel, Advanced body armor utilizing shear thickening fluids, US Patent 7226878 (issue date 5 June 2007).

[23] K.D. Sinclair, K. Webb, and P.J. Brown, The effect of various denier capillary channel polymer fibers on the alignment of NHDF cells and type I collagen, *J Biomed Mater Res A* **95** (2010), 1194–1202.

[24] J.L. Van Leeuwen, J.H. De Groot, W.M. Kier, and I.D. Walker, Evolutionary mechanics of protrusible tentacles and tongues, *Neth J Zool* **50** (2000), 113–139.

[25] D. Trivedi, C.D. Rahn, W.M. Kier, and I.D. Walker, Soft robotics: biological inspiration, state of the art, and future research, *Appl Bionics Biomech* **5** (2008), 99–117.

[26] I. Luzinov, P. Brown, G. Chumanov, M. Drews, and S. Minko, *Ultrahydrophobic fibers: lotus approach*, NTC Project Report C04-CL06 (2006), http://www.ntcresearch.org/pdf-rpts/Bref0607/C04-CL06-07.pdf (accessed 11 April 2013).

[27] K. Ramaratnam, K.S. Iyer, M.K. Kinnan, G. Chumanov, P.J. Brown, and I. Luzinov, Ultrahydrophobic textiles using nanoparticles: lotus approach, *J Eng Fiber Fabr* **3** (2008), 1–14.

[28] K.G. Kornev, I. Luzinov, P.J. Brown, V. Sa, D. Monaenkova, T. Andrukh, and J. Baker, *Deployable wet-responsive fibrous materials*, NTC Project Report M08-CL10 (2009), http://www.ntcresearch.org/pdf-rpts/AnRp09/M08-CL10-A9.pdf (accessed 11 April 2013).

[29] P. Schroeder, J. Schotter, A. Shoshi, M. Eggeling, O. Bethge, A. Hütten, and H. Brückl, Artificial cilia of magnetically tagged polymer nanowires for biomimetic mechanosensing, *Bioinsp Biomim* **6** (2011), 1–9.

[30] K. Autumn, Y.A. Liang, S.T. Hsieh, W. Zesch, W.P. Chan, T.W. Kenny, R. Fearing, and R.J. Full, Adhesive force of a single gecko foot-hair, *Nature* **405** (2000), 681–685.

[31] F. Teulé, *Genetic engineering of designed fiber proteins to study structure/function relationships in fibrous proteins*, PhD Thesis, Clemson University, (2003).

[32] D.P. Knight and F. Vollrath, Liquid crystals and flow elongation in a spider's silk production line, *Proc R Soc Lond B* **266** (1999), 519–523.

[33] R.F. Foelix, *Biology of spiders*, 2nd ed., Oxford University Press, Oxford, UK (1996).

[34] S.J. Lombardi and D.L. Kaplan, The amino acid composition of major ampullate gland silk (dragline) of Nephila clavipes (Araneae, Tetragnathidae), *J Arachnol* **18** (1990), 297–306.

[35] A. Simmons, C. Michal, and L. Jelinski, Molecular orientation and two component nature of the crystalline fraction of spider dragline silk, *Science* **271** (1996), 84–87.

[36] C.Y. Hayashi and R.V. Lewis, Evidence from flagelliform silk cDNA for the structural basis of elasticity and modular nature of spider silks, *Mol Biol* **275** (1998), 773–784.

[37] C.Y. Hayashi, N.H. Shipley, and R.V. Lewis, Hypotheses that correlate the sequence, structure, and mechanical properties of spider silk proteins, *Int J Biol Macrom* **24** (1999), 271–275.

[38] D.S. Fudge, N. Levy, S. Chiu, and J.M. Gosline, Composition, morphology and mechanics of hagfish slime, *J Exp Biol* **208** (2005), 4613–4625.

[39] D.S. Fudge, S. Hillis, N. Levy, and J. M Gosline, Hagfish slime threads as a biomimetic model for high performance protein fibres, *Bioinsp Biomim* **5** (2010), 035002.

[40] J.H. Waite, The formation of mussel byssus: anatomy of a natural manufacturing process, in *Results and problems in cell differentiation* (S.T. Case, ed.), Springer-Verlag, Berlin, Germany (1992), 27–54.

[41] J.H. Waite, X.X. Qin, and K.J. Coyne, The peculiar collagens of mussel byssus, *Matrix Biol* **17** (1998), 93–106.

[42] E. Carrington and J.M. Gosline, Mechanical design of mussel byssus: load cycle and strain rate dependence, *Am Malacol Bull* **18** (2004), 135–142.

[43] N. Holten-Andersen, G.E. Fantner, S. Hohlbauch, J. H. Waite, and F.W. Zok, Protective coatings on extensible biofibres, *Nat Mater* **6** (2007), 669–672.

[44] J.E. Herr, T.M. Weingard, M.J. O'Donnell, P.H. Yancey, and D.S. Judge, Stabilization and swelling of hagfish slime mucin vesicles, *J Exp Biol* **213** (2010), 1092–1099.

ABOUT THE AUTHOR

Michael S. Ellison is a professor in the School of Materials Science and Engineering at Clemson University. He received undergraduate and graduate degrees in Physics and a PhD in Polymer Fiber Physics (1982) from the Davis campus of the University of California. After remaining at UC Davis as a research scientist, he joined the Clemson University in 1984. He advanced to full professor in 1998 and served as Interim School Director from August 2003 until January 2005. His research embraces both natural and synthetic polymer fibers. His most active current research area is in biologically inspired materials: the study of natural systems for inspiration in new materials development. Current research considers the fibers produced by spiders as the preeminent model of an advanced material produced by a sustainable method. Elements of this program include molecular biology for gene engineering and expression; characterization of natural and synthetic protein materials; and mathematical modeling of the biomimetic materials and processes. His general research interests in natural and synthetic fiber physics include the development of structure/property relationships in melt extrusion of fibers, and in particular the on-line determination of fiber and film structure development. With colleagues in the School of Architecture, he is developing unique architectural materials. He teaches courses in physical properties of fibers and in melt extrusion production of synthetic fibers. When not holding forth as a professor, he plays traditional Irish and American fiddle music.

CHAPTER

11

Structural Colors

Natalia Dushkina[a] *and Akhlesh Lakhtakia*[b]

[a]Department of Physics, Millersville University, Millersville,
PA 17551, USA
[b]Department of Engineering Science and Mechanics, Pennsylvania State University,
University Park, PA 16802, USA

Prospectus

Structural colors originate in the scattering of light from ordered microstructures, thin films, and even irregular arrays of electrically small particles, but they are not produced by pigments. Examples include the flashing sparks of colors in opals and the brilliant hues of some butterflies such as *Morpho rhetenor*. Structural colors can be implemented industrially to produce structurally colored paints, fabrics, cosmetics, and sensors.

Keywords

Beetle, Bird, Bragg filter, Bragg phenomenon, Butterfly, Cholesteric liquid crystal, Circular Bragg phenomenon, Colloid, Cosmetics, Diffraction, Dragonfly, Electrically small particle, Fabric, Incoherence, *Infinite Color*™, Inkjet printing, Interference, Jellyfish, Layer-by-layer assembly, Latex, Multilayers, Nacre, Nacreous pigment, Opal, Optical sensing, Physical vapor deposition, Play of colors, Scattering, Sculptured thin film, Structural color, Thermal imaging

11.1 INTRODUCTION: COLORS IN NATURE

Humans, not to mention other animals, are fascinated by and make use of colors that abound in nature. A visit to a large aquarium or an extended stroll in a garden during spring reveals the extensive panoply of bright colors that are used for signaling and camouflage in the biological world. Most of the colors in the animal world are due to chemical pigmentation by dyes and pigments. With a few exceptions, floral colors are also of chemical origin [1]. Other colors—such as the glowing reds and ochres of sunsets, the blue and green hues of ocean water, the spectral colors of rainbows, and the intense colors of some minerals, animals, and plants—have purely physical origins based on optical phenomena such as refraction, reflection, scattering, interference, and polarization of light. Our eyes cannot distinguish colors based on their origin—physical, chemical, or mixed—since ocular mechanisms are entirely stimulated by rays or photons that seemingly emanate from an object but carry no information as to the mechanism of color production in that object. Moreover, mental imagery is created by electrical potentials and migrating ions in such ways that make *color*, as opposed to frequency or free-space wavelength, a partially psychological construct.

Engineered Biomimicry
http://dx.doi.org/10.1016/B978-0-12-415995-2.00011-8

Although colors of physical origin have been studied by technoscientists and artists alike, colors that have chemical origin have spawned and sustained entire industries, some for several millennia. However, physical colors are assuming technical importance nowadays because their production does not entail the release of volatile organic compounds that not only pollute the environment and are hazardous to life but also wreak havoc on endocrine systems, even when present in trace amounts.

11.1.1 Pigmental (Chemical) Colors

Chemical colors are produced by absorption of light in *pigments*, which are substances of definite chemical composition. The colors produced vary with the chemical nature of the pigment. Natural pigments can be extracted from the colored tissues of plants, woad, shellfish, or lichens by appropriate reagents. Pigments react to light in the same way, whether they are within or outside a biological object. Tissue or organisms showing only pigmental colors never have a surface gloss. Their color is not altered by immersion in any medium that does not chemically attack the pigment [2]. The extraction of chemicals from plants and animals and their conversion into dyes and pigments have a history of several millennia. Only over the last 150 years have natural dyes, pigments, and inks been replaced, gradually at first but rapidly later, by the products of modern organic chemistry.

11.1.2 Structural (Physical) Colors

Butterflies need no introduction for the brightly iridescent wing surfaces a multitude of their species display. Colors in their wings arise from either pigments or the micro and nanostructure of scales, or both. Pigments (melanins and pterins, chiefly) found in butterfly wings produce yellow, orange-yellow, red, black, and brown colors. However, there are no pigments that can produce the iridescent blue, violet, and green colors observed in some butterfly species. For instance, some species of the *Morpho* genus found in the tropical and subtropical forests of South America flash magnificent blue colors with strong iridescence and metallic glossiness, as shown in Figure 11.1. So bright is the color of its wings that a *Morpho* butterfly can be seen from a distance of a few kilometers. Although pigment granules, mainly biopterin, have been found in the scales of some *Morpho* species, their quantity is considered to be insufficient to generate color [3–5]. Instead, it has been conclusively established during the last two decades that the blue colors of *Morpho* butterflies are caused by the complex structure of the wing scales.

FIGURE 11.1 Shadow box displaying (top) *Morpho godarti* and (bottom) *Morpho didius*.

Structural colors are produced by optical phenomena, such as scattering of light from ordered microstructures, thin films, and even irregular arrays of scatterers, the key being the morphology of the substance. Striking are the luster of pearls and the iridescence of precious opals, both materials commonly used in jewelry and adored because of the *play of colors* that they display. Most fascinating are the brilliant colors of some rocks and shells, butterflies, beetles, fish, and birds, which arise from the texture or microstructure of their surfaces.

Structural colors are characterized by an intensely brilliant sheen that is almost metallic. As opposed to pigments, which scatter light diffusely, structural colors usually exhibit directional effects, i.e., they vary either with the viewing angle or with the change from reflected to transmitted light. For example, the colors of a peacock feather undergo a complete cycle of change with the viewing angle of reflected light. In addition, when the feather is held up to the light, the vivid signature colors are replaced by a dull brown or black tint.

Structural color is usually destroyed by injury to the surface or by immersion in a neutral medium whose refractive index is substantially different from that of air. For example, the colors of a peacock feather disappear when the feather is thoroughly wetted with water or oil. In contrast to pigmental colors, structural colors exhibit high stability to acids, alkali, and light and do not fade with time. Change in temperature should not significantly affect structural color unless the thermal strains are very high. According to Mason [6], a structural color is not affected by chemical treatment of the structure creating the color unless the chemical reagents swell, shrink, or destroy the structure. Certainly, pressure, distortion, swelling, or shrinking will alter the hue; however, a structural color cannot be bleached away, and no pigments can be extracted by solvents or revealed by chemical tests. Structural colors are generally iridescent and all constituents of the incident white light can be found in the scattered light. The presence of neighboring pigments may, however, change the hue of a structural color.

Structural colors arise from different types of textured surfaces. For example, the wings of the mango moth (*Bombotelia jocosatrix*) are covered with a multilayered thin film. The color changes from green to yellow to orange and purple, simply depending on the layer thicknesses and the viewing angle [7]. In some species of moths and butterflies, color arises from a diffraction grating formed by parallel ridges [8]. Both types of structures may be combined in the wings of some species, and some pigment may also contribute color [9, 10]. In addition to the blue iridescence of the *Morpho* genus, the equally spectacular iridescence of beetles and birds, illustrated in Figure 11.2, has attracted scientific attention for more than a century.

Electron microscopy has greatly assisted in the elucidation of structural color in biological structures [11, 12]. It has revealed very complicated architectures, often involving a uniformly repeating structure, which interact with light to produce color. Therefore, structural color is very difficult to reproduce with high fidelity using artificially constructed structures, even when sophisticated manufacturing techniques of nanotechnology are employed. See Chapters 14–16 on solution-based techniques, vapor-deposition techniques, and atomic layer deposition.

Very importantly, during the last two decades many researchers have been inspired by structural colors in nature to develop new and safe alternatives to the conventional pigments in order to reduce the use of hazardous and volatile chemicals. Applications have been sought in many industries such as cosmetics, textiles, and automotive paints. Our goal here is to guide the reader to a few of these applications, but let us also provide some insights into engineered biomimicry for coloration before proceeding to the applications.

FIGURE 11.2 (a) Fragment of an elytron of *Lamprocyphus augustus*. (b) The head of a male wood duck (*Aix Spousa*) displays both pigmental and structural colors. Courtesy of Joseph Grosh (Millersville University, USA). (c) The back of Anna's Hummingbird (*Calypte Anna*). Courtesy of Joseph Grosh (Millersville University, USA). (For interpretation of the references to color in this figure legend, the reader is referred to the web version of this book.)

11.2 A BRIEF HISTORY OF RESEARCH

Perhaps the first description of structural colors was given by Isaac Newton [13]. In the second book of *Opticks*, he explained the brilliant plumage of the common Indian peafowl (*Pavo cristatus*), shown in Figure 11.3, as rising from optical interference from the thin transparent part of the feathers. About 200 years later, the emergence of the Maxwell equations provided a theoretical foundation to investigate structural colors. The brilliant colors of hummingbirds, butterflies, and beetles were intensively studied thereafter.

Until about 20 years ago, research on structural colors was motivated chiefly by the need

FIGURE 11.3 Top: A male peafowl (*Pavo cristatus*) displaying its tail feathers. Middle: Close-up view of the tail feathers of the male peafowl. Bottom: Common grackle (*Quiscalus quiscula*).

to clarify the underlying physical mechanisms [14]. Rayleigh strongly supported the idea that the brilliant colors of hummingbirds, butterflies, and beetles are caused by multilayer interference. Shortly after Rayleigh, using the basic optical phenomenons such as scattering, interference, and diffraction, Bancroft explained the iridescent colors of specific insects, birds, and fish; gemstones; and natural phenomena such as sunsets, rainbows, and the blue color of the sky. Then Mason elucidated the relationship between the structural features and the colors of a variety of birds and insects and provided simple methods for distinguishing between pigment colors and structural colors when present in insects, either singly or in combination.

However, a better understanding of biological iridescence began to emerge after electron microscopes were pressed into service [15]. From the 1970s to the mid-1990s, Allyn and Downey [16–18], Ghiradella [3, 4, 19–23], and Tabata and colleagues [5, 24] described several biologically occurring arrays of uniformly repeating structures that interact with light to produce color. These developments led to another wave of extensive studies on structural colors, motivated mainly by applications for cosmetics, automobiles, and textiles of new materials to provide coloration without the use of hazardous chemicals. Notable applications-oriented research has been carried out by Kinoshita and colleagues [11, 25–27] and the group of Saito [28, 29]. Vukusic and collaborators [30–33] as well as Parker and collaborators [34–36] look for insects displaying unusual structural colors to identify natural structures that could be reproduced artificially without loss of optical functionality. Other efforts worthy of mention have been reported by Vigneron *et al.* [37], Michielsen and Stavenga [38], Galusha *et al.* [39], and Sato and colleagues [40]. During the last few years, a few books have covered a wide range of topics related to structural colors [25, 41–43]: fundamental optical processes, natural photonic structures producing structural colors, and applications of structural color.

The brilliance of structural colors and the huge potential for replication and engineering applications, as well as the big impact of pioneering Japanese research in this area, inspired the L'Oréal Art & Science Foundation in Japan to create the L'Oréal Art & Science of Color Prizes in 1997. Major monetary prizes are awarded to artists and scientists for research on color, the veritable link between art and science [44]. The 7th L'Oréal Art & Science of Color Gold Prize was awarded to Akira Saito, Shinya Yoshioka, and Shuichi Kinoshita (Osaka University) and Keiichiro Watanabe, Takayuki Hoshino, and Shinji Matsui (University of Hyogo) for their comprehensive research on the reproduction of *Morpho* blues using semiconductor lithography (the group from Osaka University) and the fabrication of individual nanoscale units with an accuracy of 10 nm (the group from University of Hyogo).

With the development of specialized techniques to make nanostructures, a new research area has opened up in the present decade: replication of biological templates such as the wings of butterflies and cicadas. Nanofabrication techniques used for bioreplication include atomic layer deposition [45], nanocasting [46], nanoimprinting [47], and physical vapor deposition [48]. (See also Chapters 14–16.)

11.3 PHYSICAL MECHANISMS FOR STRUCTURAL COLOR

The following optical phenomena have been identified as the causes of structural colors in nature: thin-film interference, multilayer interference, diffraction-grating effect, scattering from irregular assemblies of small particles, and collaborative effect in irregular structured surfaces [11, 25, 49, 50]. In general, these optical phenomena can be classified into two groups of mechanisms of interaction of light with matter: (1) scattering from electrically small particles, and (2) Bragg phenomena, exhibited by structures with periodic morphology.

11.3.1 Structural Colors Due to Scattering from Electrically Small Particles

Scattering from irregular structures is known to produce color in the biological world. Mason [51] reported that the bluish color of some jellyfish (*Cyanea lamarcki*) is that of a suspension of electrically small colloidal particles. Such a color is called a *Tyndall blue*, which varies from deep blue to pale white, depending on the particle size. The body colors of the dragonflies *Mesothemis simplicicollis* and *Libellula pulchella* arise for a similar reason [6]. Tyndall scattering is also the cause of *chatoyancy*, or the *cat's eye* effect displayed by many minerals—such as the tiger's eye shown in Figure 11.4—but even more spectacularly by the gemstone chrysoberyl (beryllium aluminum oxide). Chatoyancy, arises from the fibrous inclusions or cavities within the stone, and the luminous streak of reflected light is always perpendicular to the direction of the fibers.

Tyndall blue arises from the scattering of light from three-dimensional particles that are small (at least a tenth in maximum linear dimension) with respect to the minimum wavelength of the incident light and with a refractive index close to unity. Since the intensity of the light scattered by an electrically small particle depends directly on the fourth power of the frequency [52], light of shorter free-space wavelength (or higher frequency) is scattered more strongly than light of longer free-space wavelength (or lower frequency). Scattering by electrically small particles is called *Rayleigh scattering*. The Rayleigh

FIGURE 11.4 Chatoyancy displayed by tiger's eye, a South African mineral that is essentially quartz colored by iron oxide.

scattering of sunlight in the atmosphere is often touted as the reason for the blue color of the sky, although the response of the human eye also plays a significant role in the perceived color of sky light.

Rayleigh scattering is most prominently seen in gases and suspensions of electrically small particles. It is the elastic scattering of light from small and isolated particles scattering independently of one another. The neutral atoms or molecules in the air act as classical harmonic oscillators or dipoles. As the radiation scattered from many isolated particles adds with random phase (incoherent scattering), the intensity and not the amplitude of the scattered light is proportional to the number of scattering particles. The total scattered irradiance I_s (or the intensity integrated over a sphere of radius r completely enclosing the particle)

of each particle with volume V can be written as [52]:

$$I_s = \frac{K^2 I_i V^2}{\lambda^4 r^2}, \qquad (11.1)$$

where K is a dimensionless constant that depends on the refractive index of the particle, I_i is the excitation intensity (incident light intensity), λ is the wavelength of light in vacuum or free space, and the particle has been assumed to be spherical. This expression shows that $I_s/I_i \sim 1/\lambda^4$ or $I_s/I_i \sim f^4$, i.e., the scattered intensity is directly proportional to the fourth power of the frequency f. In addition, Rayleigh scattering is of the elastic kind, which means that the scattered light has the same frequency as the incident light. Finally, Rayleigh scattering depends on the direction of incidence in relation to the shape and orientation of the particle and therefore is anisotropic. Hence, as shown in Figure 11.5, different colors are observed in light scattered by opalescent glass in different directions.

When the electrically small particles are sprinkled in an otherwise dense material continuum, then Rayleigh scattering is also affected by the dielectric properties of the host medium. For larger particles with diameters of around one wavelength or higher, the light scattered from the different parts of the particle may interfere and result in a more complex coherent scattering, colloquially known as *Mie scattering* [52].

Many of the noniridescent blues displayed by animals result from Rayleigh scattering. Some greens such as those of parrot feathers are due to the combined effects of yellow pigment in the feather barbules and the blue from Rayleigh scattering.

When the particle size is extremely small, on the order of a few nanometers, quantum effects occur and govern the colors displayed by nanoparticles. Figure 11.6 shows suspensions of ZnS-coated CdSe nanoparticles in toluene, an organic liquid. Semiconductor nanoparticles

FIGURE 11.5 Rayleigh scattering in opalescent glass. The scattered light appears blue from the side, but orange light shines through. From www.flickr.com/photos/optick/112909824/. (For interpretation of the references to color in this figure legend, the reader is referred to the web version of this book.)

are a subclass of quantum dots. The size dependence of the luminescence upon excitation of quantum dots with ultraviolet light is manifested as follows: On increasing the diameter of the quantum dots from 3 to 8.3 nm, the

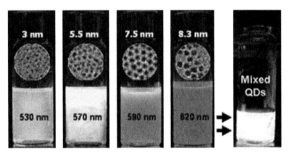

FIGURE 11.6 Quantum dots, each with a core of CdSe and a shell of ZnS, suspended in toluene. The color of a monodisperse suspension depends on the diameter of the quantum dots. A polydisperse suspension can even be white. Courtesy of Jian Xu (Pennsylvania State University, USA) and Ocean NanoTech LLC (Springdale, AR, USA). (For interpretation of the references to color in this figure legend, the reader is referred to the web version of this book.)

color of luminescent emission changes from green to red. This change is attributed to the effect of quantum confinement [53]. Since the structure of energy levels of a quantum dot is dependent on its size, the energy of the emitted photon is also determined by the diameter of the quantum dot.

11.3.2 Bragg Phenomenon

The concept of the Bragg phenomenon emerged in 1912 from studies of X-ray diffraction from crystalline solids; it also applies to electromagnetics. If white light impinges obliquely on a photonic structure that can be represented as a set of discrete and identical parallel planes, each separated from its nearest neighbors by a distance d, as shown in Figure 11.7, and the wave propagation vector of the incident light is inclined at an angle θ with respect to those planes, then light of free-space wavelength

$$\lambda = (2d \sin\theta)/m, \quad (m = 1, 2, 3, \ldots), \quad (11.2)$$

is specularly reflected due to constructive interference between neighboring planes. This angular selectivity of the Bragg phenomenon (often called *diffraction*) is a major cause of structural

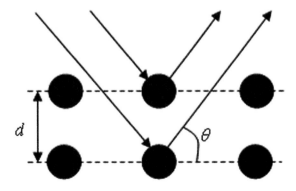

FIGURE 11.7 To explain Bragg's law.

color, as exhibited by periodic multilayered structures—which are sometimes called *Bragg filters*.

Diffraction gratings provide another simple and widely exploited manifestation of the Bragg phenomenon. Most commonly, a *diffraction grating* is a periodically corrugated sheet of a metal that is much thicker than the skin depth. When monochromatic light is incident on the grating at an angle θ with respect to the mean plane of the grating, then reflected light can be seen propagating in possibly more than one different directions. These directions are oriented at angles

$$\theta_m = \cos^{-1}(\cos\theta + m\lambda/d), \quad (m = 0, \pm 1, \pm 2, \ldots),$$

$$(11.3)$$

where d is the period of the grating. The direction for $m = 0$ is the specular direction. Only a few of the angles θ_m predicted by Eq. (11.3) being real (i.e., $|\cos\theta_m| \leq 1$), there are only a finite number of nonspecular directions.

Periodicity need not be one-dimensional. It can exist in two dimensions as well as in three dimensions. For optical purposes, we can say that the relative permittivity tensor of a periodic material has certain periodic symmetries. Nowadays these materials are called *photonic crystals*. These materials display a *photonic bandgap*, i.e., a λ-range in which propagation of light through the photonic crystal is not allowed in a certain direction [54]. An infinitely thick one-dimensional photonic crystal does not display a complete bandgap, which means that there is no λ-range, howsoever narrow, in which propagation is inhibited in all directions. Many incomplete bandgaps are exhibited by two- and three-dimensional photonic crystals. A complete bandgap is possible in some three-dimensional photonic crystals.

Most photonic crystals are fabricated artificially. The wavelength of light must be comparable to the periodicity of the photonic crystal for the Bragg phenomenon and bandgaps to be

exhibited [54]. For visible effects, the lattice constant must be in the range of 100 nm–1 μm, which can be accessed with conventional nanofabrication and self-assembly techniques [53]. A natural photonic crystal is the precious opal, which is formed by a spontaneous organization of colloidal silica spheres on a crystalline lattice. Periodic arrays of micron-scale cavities in the wings of many beetles, such as the Japanese jewel beetle *Chrysochroa fulgidissima*, function optically as inverse opals [55]. Both opals and inverse opals are made artificially as well. A splendid two-dimensional example of artificial opal is presented in Figure 11.8, which shows a

FIGURE 11.8 Photograph of a monolayer of close-packed carboxylate-modified polystyrene spheres of 510-nm diameter deposited on a silicon wafer. Courtesy of A. Shoji Hall (Pennsylvania State University, USA).

photograph of a monolayer of close-packed carboxylate-modified polystyrene spheres of 510-nm diameter deposited on a silicon wafer. Intense colors are displayed, depending on the viewing angle and the packing density.

11.4 STRUCTURAL COLORS FROM NATURAL PHOTONIC STRUCTURES

Let us now consider a few biological examples of the two main mechanisms of physical color formation. Butterflies display some of the most vivid colors in nature that result from the interaction of light with a wide diversity of micro- and nanostructures in their wings. Similarly, the exocuticles of many beetle species are brightly colored. Currently, the morphology of biological tissue can be associated with three types of photonic structures: periodic multilayered reflectors, diffraction gratings, and higher-dimensional photonic crystals.

11.4.1 Structural Colors Due to Periodic Multilayered Reflectors

Both geological and biological structures furnish numerous examples of periodic multilayered reflectors. Examples include the mineral labradorite (calcium sodium aluminum silicate), wherein a regularly stacked lamellar structure of 300-nm-thick layers creates the color [56, 57]. Another example is furnished by *Nautilus* shells that comprise alternating layers of organic and inorganic materials [58]. The luster of pearls also comes from constructive interference between adjacent unit cells of a periodic multilayered structure [59]. All three examples are shown in Figure 11.9.

Typically, the unit cell of a periodic multilayered structure in nature comprises two layers, one of a higher refractive index than the other. Interference between adjacent unit cells causes a color to be displayed on reflection.

A particularly well-studied natural structure is nacre, which is the hard, pearly iridescent substance forming the inner layer of a mollusk shell. The lamellar microstructure of natural nacre consists of polygonal aragonite platelets, approximately 5 μm across by 0.5 μm thick, wrapped in an organic matrix of conchiolin. Aragonite is a crystalline form of calcium carbonate; conchiolin is a substance of an organic nature based on polysaccharide and protein fibers.

Pearls exhibit an onion-like structure comprising alternating layers of conchiolin and aragonite. Both materials are translucent, allowing light to penetrate a large number of unit cells, and the reflected light undergoes constructive interference. The interference effect determines the fineness of the pearl luster. It depends on the number of unit cells that take part in the reflection, which in turn depends on the translucency and thinness of the constituent layers. The pearly luster rises from a bulk effect connected with the depth of light penetration in the volume, which distinguishes it from the surface luster in certain sapphires [60]. Besides luster, many pearls and *Nautilus* shells also display iridescence that can be explained with a diffraction-grating effect from the edges of the overlapping successive layers (similar to the overlapping shingles on a roof).

Interference can also be observed as the metallic reflection from the elytrons (i.e., hardened wings) of many beetles [61], as depicted in Figure 11.10. The scanning electron micrograph (SEM) in Figure 11.11 shows that the outer part of the exocuticle of longhorn beetles (*Tmesisternus isabellae*) consists of a periodic multilayered structure, and different colors are caused solely by thickness variations of the constituent layers [62]. Since the elytrons are smooth and relatively uninterrupted, the brilliant iridescent colors appear with highly metallic sheen.

Unlike butterflies, moths usually display dull brown and gray colors. However, there are exceptions, such as the Urania swallowtail moth (*Urania fulgens*), found in the Americas, or the sunset moth (*Chrysiridia rhipheus*) of

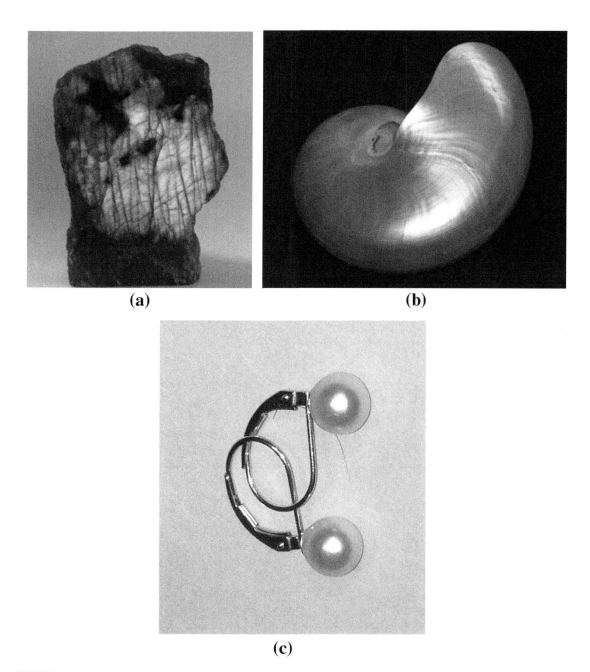

FIGURE 11.9 Structural colors due to periodic multilayered structures in (a) a labradorite specimen, (b) a *Nautilus* shell, and (c) pearls. (For interpretation of the references to color in this figure legend, the reader is referred to the web version of this book.)

(a)   **(b)**

FIGURE 11.10 Structural colors due to periodic multilayered structures in the exocuticle of beetles. (a) Japanese beetle *Popillia japonica.* Courtesy of Christopher R. Hardy (Millersville University, USA). (b) Dogbane beetle *Chrysochus auratus.* Courtesy of Gregory Hoover (Pennsylvania State University, USA). (For interpretation of the references to color in this figure legend, the reader is referred to the web version of this book.)

FIGURE 11.11 SEM of a scale of a longhorn beetle *Tmesisternus isabellae* [62]. Courtesy of Jian Zi (Fudan University, China). Reproduced with permission from the Optical Society of America.

Madagascar. Each scale of their wings comprises four or five layers of cuticle, approximately 40 nm thick, which are held apart by

tiny cuticle rods that allow the formation of 100 nm air pockets. Both the alteration of the spacing and thickness of the cuticle layers and the incorporation of different pigments within the scale create various colors [50].

Interference is also the predominant cause of iridescent colors in birds (see Figures 11.2 and 11.3) and some tropical fish. Most often, bird coloration originates from biochromes—pigments embedded within the feathers. However, no blue biochromes have been extracted from blue feathers; instead, the blue color is entirely of structural origin.

The structures creating iridescent colors in birds are diverse, but all are based on ordered arrays of melanin granules within a keratin matrix in feather barbules. For example, the feathers of grackles and other birds in the family *Icteridae* range in appearance from matte black to iridescent. Melanin is densely packed in the barbules of matte-black species, but melanin

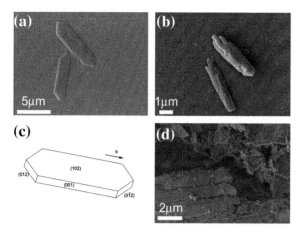

FIGURE 11.12 SEMs of (a) isolated biogenic crystals from the skin of Koi fish, and (b) *in vitro* grown anhydrous guanine crystals crystallized from an anhydrous solution of DMSO. (c) Schematic representation of the morphology of the biogenic crystals. (d) SEM showing guanine platelets from fish skin [64]. Courtesy of Lia Addadi (Weizmann Institute of Science, Israel). Reproduced with permission from the American Chemical Society.

mechanism to inhibit crystal growth in the direction of the molecular stacking. This way, the morphology of fish skin provides the best possible optical properties for reflection, due not only to constructive interference but also to birefringence of guanine.

Stacking three or four periodic multilayered structures, each tuned to a different part of the visible spectrum, leads to broadband reflectance [65–67] and multiband reflectance [66]. It may be responsible for the silvery appearance of many fish, particularly herring (*Clupea harengus*) [68]. Another way to obtain silvery appearance is by having a multilayered structure with a large number of layers with random variations of thickness over a certain range and each made of one of two materials that differ in refractive index, as demonstrated by two species of fish: the hairtail (*Trichiurus lepturus*) and the ribbonfish (*Lepidopus caudatus*) [69]. Fish use the mirror effect for camouflage.

11.4.2 Structural Colors Due to Diffraction Gratings

Precious opals best represent structural colors that arise from the diffraction-grating effect. As shown in Figure 11.13, a precious opal exhibits intense colors that change with the viewing angle, a phenomenon called the *play of colors*. Scanning electron microscopy—see the images in Figure 11.14—has revealed that precious opals contain layers of closely packed spherical particles of amorphous silica ranging from 150 to 300 nm in diameter [70]. The layers are stacked to form either a hexagonal or a cubic close-packed lattice. The luster and the play of colors in opals are thus due to the Bragg phenomenon, discussed in Section 3.2. Equation (11.2) applies, with $d = D$ for cubic close packing and $d = (2/3)^{1/2} D$ for hexagonal close packing [71], where D is the diameter of the silica spheres. With the assumption of normal incidence ($\sin\theta = 1$) and replacing λ by λ / n_{eff}, where n_{eff} is the effective refractive index, Eq. (11.2) transforms to $\lambda = 2 n_{\text{eff}} \, d / m$, with m as

granules are arranged in ordered layers around the edges of feather barbules in iridescent species. Spectrometry, transmission electron microscopy, and thin-film optical modeling have confirmed that these ordered layers typically create iridescence [63].

Periodic multilayered structures comprising guanine crystals (an organic material with high refractive index) separated by cytoplasm (a material with lower refractive index) hold the key to colored and broadband silver reflections in many fish. X-ray diffraction studies have revealed the guanine to be anhydrous. The SEMs in Figure 11.12 show isolated biogenic crystals from the skin of the koi fish (*Cyprinus carpio*) [64]. The crystals are semihexagonal thin plates (50–150 nm thick) and a few micrometers long. Because the axial component of the polarizability tensor of a guanine molecule is significantly smaller than the transverse component, guanine is anisotropic. The koi fish exhibit an efficient

FIGURE 11.13   The play of colors in a precious opal. (For interpretation of the references to color in this figure legend, the reader is referred to the web version of this book.)

a positive integer. For a hexagonal opal, $D = 160$ nm implies that $d = 130.6$ nm; additionally, with $n_{eff} \approx 1.54$ as the typical value for silica, we see that

blue color of wavelength about 402 nm would be predominantly diffracted in the visible spectrum. Increasing the value of D to 275 nm would mean that the red color of wavelength about 692 nm would be predominantly diffracted in the visible spectrum. If the silica spheres are too large ($D > 350$ nm), the play of colors would be invisible to human eyes because it would occur in the infrared regime; however, tilting the opal to reduce $\sin \theta$ would shift the Bragg phenomenon to shorter wavelengths, and the play of colors may reappear—initially with red hues.

Precious opals are rare in nature, but common opals are not. Common opals do not contain the ordered structure of precious opals and, therefore, do not diffract visible light. A typical example is the fire opal, a transparent orange variety, the vivid colors of which come from fine traces of iron oxide. Common opals are primarily built from random accumulation of silica nanograins with different effective diameters that average ~25 nm [72]. Thus, the primary cause of the lack of play of color in common opals is the asphericity in shape and polydispersity in size of the constituent nanograins.

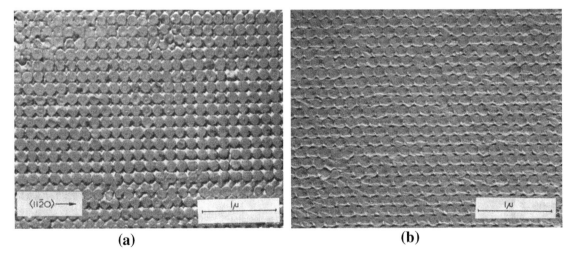

(a)   **(b)**

FIGURE 11.14   SEMs ($\times 30,000$) of the fracture surface of a precious opal reveals closely packed silica spherical particles. (a) Cubic packing. (b) Hexagonal packing [70]. Reproduced with permission of the International Union of Crystallography (http://dx.doi.org/10.1107/S0567739468000860).

The males of two butterfly species, *Lamprolenis nitida* from New Guinea and *Pierella luna* from Central America, display intriguing iridescence of the forewings due to diffraction [73, 74]. The forewings appear matte brown when illuminated from above but show rainbow coloration when illuminated along the axis from the body to the wing tip. The reflection changes dramatically in hue with the viewing angle. Males of the *L. nitida* species display green to red color when illuminated from the front and display blue to violet when illuminated from the back. This pleochroism occurs because the forewings contain two separate photonic structures in each scale, each of which causes a separate iridescent signal of different colors that can be seen in different directions [73]. The forewing of this butterfly decomposes white light, just as a diffraction grating would do. The cross-ribs of its scales, shown in Figure 11.15, can be viewed as a diffraction grating with a roughly 580-nm step lying nearly flat along the wing surface. A white beam at normal incidence diffracts, and the red diffracted beam emerges at grazing angle, whereas orange, yellow, and green diffracted beams appear at smaller viewing angles, increasingly closer to the vertical direction. The color sequence in *P. luna* is reversed, compared to *L. nitida*. Violet light exits at grazing angle, near the forewing surface; the other colors, from blue to red, emerge at viewing angles progressively closer to the direction perpendicular to the wing [74]. This effect is due to a macroscopic deformation of the entire scale, which curls to form a grating perpendicular to the wing surface, and functions in transmission but not in reflection [74].

Flashing green iridescent light from seed shrimps, which are crustaceans of class *Ostracoda*, can be also explained with the diffraction-grating effect. The reflecting surface of a seed shrimp contains fine parallel grooves similar to those on a compact disk. These grooves are used to provide a visual display during courtship [42]. Surface structures shaped like diffraction gratings have also been found in the fossils of Burgess Shale species, such as of the genuses *Canadia* and *Wiwaxia*, living 515 million years ago, and must have given rise to radiant coloration [42, 75].

11.4.3 Structural Colors Due to Collaborative Effects

In general, a structural color changes its hues according to the viewing angle. The glittering blue of South America's *Morpho* butterflies, on the contrary, exhibits a weak angular dependence, indicating that more than one physical mechanism must be responsible. Although pigment granules (mainly biopterin) have been found in the scales of some *Morpho* species, their quantity is insufficient to explain the deep radiant blues [5, 76]. Instead, these colors result from a collaboration of multilayer interference, the diffraction-grating effect, and the non uniform heights of the ridges on the scales on the wings [27, 28]. The iridescent blue and purple colors of the wings of *Hypolimnas anomala*, a Southeast Asian butterfly, also arise from complex microscopic structures on the surface of the scales.

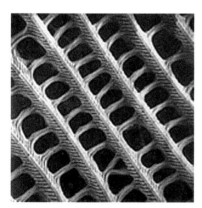

FIGURE 11.15 SEM of a wing of the butterfly *Lamprolenis nitida*. (Source: National History Museum, London, UK; www.nhm.ac.uk/about-us/news/2008/july/how-butterfly-wings-shimmer-revealed18265.html.)

FIGURE 11.16 Two details from Damien Hirst's *Incorruptible Crown*, on display at the Museum of Contemporary Art, Denver, CO, USA. The blue color of the wings of the butterflies at zero and small viewing angles are replaced by dull gray at large viewing angles. (For interpretation of the references to color in this figure legend, the reader is referred to the web version of this book.)

That the *Morpho* blue color is of physical origin can be appreciated due to the fact that it is replaced by dull gray at large viewing angles, as is evident from Figure 11.16. The *Morpho*'s brilliant sheen originates from the microscopic texture of the colorless transparent scales of about 100-μm length that cover its upper wing surfaces [3, 4]. There are about 15 scales per millimeter on the wing surface, and each scale has a set of dozens of microscopic, evenly spaced, parallel ridges. This textured structure resembles a diffraction grating with a period d between 650 and 800 nm. A ridge typically has about 15 longitudinal branches (some *Morpho* species have 6 to 10 branches on each side), all uniformly spaced and each one no thicker than a fraction of a visible wavelength and with a density of 700–2,000 lamellae per mm. This is a tree-like structure of lamellae that are self-assembled from cuticles with a refractive index of 1.56. Depending on the species, these structures effectively reflect 150- to 300-nm-wide spectral bands that peak at wavelengths ranging from 400 to 460 nm at 45–65% reflectance, thus producing the beautiful iridescent *Morpho* blue colors [50].

The physical mechanisms underlying structural color in the *Morpho* genus were extensively studied by Kinoshita *et al.* [11, 25–27]. High reflectance in the blue regime arises from multilayer interference. Diffuse reflectance is due to the small widths and irregular heights of the ridges. Blue pigment also enhances the structural color. Finally, transparent scales on top of the wing act as optical diffusers and create gloss. Thus a collaboration of four different mechanisms, illustrated in Figure 11.17, provides the brilliant *Morpho* blue.

11.4.4 Structural Colors Due to Photonic Crystals

Structural colors exhibited by periodic multilayered structures were discussed in Section 11.3.2. These structures are simple one-dimensional photonic crystals. More complicated one-dimensional photonic crystals found in certain *Coleoptera* and *Scarabaeidae* beetles are structurally chiral [77], just like chiral liquid crystals [78] and chiral sculptured thin films [79]. The helicoidal

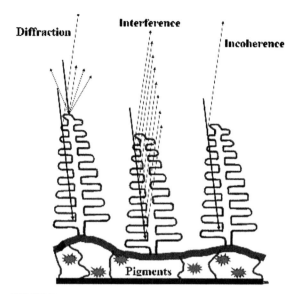

FIGURE 11.17 The complex collaboration of four mechanisms responsible for the iridescent *Morpho* blue, after Ref. 11.

structure is also found in plants and is known to produce blue color in the leaves of certain tropical understory plants [80, 81]. The reflected light is circularly polarized, often left-handed.

The exocuticle of scarab beetles is a helical arrangement of layers of parallel chitin microfibrils. The parallel microfibrils in a specific layer act as an optically anisotropic medium having a refractive index for light polarized along the microfibrils (slow axis) larger than the refractive index for perpendicularly polarized light (fast axis). In each subsequent layer, the orientation of these axes is rotated by a small angle compared to the previous layer. This helicoidal stack of layers, which usually is left-handed structurally and with a well-defined pitch (the distance perpendicular to the layers for which a 360° rotation is obtained), creates a periodicity that is responsible for circularly polarized colored reflection, depending on the pitch and the structural handedness of the exocuticle [77]. This phenomenon—called the *circular Bragg phenomenon*—has

already found practical applications in the fabrication of nano-engineered chiral reflectors for display and laser technologies [82]. Such chiral reflectors are sometimes called *circular Bragg filters* [79].

Other natural photonic crystals can be two- or three-dimensional. Thus, the green iridescence of *Lamprocyphus augustus*, which is almost independent of the viewing angle, is due to exoskeletal scales of differently oriented single-crystalline micrometer-sized domains with an interior diamond-based structure [39]. The apparently complete bandgap is collaboratively created by scales of three different orientations. Although the diamond-based lattice has been artificially engineered to function at $\lambda \sim 1,300$ nm [83], the natural structure in *L. augustus* remains the only example functioning in the visible regime. Reproducing it artificially remains a significant task [84].

Another example of collaboration in natural photonic structures is furnished by the iridescent green wings of the emerald swallowtail butterfly *Papilio palinurus*. The surface of each wing scale is a regular two-dimensional array of concave depressions that are 4–6 µm in diameter and 0.5–3 µm in depth [50]. Each depression comprises 10 unit cells, each of which is made of two constituent layers. Yellow light is preferentially reflected from the bottom and blue light is reflected by the walls of each depression. These two colors add to produce the green color perceived by the human eye. In addition, the walls, but not the bottom, are visible in a reflection-mode microscope under crossed polarizers [85].

11.5 ATTEMPTS TO MIMIC STRUCTURAL COLORS

Inspired by nature, some researchers are attempting to use the principles underlying natural structural colors in engineering practice. But engineered biomimicry is not always an easy task. Structural colors require the design and fabrication of nanostructures because the lengthscales of these structures must be on the same order as the wavelengths of visible light, but translation to even the centimeter scale appears to be a Herculean task. At the same time, the promise of avoiding organic paints and fumes is so attractive and the brilliance of structural colors is so alluring that researchers continue to strive for progress.

In particular, numerous attempts have been made to replicate the brilliant blue of the *Morpho* butterflies, most successfully by biomimetic techniques [28, 29, 86–88]. Furthermore, bioreplication techniques [89] based on nanocasting [46, 90], nanoimprinting [47, 91], physical vapor deposition [48, 92], atomic layer deposition [45], and hybrid techniques [93] are being developed. Some of these approaches and practical industrial realizations were recently summarized in a topical review by Saito [94].

The metallic reflection of koi fish was replicated using biogenic guanine crystals, which were collected from the skin beneath their scales. Guanine crystals were also grown in dimethyl sulfoxide *in vitro*. Although the two sets of crystals show resemblance in form, the biogenic crystals are exceptionally thin (~50 nm) with well-defined crystal faces, whereas the crystals grown *in vitro* are much thicker and irregularly stepped [64].

The ideal structure of precious opals inspired the synthesis of silica photonic crystals by precipitation of 10-nm grains [95] as well as by chemical self-assembly [96]. Inverse opals found on the wings of many beetles have been nanoimprinted by pressing a wing against liquid gallium that is then quickly frozen below its melting point (28.78 °C) [55]. Although this nanoimprinting technique was aimed toward the high-resolution identification of surface details, it has potential for industrial-scale production of surfaces endowed with structural color [93].

11.5.1 Structural Colors from Particle Arrays

It has long been known that beautiful and intense colors are observed in dispersions of small particles. Iridescence due to the Bragg phenomenon exhibited by three-dimensional arrays (i.e., colloidal crystals) formed in somewhat concentrated suspensions of latex particles of diameters between 150 and 500 nm has been observed, the diffraction pattern often being used for characterizing the properties of the colloidal crystals [97].

Particle diameter is an important factor for the different color-display mechanisms exhibited by multilayered stacks formed by small spheres. Latex spheres with diameters considerably smaller than the visible wavelengths (<400 nm) self-assemble in two-dimensional arrays comprising monolayers and multilayers. When illuminated with polychromatic light, a stack of nanospheres reflects intense and uniform pastel colors. Figure 11.18a shows the juxtaposition of several stacks, each of a different color in reflection [71]. These colors result from multilayer constructive interference. In contrast, colors in transmission are almost unnoticeable because each stack is just a few spherical diameters in thickness.

The interference colors in Figure 11.18a are enhanced by a gold-coated glass substrate. Each color corresponds to a stack of a specific thickness, given in Table 11.1. If monochromatic illumination is used instead of white light, the multilayers look like monocolored stripes of different intensity. By varying the monochromatic light wavelength and observing the changes in stripes' intensity, one can notice that at a given wavelength, the intensities of two neighboring stripes coincide. The value of this particular wavelength is a key factor in a theoretical model that suggests that arrays of self-assembled nanoparticles have primarily hexagonal close-packing structure. This prediction was confirmed by scanning electron microscopy and atomic force

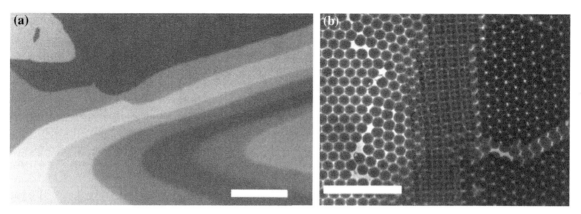

FIGURE 11.18 (a) Colored stacks of 55-nm-diameter latex spheres observed by a microscope in reflected light. To enhance the colors, a gold-coated glass substrate was used. Each color corresponds to a stack of a specific thickness. The reference bar is 100 μm. Courtesy of Late Ceco Dushkin (University of Sofia, Bulgaria) [71]. Reproduced with permission of the American Chemical Society. (b) SEM showing a tetragonal bilayer (in the center) between a hexagonal monolayer (on the right) and a hexagonal bilayer (on the left) of latex spheres of diameter 144 nm. The reference bar is 1 μm. Courtesy of Late Ceco Dushkin (University of Sofia, Bulgaria) [99]. Reproduced with permission from Springer. (For interpretation of the references to color in this figure legend, the reader is referred to the web version of this book.)

microscopy. The scanning electron micrograph shown in Figure 11.18b reveals a narrow band of tetragonal close packing between two consecutive multilayers of hexagonal close packing. This narrow band produces its own color, determined by the effective thickness of the layer with tetragonal structure. For example, a band of yellow color produced by a tetragonal four-layer could be observed between green hexagonal trilayer and red hexagonal four-layer stripes. By changing the packing geometry in an array, one can alter the color. The stepwise change in color

with each layer in such ordered stacks could be used for creating and recording of color images.

In contrast, latex spheres with diameters (1.5–10 μm) larger than the wavelength in the visible regime self-assemble as two-dimensional colloidal crystals and function as Bragg reflectors. When illuminated with white light, they produce iridescent colors that depend on the size and orientation of the domains of the hexagonal structure shown in Figure 11.19 [98, 99]. The domain size is critical for the brightness of the iridescence; indeed, polycrystalline films with

TABLE 11.1 Colors of dried multilayers of latex spheres of diameter (55 ± 4) nm assembled on gold-coated glass [71].

Number of Layers	Layer Thickness (nm)	Color in Reflected Light	Number of Layers	Thickness (nm)	Color in Reflected Light
1	47	Ochre	7	277	Magenta
2	85	Brown	8	316	Blue-purplish
3	124	Navy blue	9	354	Green
4	162	Sky blue	10	392	Yellow-green
5	201	Yellow	11	431	Orange
6	239	Orange	12	469	Red

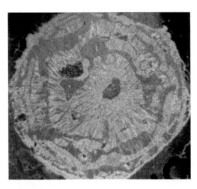

FIGURE 11.19 Optical image of a two-dimensional crystal of diameter 1.4 cm, formed by self-assembly of latex microspheres of diameter 1.696 μm on a transparent substrate [99]. Courtesy of Late Ceco Dushkin (University of Sofia, Bulgaria). Reproduced with permission from Springer.

very small domains appear white. Although the interference colors from the nanospheres can be viewed only under an optical microscope, the diffraction colors from microspheres are observable with an unaided eye. A dry two-dimensional crystal of latex microspheres is environmentally stable and exhibits colors for several years.

The process of two-dimensional crystallization of latex nanospheres on a substrate involves the evaporation of water from an aqueous suspension of monodisperse latex spheres of diameter a small fraction of visible wavelengths, as well as with microspheres of diameter ranging from 1.5 μm to 10 μm [71, 98–100]. This technique, however, does not allow for controllable growth of a single layer of particles over large areas.

In the convective-assembly technique [101], a glass substrate is pulled out from an aqueous suspension of monodisperese polystyrene spheres at an appropriate angle and with a proper speed to allow the particles to self-assemble on a regular two-dimensional lattice over a large area. This technique is applicable to a broad range of particle sizes (from about 100 nm to several micrometers in diameter), since each size results in a differently textured surface.

Identical spheres of diameters between 10 nm and 1 μm can be readily assembled into colloidal crystals and synthetic opals that possess three-dimensional order. The difference between colloidal crystals and synthetic opals is in the volume fraction of the constituent spheres. Colloidal crystals contain less than 10% v/v constituent spheres, which are highly charged and arranged on face-centered-cubic (fcc) lattice. Synthetic opals have a cubic close-packed structure (also on an fcc lattice) with a packing density close to 74%, similar to that of natural opal [102]. In addition, *inverse opals* can be fabricated by templating various kinds of precursors against crystalline arrays of colloidal spheres. Nowadays, the process of production of large quantities of exceedingly uniform spheres from silica and polymer colloids of different chemical composition (e.g., polystyrene and polymethyl methacrylate) is routine [102]. Silica colloids represent one of the best-characterized inorganic systems used for production of monodisperse spheres. The self-assembly of colloidal spheres is a well-developed technique [102, 103], and is both a faster and a cheaper approach for the fabrication of high-quality three-dimensional structures than the common microfabrication techniques (deposition, photolithography, etching, and doping) that require a clean room [104, 105]. Nanopatterning becomes easy to implement, as does the fabrication of lithography masks, templates to make three-dimensional macroporous materials, and several photonic devices—including photonic crystals and diffraction gratings [102].

11.5.2 Bioinspired and Biomimetic Reproduction of the *Morpho* Blue

The early attempts to artificially reproduce the *Morpho* blue employed self-assembled structures comprising microspheres and nanospheres [71, 98–101]. The convective-assembly technique was used to produce two-dimensional arrays of polystyrene spheres with diameter comparable to the ridge spacing on the *Morpho* wing. Such arrays were produced on large substrates and they displayed the wavelength range and the brilliant sheen of the *Morpho* blue [106]. The top half of Figure 11.20a shows the scanning electron micrograph of the parallel ridges on a scale of a *Morpho* wing, and the bottom half of the same figure has a similar micrograph of an array of 953-nm-diameter polystyrene spheres on a glass slide. An artificial wing was produced as a collage of photographs of the polystyrene-sphere array of Figure 11.20a, followed by image processing to match the real wing's shape and size. As shown in Figure 11.20b, the artificial wing displayed iridescent blue color with a brilliance competing with that of a real *Morpho* wing. However, this bioinspired attempt was successful in only partially reproducing the *Morpho*'s beauty. Due to the symmetry of the hexagonal structure and the different textured surfaces, the artificial wing displays shades of iridescent blue (appearing more like a speckle pattern), but it does not reproduce the color uniformity over a wide range of viewing angles, so characteristic of the *Morpho* blue. Although precise tuning

of the delicate blue sheen of the *Morpho* wing is really challenging, this attractive approach could be implemented to create new iridescent materials for arts, textiles, cosmetics, and paints.

The elucidation of the physical mechanisms for the *Morpho* blue [27] opened the horizons for its successful biomimetic reproduction. Saito and colleagues dry-etched a substrate using the electron-beam technique to reproduce the *Morpho* blue [28, 29, 87]. Parallel ridges, each comprising a large number of parallel pillars or tips, were etched, with the ridge width and the inter-ridge separation the same as those in the scales of a wing. The pillar heights were chosen randomly, to mimic the uneven heights of ridges in the *Morpho* wing. The fabricated substrate resembles the wing in reflectance, color, bandwidth, angular dependence, and chromatic anisotropy.

Neither of the two ways ends up mimicking the structure of the *Morpho* wing itself but only the collaboration between regularity and irregularity discussed in Section 11.4.3. Using a completely different approach, Watanabe *et al.* [88] successfully reproduced not only the color

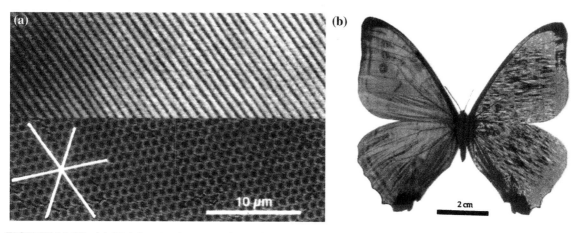

FIGURE 11.20   (a) SEM showing (top area of image) parallel ridges on a scale of a *Morpho* wing and (bottom of image) an array of polystyrene particles on a glass slide. (b) Comparison between a *Morpho stoffeli* wing (left) and an artificial wing constructed from 953-nm-diameter polystyrene spheres on a glass slide (right). The incidence and the viewing angles are about 60° and 90°, respectively, and the illumination is from the left of the figure [106]. Courtesy of Kuniaki Nagayama (National Institute for Physiological Sciences, Japan). Reproduced with permission of © SYMMETRION.

mechanism but also the structure of the ridges on the scales. They implemented a hybrid focused-ion-beam chemical vapor deposition (FIB-CVD) technique. When illuminated by white light and observed under a microscope, the artificial structure shimmers with blue to violet color, which, by means of a photonic multi channel spectral analyzer system, was proven to be similar to the reflection from real *Morpho* scales. Although indicative of technoscientific accomplishment, the hybrid technique does not appear to be industrially scalable in the foreseeable future [89].

11.5.3 Bioreplication

Perhaps the most direct way to reproduce the iridescent colors of butterfly wings is by copying the wing morphology itself with nanoscale fidelity. Several bioreplication techniques [89] are currently under development, as discussed elsewhere in this book.

Low-temperature atomic layer deposition (ALD) of alumina was used for the replication of the *Morpho* butterfly wing [45]. After the wing itself had been removed by prolonged high-temperature annealing, the negative replica left behind exhibited optical response characteristics quite similar to that of original wing. Not only the optical response

but also the surface hydrophobicity of butterfly wings is exhibited by ALD replicas [107].

The conformal evaporated film by rotation (CEFR) method, classified as a physical vapor deposition method, has also been used for bioreplication, as discussed in Chapter 15. The method employs thermal evaporation of a solid material in a low-pressure chamber to create a collimated vapor flux that is directed toward the biotemplate, which itself is rotated in a complex fashion so that the incoming vapor flux condenses and solidifies as a conformal coating of the biotemplate [48]. Details of a butterfly wing with and without the coating of a chalcogenide glass are shown in Figure 11.21, clearly demonstrating the high fidelity delivered by the CEFR method. Removal of the wing from the coating has been achieved by chemical etching to leave behind a positive replica [92].

Nanocasting methods have also been used to make replicas of butterfly wings. In the sol-gel method, replication begins by filling the void regions of a butterfly wing with a sol, drying it, and then either etching or calcining away the actual wing [46, 108–110]. This process forms a negative replica of the dried sol. Sonication may assist in the infilling step [111]. Instead of the sol-gel method, chemical vapor deposition may be used to infiltrate the void regions [112].

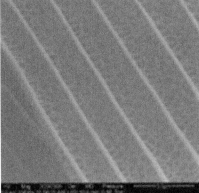

FIGURE 11.21   SEMs of (left) a butterfly wing and (right) the same wing coated with chalcogenide glass using the CEFR method [48].

A positive replica can be made by infilling the void regions of the negative replica and removing the material of the negative replica [84], but the delicate nature of butterfly wings has made it difficult to implement this step.

11.5.4 Structural Colors by Thin-Film Technologies

As mentioned in Section 11.4.4, the exocuticle of many scarab beetles has a layered helicoidal morphology and displays iridescent color when viewed with the light of one circular polarization state but not of the other. Simpler periodically multilayered structures in nature display structural color without discrimination of the circular polarization state of the incident light. A way to realize structural colors of this type is by depositing periodic arrangements of homogeneous layers of two different materials on a substrate. The refractive index of one of the two layers in the unit cell should be higher than that of the other. Deposition can be carried out using a variety of thin-film technologies.

The Langmuir–Blodgett method is used for the deposition of molecular monolayers and multilayers [53, 113]. The process of building a Langmuir–Blodgett multilayer consists of periodically dipping a substrate into an aqueous solution of amphiphilic molecules. Each amphiphlic molecule has a head and a tail. A layer is deposited during each dip. However, the molecules deposited during the removal step have their heads oriented toward the substrate, whereas the molecules deposited in the immersion step are oriented with their tails facing the substrate. The result is that the heads in the latest layer adhere to the heads in the previously deposited layer during the upstroke, and the tails of the latest layer stick to the tails in the previously deposited layer during the downstroke. Several layers of one type of molecule may be followed by several layers of another type of molecule, and then the cycle is repeated a few times, resulting in the formation of a one-dimensional photonic crystal. Although the Langmuir–Blodgett method is simple, a major shortcoming until recently has been the restricted range of materials for which it can be used.

A related technique is layer-by-layer assembly [53, 114], which is based on the alternating adsorption of positively and negatively charged species from aqueous solutions and can also be used to create one-dimensional photonic crystals. The roughness, thickness, and porosity of a multilayer film can be controlled at the molecular level by adjusting experimental parameters such as pH and ionic strength. Compared to the Langmuir–Blodgett method, layer-by-layer assembly is generally much simpler and faster, is more versatile, and usually results in more stable films [115]. The same procedure can be implemented even faster using inkjet printing of cationic and anionic polymers alternately [116].

A host of methods, collectively called *physical vapor deposition* [53, 79], are commonly used to deposit thin films. In these methods, discussed in Chapter 15, a collimated vapor flux is generated from a solid target—by heating it, by passing a current through it, or by bombarding it with electrons or ions—in a low-pressure chamber. The collimated vapor flux arrives normally on a substrate and condenses as a dense thin film. When targets of two different solid materials are used alternately, a one-dimensional photonic crystal can be formed on the substrate. With proper selection of the two materials and the thicknesses of their layers, this periodically multilayered structure can display very vivid colors [117].

If the vapor flux arrives at the substrate obliquely, the density of the film is reduced. The film is an array of parallel tilted nanowires and can be considered optically as an effectively homogeneous biaxial dielectric continuum [118]. Rotation and/or rocking of the substrate during deposition provides these nanowires with a shape. Such a film is called a *sculptured thin film* (STF) [79], which can be considered optically as a biaxial dielectric continuum that is nonhomogeneous in the thickness direction.

STFs with periodic nonhomogeneity are routinely made these days, some comprising two-dimensional nematic nanowires [119], others comprising three-dimensional helical nanowires [120]. These periodic STFs are one-dimensional photonic crystals that display the Bragg phenomenon selectively for light of a certain polarization state [79]. For example, a chiral STF is an array of parallel helical nanowires, as shown in Figure 11.22, and displays the circular Bragg phenomenon—best described as the high reflectance, within a narrow spectral regime, of circularly polarized light of the same handedness as the chiral STF of sufficient thickness, whereas circularly polarized light of the opposite handedness is reflected very little. In other words, a chiral STF is a circular Bragg filter.

Tremendous control of the circular Bragg phenomenon is possible because the helical morphology is nanoengineered during deposition. Structural color can be preserved even when a chiral STF is crushed down to sub-millimeter platelets, as shown in Figure 11.23. This fine powder could be suspended in an appropriate cosmetic fluid (lotion) or perfumed substance to form exhilarating body gels.

FIGURE 11.22   SEM of a chiral STF made of chalcogenide glass [120].

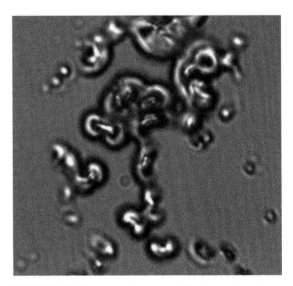

FIGURE 11.23   Optical image of platelets taken from a 3-period-thick chiral STF of magnesium fluoride deposited on a 30-nm-thick aluminum film. The pitch of the chiral STF is 450 nm, and the image was taken at 100x magnification.

11.5.5 Environmentally Responsive Structural Color

All approaches employing engineering biomimicry discussed so far in this section replicate the passive color generated by a periodic structure. Most recently, changes in structural color displayed by some fish (e.g., blue damselfish or neon tetra fish), octopuses, squids, and beetles in response to changes in the environment have inspired research on bioinspired and biomimetic materials—such as colloidal crystals—with dynamically tunable structural color [121, 122]. The fabrication of photonic materials with tunable structural color exploits the three mechanisms of reversible active structural color in nature: change in refractive index, change in spacing of a periodic structure, and change in direction of illumination. For example, the porous exocuticle of some beetles changes its refractive index by absorbing water [62]. The spacing of the periodic structure in cephalopods is tuned by change in the protein platelets' thickness (via swelling) or by varying the space between the platelets. Tilting the platelets changes the incidence angle of light and the wavelength of selective reflection [122]. Artificial structures with color tunability have been fabricated as multilayer high- and low-refractive-index organic/inorganic films (titania precursor sol and a polymer solution), in which the color can be tuned through the interlayer distance or the refractive index [122]. Tunable color has also been demonstrated by swelling of opal composites [123]. Such changes are potentially useful for colorimetric sensors of humidity, tension, and strain [122].

11.6 APPLICATIONS OF STRUCTURAL COLORS IN ART AND INDUSTRY

11.6.1 Structural Colors and Art

The applications of structural colors for art and decoration come naturally because iridescent sheen, pearly luster, and play of colors are fascinating. For thousands of years, they have been a part of our lives as decorations satisfying our aesthetic needs and signifying social status. Opals and pearls have special value in jewelry. Pearlescent laminae of ear shell or turban shell have been used to decorate lacquerware in China; the wings of jewel beetles and the feathers of birds were used for decoration in Japan more than a thousand years ago. Mica powder was used in cosmetics as well as in the Japanese art form *Ukiyo-e* in the 17th and 18th centuries [11]. Nacre or mother of pearl is still used to decorate objects, such as the box depicted in Figure 11.24. The use of paua shells, one of which is shown in Figure 11.24, for jewelry has spread from the Maori culture in New Zealand to many parts of the world. Modern art benefits from the brilliant colors of *Morpho* and other butterflies, often bred in butterfly farms [124], by incorporating their wings in works of art, as exemplified by Figure 11.16.

11.6.2 Industrial Applications of Structural Colors

The burgeoning interest in structural colors during the last two decades is fed by the possibilities of creating synthetic photonic structures that mimic natural photonic structures as well as new materials for coatings and fabrics, cosmetics, and paints [11, 94]. Industrial applications require implementation of simple, fast, efficient, and inexpensive techniques allowing production of a variety of shapes and sizes with guaranteed repeatability and quality. Most engineered-biomimicry approaches to fabricating structural colors discussed heretofore in this

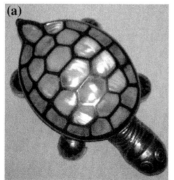 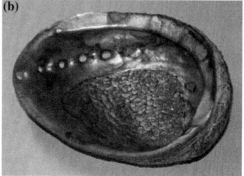

FIGURE 11.24　(a) An ornamental tortoise made of a metal and mother of pearl. (b) A paua shell from New Zealand.

chapter promise exciting applications but are still under development and not quite suitable for mass production. So far, industrial adoption is confined to the fabrication of multilayer structures and block copolymers with a periodic modulation of refractive index in one direction (one-dimensional photonic crystals) and the fabrication of three-dimensional photonic crystals based on self-assembly of colloids.

11.6.2.1 Structurally Colored Fibers and Surfaces

A very striking contemporary application of structural colors produced by a multilayer photonic structure is the structurally colored fiber and a fabric woven with that fiber. Nissan, the Japanese automobile company, claimed the world's first successful reproduction of the *Morpho* color in an engineering application: noncircular structurally colored fiber fabricated by the conjugated melt-spinning method [125, 126]. The sketch in Figure 11.25 illustrates the idea of the structurally colored fiber. As discussed in Section 11.4.1, the underlying concept is constructive interference from alternating layers of two polymers with different refractive indexes n_a and n_b, and thicknesses d_a and d_b, respectively. When the ratio of the refractive indexes obeys the condition $1.01 \leq n_b/n_a \leq 1.20$, white light at normal incidence is selectively reflected back for a wavelength

$$\lambda = 2 \left(n_a d_a + n_b d_b \right). \tag{11.4}$$

The fabrication of the fiber consists in alternative passing of molten polymers (polyester and

FIGURE 11.25 Schematic of the structurally colored fiber fabricated by Nissan Motor Co. A periodic multilayer comprising alternating layers of two polymers with different refractive indexes and different thicknesses is encased in a polyester cladding.

polyamide) through an especially designed metal mold mounted on a spinning head. The first reported fiber consisted of 61 alternate layers of polyamide (nylon 6 with mean refractive index of 1.53) and polyester (mean refractive index of 1.63) with thicknesses in the 70–90-nm range. The multilayer core was embedded in a polyester of thickness 15–17 μm and then flattened [125]. The structure was designed for high reflectance at 470-nm wavelength to correspond to the *Morpho* blue with maximum reflectance. The manufactured fiber exhibited clear color and a luxurious metallic gloss, and its color changed subtly with the viewing angle.

As it is a non pigmented fiber, its production is friendly to the environment. Due to the similar refractive indices of the alternating polymer layers, the spectral regime of high reflectance is much narrower than that of the *Morpho* wings. Although the beauty of the fiber does not match that of *Morpho* butterflies, it is quite suitable for many applications, including for dye-free upholstery and dresses, because of the flatness of the fiber; see Figure 11.26. It is also easy to diversify the color range of the manufactured fibers simply by changing the thicknesses of the constituent layers.

Dispersal of minute granules (not of pigments) in the two layers of the fiber also produces color, provided the granular substance is properly chosen, similar to the production of Tyndall blue in nature. For example, a structurally colored fiber was manufactured in the form of a yarn or filament with a coaxial structure of alternating layers of polymers with refractive indexes n_a and n_b satisfying the condition $1.1 \leq n_b/n_a \leq 1.4$ [127]. Minute granules consisting of at least one material selected from the group of calcium carbonate, zinc sulfate, zinc white, lithopone, cadmium sulfate, chrome oxide, rutile, and anatase were dispersed in the first or the second plurality of transparent layers. The granules have a size equal to or less than 1.25 μm and a refractive index n_c, which satisfies the twin conditions $n_c - n_a \geq 0.4$ and $n_c - n_b \geq 0.4$.

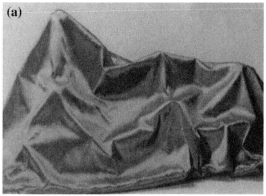

FIGURE 11.26   A fabric woven with the structurally colored fiber of Nissan Motor Co and applications thereof for upholstery and dresses. Hiroshi Tabata, personal communication to ND (1999).

The granules themselves are not structurally colored. They add the effect of scattering of white light incident on them to the selective reflection from the multilayered coaxial structure, thus producing the final color and luster of the yarn. Proper selection of the transparent and granular substances can produce reflection peaks at wavelengths of about 480 nm, 550 nm, and 650 nm and thereby exhibit vivid blue, green, and red colors. The yarn has a circular cross-section and is produced by the melt-spinning method, which involves spinning the yarn at a temperature of 285 °C and a take-up velocity of 1,000 m/min with the use of a single filament. After being cooled and caked at room temperature, the yarn has a diameter of about 40 μm.

Another technique for producing interference colors on flat surfaces or fibers involves, again, a dispersal of grains; however, the grains are structurally colored and serve as an ink—a very expensive one! A Bragg filter is deposited onto a thin substrate, which is then granulated and dispersed in a transparent polymer as ink. The idea has been implemented for color-shifting inks used on secure documents, currency, stamps, credit cards, etc. Vapor deposition techniques are commonly used for producing Bragg filters [128]. Also, layer-by-layer self-assembly [115] is used to make composite polymeric films of specific composition and thicknesses in the range of 5–1,000 nm with precision better than 1 nm [129], particularly of large areas.

Inkjet printing of Bragg filters is very attractive because it is fast and inexpensive. For instance, alternating sequences of tin oxide (SnO_2) nanoparticles (high refractive index) and silica nanoparticles (low refractive index), both overprinted with a cationic polymer (polyethyleneimine), can be inkjet printed [116]. This method is feasible for achieving bright structural colors onto already produced textile fibers and fabrics. However, further improvement is needed to allow for regular smooth layers to be built up.

These methods demonstrate that the incorporation of structural colors in textiles could be a very predictable, automated, and controlled process. The mathematical model of interference in periodic multilayered structures provides a physical coloring recipe based only on the following physical parameters: the number of layers in the unit cell and the refractive index and thickness of each layer. This recipe can precisely forecast the coloration effect of fabrics and may turn structural colors into a simple and inexpensive alternative to dyed colors. Moreover, the reflection color can be exploited for colorimetric sensing, as discussed in Section 11.5.5, thereby proffering smart textiles in the near future.

11.6.3 Nacreous Pigments

Paints containing bright materials such as aluminum flakes and mica have been on the market for quite a while. These paints are used to cover surfaces in buildings. Nacreous paints, also called pearly luster paints, exploit the multilayer-interference mechanism of labradorite's iridescence. Nowadays, most nacreous paints contain flakes of titanium-dioxide-coated mica (TiO$_2$-mica), their greatest advantages against dyes being their high stability with respect to pH (acidity and/or alkalinity), ambient light, and temperature [130, 131]. Titanium dioxide (refractive index 2.5–2.7) is the primary ingredient in white paint used for walls, since it has the highest opacity of all known pigments. Small quantities of titanium dioxide are used to lighten other pigments.

To increase the reflectance and assure wavelength selectivity, the thickness t of the titanium-dioxide coating is adjusted according to the rule

$$(2m - 1)\lambda = 4t\sqrt{n_{TiO_2}^2 - \sin^2\theta}, \quad (m = 1, 2, 3, \ldots),$$

$$(11.5)$$

where t is the thickness, n_{TiO_2} is the refractive index of titanium-dioxide coating, and θ is the angle of incidence. The color selection in transmission and reflection resulting from Eq. (11.5) for a particular wavelength is illustrated in Figure 11.27, and examples of different

thicknesses of the coating and the corresponding interference colors are given in Table 11.2. Because of the wide bandwidth and the low reflectance of each flake, a dense pile of flakes is needed for high reflectance. These paints exhibit white color with a high degree of pearly luster, and various interference colors—such as yellow, red, blue, and green—are obtained by tailoring the coating thickness.

When a paint containing titanium-dioxide-coated mica flakes is deposited on a white background, only a pearly white luster, is visible. This is because the light transmitted through the flakes onto the white background is reflected back to the eyes of the viewer, along with the light directly reflected from the flakes. When the

TABLE 11.2 Thickness and interference colors of mica coated with titanium dioxide [131].

Interference Color	Approximate Thickness (nm)
Yellow (gold)	75
Red (magenta)	90
Blue	120
Green	145
Yellow (gold)	160
Red (magenta)	185
Blue	210
Green	245

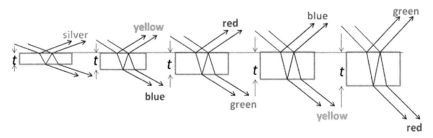

FIGURE 11.27 Reflected and transmitted colors depend on the thickness t of the titanium-dioxide coating of mica flakes, after Ref. 131. (For interpretation of the references to color in this figure legend, the reader is referred to the web version of this book.)

paint is deposited on a black ground, reflectance from the background is absent, and only the interference color reflected from the flakes can be seen. Micrometer-sized flakes produce a smooth sheen, but larger flakes give a sparkle.

Other types of nacreous pigments contain metal-coated silica flakes. They exhibit strong metallic reflectance, with the color changing with the viewing angle, and are therefore called optically variable paints [132]. Should flakes of sculptured thin films [79] be added to paints, polarization-dependent color can be expected. Nacreous pigments have found a broad range of applications in cosmetics, paints, and decorations.

11.6.4 Structural Colors in Cosmetics

11.6.4.1 Cosmetic-Grade Color Additives

Cosmetic products must not only promote healthy skin but also provide a beautiful tapestry of colors for every complexion. The state of the art in *classical cosmetics* is the use of finely crushed minerals and botanicals of the best quality. To suit even the most sensitive skin, these ingredients should be 100% pure and completely natural. However, most color additives to mass-produced cosmetic products are synthetic: indeed, unless specifically approved (such as annatto seed), all natural color additives are banned in the United States by the Food and Drug Administration to avoid irritation and hypoallergenic reactions. Cosmetic-grade color additives, either inorganic (pigments) or organic (dyes and pigments), are safe for use on humans and do not contain heavy metals such as lead, cadmium, and hexavalent chromium. The inorganic pigments include ultramarines, iron oxide, manganese, chromium oxide, titanium dioxide, and ferric ferrocyanide. Both organic and inorganic cosmetic-grade pigments are insoluble in both water and oil; however, they can be microscopically dispersed in a liquid host and made to appear as though they are dissolved. Only a few soluble dyes are used in cosmetic products because of the health hazards they may pose, especially if used in too high a concentration. For example, *lakes* are pigments manufactured by precipitating a soluble dye with a metallic salt (calcium salt, barium salt, aluminum salt, and sodium salt) because the different salts provide different colors.

Color additives can be combined with titanium dioxide and mica, bismuth oxychloride, or guanine. When combined with the titanium-dioxide-coated mica, they create very attractive dual-color pearls. Bismuth oxychloride is a soft material used as the frost in lipsticks. Guanine is derived from herring scales and is used primarily in nail polishes. In general, the mineral powders in the commercially available cosmetic products are based on titanium dioxide and zinc oxide, which, in addition to other technological reasons for their use, may also provide anti-inflammatory and calming effects.

11.6.4.2 Colored Nacreous Pigments in Cosmetics

Structural colors have also found a place in the world of cosmetic products—as glittering materials. Most of the nacreous materials exhibiting interference colors and a high degree of opacity that are used for cosmetics are titanium-dioxide-coated mica flakes. Colored nacreous pigments are obtained when colored compounds such as iron oxide (yellow); ferric ammonium ferrocyanide, known as Prussian blue (blue); chromium oxide (green); or carmine (red) have been added to those flakes [132, 133]. For example, a coating of iron oxide alone produces metallic effects similar to bronzes and coppers. The thickness of the flakes, typically between 15 and 150 μm, dictates the interference between light reflected from the substrate and that reflected from the flake and therefore controls the luster of pearlescents. Low luster—with thin flakes—gives the effect of a pearl with a smooth sheen; medium thickness creates a silky or satiny effect. The larger the flakes, the more sparkle, although they are all shiny.

For high luster, it is important that the flakes are properly dispersed in the cosmetic product to prevent agglomeration, but all have the same orientation [130]. Pearlescents reflect better when they have what is known as a *background color*.

A red pigment on a red background looks even more intense than on its own. Even though pearlescents are widely used in retail eye make-up, most eye shadows contain less than 10% pearlescent pigments; the rest is filler material. Although a wide variety of vivid interference colors and a pearly luster are produced, the addition of colored compounds to cosmetic products comprising nacreous pigments reduces the shelf life and may cause safety concerns.

11.6.4.3 The Infinite Color™ of Shiseido

Infinite Color™, the first purely physical color for cosmetic applications, was developed by the Japanese cosmetics company Shiseido after comprehensive research on nacreous paints for cosmetics [134]. Established in 1872 as the first Western-style pharmacy in Japan, Shiseido now is the leading cosmetics company in that country. *Infinite Color*™ is a photochromic formulation of titanium dioxide whose color changes with the viewing angle. The pigment consists of fine mica flakes coated with titanium dioxide and its lower oxides Ti_nO_{2n-1} ($n = 1, 2, 3, ...$) [131]. The mica flake is coated first with a black lower titanium oxide, shown in the example sketch of an *Infinite Color*™ flake in Figure 11.28 as Ti_4O_7, and then coated with titanium dioxide. As shown in Table 11.3, the lower oxides of titanium are bluish or purplish black in color and their role is to provide a black background. This black layer is very stable; it absorbs the transmitted light and causes the interference phenomenon to be greatly enhanced. Table 11.4 provides some examples of colors by *Infinite Color*™ and the corresponding optical thickness of the coated mica flakes.

In comparing the *Infinite Color*™ with the pearlescent pigments used in cosmetics, it is worth stressing that the *Infinite Color*™ is produced only by interference of light, without the addition of

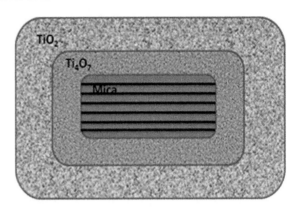

FIGURE 11.28 Schematic of a mica flake used by Shiseido in its *Infinite Color*™ line of cosmetic products, after Ref. 131.

TABLE 11.3 Color of black lower oxide of titanium in *Infinite Color*™ [131].

Composition	Color
TiO	Bronze
Ti_2O_3	Purplish black
Ti_3O_5	Bluish black
Ti_4O_7	Bluish black

TABLE 11.4 Colors of *Infinite Color*™ [131].

Interference Colors as They Appear	Base Color	Optical Thickness (nm)
Gold		210
Red	Blue purple	250
Blue		310
Green		360

coloring materials, whereas the pearlescent cosmetics also use dyes and other pigments. The *Infinite Color*™ has excellent hiding power and high flip-flop effect, i.e., change of color with the viewing angle. Similarly to pearlescent pigments, when the *Infinite Color*™ is applied on black background, both the pearly luster and the interference color show. The difference is that when pearlescent pigments are applied on a white background, the white luster shows but the interference color does not; whereas both the pearly luster and the interference color does not appear for the *Infinite Color™* even when it is applied on white background.

In addition to cosmetics, the technology of *Infinite Color™* is used to color artificial leather, construction materials, and automobile bodies. A recent review paper suggests replacing of the central mica core with a void that may result in an increased reflectance, especially in the short-wavelength range [94].

11.6.5 Nacreous Pigments in Imaging Elements

A remarkable application of nacreous pigments is in making imaging elements [135]. Such an imaging element is shaped like either a platelet or a needle. A layer of an oxide of titanium, aluminum, or barium is first deposited as a highly reflecting layer on a substrate, then that layer is coated with a layer comprising white pigment (from the group consisting of titanium oxide, zinc oxide, zinc sulfide, barium sulfate, calcium carbonate, talc, and clay), and finally a layer comprising a nacreous pigment and a polymer is deposited. The nacreous pigments contain flakes of metal-oxide-coated mica, feldspar, silicates, and quartz. The difference between the refractive indices of the second and the third layers should exceed 0.2. The thickness of the layer containing the nacreous pigment must be between 2 and 10 times the longest linear dimension of the nacreous flakes.

11.6.6 Sensor Applications Inspired by the *Morpho Sulkowskyi* Butterfly

Inspiration for the next two applications came from highly iridescent wings of the butterfly *Morpho sulkowskyi*. The reason this species was chosen from the *Morpho* genus is due to the peculiarities of the morphology of its wing scales; see Section 11.4.3. Compared to the other

species in that genus, the scales of *M. sulkowskyi* have a higher ridge density with slightly slender shapes and a more regular lamellar structure, and, most important they contain only a negligible amount of a pigment that absorbs light and decreases reflectance [26]. The reflectance of the scale changes with either thermal expansion [136] or due to the entrapment of a vapor in the void regions of the scale [137].

When a wing of *M. sulkowskyi* is illuminated with midwave infrared (3–8 μm) light, a part of the incident energy is absorbed in the wing. This causes thermal expansion of the wing, resulting in increased spacing between the ridges, expansion of the tree-like lamellar structure, and a thermally induced reduction in the refractive index of chitin (of which the wing is made); accordingly, spectral signatures of the wing shift. Doping the scales with single-walled carbon nanotubes enhanced the sensitivity and the dynamic response to the infrared exposure, and temperature changes as small as 0.02°C were detected within 25 ms [136], which is very promising for high-resolution microbolometry.

The scales trap air between the chitin structures. When that air is replaced by a vapor, the overall reflectance spectrum of a *M. sulkowskyi* wing changes. Very diverse reflectance spectra are produced with different vapors, as has been noted with water, methanol, ethanol, and dichloroethylene [137]. Replicas of these wings could therefore be used as optical sensors. Both applications—thermal imaging and optical sensing—of butterfly wings would require *standardization* before becoming industrially viable.

11.7 CONCLUDING REMARKS

Nature is not as simple as claimed by scientific reductionism. Complex structures have evolved over very long periods of time to perform one or many functions. There may be several routes to the display of the same functionality. Dyes and pigments occur widely in nature to produce color for various purposes. Nonpigmented colors also occur with some frequency in nature, both in living creatures and inanimate objects. These structural colors are due to both regular and irregular arrays of scattering elements. Rich hues of structural colors are usually produced collaboratively by several physical mechanisms. A good understanding of these mechanisms and their interplays will help us in formulating artificial structural colors for fabrics, paints, cosmetics, and other applications.

Acknowledgments

ND thanks Prof. Kuniaki Nagayama for introducing her to the field of structural colors and revealing the secrets of the blue iridescence of the beautiful *M. sulkowskyi* butterfly, as well as for leaving his collection of works and materials on the subject to her. AL thanks the Charles Godfrey Binder Endowment at Penn State for sustained support of his research activities.

The authors dedicate this chapter to ND's husband, Dr. Ceco Dushkin, who authored some of the first works on structural colors from self-assembled particle arrays and who created and financially supported the Laboratory of Nanoparticle Science & Technology at the Faculty of Chemistry, University of Sofia, in 2000, the first and pioneering laboratory in Bulgaria in this scientific area. Dr. Ceco Dushkin passed away on May 19, 2011, at the age of 52.

References

[1] D. Lee, *Nature's palette: the science of plant colors*, University of Chicago Press, Chicago, IL, USA (2007).

[2] W.D. Bancroft, The colors of colloids. VII, *J Phys Chem* **23** (1919), 365–414.

[3] H. Ghiradella, Light and color on the wing: structural colors in butterflies and moths, *Appl Opt* **30** (1991), 3492–3500.

[4] H. Ghiradella, Structure of butterfly scales: patterning in an insect cuticle, *Microsc Res Tech* **27** (1994), 429–438.

[5] K. Kumazawa and H. Tabata, Time-resolved fluorescence studies of the wings of *Morpho sulkowskyi* and *Papilio xuthus* butterflies, *Zool Sci* **13** (1996), 843–847.

[6] C.W. Mason, Structural colors in insects. I, *J Phys Chem* **30** (1926), 383–395.

[7] http://optics.org/cws/article/research/8351 (accessed 1 July 2009).

[8] S. Wickham, M.C.J. Large, L. Poladian, and L.S. Jermiin, Exaggeration and suppression of iridescence: the evolution of two-dimensional butterfly structural colours, *J R Soc Interf* **3** (2006), 99–109.

[9] P. Vukusic, Structural colour in Lepidoptera, *Curr Biol* **16** (2006), R621–R623.

[10] A.L. Ingram and A.R. Parker, A review of the diversity and evolution of photonic structures in butterflies, incorporating the work of John Huxley (The Natural History Museum, London, from 1961 to 1990), *Phil Trans R Soc Lond B* **363** (2008), 2465–2480.

[11] S. Kinoshita and S. Yoshioka, Structural colors in nature: the role of regularity and irregularity in the structure, *Chem Phys Chem* **6** (2005), 1442–1459.

[12] B.D. Wilts, H.L. Leertouwer, and D.G. Stavenga, Imaging scatterometry and microspectrophotometry of lycaenid butterfly wing scales with perforated multilayers, *J R Soc Interf* **6** (2009), S185–S192.

[13] A.R. Parker, The diversity and implications of animal structural colors, *J Exp Biol* **201** (1998), 2343–2347.

[14] N. Dushkina and A. Lakhtakia, Structural colors, cosmetics and fabrics, *Proc SPIE* **7401** (2009), 740106.

[15] T.F. Anderson and A.G. Richards Jr, An electron microscope study of some structural colors in insects, *J Appl Phys* **13** (1942), 748–758.

[16] J.C. Downey and A.C. Allyn, Wing-scale morphology and nomenclature, *Bull Allyn Museum* **31**, 1–32.

[17] A.C. Allyn and J.C. Downey, Diffraction structures in the wing scales of *Callophrys* (Mitoura) *siva siva* (Lycaenidae), *Bull Allyn Museum* **40** (1976), 1–6.

[18] A.C. Allyn Jr. and J.C. Downey, Observations on male U-V reflectance and scale ultrastructure in *Phoebis* (Pieridae), *Bull Allyn Museum* **42** (1977), 1–6.

[19] H. Ghiradella, Development of ultraviolet-reflecting butterfly scales: how to make an interference filter, *J Morphol* **142** (1974), 395–409.

[20] H. Ghiradella and W. Radigan, Development of butterfly scales. II. Struts, lattices and surface tension, *J Morphol* **150** (1976), 279–297.

[21] H. Ghiradella, Structure of iridescent lepidopteran scales: variations on several themes, *Ann Entomol Soc Am* **77** (1984), 637–645.

[22] H. Ghiradella, Structure and development of iridescent lepidopteran scales: the papilionidae as a showcase family, *Ann Entomol Soc Am* **78** (1986), 252–264.

[23] H. Ghiradella, Structure and development of iridescent butterfly scales: lattices and laminae, *J Morphol* **202** (1989), 69–88.

[24] K. Kumazawa, S. Tanaka, K. Negita, and H. Tabata, Fluorescence from wing of *Morpho sulkowskyi* butterfly, *Jpn J Appl Phys* **33** (1994), 2119–2122.

[25] S. Kinoshita, *Structural colors in the realm of nature*, World Scientific, Singapore (2008).

[26] S. Kinoshita, S. Yoshioka, Y. Fujii, and N. Okamoto, Photophysics of structural color in the *Morpho* butterflies, *Forma* **17** (2002), 103–121.

[27] S. Kinoshita, S. Yoshioka, and K. Kawagoe, Mechanisms of structural color in the *Morpho* butterfly: cooperation

of regularity and irregularity in an iridescent scale, *Proc R Soc Lond B* **269** (2002), 1417–1421.

[28] A. Saito, Y. Miyamura, Y. Ishikawa, K. Sogo, Y. Kuwahara, and Y. Hirai, Reproduction of the *Morpho* blue by nanocasting lithography, *J Vac Sci Technol B* **24** (2006), 3248–3251.

[29] A. Saito, Y. Miyamura, Y. Ishikawa, J. Murase, M. Akai-Kasaya, and Y. Kuwahara, Reproduction, mass-production, and control of the Morpho-butterfly's blue, *Proc SPIE* **7205** (2009), 720506.

[30] P. Vukusic and J.R. Sambles, Photonic structures in biology, *Nature* **424** (2003), 852–855.

[31] P. Vukusic, J.R. Sambles, and C.R. Lawrence, Structurally assisted blackness in butterfly scales, *Proc R Soc Lond B* **271** (2004), S237–S239.

[32] P. Vukusic and I. Hooper, Directionally controlled fiuorescence emission in butterflies, *Science* **310** (2005), 1151.

[33] P. Vukusic, B. Hallam, and J. Noyes, Brilliant whiteness in ultrathin beetle scales, *Science* **315** (2007), 348.

[34] A.R. Parker, Z. Hegedus, and R.A. Watt, Solar-absorber antireflector on the eye of an Eocene fly, [45 Ma], *Proc R Soc Lond B* **265** (1998), 811–815.

[35] A.R. Parker, A vision for natural photonics, *Phil Trans R Soc Lond A* **362** (2004), 2709–2720.

[36] A.R. Parker and H.E. Townley, Biomimetics of photonic nanostructures, *Nat Nanotechnol* **2** (2007), 347–353.

[37] J.P. Vigneron, J.M. Pasteels, D.M. Windsor, Z. Vértesy, M. Rassart, T. Seldrum, J. Dumont, O. Deparis, V. Lousse, L.P. Biró, D. Ertz, and V. Welch, Switchable reflector in the Panamanian tortoise beetle *Charidotel laegregia* (Chrysomelidae: Cassidinae), *Phys Rev E* **76** (2007), 031907.

[38] K. Michielsen and D.G. Stavenga, Gyroid cuticular structures in butterfly wing scales: biological photonic crystals, *J R Soc Interf* **5** (2008), 85–94.

[39] J.W. Galusha, L.R. Richey, J.S. Gardner, J.N. Cha, and M.H. Bartl, Discovery of a diamond-based photonic crystal structure in beetle scales, *Phys Rev E* **77** (2008), 050904.

[40] O. Sato, S. Kubo, and Z.-Z. Gu, Structural color films with Lotus effects, superhydrophilicity, and tunable stop- bands, *Acc Chem Res* **42** (2009), 1–10.

[41] S. Kinoshita and S. Yoshioka (eds.), *Structural colors in biological systems: principles and applications*, Osaka University Press, Osaka, Japan (2005).

[42] A. Parker, *Seven deadly colours: the genius of nature's palette and how it eluded Darwin*, Free Press, New York, NY, USA (2006).

[43] A. I. Ingram, Butterfly photonics: form and function, in *Functional surfaces in biology, Vol. 1: Little structures with big effects* (S. N. Gorb, ed.), Springer-Verlag, Heidelberg, Germany (2009).

[44] www.loreal.com/_en/_ww/loreal-art-science/loreala_sofcolor.aspx (accessed 1 February 2013).

[45] J. Huang, X. Wang, and Z.L. Wang, Controlled replication of butterfly wings for achieving tunable photonic properties, *Nano Lett* **6** (2006), 2325–2331.

[46] J. Silver, R. Withnall, T.G. Ireland, G.R. Fern, and S. Zhang, Light-emitting nanocasts formed from bio-templates: FESEM and cathodoluminescent imaging studies of butterfly scale replicas, *Nanotechnology* **19** (2008), 095302.

[47] G. Zhang, J. Zhang, G. Xie, Z. Liu, and H. Shao, Cicada wings: a stamp from nature for nanoimprint lithography, *Small* **2** (2006), 1440–1443.

[48] R.J. Martín-Palma, C.G. Pantano, and A. Lakhtakia, Biomimetization of butterfly wings by the conformal-evaporated-film-by-rotation technique for photonics, *Appl Phys Lett* **93** (2008), 083901.

[49] M. Srinivasarao, Nano-optics in the biological world: beetles, butterflies, birds, and moths, *Chem Rev* **99** (1999), 1935–1961.

[50] H. D. Wolpert, Optical filters in nature, *OSA Opt Photon News* **20** (2) (February 2009), 22–27, http://www.osa-opn.org (accessed 12 March 2013).

[51] C.W. Mason, Structural colors in feathers. I, *J Phys Chem* **27** (1923), 201–251.

[52] C.F. Bohren and D.R. Huffman, *Absorption and scattering of light by small particles*, Wiley, New York, NY, USA (1983).

[53] R.J. Martín-Palma and A. Lakhtakia, *Nanotechnology: a crash course*, SPIE Press, Bellingham, WA, USA (2010).

[54] J.D. Joannopoulos, R.D. Meade, and J.N. Winn, *Photonic Crystals*, Princeton University Press, Princeton, NJ, USA (1995).

[55] E. Adachi and K. Matsubara, Reproducibility and applicability of gallium replication as evaluated by biological specimen use, *J Electron Microsc* **49** (2000), 371–378.

[56] H.C. Bolton, L.A. Bursill, A.C. McLaren, and R.G. Turner, On the origin of the colour of labradorite, *phys stat sol (b)* **18** (1966), 221–230.

[57] Y. Miura, T. Tomisaka, and T. Kato, Experimental and theoretical approaches to iridescent labradorite, *Mem Geol Soc Jap* **11** (1974), 145–165.

[58] R. Velázquez-Castillo, J. Reyes-Gasga, D.I. García-Gutierrez, and M. Jose-Yacaman, Nanoscale characterization of nautilus shell structure: an example of natural self-assembly, *J Mater Res* **21** (2006), 1484–1489.

[59] M.S. Giridhar and S.K. Srivatsa, Pearls and shells, *Curr Sci* **76** (1999), 1324–1325.

[60] K. Lyman (ed.), *Simon & Schuster's guide to gems and precious stones*, Simon & Schuster, New York, NY, USA (1986).

[61] A.E. Seago, P. Brady, J.-P. Vigneron, and T.D. Schultz, Gold bugs and beyond: a review of iridescence and structural colour mechanisms in beetles (Coleoptera), *J R Soc Interf* **6** (2009), S165–S184.

[62] F. Liu, B.Q. Dong, X.H. Liu, Y.M. Zheng, and J. Zi, Structural color change in longhorn beetles *Tmesisternus isabellae*, *Opt Express* **17** (2009), 16183–16191.

[63] M.D. Shawkey, M.E. Hauber, L.K. Estep, and G.E. Hill, Evolutionary transitions and mechanisms of matte and iridescent plumage coloration in grackles and allies (Icteridae), *J R Soc Interf* **3** (2006), 777–786.

[64] A. Levy-Lior, B. Pokroy, B. Levavi-Sivan, L. Leiserowitz, S. Weiner, and L. Addadi, Biogenic guanine crystals from the skin of fish may be designed to enhance light reflectance, *Cryst Growth Des* **8** (2008), 507–511.

[65] F. Chiadini and A. Lakhtakia, Design of wideband circular polarization filters made of chiral sculptured thin films, *Microw Opt Technol Lett* **42** (2004), 135–138.

[66] Y. Huang, Y. Zhou, and S.-T. Wu, Broadband circular polarizer using stacked chiral polymer films, *Opt Express* **15** (2007), 6414–6419.

[67] Y.J. Park, K.M.A. Sobahan, and C.K. Hwangbo, Wideband circular polarization reflector fabricated by glancing angle deposition, *Opt Express* **16** (2008), 5186–5192.

[68] E.J. Denton and J.A.C. Nicol, Reflexion of light by external surfaces of the herring, Clupea harengus, *J Mar Biol Assoc UK* **45** (1965), 711–738.

[69] D.R. McKenzie, Y. Yin, and W.D. McFall, Silvery fish skin as an example of a chaotic reflector, *Proc R Soc Lond A* **451** (1995), 579–584.

[70] J.V. Sanders, Diffraction on light by opals, *Acta Crystallogr A* **24** (1968), 427–434.

[71] C.D. Dushkin, K. Nagayama, T. Miwa, and P.A. Kralchevsky, Colored multilayers from transparent submicrometer spheres, *Langmuir* **9** (1993), 3695–3701.

[72] E. Gaillou, E. Fritsch, B. Aguilar-Reyes, B. Rondeau, J. Post, A. Barreau, and M. Ostroumov, Common gem opal: an investigation of micro-to-nano-structure, *Am Miner* **93** (2008), 1865–1873.

[73] A.L. Ingram, V. Lousse, A.R. Parker, and J.P. Vigneron, Dual gratings interspersed on a single butterfly scale, *J R Soc Interf* **5** (2008), 1387–1390.

[74] J.P. Vigneron, P. Simonis, A. Aiello, A. Bay, D.M. Windsor, J.-F. Colomer, and M. Rassart, Reverse color sequence in the diffraction of white light by the wing of the male butterfly *Pierella luna* (Nymphalidae: Satyrinae), *Phys Rev E* **82** (2010), 021903.

[75] A. Parker, *In the blink of an eye: how vision sparked the big bang of evolution*, Perseus Publications, Cambridge, MA, USA (2003).

[76] H. Tabata, T. Hasegawa, M. Nakagoshi, S. Takikawa, and M. Tsusue, Occurrence of biopterin in the wings of Morpho butterflies, *Experientia* **52** (1996), 85–87.

[77] A.C. Neville and S. Caveney, Scarabaeid beetle exocuticle as an optical analogue of cholesteric liquid crystals, *Biol Rev* **44** (1969), 531–562.

[78] P.G. de Gennes and J. Prost, *The physics of liquid crystals*, 2nd ed., Clarendon Press, Oxford, UK (1993).

[79] A. Lakhtakia and R. Messier, *Sculptured thin films: nanoengineered morphology and optics*, SPIE Press, Bellingham, WA, USA (2005).

[80] D.W. Lee and J.B. Lowry, Iridescent blue plants: physical basis and ecologic significance, *Nature* **254** (1975), 50–51.

[81] B.J. Glover and H.M. Whitney, Structural colour and iridescence in plants: the poorly studied relations of pigment colour, *Ann Bot* **105** (2010), 505–511.

[82] S.D. Jacobs (ed.), *Selected papers on liquid crystals for optics*, SPIE Press, Bellingham, WA, USA (1992).

[83] M. Qi, E. Lidorikis, P.T. Rakich, S.G. Johnson, J.D. Joannopoulos, E.P. Ippen, and H.I. Smith, A three-dimensional optical photonic crystal with designed point defects, *Nature* **429** (2004), 538–542.

[84] J.W. Galusha, M.R. Jorgensen, and M.H. Bartl, Diamond-structured titania photonic bandgap crystals from biological templates, *Adv Mater* **22** (2010), 107–110.

[85] M. Crne, V. Sharma, J. Blair, J.O. Park, C.J. Summers, and M. Srinivasarao, Biomimicry of optical microstructures of *Papilio palinurus*, *Europhys Lett* **93** (2011), 14001.

[86] K. Nagayama and A.S. Dimitrov, Natural beauty of artificially textured surfaces: Morpho-butterfly coloring with particle arrays, *Symmetry: Cult Sci* **6** (1995), 396–399.

[87] A. Saito, S. Yoshioka, and S. Kinoshita, Reproduction of the Morpho butterfly's blue: arbitration of contradicting factors, *Proc SPIE* **5526** (2004), 188–194.

[88] K. Watanabe, T. Hoshino, K. Kanda, Y. Haruyama, and S. Matsui, Brilliant blue observation from a Morpho-butterfly-scale quasi-structure, *Jpn J Appl Phys* **44** (2005), L48–L50.

[89] D.P. Pulsifer and A. Lakhtakia, Background and survey of bioreplication techniques, *Bioinsp Biomim* **6** (2011), 031001.

[90] J.W. Galusha, L.R. Richey, M.R. Jorgensen, J.S. Gardner, and M.H. Bartl, Study of natural photonic crystals in beetle scales and their conversion into inorganic structures via a sol-gel bio-templating route, *J Mater Chem* **20** (2010), 1277–1284.

[91] P. Nagaraja and D. Yao, Rapid pattern transfer of biomimetic surface structures onto thermoplastic polymers, *Mater Sci Eng C* **27** (2007), 794–797.

[92] A. Lakhtakia, R.J. Martín-Palma, M.A. Motyka, and C.G. Pantano, Fabrication of free-standing replicas of fragile, laminar, chitinous biotemplates, *Bioinsp Biomim* **4** (2009), 034001.

[93] D.P. Pulsifer, A. Lakhtakia, R.J. Martín-Palma, and C.G. Pantano, Mass fabrication technique for polymeric replicas of arrays of insect corneas, *Bioinsp Biomim* **5** (2010), 036001.

[94] A. Saito, Material design and structural color inspired by biomimetic approach, *Sci Technol Adv Mater* **12** (2011), 064709.

[95] W. Stober, A. Fink, and E. Bohn, Controlled growth of monodisperse silica spheres in the micron size range, *J Colloid Interf Sci* **26** (1968), 62–69.

[96] R.V. Nair and B.N. Jagatap, Bragg wave coupling in self-assembled opal photonic crystals, *Phys Rev A* **85** (2012), 013829.

[97] I.M. Krieger and F.M. O'Neill, Diffraction of light by arrays of colloidal spheres, *J Am Chem Soc* **90** (1968), 3114–3120.

[98] C.D. Dushkin, H. Yoshimura, and K. Nagayama, Nucleation and growth of two-dimensional colloidal crystals, *Chem Phys Lett* **204** (1993), 455–459.

[99] C.D. Dushkin, G.S. Lazarov, S.N. Kotsev, H. Yoshimura, and K. Nagayama, Effect of growth conditions on the structure of two-dimensional latex crystals: experiment, *Colloid Polym Sci* **277** (1999), 914–930.

[100] A.S. Dimitrov, C.D. Dushkin, H. Yoshimura, and K. Nagayama, Observation of latex particle two-dimensional-crystal nucleation in wetting films on mercury, glass, and mica, *Langmuir* **10** (1994), 432–440.

[101] A.S. Dimitrov and K. Nagayama, Continuous convective assembling of fine particles into two-dimensional arrays on solid surfaces, *Langmuir* **12** (1996), 1303–1311.

[102] Y. Xia, B. Gates, Y. Yin, and Y. Lu, Monodispersed colloidal spheres: old materials with new applications, *Adv Mater* **12** (2000), 693–713.

[103] D. Dinsmore, J.C. Crocker, and A.G. Yodh, Self-assembly of colloidal crystals, *Curr Opin Colloid Interf Sci* **3** (1998), 5–11.

[104] A. Hierlemann, O. Brand, C. Hagleitner, and H. Baltes, Microfabrication techniques for chemical/biosensors, *Proc IEEE* **91** (2003), 839–863.

[105] J. Voldman, M.L. Gray, and M.A. Schmidt, Microfabrication in biology and medicine, *Annu Rev Biomed Eng* **1** (1999), 401–425.

[106] K. Nagayama and A.S. Dimitrov, Natural beauty of artificially textured surfaces: Morpho-butterfly coloring with particle arrays, *Symmetry: Cult Sci* **6** (1995), 396–399.

[107] F. Liu, Y. Liu, L. Huang, X. Hu, B. Dong, W. Shi, Y. Xie, and X. Ye, Replication of homologous optical and hydrophobic features by templating wings of butterflies *Morpho menelaus*, *Opt Commun* **284** (2011), 2376–2381.

[108] B. Li, J. Zhou, R. Zong, M. Fu, Y. Bai, L. Li, and Q. Li, Ordered ceramic microstructures from butterfly biotemplate, *J Am Ceram Soc* **89** (2006), 2298–2300.

[109] J. Silver, R. Withnall, T.G. Ireland, and G.R. Fern, Novel nano-structured phosphor materials cast from natural Morpho butterfly scales, *J Mod Opt* **52** (2005), 999–1007.

[110] Y. Chen, X. Zang, J. Gu, S. Zhu, H. Su, D. Zhang, X. Hu, Q. Liu, W. Zhang, and D. Liu, ZnO single butterfly wing scales: synthesis and spatial optical anisotropy, *J Mater Chem* **21** (2011), 6140–6143.

[111] S. Zhu, D. Zhang, Z. Chen, J. Gu, W. Li, H. Jiang, and G. Zhou, A simple and effective approach towards biomimetic replication of photonic structures from butterfly wings, *Nanotechnology* **20** (2009), 315303.

[112] G. Cook, P.L. Timms, and C. Göltner-Spickermann, Exact replication of biological structures by chemical vapor deposition of silica, *Angew Chem Int Ed* **42** (2003), 557–559.

[113] G.G. Roberts, P.S. Vincett, and W.A. Barlow, Technological applications of Langmuir-Blodgett films, *Phys Technol* **12** (1981), 69–87.

[114] M. Kim, R. Nagarajan, J.H. Snook, L.A. Samuelson, and J. Kumar, Nanostructured assembly of homopolymers for a flexible Bragg grating, *Adv Mater* **17** (2005), 631–633.

[115] A. Kumar and J. Kumar, Layer-by-layer deposition of nanoscale structures, *J Nanophoton* **3** (2009), 030306.

[116] P. Calvert, L. Skander, S. Iyenger, and P. Patra, Inkjet printing of insoluble biopolymer and polymer complexes, *PMSE Preprints* **48** (2007), 1023–1024.

[117] H.A. Macleod, *Thin-film optical filters*, 3rd ed., Institute of Physics, Bristol, UK (2001).

[118] T.G. Mackay and A. Lakhtakia, *Electromagnetic anisotropy and bianisotropy*, World Scientific, Singapore (2010).

[119] A. Lakhtakia, Y.-J. Jen, and C.-F. Lin, Multiple trains of same-color surface plasmon-polaritons guided by the planar interface of a metal and a sculptured nematic thin film. Part III: experimental evidence, *J Nanophoton* **3** (2009), 033506.

[120] D.P. Pulsifer, R.J. Martín-Palma, S.E. Swiontek, C.G. Pantano, and A. Lakhtakia, Wideband-rejection filters and reflection-hole filters of chalcogenide glass for circularly polarized IR-A and IR-B radiation, *Opt Mater Express* **1** (2011), 1332–1340.

[121] M. Kolle, *Photonic structures inspired by nature*, Springer-Verlag, Berlin, Germany (2011).

[122] H. Fudouzi, Tunable structural color in organisms and photonic materials for design of bioinspired materials, *Sci Technol Adv Mater* **12** (2011), 064704.

[123] H. Fudouzi, Optical properties caused by periodical array structure with colloidal particles and their applications, *Adv Powder Technol* **20** (2009), 502–508.

[124] P. Laufer, *The dangerous world of butterflies*, Lyons Press, Guildford, CT, USA (2009).

[125] K. Iohara, M. Yoshimura, H. Tabata, and S. Shimizu, Structurally colored fibers, *Chem Fibers Int* **50** (2000), 38–39.

[126] M. Asano, T. Kuroda, S. Shimizu, A. Sakihara, K. Kumazawa, and H. Tabata, *Fiber structure and textile using same*, US Patent 6326094 (issued 4 December 2001).

[127] K. Kumazawa and H. Tabata, *Color exhibition structure*, US Patent 6248436 (issued 19 June 2001).

[128] P.W. Baumeister, *Optical coating technology*, SPIE Press, Bellingham, WA, USA (2004).

[129] P. Calvert, M. Shah, and P. Patra, Structural colors by ionic self-assembly, *PMSE Preprints* **94** (2006), 757–758.

[130] E. F. Klenke, *Nacreous pigment compositions*, US Patent 3634119 (issued 11 January 1972).

[131] A. Kimura, personal correspondence with ND (1999). Kimura presented the data at the XIVth international federation of societies of cosmetic chemists (IFSCC) congress, held in Barcelona, Spain, (16–19 September 1986).

[132] Y. Ikuta and A. Kimura, *Color titanated mica pigment and coated-body using the same*, US Patent 6129784 (issued 10 October 2000).

[133] K. Iwano and M. Suzumeji, *Nacreous pigment containing a dye and cosmetic composition comprising the same*, US Patent 4952245 (issued 28 August 1990).

[134] http://www.shiseido.co.jp/e/story/html/sto50900.htm (accessed 18 February 2013).

[135] P.T. Aylward, N. Dontula, A.D. Camp, and R. P. Bourdelais, *Imaging element with polymer nacreous layer*, US Patent 6544713 (accessed 8 April 2003).

[136] A.D. Pris, Y. Utturkar, C. Surman, W.G. Morris, A. Vert, S. Zalyubovskiy, T. Deng, H.T. Ghiradella, and R.A. Potyrailo, Towards high-speed imaging of infrared photons with bio-inspired nanoarchitectures, *Nat Photon* **6** (2012), 195–200.

[137] R.A. Potyrailo, H. Ghiradella, A. Vertiatchikh, K. Dovidenko, J.R. Cournoyer, and E. Olson, Morpho butterfly wing scales demonstrate highly selective vapour response, *Nat Photon* **1** (2007), 123–128.

ABOUT THE AUTHORS

Natalia Dushkina received her MS in quantum electronics and lasers from the University of Sofia, Bulgaria, and a PhD in Physics from the Central Laboratory of Optical Storage and Processing of Information, Bulgarian Academy of Sciences. She was a post-doctoral fellow at the Department of Physics, University of Tokyo, and at the Mechanical Engineering Laboratory (MEL), Tsukuba, Ministry of International Trade and Industry (MITI) of Japan. Currently, she is an associate professor of physics at Millersville University, Pennsylvania, USA. Her current research interests include surface plasmon resonance, self-assembled nanostructures, color formation and structural colors, and optical properties of nanomaterials.

Akhlesh Lakhtakia received degrees from the Banaras Hindu University (BTech and DSc) and the University of Utah (MS and PhD) in electronics engineering and electrical engineering, respectively. He is the Charles Godfrey Binder (Endowed) Professor of Engineering Science and Mechanics at the Pennsylvania State University and currently serves as the editor-in-chief of the *Journal of Nanophotonics*. His current research interests include nanotechnology, forensic science, bioreplication, surface multiplasmonics, complex materials, metamaterials, and sculptured thin films. He is a Fellow of SPIE, the Optical Society of America, American Physical Society, Institute of Physics (UK), and American Association for the Advancement of Science.

CHAPTER

12

Biomimetic Antireflection Surfaces

Blayne M. Phillips and Peng Jiang

Department of Chemical Engineering, University of Florida,
Gainesville, FL 32611, USA

Prospectus

The compound eyes of moths are composed by hexagonal arrays of non-close-packed nipples that exhibit low reflectance. The outer surface of the cornea of a moth consists of periodic arrays of conical protuberances, termed corneal nipples, typically of sub-250 nm height and spacing. These arrays of subwavelength nipples generate a graded transition of refractive index, leading to minimized reflection over a broad range of wavelengths and angles of incidence. In this chapter, the fabrication, characterization, and modeling of moth-eye antireflection coatings on both transparent substrates (e.g., glass) and semiconductor wafers (such as crystalline silicon and GaAs) are discussed.

Keywords

Antiglare coating, Antireflection coating, Bioinspiration, Broadband, Close-packed, Colloidal crystal, Crystalline silicon, GaAs, GaSb, Moth-eye surface, Nanopillars, Photovoltaics, Rigorous coupled-wave analysis, Self-assembly, Self-cleaning, Solar cells, Subwavelength, Superhydrophobic coatings, Template, Thin-film multilayer model

12.1 INTRODUCTION

Antireflection coatings (ARCs) are typically optical films that are used on lenses and optoelectronic devices to reduce light reflection and mitigate its detrimental effects. ARCs are extensively utilized in applications where the transmission of light through an optical medium needs to be maximized, such as eliminating the ghost images for flat-panel displays, increasing the transmittance of optical lenses, reducing glaring from automobile dashboards, and enhancing the conversion efficiency of solar cells [1–3].

12.1.1 Traditional Quarter-Wavelength Antireflection Coatings

ARCs rely on two criteria: material refractive index and film thickness. Optical reflection from surfaces is largely governed by the difference in refractive indices at material interfaces. The amount of reflection can theoretically be calculated by using the Fresnel equation [4].

To maximize the suppression of reflection, the refractive index of a film interposed between two materials should be the geometric mean of their refractive indices. The thickness of the film should be a quarter wavelength (i.e., a quarter of the wavelength of light at the specific location where reflection reduction is desired) to take advantage of interference. Light that is reflected at the coating-substrate interface will be half of a wavelength out of phase from the incident light that is reflected from the coating-air interface, resulting in destructive interference and reduced reflectance [4].

One important application of the quarter-wavelength ARC is to improve the conversion efficiency of solar cells. Solar cells collect photons from sunlight and convert them into electric power [5]. Materials that are used to construct solar cells tend to have high refractive indices (e.g., crystalline silicon has a refractive index of $\sim$3.5), which result in high reflectance [6]. Reflected light reduces the number of photons that can be used to form the electron-hole pairs that drive the current in solar cells. As a result, one of the goals in fabricating solar cells is to reduce the amount of light that is reflected. To accomplish this goal, texturing and antireflection coatings are used. Texturing (usually at the geometrical-optics scale) reduces overall reflection by directing diffuse reflections back into the substrate. This is usually done by chemical etching using potassium hydroxide (KOH) or inorganic acids [7, 8]. Though not quite as common as wet etching, reactive ion etching (RIE) and lasers can also be used to texture silicon [9–20].

Bioinspired texturing at the multiwavelength scale has been demonstrated to improve the light-harvesting capabilities of crystalline silicon solar cells [21, 22]. Typical antireflection coatings used on silicon are silicon nitride (SiN_x) and titanium dioxide (TiO_2) films.

Quarter-wavelength SiN_x films deposited by plasma-enhanced chemical vapor deposition (PECVD) are the industrial standard for ARCs on crystalline-silicon solar cells [5, 6]. They are designed to be most effective at wavelengths $\sim$600 nm and are less effective for the near-infrared and other visible wavelengths, which are present in a large portion of the incident solar energy. These films also exhibit poor thermal stability due to the mismatch of thermal expansion coefficient between SiN_x and Si. The PECVD process is also costly.

Subwavelength structures such as pores and periodic and stochastic features can be used to realize graded refractive indices at the surface of semiconductor materials that are used in optoelectronic devices [23–30]. These structures can potentially provide broadband antireflection, but the increased surface area of these features can increase the number of defect sites where electron-hole recombination can occur, which can be detrimental to the efficiency of photovoltaic devices [31].

Glass and plastics such as polycarbonate and polymethylmethacrylate (PMMA) are important optical substrates. Transparent substrates suffer less severe reflective loss than silicon and other semiconductor materials; a loss of $\sim$4% at the air/glass interface can degrade the performance of devices with multiple components and optical interfaces. Substrates that have a refractive index of approximately 1.5 (e.g., glass) should ideally be coated with a material with a refractive index of 1.22, based on the Fresnel equation. Unfortunately, materials with this low refractive index are rare. As a result, magnesium fluoride with a refractive index of 1.38 is widely used as a single-layer ARC. But magnesium-fluoride coatings are not suitable for polymers due to the high tensile growth stress and the poor mechanical properties of fluoride thin films at low polymer-processing temperatures [32, 33]. Fluoropolymers with low refractive indices can also be used as ARCs [34–37].

To reduce reflection, expensive multilayer ARCs are typically used. The solution and temperature requirements needed for depositing these coatings make them incompatible with many substrate materials such as plastics [38].

Sol–gel techniques, discussed in Chapter 14 by Risbud and Bartl, are used for preparing ARCs on transparent surfaces [39–42]. During synthesis, the refractive index of sols can be controlled; for example, titania sol–gels can be made to have refractive indices as high as 1.8 that are useful for ARCs on materials such as indium tin oxide (ITO). Microstructure can be imprinted onto sol–gels or polymers to enhance their antireflective properties [43].

There are limitations with index-matching and quarter-wavelength coatings. First, there is a theoretical limitation on how much a single-layer coating can reduce reflection. This can be overcome by adding layers of coatings; however, high fabrication cost and limited material selection are major concerns for multilayer ARCs. Second, quarter-wavelength coatings are both narrowband and have a narrow field of vision. The coatings are narrowband because of their fixed thickness and as a result they are selective of the wavelength of light where reflection is suppressed. Also, the incident rays must be at or very close to normal incidence for the destructive interference to occur [44].

12.1.2 Bioinspired Moth-Eye Broadband Antireflection Coatings

Solutions to these shortcomings can be found in nature. Nocturnal moths possess eyes that have a microstructured cornea that exhibit excellent broadband antireflection [45, 46]. The corneas of these eyes have a hexagonal array of non-close-packed subwavelength pillars forming a grating that suppresses reflection of visible light (Figure 12.1). The dimensions and spacing of the periodic pillar structures are smaller than the wavelength of light where reflection is to be suppressed. As a result, the effective refractive index of the coating is reduced. This refractive index is the weighted average by volume fraction of the refractive index of the substrate and the space between the pillars (typically air). The tapered structure of subwavelength nipples can thus generate a

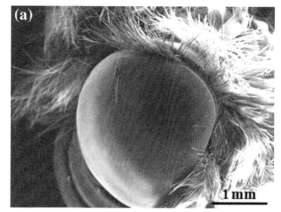

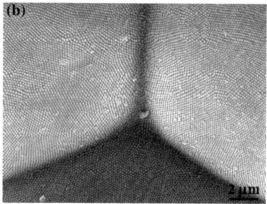

FIGURE 12.1 SEM images of a moth eye.

graded transition of refractive index, leading to minimized reflection over a broad range of wavelengths and angles of incidence [45, 47, 48].

Various top-down technologies, such as photolithography [49], electron-beam lithography [50], nanoimprint lithographyy [51, 52], and interference lithography [33, 45, 53, 54], have been developed for fabricating moth-eye ARCs. However, these techniques require sophisticated equipment and are expensive to implement [45, 55, 56]. Bottom-up self-assembly and subsequent templating nanofabrication provide a much simpler and cheaper alternative to complex nanolithography in creating subwavelength-structured moth-eye arrays [2, 57–59].

Unfortunately, the successful development of high-throughput and large-area fabrication continues to be a major challenge with these bottom-up techniques. Formation of moth-eye gratings on glass substrates typically requires different processing methods than semiconductor materials [1, 2, 33, 60–74]. Polymer or sol–gel films are usually applied to the surface and patterns are formed by either embossing them using nanoimprint lithography or etching [75, 76]. Polymers such as PMMA can have the moth-eye subwavelength structure directly patterned onto their surfaces [77–79].

12.2 SCALABLE SELF-ASSEMBLY OF COLLOIDAL PARTICLES

Various self-assembly approaches, such as evaporation-induced assembly [80–83], layer-by-layer assembly [84, 85], assembly at liquid-liquid or liquid–air interfaces [86, 87], gravity sedimentation [88, 89], electrostatic repulsion [90–93], and template-assisted assembly [94–97], have been developed for colloidal crystallization. Unfortunately, most of these bottom-up approaches are only favorable for low-volume, laboratory-scale production. They suffer from low throughput and incompatibility with mature microfabrication methods, thereby limiting the mass production of colloidal templates for fabricating moth-eye antireflection coatings. To resolve the scale-up and compatibility issues of current colloidal self-assembly, two scalable bottom-up technologies based on spin-coating [98–110] and Langmuir–Blodgett (LB) assembly of colloids at a water–air interface [111] have been developed.

12.2.1 Spin-Coating Technological Platform

Spin coating is a well-established technique in standard microfabrication for producing highly uniform thin films with tunable thickness over a large area [112]. Spin coating has been widely utilized in assembling colloidal crystals, such as colloidal masks for nanosphere lithography [59, 113–115]. However, polycrystalline samples with poor qualities are usually the results obtained from conventional spin-coating methods, due to rapid evaporation of solvents (e.g., ethanol) used to disperse colloidal particles [116]. Instead, uniform silica particles with diameters ranging from ~70 nm to over several micrometers are dispersed in various nonvolatile monomers, such as ethoxylated trimethylolpropane triacrylate (ETPTA), with 1% Darocur 1173 as photoinitiator [98, 99]. The particle volume fraction is usually tuned to ~20%. Due to refractive-index matching between silica colloids and acrylate monomers, which reduces the attractive van der Waals forces between particles, the transparent colloidal suspensions are stable for a few months.

The colloidal suspension can be dispensed on a variety of substrates, such as silicon wafers, glass microslides, and plastic plates, and spin-coated using standard spin coaters. Formation of wafer-scale colloidal crystals occurs within seconds, as indicated by the appearance of a striking diffraction star with six arms (Figure 12.2); this pattern is characteristic of long-range hexagonal ordering [98, 117]. The monomers are

2 cm

FIGURE 12.2 Photograph of a 3D ordered colloidal crystal-polymer nanocomposite film consisting of 325 nm silica spheres on a 4-in. silicon wafer illuminated with white light. Adapted with permission from *Am Chem Soc* **126** (2004), 13778–13786. Copyright 2004, American Chemical Society.

then rapidly polymerized to form 3D ordered polymer nanocomposites (Figure 12.2). A scanning electron microscope (SEM) image (Figure 12.3a) and its Fourier transform (inset of Figure 12.3a) demonstrate the highly ordered structures with hexagonal packing on the film surface.

At higher magnification (Figure 12.3b), another interesting feature is evident, i.e., the spheres of the top layer are not contacting with each other, but they exhibit center-to-center distance around $1.41D$, where D is the diameter of colloids. The non-close-packing of colloids and the specific spacing between intra-layer spheres are much more apparent after the polymer matrix is selectively removed by conventional oxygen plasma etching. As shown in Figure 12.3c, the top-layer spheres only fill in the triangularly arranged crevices made by the non-touching spheres of the underneath layer. This non-close-packed structure is indeed exhibited by all layers of the spin-coated colloidal crystal. The ordering perpendicular to the substrate surface is apparent in the cross-sectional SEM image (Figure 12.3d).

Besides the unusual non-close-packed crystalline structure, the spin-coating technology enables rapid production of wafer-sized colloidal crystals with highly uniform and tunable thickness ranging from a single monolayer to hundreds of monolayers [98, 99]. The crystal thickness can be easily adjusted by tuning the spin speed and duration. The crystal thickness is inversely proportional to the spin speed and the square root of the spin duration. For instance, colloidal crystals consisting of 325-nm-diameter silica spheres with 2, 5, and 41 colloidal multilayers can be fabricated by spin coating at 6,000 rpm for 900, 170, and 120 s, respectively. The typical spin-coating condition to assemble submicrometer-sized particles into monolayer, non-close-packed colloidal arrays (Figure 12.4) is 8,000 rpm for 5–6 min.

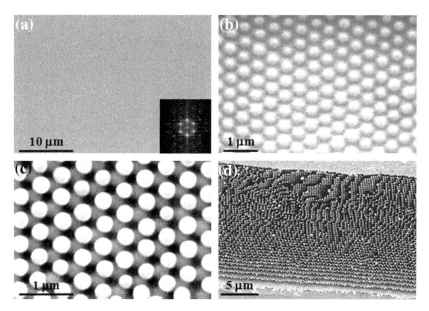

FIGURE 12.3 (a) Typical top-view SEM image of a spin-coated colloidal crystal-polymer nanocomposite film. The inset shows a Fourier transform of a $40 \times 40 \ \mu m^2$ region. (b) Higher-magnification image. (c) Typical top-view SEM image of a released multilayer colloidal crystal after removing polymer matrix. (d) Cross-sectional SEM image of the released colloidal crystal. Adapted from Ref. 98.

FIGURE 12.4 (a) Photograph of a non-close-packed monolayer colloidal crystal made by spin coating. (b) Typical top-view SEM image of the monolayer sample. Inset showing a Fourier transform of a $40 \times 40 \ \mu m^2$ region. Reprinted with permission from *Appl Phys Lett* **89** (2006), 011908. Copyright 2006, American Institute of Physics.

The wide range of particle diameters achievable with the spin-coating technique is another advantage over older self-assembly methods (e.g., convective self-assembly), since quick gravitational sedimentation of large silica spheres (>400 nm) causes serious problems in making high-quality crystals [82]. Given the short spin-coating and photopolymerization time (in minutes), the formidable challenge of sedimentation can be completely avoided. With the spin-coating technique, monodispersed silica colloids with a wide diameter range from 70 nm to 2 μm can be assembled to form 3D ordered colloidal crystals [98, 99, 106].

Two samples made from 70-nm (Figure 12.5a) to 1,320-nm (Figure 12.5b) diameter silica particles show similar long-range hexagonal ordering and center-to-center distance (around $1.41D$) to those of submicrometer-sized spheres (Figures 12.3 and 12.4). The well-established microemulsion method was used to synthesize monodispersed silica nanoparticles with 70-nm diameter [118, 119]. Higher spin-coating speed and longer coating durations are needed to align 70-nm particles than their submicrometer-sized counterparts [106]. This is not surprising, since the

FIGURE 12.5 Typical top-view SEM images of spin-coated colloidal crystals with different sphere sizes. (a) A sample made from 70-nm-diameter spheres and coated at 10,000 rpm for 10 min. (b) A sample made from 1,320-nm-diameter spheres and coated at 600 rpm for 120 s. Adapted with permission from *J Am Chem Soc* **126** (2004), 13778–13786. Copyright 2004, American Chemical Society. Adapted from Ref. 106.

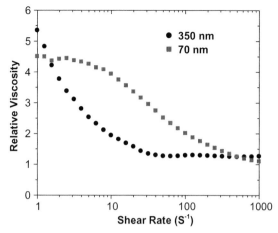

FIGURE 12.6 Shear-thinning behavior of silica-ETPTA suspensions using 70- and 350-nm particles at a volume fraction of 0.2. Adapted from Ref. 106.

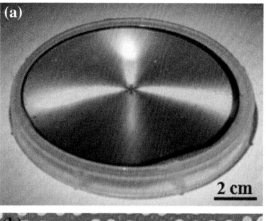

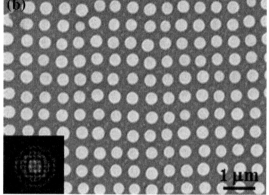

FIGURE 12.7 (a) Photograph of a spin-coated monolayer, non-close-packed colloidal crystal with metastable square lattice. (b) SEM image of the sample. The inset shows a Fourier transform of a low-magnification image. Reprinted with permission from *J Vac Sci Technol B* **27** (2009), 1043–1047. Copyright 2009, American Institute of Physics.

rheology measurements (Figure 12.6) show that much higher shear rate is required to achieve low relative viscosity for 70-nm particles compared to 350-nm spheres with the same particle volume fraction (0.20), indicating higher spin speed is necessary to crystallize smaller particles. The shear-thinning behavior exhibited by the colloidal suspensions is due to the formation of hexagonally packed colloidal layers caused by the coupling of the centrifugal and viscous forces [91, 117, 120–124] and the reduced resistance when layers of ordered spheres glide over one another [120, 124].

Besides energy-favorable hexagonally ordered colloidal crystals, the spin-coating technology also enables the formation of wafer-scale, non-close-packed crystals with unusual square ordering (Figure 12.7) [110] that are consistent with the industry-standard rectilinear coordinate system for simplified addressing and circuit interconnection [125]. The alternate formation of hexagonal and square diffraction patterns when the spin speed is higher than 6,000 rpm is observed. The spin-coating process can be stopped, once a strong four-arm diffraction pattern is formed on the wafer surface.

Figure 12.7a shows a photograph of a 4-in. colloidal monolayer sample consisting of 380 nm silica spheres and spin-coated at 8,000 rpm for 150 s. The sample exhibits a distinctive four-arm diffraction pattern under white-light illumination, and the angles between the neighboring diffraction arms are 90°. This pattern is characteristic of long-range square ordering. This is confirmed by the SEM image in Figure 12.7b and is further evidenced by the squarely arranged peaks in the Fourier transform of a low-magnification SEM image. The interparticle distance of the squarely ordered crystal is determined to be $\sim 1.46D$ by the first peak of the pair correlation function (PCF), which is calculated from a low-magnification SEM image [110].

12.2.2 Langmuir–Blodgett Particle Assembly at Air–Water Interface

The spin-coating technology described in Section 12.2.1 is suitable for large-scale production of highly ordered colloidal crystals with unusual non-close-packed crystalline structure. However, spin coating can only be performed on flat substrates with low surface roughness. It is almost impossible for spin coating to create colloidal crystals on curved surfaces (e.g., optical lenses) and substrates with rough surface, such as a solar-grade multicrystalline silicon (mc-Si) wafer. To enable the formation of colloidal crystals on curved and/or rough surfaces, a simple but scalable Langmuir–Blodgett particle-assembly technology has been developed [111].

Monodispersed silica particles synthesized by the standard Stöber method [126] are purified by repeated centrifugation/redispersion cycles in ethanol and then redispersed in ethylene glycol with particle volume fraction of 0.20. With a clamp attached to a syringe pump, various substrates (e.g., mc-Si wafer) can be vertically immersed in a crystallizing dish containing deionized water. The silica/ethylene glycol suspension is then added dropwise to the surface of the water. The suspension is spread to form a thin layer floating on the surface of the water. With the gradual dissolving of ethylene glycol in water, silica microspheres accumulate at the water–air interface due to the high surface tension of water (72.75 mN/m at 20°C). The capillary action between neighboring silica microspheres can then organize the floating particles into close-packed monolayer colloidal crystals that exhibit striking iridescence caused by light diffraction [80]. The substrate is then slowly withdrawn at a rate of ~ 0.5 mm/min

controlled by the syringe pump. As the wafer is withdrawn, the floating monolayer colloidal crystal is transferred onto the substrate.

Figure 12.8a shows a photograph of a 5-in. solar-grade (mc-Si) wafer with the right half (yellowish region) covered by a uniform monolayer of 250-nm silica particles. The typical top-view SEM images in Figure 12.8b illustrate the right part of the wafer. The high surface roughness of the wafer as evidenced by the randomly distributed, micrometer-sized pits, and the uniform coverage of the rough surface by hexagonally close-packed silica particles is clearly evident. Figure 12.8c shows a photograph of a 5-in. single-crystal silicon (sc-Si) wafer covered by a high-quality monolayer of 200-nm silica spheres assembled using the LB method. The long-range ordering of the colloidal array is demonstrated by the top-view SEM image in Figure 12.8d. This simple colloidal self-assembly technology does not require sophisticated equipment (e.g., a Langmuir–Blodgett trough) [127] to organize silica microspheres with diameter ranging from ~ 70 nm to ~ 30 μm over wafer-sized areas. In addition, extensive experimental results show that this technique is compatible with roll-to-roll processing, promising for scaling up to large-volume production.

12.3 TEMPLATED BROADBAND MOTH-EYE ANTIREFLECTION COATINGS ON SEMICONDUCTOR WAFERS

The self-assembled colloidal arrays can be used as structural templates to create broadband moth-eye ARCs on a large variety of technologically important inorganic semiconductor wafers. The high refractive indices of semiconductors lead to severe surface reflection, greatly impeding the efficiencies of many optoelectronic devices ranging from solar cells to photodiodes.

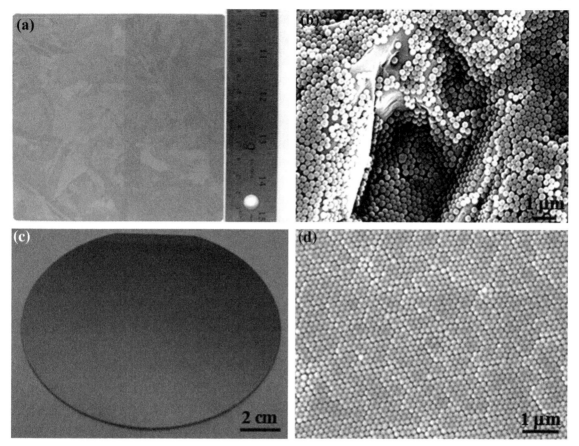

FIGURE 12.8   (a) Photograph of a commercial mc-Si wafer with the right half covered by a close-packed monolayer of 250-nm silica particles assembled using the LB method. (b) Typical top-view SEM image of the right part of the wafer in (a). (c) Photograph of a silicon wafer covered with a close-packed monolayer of 200-nm silica particles. (d) Top-view SEM image of the sample in (c). Reprinted with permission from *Appl Phys Lett* **99** (2011), 191103. Copyright 2011, American Institute of Physics. (For interpretation of color in this figure, the reader is referred to the web version of this book.)

12.3.1 Crystalline Silicon Moth-Eye Antireflection Coatings

Crystalline silicon is widely used in fabricating solar cells [5]. The production of photovoltaic panels is dominated by crystalline-silicon solar cells with 98% of the market share. More specifically, 36% of the 2004 production is based on sc-Si, 58% on mc-Si, and 4% on thin-film amorphous silicon (a-Si) [128]. Ideally, a solar cell should absorb all available photons.

However, due to the high refractive index of silicon, more than 35% of incident light is reflected back from the surface [3, 6, 53]. To lower manufacturing costs and increase conversion efficiency of solar cells, it is highly desirable to develop inexpensive nanofabrication techniques that enable large-scale production of broadband ARCs with high coating stability and durability for reducing reflection over a broad range of wavelengths and angles of incidence.

12.3.1.1 Wet-Etched Single-Crystalline Silicon Moth-Eye Antireflection Coatings

The first approach to fabricate nanostructured ARCs on sc-Si wafers is similar to the conventional anisotropic etching processes used in standard crystalline-silicon solar-cell production. A schematic illustration of the fabrication procedures is shown in Figure 12.9 [102]. Non-close-packed colloidal monolayers on (1 0 0) silicon wafers are first assembled by the spin-coating technology [98, 99]. The non-close-packed silica particles function as shadow masks during an electron-beam evaporation process for depositing a 30-nm-thick chromium layer. After lifting off the templating silica particles, a periodic array of nanoholes (Figure 12.10a), the diameter of which is determined by the size of the templating silica spheres, can be formed [129]. These circular nanoholes can then be used as etching masks during a KOH anisotropic etching process to create wafer-scale inverted pyramidal arrays (Figure 12.10b) on the wafer surface [101].

The specular optical reflectivity of the replicated pyramidal arrays is evaluated using visible-near-IR reflectivity measurement at normal incidence. Figure 12.10c shows the measured specular reflection spectra from a polished (1 0 0) silicon wafer and an inverted pyramidal array with 360-nm pits. The flat silicon substrate exhibits high reflectance (>35%) for visible and near-infrared wavelengths, whereas the templated gratings show reduced reflectance of ~10% for long wavelengths (>600 nm). The reflectance is further reduced to ~2% for

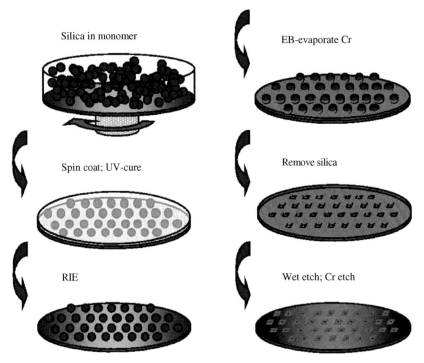

FIGURE 12.9 Schematic illustration of the templating nanofabrication procedures for creating inverted pyramidal gratings on single-crystalline (1 0 0) silicon wafers. Adapted with permission from *Chem Mater* **19** (2007), 4551–4556. Copyright 2007, American Chemical Society.

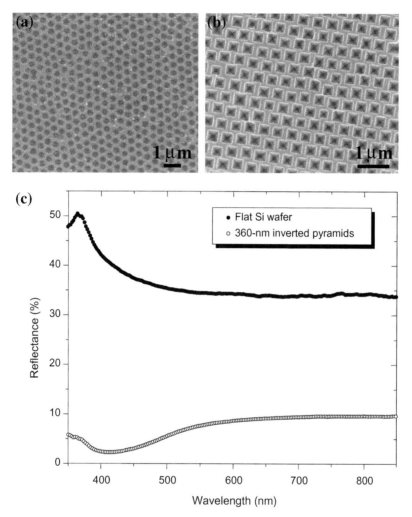

FIGURE 12.10 (a) SEM image of a templated chromium nanohole array using spin-coated non-close-packed monolayer colloidal crystal as deposition mask. **(b)** SEM image of a 360-nm-sized inverted pyramidal array templated from 320-nm-diameter silica spheres. (c) Specular optical reflectivity spectra at normal incidence. Black: bare (1 0 0) silicon wafer. Red: the sample in (b). Reprinted with permission from *Appl Phys Lett* **91** (2007), 231105. Copyright 2007, American Institute of Physics. (For interpretation of the references to color in this figure legend, the reader is referred to the web version of this book.)

wavelengths around 400 nm. This antireflection performance is good but not outstanding. The optical simulations based on a rigorous coupled-wave approach (RCWA) [130, 131] indicate that the limited antireflection performance of the inverted pyramidal arrays is caused by the limited optical depth of the inverted pyramids. Due

to the characteristic 54.7° sidewalls, the depth of the V-shaped inverted pyramids created by the anisotropic wet etch is affected by the size and separation of the templating spheres. To greatly improve the antireflection performance of the templated ARCs, the optical depth of the gratings needs to be increased.

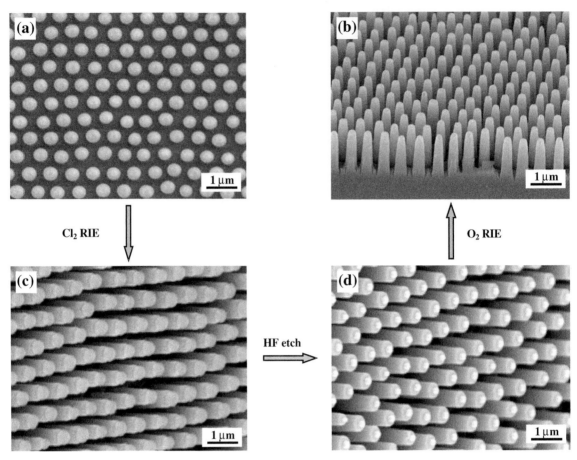

FIGURE 12.11 Outline of the templating procedures for fabricating antireflective silicon pillar arrays by using spin-coated, non-close-packed, monolayer colloidal crystal as template. Adapted from Ref. 105.

12.3.1.2 Dry-Etched Single-Crystalline Silicon Moth-Eye Antireflection Coatings

To fabricate periodic arrays of moth-eye pillars with high aspect ratio, a versatile dry-etching-based templating nanofabrication technology has been developed [105, 109]. The outline of the templating procedures for patterning moth-eye pillar arrays on single-crystalline silicon substrates is shown in Figure 12.11. The spin-coating technique [99] is first used to generate non-close-packed colloidal monolayers of hexagonally ordered silica particles on silicon wafers (Figure 12.11a). These particles are then used as etching masks during a chlorine-RIE process (5 mTorr pressure, 20 sccm chlorine flow rate, and 80 W) or a SF_6-RIE process (40 mTorr pressure, 26 sccm SF_6, 5 sccm O_2, and 25 W). Because the etching rate of silica is much lower than that of silicon under above-RIE conditions [112], silica particles protect silicon immediately underneath them from being etched, resulting in the formation of pillar arrays directly on the silicon surface (Figure 12.11b). Once the silicon pillars are deep enough, the templating silica spheres can be removed by dissolving in a 2% hydrofluoric-acid aqueous solution.

Interestingly, arrays of *micro-candles*, consisting of silicon columns as candle bodies and polymer dots as candle wicks, are clearly evident after removing the templating silica spheres (Figure 12.11c). The polymer dots are unetched residues of the thin polymer wetting layer (~100 nm thick) between the spin-coated colloidal monolayer and the silicon substrate. These dots can be easily removed by brief oxygen RIE to generate clean silicon pillar arrays (Figure 12.11d).

The resulting silicon moth-eye ARCs with high aspect ratio show excellent broadband antireflection properties, as illustrated by the specular reflectance spectra shown in Figure 12.12. Very low reflectance (<2.5%) over a wide range of wavelengths is obtained. The experimental reflectance measurements are complemented by theoretical calculations using the RCWA. The experimental spectra match reasonably well with the simulated spectra.

Besides optical depth, the crystal structure of the moth-eye ARCs also affects their antireflection performance. As shown in Figure 12.7, the spin-coating technology enables wafer-scale assembly of non-close-packed colloidal crystals with meta-stable square ordering. This allows us to create moth-eye ARCs with square arrays using the same templating technique described previously and then compare their antireflection performance with the nature-inspired hexagonal arrays [110]. Figures 12.13a and 12.13b show side-view SEM images of a square and a hexagonal moth-eye array fabricated using the same templating conditions. The antireflection performance of the square array is apparently better than that of the hexagonal array (Figure 12.13c).

The pillar pitch of moth-eye ARCs also affects the final antireflection performance. Figure 12.14a shows a moth-eye ARC fabricated using 70-nm silica spheres (see Figure 12.5a) as template. Figure 12.15b compares the specular reflection from a commercial crystalline silicon solar cell with PECVD-deposited SiN_x ARC, and the templated nanopillar array [106]. It is apparent

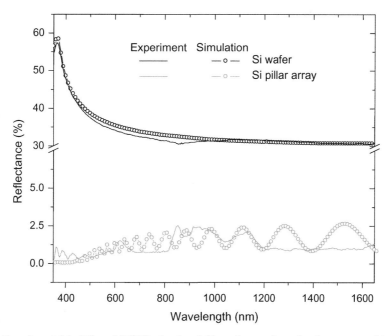

FIGURE 12.12 Experimental (solid) and RCWA-simulated (dotted) specular reflection at normal incidence from a flat silicon wafer and a 60-min Cl_2-RIE-processed silicon pillar array. Adapted from Ref. 105.

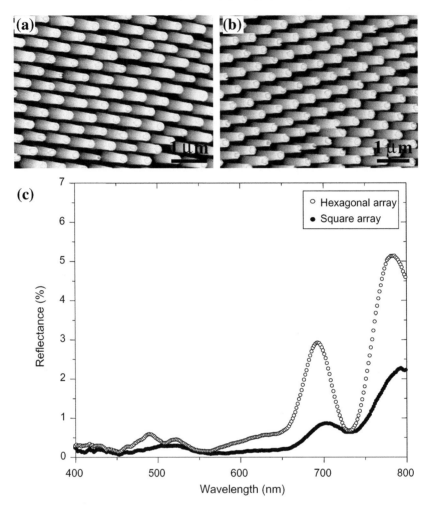

FIGURE 12.13 Comparison of the specular optical reflection at normal incidence between (a) a square silicon array and (b) a hexagonal silicon array patterned under the same conditions. Reprinted with permission from *J Vac Sci Technol B* **27** (2009), 1043–1047. Copyright 2009, American Institute of Physics.

that the nanopillar ARC exhibits excellent broadband antireflection property, better than commercial ARC on the crystalline silicon solar cell, which shows reduced reflection only around 600–800 nm [6].

12.3.1.3 Dry-Etched Multicrystalline Silicon Moth-Eye Antireflection Coatings

The same dry-etch approach can also be utilized to create broadband moth-eye ARCs on commercial mc-Si wafers covered with close-packed silica particles (Figure 12.8b) [111]. Figure 12.15a shows the typical top-view SEM image of a templated moth-eye grating using LB-assembled silica spheres as etching mask during a Cl$_2$-RIE process. Figure 12.15b compares the specular hemispherical reflectance obtained from a polished sc-Si wafer, a solar-grade mc-Si wafer, and the templated mc-Si grating. The flat sc-Si wafer exhibits 30–50%

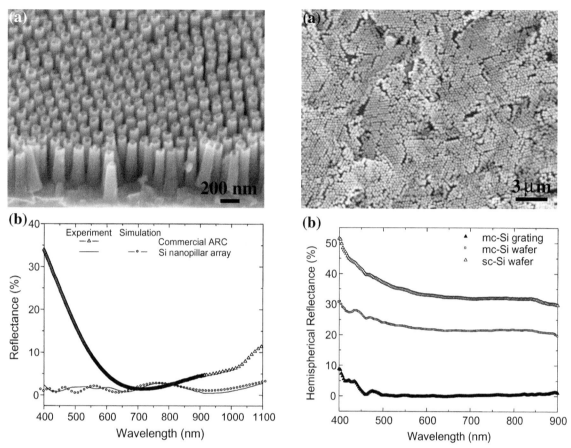

FIGURE 12.14 (a) Cross-sectional SEM image of 20-min-etched silicon nanopillars templated from 70-nm silica spheres. (b) Experimental and simulated specular reflection at normal incidence from a commercial single-crystalline silicon solar cell with SiN$_x$ ARC, and the templated silicon nanopillar array as shown in (a). Adapted from Ref. 106.

FIGURE 12.15 (a) Typical top-view SEM image of a templated moth-eye grating on multicrystalline silicon. (b) Comparison of hemispherical reflectance obtained from a polished sc-Si wafer, a commercial mc-Si wafer, and a templated mc-Si grating. Reprinted with permission from *Appl Phys Lett* **99** (2011), 191103. Copyright 2011, American Institute of Physics.

hemispherical reflectance for wavelengths from 400 to 900 nm, matching early measurements in the literature [132, 133]. The rough surface of the commercial mc-Si facilitates to reduce the hemispherical reflectance to 20–30%. By contrast, the templated mc-Si grating shows excellent broadband antireflection property, and the hemispherical reflectance is near zero for a wide range of wavelengths from ∼500 nm to ∼850 nm.

12.3.2 Templated Moth-Eye Antireflection Coatings on Other Semiconductor Wafers

The templating nanofabrication technology developed for making moth-eye ARCs on crystalline silicon substrates can be easily extended to many other technologically important semiconductor wafers, such as GaAs and GaSb, which have been widely utilized in making high-efficiency

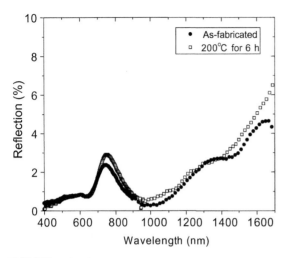

FIGURE 12.16 Comparison of the normal-incidence specular reflectance between an as-fabricated GaSb moth-eye grating and the same sample after annealing at 200°C for 6 h. Reprinted with permission from *Appl Phys Lett* **92** (2008), 141109. Copyright 2008, American Institute of Physics.

reflective loss from glass is not as severe as that from semiconductors, a ~4% optical reflection from each air–glass interface could still degrade the performance of optical devices, especially when multiple components are involved [2, 33]. To generate transparent ARCs on glass substrates, various bottom-up self-assembly techniques have been exploited [1, 3, 53, 60, 61, 137–142]. For instance, layer-by-layer assembly of polyelectrolyte or polyelectrolyte-colloid multilayers has been demonstrated as an efficient means of creating ARCs on glass [2, 60, 138, 140]. Unfortunately, traditional bottom-up techniques suffer from low throughput and incompatibility with standard microfabrication, limiting the mass production of practical coatings. By contrast, the spin-coating technological platform enables scalable production of transparent moth-eye ARCs with tunable structural parameters on glass substrates.

concentrating solar cells and thermophotovoltaic cells [104, 108]. The surface temperature of these cells is usually higher than that of a conventional cell [134, 135]. Therefore, ARCs with high thermal stability are highly preferred. Fortunately, the templated moth-eye ARCs exhibit excellent thermal stability because the resulting coatings are directly patterned on the wafer surface and no foreign materials, as in conventional quarter-wavelength design, need to be deposited on the substrates. Figure 12.16 compares the normal-incidence specular reflectance spectra of a templated GaSb moth-eye ARC prior to and after annealing at 200°C for 6 h. The change in reflectance is very small. This is in sharp contrast to the conventional quarter-wavelength ARCs that exhibit significant antireflection degradation, even at temperature as low as 100°C [136].

12.4 TEMPLATED TRANSPARENT MOTH-EYE ANTIREFLECTION COATINGS

Transparent substrates, such as glass, are widely used in our daily life. Although the

12.4.1 Templated Polymer Moth-Eye Antireflection Coatings

A schematic outline of the templating procedures for fabricating polymer moth-eye ARCs on glass is shown in Figure 12.17 [100, 107]. The polymer matrix of spin-coated colloidal crystal-polymer nanocomposites can be plasma-etched (40 mTorr oxygen pressure, 40 sccm flow rate, and 100 W) to adjust the height of the protruded portions of silica spheres. The long-range periodic surface protrusions of the exposed silica spheres can be easily transferred to a poly(dimethylsiloxane) (PDMS) mold. The solidified PDMS mold can then be peeled off and put on top of ETPTA monomer supported by a glass slide with spacers in between. After polymerization of ETPTA and peeling off PDMS mold, polymer moth-eye nipple arrays with tunable depth can be easily generated. The flexible PDMS mold enables the creation of polymer moth-eye ARCs on both planar and curved surfaces.

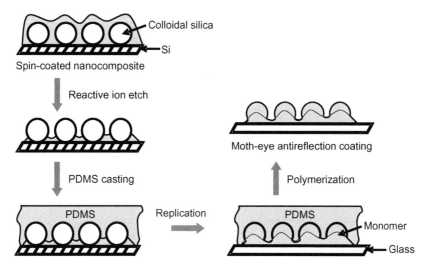

FIGURE 12.17 Schematic of the templating procedures for making polymer moth-eye ARCs on glass. Reprinted with permission from *Appl Phys Lett* **91** (2007), 101108. Copyright 2007, American Institute of Physics.

Figures 12.18a and 12.18b show atomic force microscope (AFM) images of two polymer moth-eye nipple arrays replicated from the same nanocomposite sample consisting of 360-nm silica spheres after 20 s and 45 s oxygen-plasma etching, respectively. The shape of nipples in the latter sample is close to hemispherical, as revealed by the AFM depth profile. Figure 12.18c compares the specular reflectance spectra obtained from a flat polymer control sample and two polymer moth-eye arrays with different nipple depths. It is apparent that the hemispherical nipple array shows significantly lower reflectance than that of the flat control sample and the moth-eye grating with shallower spherical caps for the whole visible spectrum.

The antireflection performance of polymer moth-eye ARCs can be further improved by using fluoropolymers [107]. A few perfluoroether acrylate monomers have been synthesized and used in creating moth-eye gratings using the same templating procedures as described previously. The utilization of fluoropolymers as the materials for templating moth-eye ARCs has at least three advantages over nonfluorinated

polymers. First, the low refractive indices of fluoropolymers, coupled with the templated moth-eye microstructures, improve the antireflective performance. The normal-incidence reflectance from a perfluoroether acrylate hemispherical nipple array is below 0.5% for most of the visible spectrum (400–700 nm). Second, the elastomeric properties of fluoropolymers with low glass temperature enable the creation of high-performance ARCs on curved surfaces. Third, fluoropolymers are intrinsically hydrophobic, facilitating the realization of self-cleaning ARCs.

The described soft-lithography-like templating technology enables wafer-scale fabrication of polymer moth-eye ARCs with periodic arrays of unitary hemispherical nipples. However, the height of the resulting nipples is at most the radius of the templating spheres, limiting the optical depth and the final antireflection performance of the unitary ARCs.

The spin-coating technological platform has therefore been extended one step further by demonstrating the production of binary dimple-nipple moth-eye ARCs on glass substrates [103]. The optical depth of the resulting binary

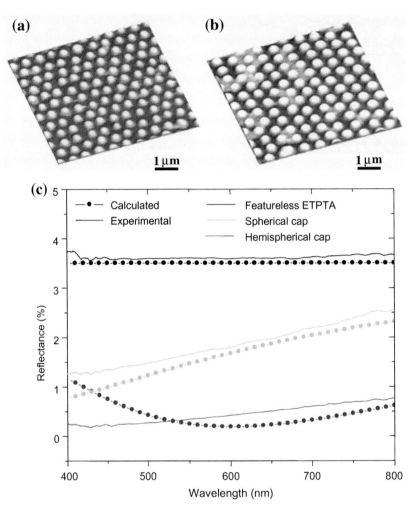

FIGURE 12.18 (a, b) AFM images of templated polymer moth-eye arrays with different nipple depths. (c) Experimental (solid) and simulated (dotted) specular optical reflectance at normal incidence obtained from a flat ETPTA control sample, a moth-eye nipple array with 110-nm spherical caps, and a nipple array with 180-nm hemispherical caps. Reprinted with permission from *Appl Phys Lett* **91** (2007), 101108. Copyright 2007, American Institute of Physics.

ARCs can almost be double that of their unitary counterparts. To generate binary ARCs, double-layer, non-close-packed colloidal crystals are first assembled by using the spin-coating technology. The polymer matrix is then partially removed and the exposed silica spheres are dissolved in a 2% hydrofluoric-acid aqueous solution, leaving behind a polymer dimple-nipple array, as shown by the side-view SEM image in Figure 12.19a. The resulting periodic binary structure can then be replicated into a thin layer of polymer (e.g., ETPTA) on a glass substrate using the same templating approach as described previously. From the reflectance measurements in Figure 12.19b, it is evident that the binary ARC shows improved antireflective performance over the unitary hemispherical ARC for most of the visible wavelengths.

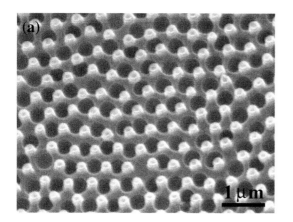

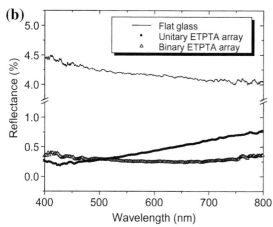

FIGURE 12.19 (a) Cross-sectional SEM image of a polymer moth-eye grating with binary structure. (b) Normal-incidence specular reflectance spectra from a bare glass substrate, a glass slide covered with a templated ETPTA binary dimple-nipple array, and a glass slide coated with an ETPTA unitary hemispherical nipple array. Adapted with permission from *J Phys Chem C* **112** (2008), 17586–17591. Copyright 2008, American Chemical Society.

12.4.2 Templated Sol–Gel Glass Moth-Eye Antireflection Coatings

The templated silicon pillars with high aspect ratio (Figure 12.11d) can be used as second-generation templates to replicate sol–gel glass moth-eye ARCs on transparent substrates [105]. A PDMS mold is first cast over the silicon template and then put on top of a sol–gel glass precursor supported by a glass slide [143]. The precursor

is then solidified by baking at 120°C for 5 min. Glass pillar arrays with high aspect ratio can then be made after peeling off the PDMS mold.

Figure 12.20a shows a side-view SEM image of a templated sol–gel glass pillar array. The size and depth of the glass pillars are reduced by ~10% over those of the templating silicon pillars due to the volume shrinkage during the solidification of sol–gel precursor. The templated glass pillar arrays exhibit excellent antireflective properties over the whole visible spectrum (Figure 12.20b).

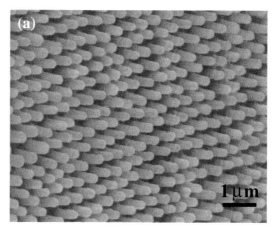

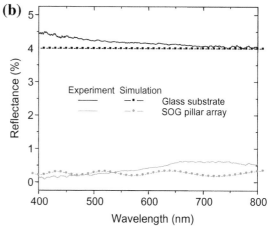

FIGURE 12.20 (a) Cross-sectional SEM image of a templated sol–gel glass moth-eye pillar array. (b) Experimental (solid) and RCWA-simulated (dotted) specular reflectance at normal incidence from a flat glass substrate and a sol–gel glass pillar array. Adapted from Ref. 105.

12.5 NANOSTRUCTURED SUPERHYDROPHOBIC COATINGS

Besides the broadband antireflection properties described in Sections 12.3 *and* 12.4, the templated nanopillar arrays with high aspect ratio can also significantly enhance the hydrophobicity of the substrate surface due to the high fraction of air trapped in the trough area between pillars [144–146]. Indeed, periodic arrays of nanopillars have been observed on the wings of cicada to render superhydrophobic surfaces for self-cleaning functionality [146]. The hydrophobicity of the templated silicon and glass nanopillars can be improved by functionalizing them with fluorosilane through the well-established silane coupling reaction [147].

Figures 12.21a and 12.21b show water-drop profiles on fluorosilane-modified silicon and glass pillar arrays, respectively. Both coatings

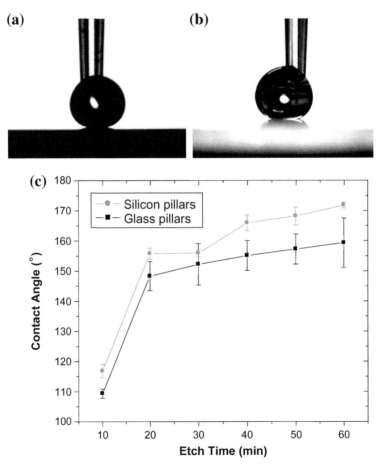

FIGURE 12.21 Superhydrophobic coatings achieved on both templated silicon and sol–gel glass nanopillar arrays. (a) Water-drop profile on a fluorosilane-modified silicon pillar array. (b) Water-drop profile on a fluorosilane-modified glass pillar array templated from the sample in (a). (c) Apparent water contact angle of templated silicon and glass pillar arrays etched at different reactive ion etching durations. Adapted from Ref. 105.

are superhydrophobic and the measured apparent water contact angle (CA) is ~172° for the former and ~160° for the latter, significantly enhanced from ~108° to ~105° on fluorinated flat silicon and glass substrates.

Figure 12.21c shows the dependence of the measured water CA on RIE duration. It is apparent that longer etching duration (i.e., larger aspect ratio for the resulting nanopillars) leads to a more hydrophobic surface. This agrees well with previous studies on microstructure-induced dewetting [144–146]. The volume shrinkage during the sol–gel glass solidification process could explain the reduced CA for glass pillar arrays over the corresponding silicon arrays.

12.6 CONCLUSIONS

Biomimetic moth-eye ARCs exhibit much-improved broadband antireflection and thermal stability than traditional quarter-wavelength ARCs. Colloidal lithography-based templating nanofabrication provides a much simpler, cheaper, and faster alternative to complex top-down nanolithography in creating subwavelength-structured moth-eye ARCs.

To improve the scalability and compatibility of current colloidal self-assembly, a versatile spin-coating technological platform and a Langmuir–Blodgett assembly technology that enable large-scale production of colloidal template for patterning moth-eye ARCs have been developed. The structural parameters of the resulting antireflection gratings, such as crystalline structure, pitch size, pillar depth, aspect ratio, and shape, can be easily adjusted by tuning the templating conditions. Excellent broadband antireflection has been achieved on a large variety of substrates for important technological applications ranging from improving the efficiency of PV cells to reducing glare from optoelectronic devices. The templated periodic arrays of nanopillars with high aspect ratios can also facilitate the achievement of superhydrophobic surface

state, which is promising for developing self-cleaning antireflection coatings.

The bioinspired moth-eye ARCs can be directly used in improving the performance of many optical and optoelectronic devices (e.g., optical lenses and photodiodes). However, on some occasions, the low reflection rendered by the moth-eye structure may not be sufficient in enhancing the final device performance. A prominent example is furnished by crystalline-silicon solar cells. One key technical risk that could limit the conversion efficiency of solar cells with integrated moth-eye ARCs is the surface recombination of charge carriers [5]. The high surface area enabled by nanostructured moth-eye gratings could greatly increase the surface recombination rate of electrons and holes, impacting the final conversion efficiency of the solar cells. The moth-eye nanostructure (e.g., size, shape, and depth of nanopillars) therefore needs to be optimized to balance the optical and optoelectronic properties of the resulting coatings.

Acknowledgments

This work was supported by the U.S. National Science Foundation (NSF) under Grants Nos. CBET-0744879 and CMMI-1000686.

References

[1] S. Walheim, E. Schaffer, J. Mlynek, and U. Steiner, Nanophase-separated polymer films as high-performance antireflection coatings, *Science* **283** (1999), 520–522.

[2] J. Hiller, J.D. Mendelsohn, and M.F. Rubner, Reversibly erasable nanoporous anti-reflection coatings from polyelectrolyte multilayers, *Nat Mater* **1** (2002), 59–63.

[3] B.G. Prevo, E.W. Hon, and O.D. Velev, Assembly and characterization of colloid-based antireflective coatings on multicrystalline silicon solar cells, *J Mater Chem* **17** (2007), 791–799.

[4] G. Chartier, *Introduction to optics*, Springer-Verlag, Heidelberg, Germany (2005).

[5] A. Luque and S. Hegedus, *Handbook of photovoltaic science and engineering*, Wiley, Chicester, UK (2003).

[6] P. Doshi, G.E. Jellison, and A. Rohatgi, Characterization and optimization of absorbing plasma-enhanced chemical vapor deposited antireflection coatings for silicon photovoltaics, *Appl Opt* **36** (1997), 7826–7837.

[7] R. Barrio, N. González, J. Cárabe, and J.J. Gandía, Optimisation of NaOH texturisation process of silicon wafers for heterojunction solar-cells applications, *Sol Energy* **86** (2012), 845–854.

[8] L. Forbes, Texturing, reflectivity, diffuse scattering and light trapping in silicon solar cells, *Sol Energy* **86** (2012), 319–325.

[9] P.O. Caffrey, B.K. Nayak, and M.C. Gupta, Ultrafast laser-induced microstructure/nanostructure replication and optical properties, *Appl Opt* **51** (2012), 604–609.

[10] W.T. Hsiao, S.F. Tseng, K.C. Huang, Y.H. Wang, and M.F. Chen, Pulsed Nd:YAG laser treatment of monocrystalline silicon substrate, *Int J Adv Manufact Technol* **56** (2011), 223–231.

[11] V.V. Iyengar, B.K. Nayak, K.L. More, H.M. Meyer, M.D. Biegalski, J.V. Li, and M.C. Gupta, Properties of ultrafast laser textured silicon for photovoltaics, *Sol Energy Mater Sol Cells* **95** (2011), 2745–2751.

[12] Z. Li, B.K. Nayak, V.V. Iyengar, D. McIntosh, Q.G. Zhou, M.C. Gupta, and J.C. Campbell, Laser-textured silicon photodiode with broadband spectral response, *Appl Opt* **50** (2011), 2508–2511.

[13] D.A. Zuev, O.A. Novodvorsky, E.V. Khaydukov, O.D. Khramova, A.A. Lotin, L.S. Parshina, V.V. Rocheva, V.Y. Panchenko, V.V. Dvorkin, A.Y. Poroykov, G.G. Untila, A.B. Chebotareva, T.N. Kost, and M.A. Timofeyev, Fabrication of black multicrystalline silicon surface by nanosecond laser ablation, *Appl Phys B* **105** (2011), 545–550.

[14] J.H. Kim, S.M. Chun, and H.J. Lee, Study on the plasma texturing for increasing the conversion efficiency of a solar cell with a DC arc plasmatron, *J Korean Phys Soc* **57** (2010), 1218–1223.

[15] K.S. Lee, M.H. Ha, J.H. Kim, and J.W. Jeong, Damage-free reactive ion etch for high-efficiency large-area multi-crystalline silicon solar cells, *Sol Energy Mater Sol Cells* **95** (2011), 66–68.

[16] M. Moreno, D. Daineka, and P.R.I. Cabarrocas, Plasma texturing for silicon solar cells: from pyramids to inverted pyramids-like structures, *Sol Energy Mater Sol Cells* **94** (2010), 733–737.

[17] J.M. Shim, H.W. Lee, K.Y. Cho, J.K. Seo, J.S. Kim, E.J. Lee, J.Y. Choi, D.J. Oh, J.E. Shin, J.H. Kong, S.H. Lee, and H.S. Lee, 17.6% conversion efficiency multicrystalline silicon solar cells using the reactive ion etching with the damage removal etching, *Int J Photoenergy* (2012), 248182.

[18] J. Yoo, Reactive ion etching (RIE) technique for application in crystalline silicon solar cells, *Sol Energy* **84** (2010), 730–734.

[19] J. Yoo, K. Kim, M. Thamilselvan, N. Lakshminarayan, Y.K. Kim, J. Lee, K.J. Yoo, and J. Yi, RIE texturing optimization for thin c-Si solar cells in SF(6)/O(2) plasma, *J Appl Phys D* **41** (2008), 125205.

[20] J. Yoo, G. Yu, and J. Yi, Large-area multicrystalline silicon solar cell fabrication using reactive ion etching (RIE), *Sol Energy Mater Sol Cells* **95** (2011), 2–6.

[21] F. Chiadini, V. Fiumara, A. Scaglione, and A. Lakhtakia, Simulation and analysis of prismatic bioinspired compound lenses for solar cells, *Bioinsp Biomim* **5** (2010), 026002.

[22] F. Chiadini, V. Fiumara, A. Scaglione, and A. Lakhtakia, Simulation and analysis of prismatic bioinspired compound lenses for solar cells: II. Multifrequency analysis, *Bioinsp Biomim* **6** (2011), 014002.

[23] B.S. Kim, D.H. Lee, S.H. Kim, G.H. An, K.J. Lee, N.V. Myung, and Y.H. Choa, Silicon solar cell with nanoporous structure formed on a textured surface, *J Am Chem Soc* **92** (2009), 2415–2417.

[24] M. Malekmohammad, M. Soltanolkotabi, R. Asadi, M.H. Naderi, A. Erfanian, M. Zahedinejad, S. Bagheri, and M. Khaje, Combining micro- and nano-texture to fabricate an antireflective layer, *J Micro-Nanolith* **11** (2012), 013011.

[25] N. Marrero, R. Guerrero-Lemus, B. Gonzalez-Diaz, and D. Borchert, Effect of porous silicon stain etched on large area alkaline textured crystalline silicon solar cells, *Thin Solid Films* **517** (2009), 2648–2650.

[26] A. Ramizy, Z. Hassan, K. Omar, Y. Al-Douri, and M.A. Mahdi, New optical features to enhance solar cell performance based on porous silicon surfaces, *Appl Surf Sci* **257** (2011), 6112–6117.

[27] X. Chen, Z.C. Fan, J. Zhang, G.F. Song, and L.H. Chen, Pseudo-rhombus-shaped subwavelength crossed gratings of GaAs for broadband antireflection, *Chin Phys Lett* **27** (2010), 124210.

[28] K.Y. Lai, Y.R. Lin, H.P. Wang, and J.H. He, Synthesis of anti-reflective and hydrophobic Si nanorod arrays by colloidal lithography and reactive ion etching, *Cryst Eng Comm* **13** (2011), 1014–1017.

[29] Y. Li, J. Zhang, S. Zhu, H. Dong, Z. Wang, Z. Sun, J. Guo, and B. Yang, Bioinspired silicon hollow-tip arrays for high performance broadband anti-reflective and water-repellent coatings, *J Mater Chem* **19** (2009), 1806–1810.

[30] D. Qi, N. Lu, H. Xu, B. Yang, C. Huang, M. Xu, L. Gao, Z. Wang, and L. Chi, Simple approach to wafer-scale self-cleaning antireflective silicon surfaces, *Langmuir* **25** (2009), 7769–7772.

[31] Y.P. Liu, T. Lai, H.L. Li, Y. Wang, Z.X. Mei, H.L. Liang, Z.L. Li, F.M. Zhang, W.J. Wang, A.Y. Kuznetsov, and

X.L. Du, Nanostructure formation and passivation of large-area black silicon for solar cell applications, *Small* **8** (2012), 1392–1397.

[32] C.M. Kennemore III and U.J. Gibson, Ion beam processing for coating MgF$_2$ onto ambient temperature substrates, *Appl Opt* **23** (1984), 3608–3611.

[33] U. Schulz, Review of modern techniques to generate antireflective properties on thermoplastic polymers, *Appl Opt* **45** (2006), 1608–1618.

[34] J.Y. Kim, Y.K. Han, E.R. Kim, and K.S. Suh, Two-layer hybrid anti-reflection film prepared on the plastic substrates, *Curr Appl Phys* **2** (2002), 123–127.

[35] H. Kondo, L. Sungkil, and H. Hanaoka, Durable antismudge materials for display terminals, *Tribol Lubr Technol* **65** (2009), 54–61.

[36] B.T. Liu, W.D. Yeh, and W.H. Wang, Preparation of low refractive index fluorinated materials for antireflection coatings, *J Appl Polym Sci* **118** (2010), 1615–1619.

[37] H. Mori, C. Sada, T. Konno, and K. Yonetake, Synthesis and characterization of low-refractive-index fluorinated silsesquioxane-based hybrids, *Polymer* **52** (2011), 5452–5463.

[38] M.S. Park, Y. Lee, and J.K. Kim, One-step preparation of antireflection film by spin-coating of polymer/solvent/nonsolvent ternary system, *Chem Mater* **17** (2005), 3944–3950.

[39] P.W. de Oliveira, C. Becker-Willinger, and M.H. Jilavi, Sol–gel derived nanocomposites for optical applications, *Adv Eng Mater* **12** (2010), 349–361.

[40] J. Pilipavicius, I. Kazadojev, A. Beganskiene, A. Melninkaitis, V. Sirutkaitis, and A. Kareiva, Hydrophobic antireflective silica coatings via sol–gel process, *Mater Sci* **14** (2008), 283–287.

[41] R. Prado, G. Beobide, A. Marcaide, J. Goikoetxea, and A. Aranzabe, Development of multifunctional sol–gel coatings: anti-reflection coatings with enhanced self-cleaning capacity, *Sol Energy Mater Sol Cells* **94** (2010), 1081–1088.

[42] K. Tadanaga, N. Yamaguchi, Y. Uraoka, A. Matsuda, T. Minami, and M. Tatsumisago, Anti-reflective properties of nano-structured alumina thin films on poly(methyl methacrylate) substrates by the sol–gel process with hot water treatment, *Thin Solid Films* **516** (2008), 4526–4529.

[43] K.-S. Han, H. Lee, D. Kim, and H. Lee, Fabrication of anti-reflection structure on protective layer of solar cells by hot-embossing method, *Sol Energy Mater Sol Cells* **93** (2009), 1214–1224.

[44] S. Chattopadhyay, Y.F. Huang, Y.J. Jen, A. Ganguly, K.H. Chen, and L.C. Chen, Anti-reflecting and photonic nanostructures, *Mater Sci Eng R* **69** (2010), 1–35.

[45] P.B. Clapham and M.C. Hutley, Reduction of lens reflection by moth eye principle, *Nature* **244** (1973), 281–282.

[46] D.G. Stavenga, S. Foletti, G. Palasantzas, and K. Arikawa, Light on the moth-eye corneal nipple array of butterflies, *Proc R Soc Lond B* **273** (2006), 661–667.

[47] S.A. Boden and D.M. Bagnall, Tunable reflection minima of nanostructured antireflective surfaces, *Appl Phys Lett* **93** (2008), 133108.

[48] R. Dewan, S. Fischer, V.B. Meyer-Rochow, Y. Ozdemir, S. Hamraz, and D. Knipp, Studying nanostructured nipple arrays of moth eye facets helps to design better thin film solar cells, *Bioinsp Biomim* **7** (2012), 016003.

[49] C. Heine and R.H. Morf, Submicrometer gratings for solar-energy applications, *Appl Opt* **34** (1995), 2476–2482.

[50] Y. Kanamori, E. Roy, and Y. Chen, Antireflection subwavelength gratings fabricated by spin-coating replication, *Microelectron Eng* **78–79** (2005), 287–293.

[51] Z.N. Yu, H. Gao, W. Wu, H.X. Ge, and S.Y. Chou, Fabrication of large area subwavelength antireflection structures on Si using trilayer resist nanoimprint lithography and liftoff, *J Vac Sci Technol B* **21** (2003), 2874–2877.

[52] K.S. Han, J.H. Shin, K.I. Kim, and H. Lee, Nanosized structural anti-reflection layer for thin film solar cells, *Jpn J Appl Phys* **50** (2011), 020207.

[53] A. Gombert, W. Glaubitt, K. Rose, J. Dreibholz, B. Blasi, A. Heinzel, D. Sporn, W. Doll, and V. Wittwer, Subwavelength-structured antireflective surfaces on glass, *Thin Solid Films* **351** (1999), 73–78.

[54] K.M. Baker, Highly corrected close-packed microlens arrays and moth-eye structuring on curved surfaces, *Appl Opt* **38** (1999), 352–356.

[55] Q. Chen, G. Hubbard, P.A. Shields, C. Liu, D.W.E. Allsopp, W.N. Wang, and S. Abbott, Broadband moth-eye antireflection coatings fabricated by low-cost nanoimprinting, *Appl Phys Lett* **94** (2009), 263118.

[56] Y. Kanamori, M. Sasaki, and K. Hane, Broadband antireflection gratings fabricated upon silicon substrates, *Opt Lett* **24** (1999), 1422–1424.

[57] N.H. Finkel, B.G. Prevo, O.D. Velev, and L. He, Ordered silicon nanocavity arrays in surface-assisted desorption/ionization mass spectrometry, *Anal Chem* **77** (2005), 1088–1095.

[58] D.G. Choi, H.K. Yu, S.G. Jang, and S.M. Yang, Colloidal lithographic nanopatterning via reactive ion etching, *J Am Chem Soc* **126** (2004), 7019–7025.

[59] C.L. Haynes and R.P. Van Duyne, Nanosphere lithography: a versatile nanofabrication tool for studies of size-dependent nanoparticle optics, *J Phys Chem B* **105** (2001), 5599–5611.

[60] F.C. Cebeci, Z.Z. Wu, L. Zhai, R.E. Cohen, and M.F. Rubner, Nanoporosity-driven superhydrophilicity: a means to create multifunctional antifogging coatings, *Langmuir* **22** (2006), 2856–2862.

[61] M. Ibn-Elhaj and M. Schadt, Optical polymer thin films with isotropic and anisotropic nano-corrugated surface topologies, *Nature* **410** (2001), 796–799.

[62] S.H. Baek, S.B. Kim, J.K. Shin, and J.H. Kim, Preparation of hybrid silicon wire and planar solar cells having ZnO antireflection coating by all-solution processes, *Sol Energy Mater Sol Cells* **96** (2012), 251–256.

[63] J.Y. Chen, W.L. Chang, C.K. Huang, and K.W. Sun, Biomimetic nanostructured antireflection coating and its application on crystalline silicon solar cells, *Opt Express* **19** (2011), 14411–14419.

[64] N. Kadakia, S. Naczas, H. Bakhru, and M.B. Huang, Fabrication of surface textures by ion implantation for antireflection of silicon crystals, *Appl Phys Lett* **97** (2010)

[65] J.W. Leem, Y.M. Song, and J.S. Yu, Broadband wide-angle antireflection enhancement in AZO/Si shell/core subwavelength grating structures with hydrophobic surface for Si-based solar cells, *Opt Express* **19** (2011), A1155–A1164.

[66] Y.S. Lin, W.C. Hsu, K.C. Huang, and J.A. Yeh, Wafer-level fabrication and optical characterization of nanoscale patterned sapphire substrates, *Appl Surf Sci* **258** (2011), 2–6.

[67] Y. Liu and M.H. Hong, Ultralow broadband optical reflection of silicon nanostructured surfaces coupled with antireflection coating, *J Mater Sci* **47** (2012), 1594–1597.

[68] Y. Liu, S.H. Sun, J. Xu, L. Zhao, H.C. Sun, J. Li, W.W. Mu, L. Xu, and K.J. Chen, Broadband antireflection and absorption enhancement by forming nanopatterned Si structures for solar cells, *Opt Express* **19** (2011), A1051–A1056.

[69] J.N. Munday and H.A. Atwater, Large integrated absorption enhancement in plasmonic solar cells by combining metallic gratings and antireflection coatings, *Nano Lett* **11** (2011), 2195–2201.

[70] E. Osorio, R. Urteaga, L.N. Acquaroli, G. García-Salgado, H. Juaréz, and R.R. Koropecki, Optimization of porous silicon multilayer as antireflection coatings for solar cells, *Sol Energy Mater Sol Cells* **95** (2011), 3069–3073.

[71] H. Park, D. Shin, G. Kang, S. Baek, K. Kim, and W.J. Padilla, Broadband optical antireflection enhancement by integrating antireflective nanoislands with silicon nanoconical-frustum arrays, *Adv Mater* **23** (2011), 5796.

[72] J.H. Selj, T.T. Mongstad, R. Sondena, and E.S. Marstein, Reduction of optical losses in colored solar cells with multilayer antireflection coatings, *Sol Energy Mater Sol Cells* **95** (2011), 2576–2582.

[73] N. Yamada, T. Ijiro, E. Okamoto, K. Hayashi, and H. Masuda, Characterization of antireflection moth-eye film on crystalline silicon photovoltaic module, *Opt Express* **19** (2011), A118–A125.

[74] T.C. Yang, T.Y. Huang, H.C. Lee, T.J. Lin, and T.J. Yen, Applying silicon nanoholes with excellent antireflection for enhancing photovoltaic performance, *J Electrochem Soc* **159** (2012), B104–B108.

[75] B.J. Bae, S.H. Hong, E.J. Hong, H. Lee, and G.Y. Jung, Fabrication of moth-eye structure on glass by ultraviolet imprinting process with polymer template, *Jpn J Appl Phys* **48** (2009), 010207.

[76] S.H. Hong, B.J. Bae, K.S. Han, E.J. Hong, H. Lee, and K.W. Choi, Imprinted moth-eye antireflection patterns on glass substrate, *Electron Mater Lett* **5** (2009), 39–42.

[77] H.S. Jang, J.H. Kim, K.S. Kim, G.Y. Jung, J.J. Lee, and G.H. Kim, Improvement of transmittance by fabricating broadband subwavelength anti-reflection structures for polycarbonate, *J Nanosci Nanotechnol* **11** (2011), 291–295.

[78] I. Saarikoski, M. Suvanto, and T.A. Pakkanen, Modification of polycarbonate surface properties by nano-, micro-, and hierarchical micro-nanostructuring, *Appl Surf Sci* **255** (2009), 9000–9005.

[79] H.-Y. Tsai and C.-J. Ting, Optical characteristics of moth-eye structures on poly(methyl methacrylate) and polycarbonate sheets fabricated by thermal nano-imprinting processes, *Jpn J Appl Phys* **48** (2009), 06FH19.

[80] N.D. Denkov, O.D. Velev, P.A. Kralchevsky, I.B. Ivanov, H. Yoshimura, and K. Nagayama, 2-Dimensional crystallization, *Nature* **361** (1993), 26.

[81] P. Jiang, J.F. Bertone, K.S. Hwang, and V.L. Colvin, Single-crystal colloidal multilayers of controlled thickness, *Chem Mater* **11** (1999), 2132–2140.

[82] Y.A. Vlasov, X.Z. Bo, J.C. Sturm, and D.J. Norris, On-chip natural assembly of silicon photonic bandgap crystals, *Nature* **414** (2001), 289–293.

[83] S. Wong, V. Kitaev, and G.A. Ozin, Colloidal crystal films: advances in universality and perfection, *J Am Chem Soc* **125** (2003), 15589–15598.

[84] A.A. Mamedov, A. Belov, M. Giersig, N.N. Mamedova, and N.A. Kotov, Nanorainbows: graded semiconductor films from quantum dots, *J Am Chem Soc* **123** (2001), 7738–7739.

[85] S. Srivastava and N.A. Kotov, Composite layer-by-layer (LBL) assembly with inorganic nanoparticles and nanowires, *Acc Chem Res* **41** (2008), 1831–1841.

[86] Y. Lin, H. Skaff, T. Emrick, A.D. Dinsmore, and T.P. Russell, Nanoparticle assembly and transport at liquid–liquid interfaces, *Science* **299** (2003), 226–229.

[87] A.G. Dong, J. Chen, P.M. Vora, J.M. Kikkawa, and C.B. Murray, Binary nanocrystal superlattice membranes self-assembled at the liquid–air interface, *Nature* **466** (2010), 474–477.

[88] A. Blanco, E. Chomski, S. Grabtchak, M. Ibisate, S. John, S.W. Leonard, C. Lopez, F. Meseguer, H. Miguez, J.P. Mondia, G.A. Ozin, O. Toader, and

H.M. van Driel, Large-scale synthesis of a silicon photonic crystal with a complete three-dimensional bandgap near 1.5 micrometres, *Nature* **405** (2000), 437–440.

[89] R. Mayoral, J. Requena, J.S. Moya, C. López, A. Cintas, H. Miguez, F. Meseguer, L. Vázquez, M. Holgado, and A. Blanco, 3D long-range ordering in an SiO$_2$ submicrometer-sphere sintered superstructure, *Adv Mater* **9** (1997), 257–260.

[90] J.M. Jethmalani, W.T. Ford, and G. Beaucage, Crystal structures of monodisperse colloidal silica in poly(methyl acrylate) films, *Langmuir* **13** (1997), 3338–3344.

[91] P. Pieranski, Colloidal crystals, *Contemp Phys* **24** (1983), 25–73.

[92] J.M. Weissman, H.B. Sunkara, A.S. Tse, and S.A. Asher, Thermally switchable periodicities and diffraction from mesoscopically ordered materials, *Science* **274** (1996), 959–960.

[93] A. Yethiraj and A. van Blaaderen, A colloidal model system with an interaction tunable from hard sphere to soft and dipolar, *Nature* **421** (2003), 513–517.

[94] Y. Cui, M.T. Bjork, J.A. Liddle, C. Sonnichsen, B. Boussert, and A.P. Alivisatos, Integration of colloidal nanocrystals into lithographically patterned devices, *Nano Lett* **4** (2004), 1093–1098.

[95] G.A. Ozin and S.M. Yang, The race for the photonic chip: colloidal crystal assembly in silicon wafers, *Adv Funct Mater* **11** (2001), 95–104.

[96] A. van Blaaderen, R. Ruel, and P. Wiltzius, Template-directed colloidal crystallization, *Nature* **385** (1997), 321–324.

[97] Y.D. Yin, Y. Lu, B. Gates, and Y.N. Xia, Template-assisted self-assembly: a practical route to complex aggregates of monodispersed colloids with well-defined sizes, shapes, and structures, *J Am Chem Soc* **123** (2001), 8718–8729.

[98] P. Jiang and M.J. McFarland, Large-scale fabrication of wafer-size colloidal crystals, macroporous polymers and nanocomposites by spin-coating, *J Am Chem Soc* **126** (2004), 13778–13786.

[99] P. Jiang, T. Prasad, M.J. McFarland, and V.L. Colvin, Two-dimensional nonclose-packed colloidal crystals formed by spincoating, *Appl Phys Lett* **89** (2006), 011908.

[100] N.C. Linn, C.H. Sun, P. Jiang, and B. Jiang, Self-assembled biomimetic antireflection coatings, *Appl Phys Lett* **91** (2007), 101108.

[101] C.H. Sun, N.C. Linn, and P. Jiang, Templated fabrication of periodic metallic nanopyramid arrays, *Chem Mater* **19** (2007), 4551–4556.

[102] C.H. Sun, W.L. Min, N.C. Linn, P. Jiang, and B. Jiang, Templated fabrication of large area subwavelength antireflection gratings on silicon, *Appl Phys Lett* **91** (2007), 231105.

[103] W.H. Huang, C.H. Sun, W.L. Min, P. Jiang, and B. Jiang, Templated fabrication of periodic binary nanostructures, *J Phys Chem C* **112** (2008), 17586–17591.

[104] W.L. Min, A.P. Betancourt, P. Jiang, and B. Jiang, Bioinspired broadband antireflection coatings on GaSb, *Appl Phys Lett* **92** (2008), 141109.

[105] W.L. Min, B. Jiang, and P. Jiang, Bioinspired self-cleaning antireflection coatings, *Adv Mater* **20** (2008), 3914–3918.

[106] W.L. Min, P. Jiang, and B. Jiang, Large-scale assembly of colloidal nanoparticles and fabrication of periodic subwavelength structures, *Nanotechnology* **19** (2008), 475604.

[107] C.H. Sun, A. Gonzalez, N.C. Linn, P. Jiang, and B. Jiang, Templated biomimetic multifunctional coatings, *Appl Phys Lett* **92** (2008), 051107.

[108] C.H. Sun, B.J. Ho, B. Jiang, and P. Jiang, Biomimetic subwavelength antireflective gratings on GaAs, *Opt Lett* **33** (2008), 2224–2226.

[109] C.H. Sun, P. Jiang, and B. Jiang, Broadband moth-eye antireflection coatings on silicon, *Appl Phys Lett* **92** (2008), 061112.

[110] C.H. Sun, W.L. Min, N.C. Linn, P. Jiang, and B. Jiang, Large-scale assembly of periodic nanostructures with metastable square lattices, *J Vac Sci Technol B* **27** (2009), 1043–1047.

[111] B.M. Phillips, P. Jiang, and B. Jiang, Biomimetic broadband antireflection gratings on solar-grade multicrystalline silicon wafers, *Appl Phys Lett* **99** (2011), 191103.

[112] M.J. Madou, *Fundamentals of microfabrication: the science of miniaturization*, 2nd ed., CRC Press, New York, NY, USA (2002).

[113] H.W. Deckman and J.H. Dunsmuir, Natural lithography, *Appl Phys Lett* **41** (1982), 377–379.

[114] J.C. Hulteen and R.P. Van Duyne, Nanosphere lithography—a materials general fabrication process for periodic particle array surfaces, *J Vac Sci Technol A* **13** (1995), 1553–1558.

[115] A. Kosiorek, W. Kandulski, P. Chudzinski, K. Kempa, and M. Giersig, Shadow nanosphere lithography: simulation and experiment, *Nano Lett* **4** (2004), 1359–1363.

[116] Y. Xu, G.J. Schneider, E.D. Wetzel, and D.W. Prather, Centrifugation and spin-coating method for fabrication of three-dimensional opal and inverse-opal structures as photonic crystal devices, *J Microlith Microfab Microsys* **3** (2004), 168–173.

[117] R.L. Hoffman, Discontinuous and dilatant viscosity behavior in concentrated suspensions. 1. Observation of a flow instability, *Trans Soc Rheol* **16** (1972), 155–165.

[118] S.A. Johnson, P.J. Ollivier, and T.E. Mallouk, Ordered mesoporous polymers of tunable pore size from colloidal silica templates, *Science* **283** (1999), 963–965.

[119] F.J. Arriagada and K. Osseo-Asare, Synthesis of nano-size silica in a nonionic water-in-oil microemulsion: effects of the water/surfactant molar ratio and ammonia concentration, *J Colloid Interf Sci* **211** (1999), 210–220.

[120] B.J. Ackerson, Shear induced order of hard-sphere suspensions, *J Phys: Condens Matter* **2** (1990), SA389–SA392.

[121] B.J. Ackerson and P.N. Pusey, Shear-induced order in suspensions of hard-spheres, *Phys Rev Lett* **61** (1988), 1033–1036.

[122] R.M. Amos, T.J. Shepherd, J.G. Rarity, and P. Tapster, Shear-ordered face-centred cubic photonic crystals, *Electron Lett* **36** (2000), 1411–1412.

[123] L.B. Chen, B.J. Ackerson, and C.F. Zukoski, Rheological consequences of microstructural transitions in colloidal crystals, *J Rheol* **38** (1994), 193–216.

[124] J. Vermant and M.J. Solomon, Flow-induced structure in colloidal suspensions, *J Phys: Condens Matter* **17** (2005), R187–R216.

[125] C. Tang, E.M. Lennon, G.H. Fredrickson, E.J. Kramer, and C.J. Hawker, Evolution of block copolymer lithography to highly ordered square arrays, *Science* **322** (2008), 429–432.

[126] W. Stober, A. Fink, and E. Bohn, Controlled growth of monodisperse silica spheres in micron size range, *J Colloid Interf Sci* **26** (1968), 62–68.

[127] C.M. Hsu, S.T. Connor, M.X. Tang, and Y. Cui, Wafer-scale silicon nanopillars and nanocones by Langmuir–Blodgett assembly and etching, *Appl Phys Lett* **93** (2008), 133109.

[128] J. Poortmans and V. Arkhipov, *Thin films solar cells: fabrication, characterization and applications*, Wiley, Chichester, UK (2006).

[129] P. Jiang and M.J. McFarland, Wafer-scale periodic nanohole arrays templated from two-dimensional nonclose-packed colloidal crystals, *J Am Chem Soc* **127** (2005), 3710–3711.

[130] M.G. Moharam and T.K. Gaylord, Rigorous coupled-wave analysis of planar-grating diffraction, *J Opt Soc Am* **71** (1981), 811–818.

[131] M.G. Moharam, D.A. Pommet, E.B. Grann, and T.K. Gaylord, Stable implementation of the rigorous coupled-wave analysis for surface-relief gratings: enhanced transmittance matrix approach, *J Opt Soc Am A* **12** (1995), 1077–1086.

[132] S.H. Zaidi, D.S. Ruby, and J.M. Gee, Characterization of random reactive ion etched-textured silicon solar cells, *IEEE Trans Electron Devices* **48** (2001), 1200–1206.

[133] Y.F. Huang, S. Chattopadhyay, Y.J. Jen, C.Y. Peng, T.A. Liu, Y.K. Hsu, C.L. Pan, H.C. Lo, C.H. Hsu, Y.H. Chang, C.S. Lee, K.H. Chen, and L.C. Chen, Improved broadband and quasi-omnidirectional anti-reflection properties with biomimetic silicon nano-structures, *Nat Nanotech* **2** (2007), 770–774.

[134] E. Oliva, F. Dimroth, and A.W. Bett, GaAs converters for high power densities of laser illumination, *Prog Photovolt: Res Appl* **16** (2008), 289–295.

[135] S. Luca, J.L. Santailler, J. Rothman, J.P. Belle, C. Calvat, G. Basset, A. Passero, V.P. Khvostikov, N.S. Potapovich, and R.V. Levin, GaSb crystals and wafers-for photovoltaic devices, *J Sol Energy Eng* **129** (2007), 304–313.

[136] Y. Kanamori, K. Kobayashi, H. Yugami, and K. Hane, Subwavelength antireflection gratings for GaSb in visible and near-infrared wavelengths, *Jpn J Appl Phys* **42** (2003), 4020–4023.

[137] A. Gombert, B. Blasi, C. Buhler, P. Nitz, J. Mick, W. Hossfeld, and M. Niggemann, Some application cases and related manufacturing techniques for optically functional microstructures on large areas, *Opt Eng* **43** (2004), 2525–2533.

[138] D. Lee, M.F. Rubner, and R.E. Cohen, All-nanoparticle thin-film coatings, *Nano Lett* **6** (2006), 2305–2312.

[139] B.G. Prevo, Y. Hwang, and O.D. Velev, Convective assembly of antireflective silica coatings with controlled thickness and refractive index, *Chem Mater* **17** (2005), 3642–3651.

[140] X.T. Zhang, O. Sato, M. Taguchi, Y. Einaga, T. Murakami, and A. Fujishima, Self-cleaning particle coating with antireflection properties, *Chem Mater* **17** (2005), 696–700.

[141] H. Jiang, K. Yu, and Y.C. Wang, Antireflective structures via spin casting of polymer latex, *Opt Lett* **32** (2007), 575–577.

[142] Y. Zhao, J.S. Wang, and G.Z. Mao, Colloidal subwavelength nanostructures for antireflection optical coatings, *Opt Lett* **30** (2005), 1885–1887.

[143] Y.N. Xia and G.M. Whitesides, Soft lithography, *Angew Chem Int Ed* **37** (1998), 551–575.

[144] C.W. Guo, L. Feng, J. Zhai, G.J. Wang, Y.L. Song, L. Jiang, and D.B. Zhu, Large-area fabrication of a nano-structure-induced hydrophobic surface from a hydrophilic polymer, *Chem Phys Chem* **5** (2004), 750–753.

[145] H. Nakae, R. Inui, Y. Hirata, and H. Saito, Effects of surface roughness on wettability, *Acta Mater* **46** (1998), 2313–2318.

[146] T.L. Sun, L. Feng, X.F. Gao, and L. Jiang, Bioinspired surfaces with special wettability, *Acc Chem Res* **38** (2005), 644–652.

[147] Y. Coffinier, S. Janel, A. Addad, R. Blossey, L. Gengembre, E. Payen, and R. Boukherroub, Preparation of superhydrophobic silicon oxide nanowire surfaces, *Langmuir* **23** (2007), 1608–1611.

ABOUT THE AUTHORS

Blayne M. Phillips is a doctoral candidate in the Chemical Engineering Department at the University of Florida in Gainesville, Florida. Blayne grew up in Trinidad and Tobago, where he attended high school at St. Mary's College, Port of Spain. He attended Northeastern University, Boston, MA, where he was awarded the Reggie Lewis scholarship and graduated with a BS in Chemical Engineering. He was admitted to the University of Florida in 2010 with the NSF Bridge to the Doctorate Fellowship. His current research interests include colloidal self-assembly, polymer membranes, antireflection coatings, and photovoltaics.

Peng Jiang is currently an associate professor in the Chemical Engineering Department at the University of Florida. He obtained his PhD in materials chemistry at Rice University and was a postdoctoral fellow in the Chemical Engineering Department at Princeton University. After gaining industrial R&D experiences in Corning and General Electric, he started his academic career in 2006. His current research interests include nanophotonics, material self-assembly, plasmonic biosensors, biomimetic materials, and scalable nanomanufacturing. He is a recipient of the NSF CAREER Award.

13

Biomimetic Self-Organization and Self-Healing

Torben Lenau[a] *and Thomas Hesselberg*[b]

[a]Department of Mechanical Engineering, Technical University of Denmark,
DK2800 Lyngby, Denmark
[b]Department of Zoology, University of Oxford,
Oxford OX1 3PS, United Kingdom

Prospectus

Self-organization and self-healing appeal to humans because difficult and repeated actions can be avoided through automation via bottom-up nonhierarchical processes. This is in contrast to the top-level controlled manner we normally apply as an action strategy in manufacturing and maintenance work. This chapter presents eight different self-organizing and self-healing approaches in nature and takes a look at realized and potential applications. Furthermore, the core principles for each approach are described using simplified drawings in order to make the ideas behind the self-organizing and self-healing principles more accessible to design practitioners.

Keywords

Adaptive growth, Animal behavior, Bio-inspired design, Breaking plane, Collective decision making, Decentralized control, Emergent structures, Extensibility, Flock coordination, Multilayer formation, Self-healing, Self-organizing, Social insects, Stigmergy, Swarm intelligence, Tensairity

13.1 INTRODUCTION

We humans normally apply a very controlled approach to organizing activities and producing goods. For instance, a sculptured metal part for a car is produced by shaping bulk material into thin sheets and then pressing them into shape in a die with a predefined geometry. In contrast, nature constructs geometry in seashells, insects, and plants without a guiding mold or die. Instead, the geometrical instructions are inherent in the single building blocks, namely, in the cells. The traffic movements of trains and airplanes are controlled through time schedules and centralized command structures. In nature, bird flocks, fish schools, ants, and penguins move nimbly and purposefully without an overall controlling body. In many situations, it could be advantageous to emulate nature and move some of the planning capacity away from the central hierarchical level down to the single building blocks. This chapter examines how nature uses self-organization and self-healing and identifies the underlying principles. Furthermore, it describes how the principles have been used and can be applied by humans.

The concepts of self-organization are often explained by looking at emergent structures

Engineered Biomimicry

http://dx.doi.org/10.1016/B978-0-12-415995-2.00013-1

and properties [1, 2]. Emergent structures are more than the sum of the constituents in the sense that an emergent structure has characteristics that are not expected from looking at the single elements brought together to make it. Emergence is used to describe macroscopic phenomena that are not seen on the microscopic scale. An emergent structure is governed by a combination of rules and principles at the microscopic scale with the interaction of the macroscopic structure and its surroundings. Emergent properties are complex patterns or processes derived from simple interactions between multiple agents. There are many examples of emergent properties in biological systems, ranging from the construction of ant and termite nests from simple individual behavioral rules to the coordinated movements and patterns of large bird flocks.

The eight phenomena described in this chapter include the agile movement and navigation in large groups like bird flocks and fish schools. Another phenomenon is how groups of animals coordinate the building of large constructions such as termite mounds. In swarm intelligence, seemingly intelligent behavior emerges from the collective action of a large number of autonomous agents such as individual ants and termites. Another impressive phenomenon in large groups is the collective decision making involved when ants find the shortest path to food sources and honeybees determine where to build a new nest.

Self-sealing is the first step in the healing of liquid transporting vessels and is seen in blood clotting and in liana plants. Lizards and spiders sometimes apply a drastic defense action when they self-amputate limbs. The byssus threads that mussels use to fasten themselves to rocks have a remarkable self-healing capability whereby fractures can be healed. Slime molds are very resource efficient since they not only build up and expand food vessels but also break down the ones that are not needed. Finally, the construction of a composite layered structure in mussel nacre is examined at the end of the chapter.

13.2 NAVIGATION IN LARGE GROUPS

Some animals aggregate in very large groups consisting of thousands of individuals that nonetheless act in a coordinated manner; examples include fish schools, bird flocks, and building and foraging tasks in social insect communities. The sheer number of individuals and the reliance on relatively simple behavioral rules suppress the effects of individual variations in behaviors to such a degree that collective behaviors of large groups are similar to the physical behaviors of inanimate objects (for example, the tide of people exiting a busy commuter train follows the path of least resistance and can be likened to water flowing downhill) and can therefore be analyzed by mathematical and statistical models employed to describe physical phenomena [3, 4].

Many birds have a tendency to fly in flocks, and for common starlings (*Sturnus vulgaris*) the flocks become extremely huge, as many as 100,000 birds in a single flock. Despite the fact that the birds fly very close to each other and at high speeds (about 70 km/h), it is rarely seen that the birds hit each other. A spectacular phenomenon called *murmuration* (aptly named *Sort sol* in Danish, meaning *black sun*) involving such starling flocks can be witnessed on the west coast of Denmark during the spring and autumn migrations (Figure 13.1). Around sunset, large flocks are seen flying around looking like quickly moving black clouds. Hunting falcons attack the flock, causing rapid changes locally in the flock so that the "cloud" changes direction very quickly and abruptly. When the flock is landing, a similar and apparently very controlled behavior can be witnessed. Despite the huge numbers of birds in the flock, all birds land within a short distance of each other without colliding.

FIGURE 13.1 Murmuration involving large starling flocks seen on the west coast of Denmark. Photograph reproduced with kind permission from www.sortsafari.dk.

Researchers in the computer science community have long been interested in simulating this flock behavior. One application area is realistic animations of animal flocks on film; feasible explanations of the animal behavior can therefore be found in the computer-science literature [5, 6]. Reynolds describes one explanation of the mechanism behind the starlings' agile navigation in terms of three simple rules that each bird obeys: the separation rule, the alignment rule, and the cohesion rule, as shown in Figure 13.2 [5, 6]. The *separation rule* makes sure that the birds avoid crowding within the flock. If the number of birds within a local region becomes too big, the outsiders avoid getting into this

region. This gives rise to short-distance repulsion between individual birds, which in turns results in an exclusion zone around each bird. In one empirical study, the exclusion zone was found to be 0.38 m in radius and independent of the flock size [7]. The *alignment rule* causes each bird to fly in the same direction as the adjacent birds. Finally, the *cohesion rule* steers the birds on the periphery to position themselves so that the flock does not get separated or broken. The distance and the direction determine the vicinity of every bird. Flock mates outside the local vicinity are ignored. The three rules are reported to give reasonable and realistic simulation results, but more accurate models also include the avoidance of obstacles [5].

The ability to act in a coordinated fashion and retain group adherence during predatory events by a bird of prey is likely to have significant fitness advantages for individual birds as the raptor encounters problems with picking out individual birds to attack. In general, a collective response is the hallmark of self-organized distributive order as opposed to the centralized order present in hierarchical systems where individuals follow a leader [8]. The ability to act collectively arises from the simple individual behavioral rules outlined in the previous paragraph: Each bird typically interacts with up to seven of its closest neighbors (the interaction distance). The degree to which non-interacting birds correlate their behavior depends on how well the

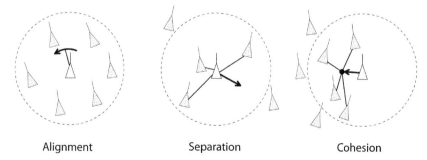

Alignment   Separation   Cohesion

FIGURE 13.2 Three local rules birds use to navigate within a flock: the alignment rule (left), the separation rule (middle), and the cohesion rule (right). Drawing based on description in Ref. 5.

information is transmitted via individual interactions. In most biological and physical systems, the correlation distance is significantly longer than the interaction distance but shorter than the size of the system. However, in starling flocks the correlation distance is as large as the group, independent of the number of individuals in the group; the behavior of all birds is correlated, which is known as *scale-free correlation* [8].

Computer-automated measurements of individual starlings in flocks ranging in size from 500 to 2,500 birds have revealed that the speed of the entire flock (center of mass) averages between 9 and 15 m/s. Perhaps somewhat surprisingly, the flocks are not as dense as they appear to a human observer standing on the ground. The density is 0.04–0.8 birds per cubic meter (compared to molecular density of materials, this would be closer to a gas than a liquid) and the distance from a bird to its nearest neighbor ranges from 0.7 to 1.5 m [7].

The nearest-neighbor distance in fish schools is not independent of the size of the group, as it appears to be in bird flocks [5]. Apart from this, the collective behavior of fish schools closely follows the behavior outlined for bird flocks. The individual fish interacts with neighbors by showing behavioral matching (i.e., it matches its behavior with that of its neighbors) and positional preference (i.e., it maintains a constant distance or position relative to its neighbors) [1]. Similarly to bird flocks, this is thought to confer fitness advantages to individual fish through protection from predation.

This insight has potential for many applications. Computer animations use swarm algorithms for simulating flock behavior among birds, bats, herds, and fish at sea [4]. An interesting observation is that human beings also exhibit similar flock behavior when we crowd together—for instance, when driving cars. The distributed algorithms are therefore also usable for predicting and controlling car traffic [4].

Similarly, the field of robotics benefits from studies of insects and other animals. Especially,

insects attract interest since they perform difficult tasks even though they only have very small brains. This is important for designers of mobile robots, because processing capacity is restricted by weight and limited energy access. The interest also goes in the opposite direction, where robots can be used by biologists and engineers in collaboration to simulate the behavior of animals in order to test a hypothesis explaining that behavior [9, 10]. Robot researchers study and mimic various capabilities in insects by looking at chemical, visual, and auditory sensing and on how complex motor control is performed [11]. Robots are made that navigate like bees and ants, which use visual landmarks to remember the trail back to the nest. One example is the mobile robot Sahabot 2 that uses a 360° camera that captures how the light intensity in a 360° view changes during movement [11]. Other examples are provided in the accompanying chapters on microflyers (Chapter 5) and biomimetic vision sensors (Chapter 1).

Biological inspiration from flock behavior is also valued in bigger robots such as automated cars. The decentralized control of unmanned vehicles to form convoys, e.g., on highways, is called *platooning*. Inspiration comes from birds such as geese that fly in formation. The advantage of platooning is that car drivers can relax and do activities other than driving. Furthermore, it makes traffic safer and allows for higher capacity on the roads since the distance between the cars can be reduced and the risk of queues due to variation in speed can be lowered. The European Sartre project has recently demonstrated platooning on a public road with eight cars where only the front car had a human driver [12].

Finally, the flock behavior in birds and fish attracts attention from researchers developing unmanned aerial vehicles (UAV), i.e., small airplanes and helicopters [13–15]. As discussed in Chapter 5 on microflyers, effort has so far been invested in the development of hardware and software platforms to make the vehicles fly and sense their surroundings, but it is a challenge to sense the surroundings using cameras to accurately detect other vehicles. The very low weights of these vehicles also set severe limitations to the on-board computer handling.

13.3 COLLECTIVE DECISION MAKING

A person with limited biological knowledge could think that decisions in large insect societies follow the classical hierarchical structure known from pre-democratic human cultures and the business world, where a leader individual organizes and controls the behavior of the follower individuals. Thus larger strategic decisions such as how to allocate tasks to individual workers, foraging priorities, and new nest selection would be under the direct control of the ant or bee queen via pheromones. However, this is not the case.

Such decisions are reached by the collective actions of individuals following simple behavioral rules. Decisions on choosing an optimal path to a food source in ants is, for instance, primarily determined by the chemical concentration of recruitment pheromones left by other foraging workers [2]. The simple behavioral rule of following the path with the highest pheromone concentrations can explain the correct choice between a shorter path and a longer path leading to the same food. If we imagine two foragers that continuously deposit pheromones as they follow the paths, with one ant collecting food by taking the short path and the other ant taking the longer path, then the ant following the shorter path will take shorter time and thus pass by a given point on the path more often than the ant on the longer path. Since the foragers continuously leave a pheromone trail as they move, the concentration of pheromone on the shorter path will automatically become higher as time passes (Figure 13.3). New foragers will therefore choose the shorter path, which will then get an even higher pheromone concentration.

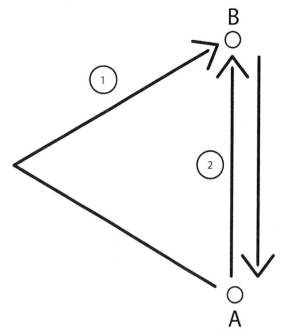

FIGURE 13.3 The basic principle used by ants to find the shortest route to a food source. Ants leave odorant tracks (pheromones) that quickly evaporate. The more often the track is passed, the stronger the odorant is and is therefore preferred by ants that follow.

However, in this section we focus on another well-studied behavior: swarming and nest-site selection in honeybees. Honeybees live in large colonies with one queen and up to 100,000 workers. In the spring, the queen begins to lay eggs destined to become new queens. Soon after these new queens emerge, the old queen, along with a sizable number of the workers, forms a swarm and leaves the old nest. The swarm usually settles on a tree for a couple of days while scouts are sent out to search for new nest sites. Nest-site selection and swarm lift-off are determined by at least four different communication methods, as illustrated in Figure 13.4 [16]: (1) In the waggle dance, scouts advertise the position of possible nest sites to other scouts. (2) Vibration signals, which consist of scout bees vibrating their bodies rapidly, are used to stimulate other bees and

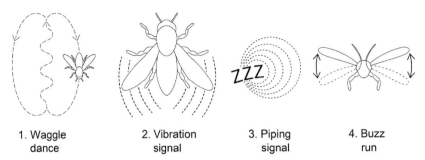

1. Waggle dance   2. Vibration signal   3. Piping signal   4. Buzz run

FIGURE 13.4 Four communication methods used by bees in the decision-making process when selecting a new nest site.

make them more receptive to other signals and thus play a role both in nest-site selection and in decisions on when the swarm should take off. (3) A piping signal appears after a decision on a nest site has been reached and consists of a high-pitched noise produced by pressing the bee's body to a surface or to another bee. It is a signal for other bees to warm up their flight muscles. (4) A buzz run occurs shortly before lift-off and is done by the same bees that performed the piping signals. It consists of running across the surface of the swarm while buzzing the wings every second and eventually triggers the lift-off.

Let us now focus on what happens during the decision-making process, i.e., before piping signals and buzz runs occur. A number of scouts leave the swarm to fly off in different directions to search for suitable nest sites. Once located, they return to the swarm to advertise for the nest sites with waggle dances. The better the site, the longer and more vigorous the dance. In addition, scouts visiting good sites use a stop signal toward scouts waggle dancing for other nest sites [17]. The signal consists of the bee touching its head against the dancing bee while emitting a high-frequency vibration signal. Although one stop signal is not enough to stop a waggle dance, the more stop signals a dancing bee receives, the more likely she is to abandon her waggle dance. The process of advertisement and inhibitory signals means that the preference for the different sites (measured as the number of scouts waggle dancing for them) will change

over time. Eventually, usually within a day, a consensus is reached such that the majority of scouts advertise for the same location, as illustrated in Figure 13.5 [18]. When this consensus is reached, some scouts start emitting the piping signal, followed by the buzz run and migration of the swarm to the chosen nest site. Thus the queen is not involved in the decision of which location is going to be the site of the new nest. Instead, it is determined by a *democratic* process of a convergence of preferences by the majority of the scouting honeybees.

One example of a planned human behavior with similarities to the bee decision-making behavior is described by Hansen and Lenau [19]. The behavior of a team of engineering design students is influenced so they can easily follow the work of other teams. The students make extensive use of worksheets on which they

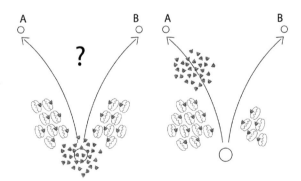

FIGURE 13.5 Part of the decision-making process when the bee swarm determines which nest site to choose.

graphically describe how they perceive the design problem and propose conceptual solutions. Each worksheet is analogous to a bee doing its waggle dance: Other students either agree and maybe get inspired or actively go into a discussion where they argue against the proposals on the worksheet. This results in more distributed learning and faster dissemination of knowledge. The students are confronted with many partial solutions to design problems similar to their own and can readily determine which ones to be inspired by. The average quality of the design solutions is better than if the student groups had worked alone without insight into the work of the other groups. Instead of fearing that the other students will steal their ideas, Hansen and Lenau found that everyone performed better.

13.4 COORDINATION OF LARGE CONSTRUCTION WORK

Several different species across the animal kingdom build structures either from internally secreted building materials (such as silk used by caterpillars to construct cocoons or by spiders in their impressive aerial webs) or from externally collected materials (such as stones and leaves in the case of caddisfly larvae or twigs in bird's nests). The majority of animal constructions are built by individual animals for shelter, protection, mate attraction, or capturing prey. However, the most impressive examples of animal architecture are the result of the collaboration of a large number of individuals. For instance, termite nests (also known as *mounds*) can reach heights of several meters on the African savannah, the heights exceeding more than 200 times the size of the termite workers. In human terms, a termite mound is as high as the Empire State Building in New York.

Termite mounds are not only impressive when viewed from the outside, but they also contain many sophisticated internal adaptations to maintain a constant indoor climate [20].

Complex ventilation systems run within the nest either as (1) open systems with chimneys and ventilation shafts or (2) closed systems, wherein gas exchange occurs in special galleries close to the outer surface. The interaction among temporal variations in wind speed, wind direction, and turbulence caused by the morphology of the ventilation system in the termite mound has been suggested to cause a tidal gas exchange in the mound similar to that found in human lungs [21]. Such passive ventilation systems have significant biomimetic potential for use in managing the internal climate of buildings. Similar wind-induced ventilation systems have been found in large underground nests of leaf-cutter ants, comprising millions of individual workers [22].

The structure of these giant nests is impressive, but equally fascinating is the question of how such elaborate structures can be constructed by individual ants working without any centralized and hierarchical supervision. Recent research shows that the nest structure appears as an emergent property from the individual actions of thousands of workers that indirectly communicate through modifications of the environment [20, 23, 24]. The concept that the actions of individuals modify the environment, which in turn modifies the behavior of other individuals, is known as *stigmergy*. One example of how stigmergy can result in self-organization can be seen in the ant "cemeteries" found outside the nests of many species. The large aggregations of dead ants result from the corpse-removal behavior of individual ants, who pick up corpses and then drop them as a function of the density of corpses in the vicinity [23]. This means that once a pile of corpses is starting to build up, the probability that further corpses are dropped on it increases, which in turns results in all corpses being dropped in a few large aggregations only.

A similar example of stigmergy is seen in wall building by the ant *Leptothorax tuberointerruptus*. These ants construct a simple circular wall around the colony at a certain distance

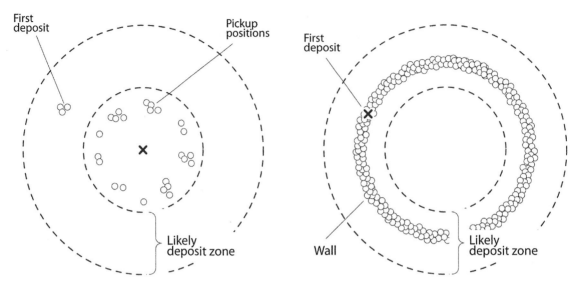

FIGURE 13.6 Collective decision making using stigmergy, i.e., the concept whereby individuals modify the environment, which in turn modifies the behavior of the individuals. The drawing illustrates the principle of how ants build a wall at a position determined by the distance from the pickup place and the positions where the other ants have placed their building blocks. The very simplified drawing is based on the description in Ref. 24.

from the brood, as shown in Figure 13.6. The worker ants collect grains and deposit them based on the distance from the brood and the local density of grains [24]. The presence of grains generally acts as a positive reinforcement, and ants are most likely to drop a grain if the distance to the brood is right and there already are other grains present at the site.

The collective nonhierarchical behavior of social insects has significant biomimetic potential in management, manufacturing, and computer science. Examples include robot coordination, flow shop scheduling, and comparisons between collective decision-making strategies in house-hunting ants and those used in internet search engines [25].

The concept of stigmergy is used in open-source user-driven software on the internet, such as the encyclopedia Wikipedia (wikipedia. org). Here is the basic principle that anyone can write what they like, but other users can alter the content if they do not agree or if they have supplementary information. When someone starts to describe a topic, others will tend to expand it—so the mere existence of elements of a description will stimulate further writing on the topic. However, an important difference between the way human beings and ants approach stigmergy is that humans sometime counteract and write things that are not in the common interest of the software-user community. Wikipedia therefore has editors who can intervene and remove undesired material.

Human building construction is a complicated task involving many different actors who are mutually dependent on the outcome and timing of each other's work. Part of the planning and control is done by bringing many of the actors together at meetings, so that they together can experience the present state of the building project and determine if plans should be changed. This coordination activity can be seen as stigmergy, since the planning of the building construction relies on actors monitoring the progress of the work and changing plans according to those observations. Perhaps, the building industry could learn from the different ways that stigmergy is applied in insect colonies.

Graffiti and littering by human beings also have stigmergy characteristics. Places where graffiti and litter emerge seem to attract more of the same. Railway companies therefore have the anti-graffiti strategy to remove the graffiti as quickly as possible. By removing the traces of other *members of the tribe*, it is believed that the undesired decorations will be kept at a minimum.

13.5 SWARM INTELLIGENCE

Swarm intelligence is the seemingly intelligent behavior that emerges from the collective behavior of a large number of autonomous agents. In biology, this term is most widely used with reference to the colony-level behaviors seen in social insects. For instance, whereas individual fire ants (*Solenopsis invicta*) struggle and relatively rapidly drown after falling into a pool of water, groups of fire ants link together to form rafts that can float on the water surface for days [26]. Similarly, army ants in their migratory phase will fabricate temporary nests (bivouacs) consisting of several hundred thousand workers linking their bodies together to protect the queen and the brood [27]. However, where this collective behavior is perhaps the most impressive and the most biomimetically relevant is in the large-scale traffic control of foraging ants.

Most ant species use specific paths that lead from the nest to the foraging site, which, due to attraction of previous workers' pheromone, is chosen by almost all foraging ants. Thus at any given moment, a large number of foragers are moving toward the food source and simultaneously a similar number are moving in the opposite direction toward the nest. In one group of ants, the army ants, the potential for collisions and traffic jams is particularly large. Colonies of army ants, *Eciton burchelli*, have more than 200,000 foragers and carry over 3,000 prey items

per hours on paths over 100 m in length [28]. These ants solve the traffic problem in a way that is similar to how we structure our road systems. They form distinct lanes, where the inbound traffic of foragers returning to the nest with prey takes place in a central lane and outbound foragers move in two bands on either side of the central lane. This has the added benefit that the ants carrying valuable prey items are better protected from predators. The separation of flow is formed by individual workers following simple interaction rules and having asymmetric turning rates, where returning workers with prey have a lower turning rate than outgoing ants, as illustrated in Figure 13.7 [28].

However, most ants do not separate flow into specific lanes like the army ants. Instead they rely on other traffic-control rules. For the European black garden ant, *Lasius niger*, one path is used at low densities where collision-avoidance maneuvers of the individual ants end up avoiding the formation of traffic jams. However, as the density of workers in the path builds up, a new separate path is formed before the density gets so high that significant delays from traffic jams can occur [29]. Again, this bifurcation arises as a result of the behavioral rules of the individual worker ants in that at higher densities ants start to push oncoming ants onto another path.

Swarm-intelligence principles inspired by the collective insect societies are used for developing computer algorithms and motion control principles for robotics. The basic idea is that a swarm of individuals can coordinate and behave as a single entity that performs better than the individuals. Using cooperative behavior, the individuals help each other and solve problems that cannot be handled by the single individuals. Such a collection of cooperating robots is referred to as a *swarm-bot* [30–32]. Because the individual robots in the swarm can communicate with simple sensors such as light and sound, they stay together and avoid barriers such as walls and holes [31, 32].

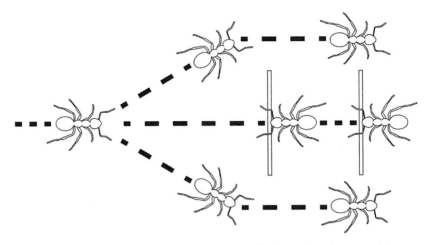

FIGURE 13.7   Basic principle used by army ants to form distinct traffic lanes. Outgoing ants without cargo have the duty to give way to the returning ants carrying prey. Drawing based on the description in Ref. 28.

Nouyan and colleagues [33] performed an experiment with up to 12 small physical robots that were given the foraging task to move a prey to a home position called the *nest*. Movement of the prey required concurrent physical handling from more than one robot. The single robots physically attached to each other in order to pull the prey. The robots only had a small perceptual range and would therefore not be able to find their way back to the nest on their own. They relied on contact with other robots that were within their range. In most of their experiments, where the group size was sufficiently large, the robot swarm group succeeded in retrieving the prey to the nest.

13.6 SELF-SEALING

Both animals and plants have mechanisms that heal wounds rapidly so that only limited amounts of liquid are lost. First the body makes sure to close and seal the wound, and thereafter the healing takes place. Repairing damage in blood vessels involves three steps: (1) forming a plug in the hole (called *primary hemostasis*),

(2) sealing the damage (called *secondary hemostasis*) so that the plug will last until the wound is healed, and (3) healing the wound. In humans (and other animals), specialized cells cover the damage in a blood vessel through a cascade of events, as explained by Purves and colleagues [34]. When the inner membrane of the vessel (*endothelium*) is damaged, the blood comes in contact with collagen fibers in the tissue. This activates small cell fragments in the blood called *platelets* (thrombocytes), which swell become sticky, and then release several clotting factors. The clotting factors activate more platelets, which together form a plug. Furthermore, they initiate the formation of fibrin fibers that form a cloth which seals the vessel and acts as a scaffold for the healing, where scar tissue is built up. Fibrin formation involves a sequence of actions. The clotting factors activate the pro-enzyme prothrombin that circulates in the blood so that it is changed into the enzyme thrombin. Thrombin causes the plasma protein fibrinogen to polymerize fibers of fibrin. The steps in the sealing process are illustrated in Figure 13.8.

The clotting process is very complex, and from a biomimetic point of view, it would be attractive if a similar effect could be achieved with simpler

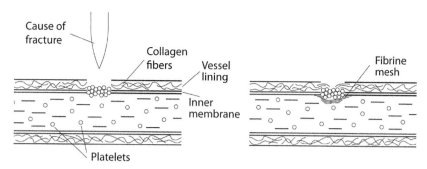

FIGURE 13.8 The plug formation and sealing process in blood coagulation. Left: Platelets in contact with collagen fibers for the plug. Right: A hole is sealed using a clot of fibrin fibers.

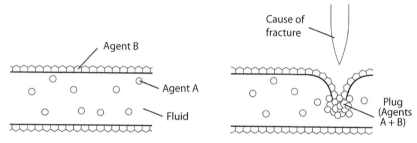

FIGURE 13.9 Simplified principle of the first steps in coagulation. When the two agents A and B are in contact with each other, they form a plug.

means, as shown in Figure 13.9. A central part of the sequence of actions in clotting is when a substance (platelets = agent A) is being exposed to another substance (collagen fibers = agent B) that causes a reaction (the platelets get sticky and plug the hole). A possible application of such a sealing process could be sealing bicycle and car tires. The tire could be made as a double-layer structure. A liquid agent placed between the layers would react and *cure* on exposure to air. Another possible application could be sealing of leaking gas pipes in which an external coating could swell when exposed to the gas.

Researchers at the University of Illinois work with self-healing of polymers and composites that is inspired by blood clotting in animals [35, 36]. Three different approaches are used in this research: (1) capsule based, (2) vascular, and

(3) intrinsic. The basic idea in the capsule-based and the vascular approaches is to include two different and isolated chemical agents in the material; these agents can interact with each other at a microscopic level when damage occurs. In the first approach, the two chemical agents are contained in microscopic capsules or spheres. In the vascular approach, the agents are held in separate hollow channels or capillaries. The intrinsic approach designates materials that have latent self-healing functionality that is activated by the damage or an external stimulus. Examples of basic mechanisms are reversible polymerizations, melting of thermoplastic phases, hydrogen bonding, and ionic interactions. Toohey and colleagues reported making a coated material that repeatedly can repair cracks in a surface coating [37]. A ductile base material with a three-dimensional

network of hollow microchannels is built up using three-dimensional printing. The channels are filled with a liquid healing agent and the material is coated with an epoxy material with embedded solid catalyst particles. When the surface coating cracks under load, the liquid healing agent wicks from the microchannels and cures due to contact with the catalyst. The process can be repeated.

Companies from Germany and Switzerland, in collaboration with the University in Freiburg in Germany, have developed a self-sealing mechanism that can be used in pneumatic structures such as rubber boats and air-filled load-bearing constructions [38, 39]. One example of the latter is a bridge in the Alps that spans 52 m, as shown in Figure 13.10 [40]. The bridge has two long, tubular, low-pressure air beams (air-filled balloons) with walls made of a heavy-duty flexible membrane. The air beams are part of a *tensairity* structure in which the top side of the balloon stabilizes the compression bar against buckling. Wires are swept around each balloon to keep it at a distance from the compression bar and thereby add to the strength and stiffness. The tensairity principle makes it possible to significantly reduce the weight of the load-bearing construction [39, 41]. A concern, however, is the risk of failure due to puncture of the balloons, but that risk can be reduced thanks to the self-sealing mechanism. At the present time the self-sealing mechanism is not applied in the bridge but has been tested on demonstrator constructions.

The self-sealing mechanism is inspired by the twining liana *Aristolochia macrophylla*. During growth, the stem increases its diameter, thereby causing circumferential rings of sclerenchyma fibers to break and split into segments. To prevent the fissures between the segments from reaching the stem surface, the liana uses a rapidly sealing operation as the first element in the self-healing mechanism. The underlying cortex parenchymatous cells have a high internal pressure called *turgor* that presses them into and closes the fissure. This principle is mimicked using soft

FIGURE 13.10 The Lanselevillard bridge in the Alps spanning 52 m. A very low-weight structure is achieved using the tensairity principle where airbeams stabilize compression and tension elements. Such a structure would benefit from a self-sealing mechanism using a principle inspired by the twining liana *Aristolochia macrophylla*. Currently, the self-sealing mechanism has only been tested on demonstrator constructions. Photograph reproduced with kind permisssion from Dr. Rolf Luchsinger, Empa—Center for Synergetic Structures.

closed-cell polymeric foam on the inside of the membrane. Internal stress is introduced into the cells by curing the foam under pressure [42].

Rampf *et al.* [38] explained the mechanics and working principles of the self-sealing membrane. Figure 13.11 illustrates the principle. A base membrane made from PVC and polyester was coated on the inside with polyurethane polymer foam. The polyurethane is made in a chemical reaction between isocyanate and polyol. Mixing the two components results in open-cell foam polyurethane, but applying pressure to the foam while it cures gives a closed-cell structure. Two-bar overpressure resulted in a solid coating almost without air cells and therefore no sealing effect. An overpressure of one bar gave the best repair efficiency. The repair efficiency was calculated as the relation between airflow from a punctured coated membrane and from a reference uncoated membrane. A completely closed fissure would result in an efficiency of 1; no reduction of airflow would result in efficiency 0. Rampf *et al.* achieved a repair efficiency of 0.999.

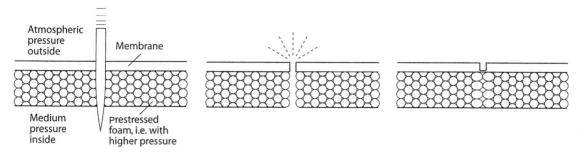

FIGURE 13.11 A self-sealing principle. The higher pressure within the foam cells causes the foam to expand into puncturing fissures. Drawing based on Ref. 39.

To illustrate the practical implication of the repair efficiency, Rampf and colleagues provide the following example. If a medium-sized rubber boat with a conventional membrane is punctured with a 2.5 mm spike, it would take about 26 min for the pressure to drop from 500 to 100 mbar, which is considered the limit for safe use. Using a foam-coated membrane with a repair efficiency of 0.99 instead would increase the time for the pressure drop to about 43 h, i.e., by a factor of 100.

Some animals employ a drastic antipredator strategy that also involves a self-sealing action. They self-amputate parts of themselves (also called *autotomy*) as a last resort. The best-known example comes from lizards that readily let go of their tails when the tails are grabbed by predators. However, most of the animals that display autotomy are invertebrates from almost all taxonomic groups [43]. Autotomy has evolved independently many times because the costs associated with the loss of an appendage are far outweighed by the advantage of staying alive. However, common to all groups is that autotomy occurs along predefined breakage lines, such as between specific leg joints in spiders or at the tail root in lizards [43]. The predefined breaking plane is particularly weak and gives a clean break with minimal external force, which ensures that the amputation is followed by a minimal loss of blood and in many cases also facilitates the ensuing regeneration. Sometimes,

autotomy in lizards does not happen at the joints between two vertebrae but by bone fracture [44]. The core principles are illustrated in Figure 13.12 based on the description in Ref. 44. A fracture along a breaking plane in one of the

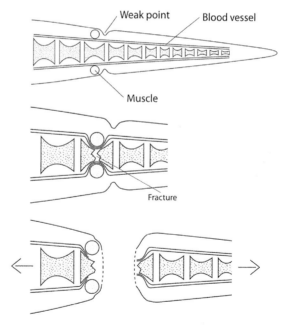

FIGURE 13.12 Simplified sketch of a basic principle in autotomy, i.e., self-amputation, seen in some lizards. When certain muscles contract, one of the vertebrae breaks along a predefined fracture line. When the predator pulls the tail, the tissue is torn along weak planes. The drawing is based on the description of autotomy in Ref. 44.

vertebrae is activated by contraction of tail muscles. An external pull of the tail causes the remaining tissue to break along weak planes.

The principles of autotomy could be used in many places—for example, in fire protection of buildings and in flow control of sewer systems to avoid massive central overflows during heavy rains. Disassembly of industrial products such as cars and washing machines can also benefit from these principles. When the products are disposed at the end of the use period, it is important that the different material fractions can be separated easily. One way to do this is to build in fracture lines at strategic places in the product so that it is easy and fast to remove panels, windows, and other parts.

13.7 SELF-HEALING

Mussels such as *Mytilus californianus* or *Mytilus edulis* can attach to rocks and other surfaces using thin byssus threads, as shown in Figure 13.13. The mussels use 50–100 individual byssus threads to fasten themselves to accessible surfaces of the rocky seashore [45]. The byssus thread has remarkable properties: It can repeatedly extend up to 100% strain and has good antideterioration properties against microbial attack and abrasion from suspended sand particles, thanks to a hard surface [46]. Extensibility can

FIGURE 13.13 Anchor thread byssus from the blue mussel *Mytilus edulis.*

normally not be combined with a hard surface, and the mussel byssus has therefore attracted scientific interest.

The byssus threads are silky filaments composed of a fibrous core and a rough cuticle coating [45, 47]. The core exhibits remarkable toughness and self-healing that is explained by sacrificial bonds [45]. Instead of the normal cross-links seen in collagen structures, the byssus core uses ligand–metal complexes as bonds. These bonds are only half as strong as covalent bonds but are reversibly breakable many times. The collagen core is coated with a 5-μm-thin composite structure consisting of equal amounts of matrix and dispersed granules [45, 47]. The cuticle is about five times harder than the core of the byssus. Within the cuticle, a large number of small granules with a size in the order of 200–800 nm (depending on the mussel species) are evenly distributed in a matrix material. The granules play a central role in preventing crack propagation.

Both the granules and the matrix material consist mainly of the curly shaped mussel foot protein (mfp-1), but the concentration of mfp-1 is higher within the granules than in the matrix material. Part of the mfp-1 protein is a ligand called *dopa* that forms ligand–metal complexes with very small amounts of metals like iron and calcium (metal content less than 1%). These complexes function as bonds between the protein chains. Byssus extension up to 30% is explained by the straightening out of the curly protein chains.

At higher extensions, the bonds within the matrix will break, but in a distributed way to form many small micro cracks in between the granules. The granules prevent the cracks from extending further. A simplified sketch illustrates the principle in Figure 13.14.

The potential for self-healing polymers is significant. Self-healing paint, in which small scratches repair themselves, will have an obvious advantage on cars and on many industrial products and consumer goods. Not only will it

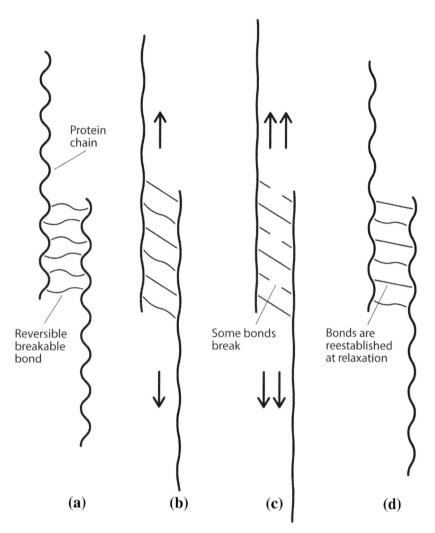

Protein chain

Reversible breakable bond

Some bonds break

Bonds are reestablished at relaxation

(a)   **(b)**   **(c)**   **(d)**

FIGURE 13.14 The principle in part of the self-healing mechanism within the matrix in the core of the byssus thread. First extension is reached by straightening the curly protein chains. Further extension is then achieved by breaking the reversibly breakable metal–ligand bonds. The drawing is based on the description in Ref. 45.

increase the esthetic value over time, but it may also contribute to reduced corrosion and improved hygienic properties, since dirt and microbes will have fewer areas in which to hide. Another potential area of use is in prostheses such as artificial hips and heart valves.

Desired properties include flexibility and strength combined with self-healing. Flexible materials are exposed to dynamic movements that, over time, cause fatigue and cracks that propagate through the material. In living systems these undesired effects are dealt with by regenerating tissue, but this is not an option in artificial materials. The self-healing properties seen in the mussel threads are therefore desirable, in particular for parts that are surgically inserted into the human body, which are generally difficult and problematic to replace.

Holten-Andersen and colleagues are pursuing ways of synthetically creating polymer networks with ligand–metal bonds in order to achieve self-healing properties [48]. They have developed a strategy for introducing bis- and tris-catechol-Fe^{3+} cross-links in polymeric networks and have demonstrated that they can produce a polymer network with high elastic moduli and self-healing properties—much like the mussel thread cuticle.

The formation of catechol-Fe^{3+} complexes requires an alkaline environment, i.e., pH higher than 7, but Fe^{3+} ion solubility is very low under such conditions. This challenge is overcome by prebinding Fe^{3+} in mono-dopa- Fe^{3+} complexes by mfp-1 at pH ≤ 5. Release into seawater (pH 8) causes the material to spontaneously cross-link. The resulting substance has the desired properties, but it is a model gel that requires further research and development before becoming useful for practical purposes.

13.8 ADAPTIVE GROWTH

Earlier in this chapter, we discussed how the behavior of individuals following simple rules gives rise to emergent optimal properties such as collective behavioral decisions on optimal foraging and nest-site selection. Similar outcomes are found if we look at growth in uni- and simple multicellular organisms, where local growth rules can result in optimal overall growth patterns without the necessity for gathering any global information.

Slime molds are eukaryotic (i.e., a cell containing complex structures enclosed in membranes, such as the cell nucleus and the mitochondria) unicellular organisms that were previously thought to belong to the set of fungi but are now classified under *Protista* along with algae. The slime mold begin life as individual amoebas, which then mate and fuse together to form large colonies that usually are up to several centimeters in length but can grow up to a meter.

They feed on microorganisms inhabitating dead plant material and reproduce with spores. Researchers from the Hokaido University in Japan have found inspiration for the simulation of traffic networks in the slime mold *Physarum polycephalum* [49, 50]. The researchers have developed a biologically inspired mathematical model that captures the core mechanisms in the slime mold's network design [51, 52].

Conventional network planning requires centralized control and globally available information. In contrast, the slime mold is self-organized and bases its optimization only on locally available information. The researchers compared the network performance in the Tokyo rail network with a network created by the slime mold. The Tokyo rail network was mimicked by placing oat flakes at positions matching the rail stations and by restricting the growth area using light (Figure 13.15). The slime mold avoids light and will therefore not grow into exposed areas. The remarkable results show that the slime mold has a network performance that is comparable to the one of the railway network.

Network performance can be measured by looking at the total cost (total length of connections), the transport efficiency (average minimum distance between nodes), and resilience, which is the fault tolerance to accidental disconnection. A relative number for each of these three measures can be found by comparing with a minimal *spanning tree*, i.e., the most cost-effective network that uses as few connections as possible to link all nodes. A spanning tree is a graph where all nodes are connected, either directly or through other nodes. The relative cost of the slime mold network is about 175% and the railway network is about 180% of the minimum spanning tree. The transport efficiency is comparable for the two, since the minimum distance between nodes is only about 85% of the minimum spanning tree. The resilience is best for the railway network since only 4% of faults in the network would lead to isolation of any part of the network. The same number for

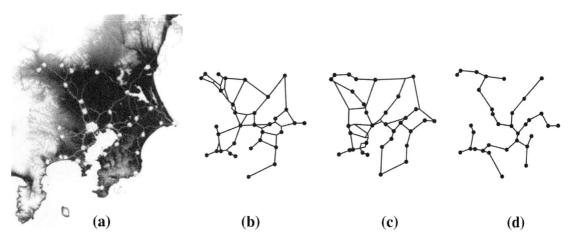

FIGURE 13.15 The Tokyo railway network simulated by the slime mold *P. polycephalum*. (a) and (b) together show how the slime mold has formed food corridors between food sources following a route similar to the existing Tokyo railway system (c). (d) shows the *minimum spanning tree*, i.e., the shortest graph that connects all nodes. Reprinted from Ref. [49] with permission from Prof. T. Nakagaki and the American Association for the Advancement of Sciences.

the slime mold was about 14%. However, the slime mold reached these results without having global knowledge of the area. It only used local rules for either reinforcing preferred routes or removing redundant ones. This behavior in the slime mold can be explained by a change of the tube diameter depending on the flow rates of

the liquid in the tube, as illustrated in Figure 13.16. When a tube during growth reaches a food source, the flow rate of liquid increases. This can be formulated as a functional principle whereby the capacity of a connection is either increased or decreased depending on the amount of traffic.

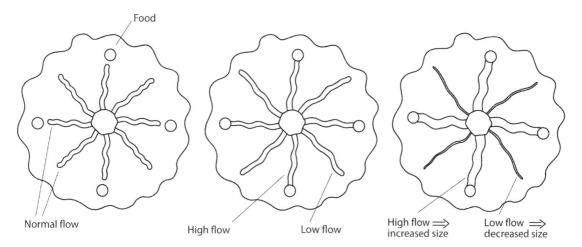

FIGURE 13.16 The basic principle in adapting size of food channels to the location with the highest need. The drawing is based on the description in Ref. 49.

Tero and colleagues developed a mathematical model that uses the same principles [49, 51], i.e., increases or decreases in the capacity of a connection between two nodes depend on the flux. When used to generate a network for the Tokyo station layout, the model performed slightly better than both the actual Tokyo railway network and the slime mold networks. The decentralized planning model can be used for many types of infrastructure networks, ranging from railway systems to power grids, financial systems, information networks, and supply networks.

13.9 MULTILAYER FORMATION

Many molluscs (Mollusca) produce hard shells to cover themselves. The beautiful nacre (see Figure 13.17) on the inside of the bivalve shell is a composite layered structure that is produced layer by layer. Two different types of layers alternate in nacre. The relatively thick layers ($\sim$500 nm) of aragonite crystals (calcium carbonate) are separated by thinner layers ($\sim$30 nm) of highly cross-linked protein [53]. The aragonite layers are furthermore split into small units and separated vertically by protein walls.

Nacre has remarkable mechanical properties. It is extremely tough, considering that the majority of the material content is chalk, which is very brittle. Nacre is 95% aragonite by volume, which essentially is chalk, but the toughness is 1,000 times greater than that of monolithic calcium carbonate [54–56]. Toughness can be measured as the specific work of fracture (R/ρ), which for aragonite is 0.0002 kJ m^{-2} and for nacre is 0.4 kJ m^{-2} [55].

The explanation of this behavior is found in the layered composite structure, also referred to as *brick and mortar*, shown in Figure 13.18. See also Chapter 3 on biomimetic hard materials. A popular explanation of the phenomenon is the prevention of crack propagation. The cracks are stopped when they meet the more ductile proteinaceous layers. A more detailed explanation can be gleaned by looking at the energy conversions, which also reveal that self-healing mechanisms are involved [54]. When a predator such as a crab attacks a mollusc, it is vital that the mollusc shell can absorb the energy from the repeated bite of the crab claw. This is done through the formation of many small cracks in

FIGURE 13.17   Nacre ("mother-of-pearl") on the inside of an abalone shell (genus *Haliotis*).

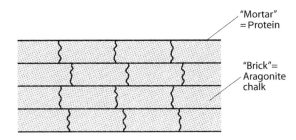

FIGURE 13.18   The layered "brick-and-mortar" structure in the mussel shell with very thin layers of protein between aragonite sections.

the aragonite crystals. In a more homogeneous material the cracks would propagate and cause a failure, but the proteinaceous layers absorb energy by deforming elastically and distribute part of the energy as crack formation in many other aragonite crystals. The result is that not only is a failure prevented, but many small microcracks are created. However, to avoid that repeated attacks from predators break the shell, the microcracks need to heal. This is assumed to happen as self-healing caused by a combination of the very small dimensions of the cracks and the energy being released from the proteinaceous layers.

Addadi et al. [53] describe a model for how the layers are produced and aragonite crystals are formed. In order to make the layered structure, a biological cascade with several steps is followed. The mollusc mantle secretes the highly cross-linked protein layer that is called the *periostracum* from its ectodermic (epithelial) cells. In the space between the mantle and the periostracum, the epithelial cells in the mantle form a gel-like framework matrix of various macromolecules. It is in this matrix that the aragonite mineral forms. The matrix consists of hydrophobic proteins (polysaccharide β-chitin) and hydrophilic proteins that are rich in aspartic acid. Within the matrix, colloidal particles of chemically unstable amorphous calcium carbonate (ACC) may also be present. Possibly, they are isolated from the aqueous environment by vesicle lipid membranes. Definitely at the first periostracum layer, nucleation sites are laid out. Each nucleation site equals an aragonite crystal that has a typical size of $10 \times 10 \times 0.5 \ \mu m^3$. The crystals grow from the nucleation site upward, as shown in Figure 13.19. As the growing crystal reaches the upper periostracum layer, the growth continues laterally until the crystal meets other crystals. Between the crystals, β-chitin is trapped. Maybe new nucleation sites are being laid out on each periostracum layer or the crystallization continues through small holes in the periostracum.

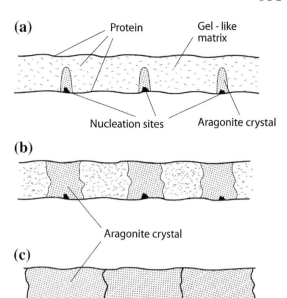

FIGURE 13.19 The formation of the aragonite-protein structures. 500-nm-thick layers of gel are laid out between 30-nm-thin protein layers. Aragonite crystals grow from nucleation sites, first upward (a) and then sideward (b). Protein particles in the gel are "pushed" in front of the crystal and form the vertical protein borders between cells (c). Drawing based on the description in Ref. 53.

The micro- and nanostructure in nacre is interesting for material development in many areas. Much effort has gone into making materials with better strength and toughness properties as well as materials for biomedical applications such as bone and dental analogs. A less explored area is of photonic multilayer structures, where selective reflection of narrow or broadband light through interference could make possible coatings that reflect light like metal without having the thermal or electric conducting properties typical of metals [57, 58]. A process of layer formation and biomineralization similar to nacre production would be attractive if the translucent calcium carbonate could be replaced with a transparent substitute. See Chapter 11 on structural colors.

Munch and colleagues describe a technique called *ice templating* of ceramic alumina and polymethyl methacrylate (PMMA) into a very tough, strong composite material [54, 59]. The hybrid ceramic material has a yield strength of 200 MPa and a fracture toughness of 30 MPa m$^{1/2}$ and is thus comparable to aluminum. The fracture toughness measured as fracture energy is more than 300 times what is known for any of the constituents. The material is produced by first freezing a slurry of alumina in water [59, 60] under controlled conditions. When water freezes, ice crystals grow vertically from the cooling surface, as shown in Figure 13.20. The ceramic particles are repelled by the ice crystals and become trapped in the space between the ice dendrites. In this way, vertical layers of ceramic particles are made. The ice is removed by freeze-drying and the ceramic particles are sintered. Compacting the structure can reduce the size of the void regions, and then infiltrating the ceramic structure with the polymer makes the composite.

Remarkable properties have been achieved, but so far it has not been possible to fabricate structures as small as in nacre. *Bricks* 5–10 μm wide and 20–100 μm long are separated by polymer layers of 1–2 μm thickness, although submicrometer structures are found in large areas. The layer formation is self-organized, apart from the necessary compacting required to achieve the narrow spaces for the polymer.

A remarkable feature of nacre that adds to the desirable combination of strength and toughness is the almost defect-free structure where all space is occupied by either aragonite or protein. In contrast, a high density of defects in common cement is the reason for its relatively poor structure properties. Even the compressive strength of about 40 MPa is much less than what can be achieved with polymers, ceramics, and metal [61]. The reason is that the many trapped-air cavities in cement function as crack initiators. The cavities originate from trapped air and dehydrated areas.

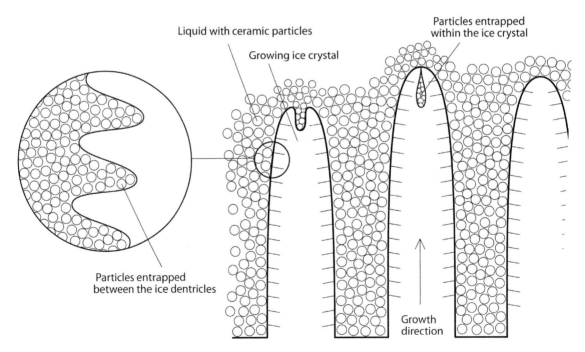

FIGURE 13.20   The principle of freeze casting, based on the description in Ref. 59.

Kendall and colleagues have described how they made Macro Defect-Free cement (MDF-cement) [61]. They mixed the cement with a plasticizer (a water-soluble polymer) that allowed them to use less water and to more easily remove trapped air cavities. The properties were significantly improved, e.g., the compressive strength improved by a factor of between two and seven.

According to Luz and Mano [62], nacre integrates well into bone tissue, and nacre analogs therefore are potentially interesting for orthopedic and dental applications. They have described strategies, including the ice-templating technique, for creating similar brick-and-mortar structures. However, these structures do not have the self-healing property that would be desirable to ensure a long lifetime for the prosthetic device.

13.10 DISCUSSION AND PERSPECTIVES

In contrast to the human way of controlling, producing, and repairing objects in a top-down controlled manner, extensive use of decentralized self-organization and self-healing occurs in nature. It is attractive to adopt the principles and methods from nature for manmade products and the way we control the activities around us. It is particularly recommended for micro- and nanoproduction of delicate structures, which are difficult to produce in a top-down way.

Another significant area is of the computer software systems that control many objects and activities. Controlling procedures are often designed in a top-down hierarchical manner to make sure that things happen the way they are intended to and on time. The benefit of the top-down control principle is the predictability, but the downside is the administrative cost of nonproductive control resources and the vulnerability of the system. In hierarchical control systems, the strength is not better than that of the weakest element. Decentralized self-organized systems,

on the other hand, have the advantages of very low administrative costs and robustness, since there are no centralized command structures that can be disturbed. However, the intended goals of the decentralized system need to be built in as emergent properties within the single units of the self-organized system.

The principles for decentralized control can be applied in a very wide array of activities in our daily life: production and maintenance of products, computer programs, organization of service systems, and organization of social structures. It is therefore very relevant and interesting to study how nature self-organizes and self-heals in order to understand the core principles and mimic them when we're designing manmade artefacts.

Since design processes are carried out in collaboration between experts from different disciplines, the communication of the core biomimetic principles is a vital element in the transfer of knowledge. Many of the self-organizing and self-healing processes are well understood within the biological community, but the knowledge is not accessible in a form that is understandable and useful for design practitioners.

The aims of this chapter have therefore been to collect information about eight phenomena within self-organization and self-healing, to outline existing and possible applications within the manmade world, and to describe the simplified core principles for each of the phenomena. To facilitate description, simple drawings of the principles were included for the phenomena. This way of communication and documentation is appreciated among design practitioners [19].

The eight phenomena were selected based on their remarkable and effective characteristics. Section 13.2 discussed how the coordinated movement of large groups can be inspired from the way birds fly in flocks and fish swim in schools. The basic principles within the two natural phenomena are similar, since the individual animals follow simple and local rules. Basically, they move in the same direction as their neighbors, and they make sure to keep the right distance from one another—not too close and not too far away. Actually, this behavior is not that different from the way human beings behave when we are in a flock, e.g., when traveling by cars on highways. However, when cars become robotic, they need control systems that ensure a smooth and effective traffic flow. The study of the behavior of birds and fish in flocks gives input to a decentralized control mode that is expected to increase the density of automatic vehicles. Another application is realistic computer animations of bird flocks, fish flocks, and grazing animal herds.

Section 13.3 looked at how collective decision making is found in insect societies, where many individuals live together. Ants leave volatile pheromones on their trails, which help them determine the shortest paths to food sources. The same principle can be used to make computer programs that solve logistic problems. Honeybees follow an almost democratic procedure whereby new nests are found based on discussions between scouts that advertise in favor or against a site. This principle is also useful for decision making in computer programs and it is seen in social media on the Internet, where user recommendations determine the sales success of products. Political polls increasingly control public decision making. However, a difference is that scouting bees have clear criteria for the selection of nest sites, in contrast to political polls, where selection criteria are unclear.

Section 13.4 treated the building of large structures such as the termite towers that are not only spectacular because of relative size. Just as impressive is the apparent lack of a master plan and centralized structure. Part of the explanation lies in the concept of stigmergy, whereby the actions of individuals modify the environment and the modification affects the behavior of other individuals. Stigmergy is also seen in the behavior of human beings, for example, in the unwanted concentration of litter at highway pull-up areas. Once someone has left their litter on the side of the road, others are more willing to do the same. Attention to this type of behavior can be used in designing public places such as playgrounds and stadiums.

A positive use of similar principles is seen when temporary findings are put on display in design work. Colleagues notice and follow the work, providing criticism that can be beneficial for the creators, but the critics also pick up good ideas. The graphical display of works in progress is of mutual benefit and certainly shapes the designs made by all the involved parties.

Section 13.5 looked at swarm intelligence that underlies the clever collective behavior seen when a large number of autonomous agents work together. The swarm can achieve better results than is possible for the individual members. Strangely enough, the behavior in bee and ant swarms is controlled solely on a local level. Each member knows how to deal with problems when they occur, without consulting a higher-level control authority. Similar strategies for distributed local control are used to design robotic swarms that behave purposefully without any interference from a central controlling computer. Such robotic swarms can be much more robust than centralized controlled systems and much less vulnerable to communication problems.

Sealing problems was the topic of Section 13.6. A general problem for all types of containers that hold gas or liquids is the risk of leakage. Both animals and plants have mechanisms that quickly close leakages and seal them before the actual healing takes place. The two-component plugging mechanism that is the first step in blood clotting could potentially be used for self-sealing pneumatic structures such as car tires. However, the question is whether the process can be made to react quickly enough to preserve a high enough working pressure. Another sealing mechanism found in plants that uses internal pressure in foam-like structures has proved workable for low-pressure pneumatic structures.

The need to isolate, cut off, and seal the connection to a certain area is a desired function for flooded sewers, fire protection in buildings, stabilization of leaking ships, and minimizing the spreading of contamination or diseases. Inspiration can be found in the way reptiles and invertebrates self-amputate limbs.

Section 13.7 looked at automatic healing phenomena. Most surfaces are exposed to damage from wear and tear. In many cases, damage reduces the functionality of the surface, which then has to be repaired or replaced. If the surface could repair or heal itself, improved functionality and reduced cost could be achieved. Self-healing phenomena in nature therefore attract interest, and one of these is the byssus thread in sea mussels. The self-healing mechanisms rely on at least two factors: (1) a distribution of the impact so that it results in many small cracks, and (2) metal–ligand bonds that are reversibly breakable. Today artificial substances that mimic the byssus-thread behavior have been made at a laboratory scale.

Section 13.8 treated adaptive growth, which also is a desirable property for many structures. Growth normally expands the object while consuming energy and resources, but when a branch is made it will stay there, even when it is no longer needed. Slime molds have the remarkable property that they can reabsorb parts of their bodies when no longer needed and recycle the material in other parts of the body. Apparently, the basic principle used to determine if a part of the body should be strengthened or broken down is fairly simple. It is similar to the principle used by ants to find the shortest path to a food source. The slime mold has a large number of vessels that transport liquid to and from food sources. The vessels that are connected to food sources have a larger flow than the other vessels. The difference in flow level determines the fate of the vessel. The principles from the slime mold are used to make efficient computer programs for planning networks such as railroad systems.

Section 13.9 looked at composite nanostructured materials that are desirable in a broad range of applications. One area is improvement of mechanical properties so that stiffness and hardness are combined with ductility. This could be for instance, to combine the wear resistance found in ceramic materials with elastic properties in metals. However, combining such different materials is not an easy manufacturing task. Therefore, the multilayer structures found in mussels and snails attract attention. Thin layers of gel are spread in mussels to act as scaffolds for a biomineralization process that forms stiff chalk layers and thin elastic layers. The biological process has not yet been mimicked, but other processes such as freeze casting have been used on a laboratory scale to make structures that emulate natural ones.

Apart from improved material properties, another perspective in producing the nanostructured composites is to be able to make the material for biological spare parts. Nacre is known to have good compatibility with human tissue. Another perspective from producing the delicate multilayer structure is to make photonic architectures that selectively reflect certain wavebands through interference. The pearlescent appearance of nacre is actually an example of this, as discussed in Chapter 11.

The aim of this chapter was to show how self-organization and self-healing take place in nature and can be mimicked in manmade applications. The eight approaches show that this is achieved through more localized control of activities. Key principles for each of the approaches have been described in simple graphical illustrations facilitating the further use of new design activities.

Acknowledgments

Thanks are due to Çağrı Mert Bakirci and Silviu Iosif for their contributions to the initial literature research and discussions on self-organizing principles, and to Tomas Benzon for drawing the illustrations.

References

[1] J.K. Parrish, S.V. Viscido, and D. Grunbaum, Self-organized fish schools: an examination of emergent properties, *Biol Bull* **202** (2002), 296–305.

[2] J.M. Pasteels, J.-L. Deneubourg, and S. Goss, Self-organisation mechanisms in ant societies. I. The example of food recruitment, in *From individual to collective behavior in social insects* (J.M. Pasteels and J.-L. Deneubourg, eds.), Birkhäuser, Basel, Switzerland (1987), 177–196.

[3] W. Bialek, A. Cavagna, I. Giardina, T. Morad, E. Silvestrib, M. Vialeb, and A.M. Walczake, Statistical mechanics for natural flocks of birds, *Proc Natl Acad Sci* **109** (2012), 4786–4791.

[4] D.J.T. Sumpter, The principles of collective animal behavior, *Phil Trans R Soc Lond B* **361** (2006), 5–22.

[5] C. Reynolds, Boids—background and update, http://www.red3d.com/cwr/boids/ (accessed 30 April 2012).

[6] C.W. Reynolds, Flocks, herds, and schools: a distributed behavioral model, *Comput Graph* **21** (1987), 25–34.

[7] M. Ballerini, N. Cabibbo, R. Candelier, A. Cavagna, E. Cisbani, I. Giardina, A. Orlandi, G. Parisi, A. Procaccini, M. Viale, and V. Zdravkovic, Empirical investigation of starling flocks: a benchmark study in collective animal behavior, *Anim Behav* **76** (2008), 201–215.

[8] A. Cavagna, A. Cimarelli, I. Giardina, G. Parisi, R. Santagati, F. Stefanini, and M. Viale, Scale-free correlations in starling flocks, *Proc Natl Acad Sci* **107** (2010), 11865–11870.

[9] B. Webb, Can robots make good models of biological behavior? *Behav Brain Sci* **24** (2001), 1033–1050.

[10] S. Garnier, From ants to robots and back: how robotics can contribute to the study of collective animal behavior, in *Bio-inspired self-organizing robotic systems*, (Y. Meng and Y. Jin, eds.), Springer-Verlag, Heidelberg, Germany (2011), 105–120.

[11] B. Webb and T.R. Consi, *Biorobotics—methods and application*, MIT Press, Menlo Park, CA, USA (2001).

[12] http://www.sartre-project.eu (accessed 31 May 2012).

[13] A. Bürkle, F. Segor, and M. Kollmann, Towards autonomous micro UAV swarms, *J Intell Robot Syst* **61** (2011), 339–353.

[14] O. Holland, R. De Nardi, J. Woods, and A. clark, Beyond swarm intelligence: the ultraswarm, *Proceedings—2005 IEEE swarm intelligence symposium*, Pasadena, CA, USA (8–10 June 2005), 225–232.

[15] R. De Nardi and O. Holland, UltraSwarm: a further step towards a flock of miniature helicopters, *Swarm Robotics* **4433** (2007), 116–128.

[16] S. Gilbert, L.A. Lewis, and S.S. Schneider, The role of the vibration signal during nest-site selection by honey bee swarms, *Ethology* **117** (2011), 254–264.

[17] T.D. Seeley, P.K. Visscher, T. Schlegel, P.M. Hogan, N.R. Franks, and J.A.R. Marshall, Stop signals provide cross-inhibition in collective decision-making by honeybee swarms, *Science* **335** (2012), 108–111.

[18] T.D. Seeley, *Honeybee democracy*, Princeton University Press, Princeton, NJ, USA (2010).

[19] C.T. Hansen and T.A. Lenau, Developing engineering design core competences through analysis of industrial products, *Proceedings of the 7th International CDIO conference*, Technical University of Denmark, Copenhagen, Denmark (20–23 June 2011).

[20] M. Worall, Homeostasis in nature: nest building termites and intelligent buildings, *Intell Buildings Int* **3** (2011), 87–95.

[21] J.S. Turner and R.C. Soar, Beyond biomimicry: what termites can tell us about realizing the living building, *Proceedings of the 1st International conference on industrialized, intelligent construction (I3CON)* (T. Hansen and J. Ye, eds.), Loughborough University, UK (14–16 May 2008), 221–237.

[22] C. Kleineindan, R. Ernst, and F. Roces, Wind-induced ventilation of the giant nests of the leaf-cutting ant *Atta vollenweideri*, *Naturwissen* **88** (2001), 301–305.

[23] S. Garnier, J. Gautrais, and G. Theraulaz, The biological principles of swarm intelligence, *Swarm Intell* **1** (2007), 3–31.

[24] N.R. Franks and J.-L. Deneubourg, Self-organizing nest construction in ants: the behavior of individual workers and the properties of the nest's dynamics, *Anim Behav* **54** (1997), 779–796.

[25] C. Tate-Holbrook, R.M. Clark, D. Moore, R.P. Overson, C.A. Penick, and A.A. Smith, Social insects inspire human design, *Biol Lett* **6** (2010), 421–433.

[26] N.J. Mlot, C.A. Tovey, and D.L. Hu, Fire ants self-assemble into waterproof rafts to survive floods, *Proc Natl Acad Sci* **108** (2011), 7669–7673.

[27] T.C. Schneirla, R.Z. Brown, and F.C. Brown, The bivouac or temporary nest as an adaptive factor in certain terrestrial species of army ants, *Ecol Monogr* **24** (1954), 269–296.

[28] I.D. Couzin and N.R. Franks, Self-organized lane formation and optimized traffic flow in army ants, *Proc R Soc Lond B* **270** (2003), 139–146.

[29] A. Dussotour, V. Fourcassie, D. Helbing, and J.-L. Deneubourg, Optimal traffic organization in ants under crowded conditions, *Nature* **428** (2004), 70–73.

[30] D. Floreano, P. Husbands, and S. Nolfi, Evolutionary robotics, in *Handbook of robotics* (B. Siciliano and O. Khatib, eds.), Springer-Verlag, Berlin, Germany (2008), 1423–1451.

[31] V. Trianni, S. Nolfi, and M. Dorigo, Cooperative hole avoidance in a swarm-bot, *Robot Auton Syst* **54** (2006), 97–103.

[32] G. Baldassarre, S. Nolfi, and D. Parisi, Evolving mobile robots able to display collective behaviors, *Artif Life* **9** (2003), 255–267.

[33] S. Nouyan, R. Gross, M. Bonani, F. Mondada, and M. Dorigo, Teamwork in self-organized robot colonies, *IEEE Trans Evol Comput* **13** (2009), 695–711.

[34] W.K. Purves, D. Sadava, G.H. Orians, and H.C. Heller, *Life—the science of biology*, Sinauer, Sunderland, MA, USA (2004).

[35] B.J. Blaiszik, S.L.B. Kramer, S.C. Olugebefola, J.S. Moore, N.R. Sottos, and S.R. White, Self-healing polymers and composites, *Annu Rev Mater Res* **40** (2010), 179–211.

[36] A.R. Hamilton, N.R. Sottos, and S.R. White, Self-healing of internal damage in synthetic vascular materials, *Adv Mater* **22** (2010), 5159–5163.

[37] K.S. Toohey, N.R. Sottos, J.A. Lewis, J.S. Moore, and S.R. White, Self-healing materials with microvascular networks, *Nat Mater* **6** (2007), 581–585.

[38] M. Rampf, O. Speck, T. Speck, and R. Luchsinger, Self-repairing membranes for inflatable structures inspired by a rapid wound sealing process of climbing plants, *J Bionic Eng* **8** (2011), 242–250.

[39] R. Luchsinger, M. Pedretti, and A. Reinhard, Pressure induced stability: from pneumatic structures to tensairity, *J Bionic Eng* **1** (2004), 141–148.

[40] R. Luchsinger and A. Schmid, Bionische selbstreparieerende Membranen für pneumatische Strukturen, *Biomimetic convention*, Biokon, Berlin, Germany (March 2011).

[41] R.H. Luchsinger, A. Pedretti, M. Pedretti, and P. Steingruber, The new structural, in *Progress in structural engineering mechanics and computation* (A. Zigoni, ed.), Taylor & Francis, London, UK (2004).

[42] M. Rampf, O. Speck, T. Speck, and R.H. Luchsinger, Structural and mechanical properties of flexible polyurethane foams cured under pressure, *J Cell Plast* **48** (2012), 49–65.

[43] P. Fleming, D. Muller, and P.W. Bateman, Leave it all behind: a taxonomic perspective of autotomy in invertebrates, *Biol Rev* **82** (2007), 481–510.

[44] A. Bellairs and S.V. Bryant, Autotomy and regeneration in reptiles, in *Biology of the reptilia*, vol. 15B (G.C.F. Billet and P.F.A. Maderson, eds.), Wiley, New York, NY, USA (1985).

[45] M.J. Harrington, A. Masic, N. Holten-Andersen, J.H. Waite, and P. Fratzl, Iron-clad fibers: a metal-based biological strategy for hard flexible coatings, *Science* **328** (2010), 216–220.

[46] N. Holten-Andersen, H. Zhao, and J.H. Waite, Stiff coatings on compliant biofibers: the cuticle of mytilus californianus byssal threads, *Biochemistry* **48** (2009), 2752–2759.

[47] N. Holten-Andersen and J.H. Waite, Mussel-designed protective coatings for compliant substrates, *J Dent Res* **87** (2008), 701–709.

[48] N. Holten-Andersen, Ph-induced metal-ligand crosslinks inspired by mussel yield self-healing polymer networks with near-covalent elastic moduli, *Proc Natl Acad Sci* **108** (2010), 2651–2655.

[49] A. Tero, S. Takagi, T. Saigusa, K. Ito, D.P. Bebber, M.D. Fricker, K. Yumiki, R. Kobayashi, and T. Nakagaki, Rules for biologically inspired adaptive network design, *Science* **327** (2010), 439–442.

[50] T. Nakagaki and R.D. Guy, Intelligent behaviors of amoeboid movement based on complex dynamics of soft matter, *Soft Matter* **4** (2008), 57–67.

[51] A. Tero, K. Yumiki, R. Kobayashi, T. Saigusa, and T. Nakagaki, Flow-network adaptation in *Physarum* Amoebae, *Theor Biosci* **127** (2008), 89–94.

[52] A. Tero, R. Kobayashi, and T. Nakagaki, A mathematical model for adaptive transport network in path finding by true slime mold, *J Theor Biol* **244** (2007), 553–564.

[53] L. Addadi, D. Joester, F. Nudelman, and S. Weiner, Mollusk shell formation: a source of new concepts for understanding biomineralization processes, *Chem Eur J* **12** (2006), 980–987.

[54] D.M. Williamson and W.G. Proud, The conch shell as a model for tougher composites, *Int J Mater Eng Innov* **2** (2011), 149–164.

[55] A.P. Jackson, J.F.V. Vincent, and R.M. Turner, Comparison of nacre with other ceramic composites, *J Mater Sci* **25** (1990), 3173–3178.

[56] A.P. Jackson, J.F.V. Vincent, and R.M. Turner, The mechanical design of nacre, *Proc R Soc Lond B* **234** (1988), 415–440.

[57] T. Lenau, Nature inspired structural colour applications, in *Biomimetic photonics* (Olaf Karthaus, ed.), Taylor & Francis, London, UK (2012).

[58] T. Lenau and M. Barfoed, Colours and metallic sheen in beetle shells- a biomimetic search for material structuring principles causing light interference, *Adv Eng Mater* **10** (2008), 299–314.

[59] E. Munch, M.E. Launey, D.H. Alsem, E. Saiz, A.P. Tomsia, and R.O. Ritchie, Tough bio-inspired hybrid materials, *Science* **322** (2008), 1516–1520.

[60] S. Deville, E. Saiz, R.K. Nalla, and A.P. Tomsia, Freezing as a path to build complex composites, *Science* **311** (2006), 515–518.

[61] K. Kendall, A.J. Howard, J.D. Birchall, P.L. Pratt, B.A. Proctor, and S.A. Jefferis, The relation between porosity, microstructure and strength, and the approach to advanced cement-based materials, *Phil Trans R Soc Lond A* **310** (1983), 139–153.

[62] G.M. Luz and J.F. Mano, Biomimetic design of materials and biomaterials inspired by the structure of nacre, *Phil Trans R Soc Lond A* **367** (2009), 1587–1605.

ABOUT THE AUTHORS

Torben Lenau is associate professor of design methodology, material selection, and biomimetics in the Department of Mechanical Engineering, Technical University of Denmark. His research interests are creative methods in product design with focus on materials, manufacturing, and biomimetics (inspiration from nature). He has conducted several industrial case studies on how to integrate biomimetics in product development and has developed biocards used to communicate design principles found in nature. Furthermore, he studies naturally occurring photonic structures in order to develop new surface coatings based on structural colors.

Thomas Hesselberg is a research associate in the Department of Zoology, University of Oxford, United Kingdom. His research interests include animal behavior, spider web building, comparative biomechanics, and biomimetics. He holds a PhD in biomimetics from the University of Bath and has been involved in several biomimetic projects, including the development of robotic endoscopes based on ragworm locomotion and morphology, and the study of insect flight-control strategies and their application to micro-air vehicles.

CHAPTER

14

Solution-Based Techniques for Biomimetics and Bioreplication

Aditi S. Risbud[a] and Michael H. Bartl[b,c]

[a]Lawrence Berkeley National Laboratory,
MS 67R3110, Berkeley, CA 94720, USA
[b]Department of Chemistry, University of Utah,
Salt Lake City, UT 84112, USA
[c]HRL Laboratories, LLC, Sensors and Materials Laboratory,
Malibu, CA 90265, USA

Prospectus

Nature generates structurally complex architectures with feature sizes covering several length scales under rather simple environmental conditions and with limited resources. Today, researchers understand how many of these structures look and behave, but, in many instances, we still lack nature's ability to marry elegant structures with complex functionality. By unraveling the wonders of nature's design, scientists have developed biomimetic and biotemplated materials with entirely new functions and behaviors. In particular, solution-based methods provide simple, inexpensive routes to generating bioreplicated structures. In this chapter, we survey solution-based bioreplication methods and provide an example for generating three-dimensional photonic crystal structures based on colored weevil scales. This example illustrates how structural engineering in biology can be replicated using sol–gel chemistry and results in an entirely new optical material with fascinating properties.

Keywords

Band gap, Band structure, Beetle, Bioinspiration, Biomaterials, Bioreplication, Biotemplated, Butterfly, Colloids, Electrochemistry, Functional materials, Gecko, Lotus leaf, Nanoparticle, Optical properties, Photonic crystal, Polymer, Replica, Scaffold, Soft chemistry, Sol–gel, Solution-based synthesis, Structural color, Structure engineering, Templating, Three-dimensional structure, Weevil

14.1 INTRODUCTION

Millions of years before researchers engineered bioinspired materials, biological systems were using nanometer-scale architectures to produce remarkable functionalities. Today, scientists are mimicking biological systems to enable a wide range of technologies. Advanced functional materials such as superhydrophobic surfaces have been fabricated based on structures designed by living creatures, such as the pristine lotus leaf.

The lotus leaf has long been viewed as a symbol of purity due to its self-cleaning properties. Dewy rainwater drops roll off a lotus leaf's surface, taking dirt with them. However, this *cleanliness* is actually due to a

http://dx.doi.org/10.1016/B978-0-12-415995-2.00014-3

complex micro- and nanoscale surface architecture that minimizes adhesion and results in the lotus leaf's ability to pick up dirt in water droplets (Figure 14.1) [1].

This self-cleaning behavior, called *superhydrophobicity*, is useful for many modern applications, including stain-resistant paints and roof tiles as well as coatings for fabrics and other surfaces that need to stay dry and repel dirt. Scientists are also studying this effect for lab-on-a-chip applications, in which hydrophobic and hydrophilic materials can be used to control the flow of liquids through microfluidic components [2].

Although the self-cleaning properties of lotus leaves have been documented for millennia and are being implemented in technological applications today, there are many other structures,

especially in the animal kingdom, which we have only recently begin to understand.

Geckos are able to climb on walls or crawl upside down on ceilings without adhesives. This remarkable behavior originates on a gecko's foot. Although smooth in appearance to the naked eye, this reptile's toes are covered in tiny structures that allow it to attach to a wide variety of surfaces (Figure 14.2). Indeed, gecko toes have fine, micrometer-width hairs called *setae* in the lamellae of their toes. Each seta has about 400–1,000 branches ending in a spatula-like structure about 0.2–0.5 μm long [3, 4].

When a gecko crawls up a surface, these tiny spatulae "stick" with weak van der Waals forces to help the gecko's toes temporarily adhere to nanoscale undulations on a given surface. These forces are relatively weak compared to normal

FIGURE 14.1   The surface of a lotus leaf has a complex micro- and nanostructured surface to help minimize adhesion, as shown in (a) the SEM image of a lotus leaf surface and (b) a higher-magnification image with hierarchical structures on the leaf's surface. A drop of water (c) on the surface of the lotus leaf forms nearly a perfect sphere. Adapted from Ref. 1, with permission from SPIE.

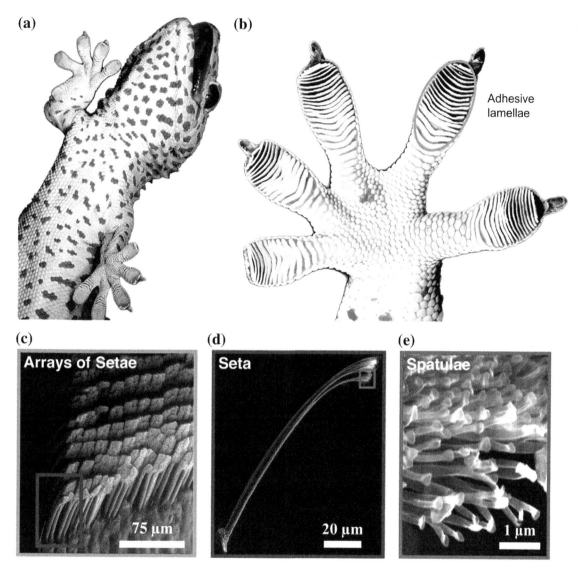

FIGURE 14.2 A hierarchy of structures on a gecko's foot helps it navigate a wide variety of surfaces. A tokay gecko (*G. gecko*) climbs vertical glass (a); the adhesive lamellae on this gecko's foot (b) serve as overlapping pads. Each seta in this array (c) is approximately 5 μm in diameter; a single seta (d) with branched structure terminates in hundreds of spatular tips (e). Adapted from Ref. 3. Copyright (2005) National Academy of Sciences, USA.

bonding forces, but because of the large surface area of contact between the spatulae and a surface, the forces are strong enough to hold up a gecko—upside down or otherwise. Researchers are actively pursuing materials that could mimic this behavior for use in fabrics and adhesive tapes and for the feet of robots and unmanned vehicles [5].

The colorful world of insects, birds, and marine animals demonstrates myriad hues of color and optical effects and provides another example of the elaborate architectures found in nature [6–8]. These colors and effects arise from a delicate interplay of light and periodically organized architectures with feature sizes of a few hundreds of nanometers and are in part the result of *structural* colors [9], discussed in detail in Chaper 11 by Dushkina and Lakhtakia. For example, it is the periodic variation of biopolymeric compounds embedded into wings and exoskeletons that lends many butterflies, birds, beetles, and marine animals their iridescent appearance (Figure 14.3). Mimicking or replicating the structure in these compounds can result in entirely new materials with fascinating properties, as we discuss in detail in Section 14.3.

Nature's ability to generate structurally complex architectures with feature sizes covering several length scales under rather simple environmental conditions and with limited resources is still largely unmatched by our synthetic abilities. However, by unraveling the wonders of nature's design, scientists have developed biomimetic and biotemplated materials with entirely new capabilities.

Biomimetic materials research draws inspiration from nature to address technological issues or, more fundamentally, to reveal knowledge about a biological structure of interest for a specific application [10]. As such, biomimicry relies on expertise from chemists, biologists, physicists, materials scientists, and engineers to unlock mechanisms and design principles in nature.

Nature creates fine structures based on the self-assembly of component materials. For biomimetic materials, self-assembly processes seen in biological systems are leveraged for the fabrication of advanced materials. These bottom-up self-assembly routes use soft chemistry-based techniques to generate hybrid materials. Among these methods, the sol–gel process is a versatile technique to express organic or even biological species in new materials derived from precursors.

Today researchers understand how many of these structures look and behave but, in many cases, still lack nature's synthetic capabilities in

marrying elegant structures with complex functionality. Unlike biomimetic structures, biotemplated materials borrow from nature's blueprints for our own technological needs, quite often resulting in new materials with altogether different chemical composition and function [11–14]. In biotemplating, natural systems are used as scaffolds to combine complex structural characteristics with specific functions. Biotemplating is achieved by duplicating a specific structure or by extracting design principles encoded in natural structures.

Using templates such as organic molecules, supramolecular aggregates, colloids, nanoparticles, and their assemblies, biotemplated materials can be made with technologically significant structures at nanometer length scales. Depending on the template's structural properties and desired dimensionality, this can be accomplished by imprinting, casting, molding, infiltration, coating, and several other techniques. These materials have ordered pores, reactive sites, and other attractive features advantageous for applications in catalysis, drug delivery, photonics, and molecular electronics.

Regardless of the synthetic technique employed for bioreplication, there are a few key requirements: simultaneous replication of large and small feature sizes; preservation of framework geometry and lattice parameters; and avoidance of crack formation, structural damage, and loss of long-range features. In general, inorganic materials with hierarchical structures based on the biotemplating of plants or animals are promising for lightweight structural materials, filters, and catalysts, or photonic devices.

14.2 BIOREPLICATION TECHNIQUES AND PROCESSES

14.2.1 Some Definitions

In bioreplication, terms such as *hollow and solid*, *inverse*, *true*, *negative* and *positive replicas* are typically chosen to describe the structural

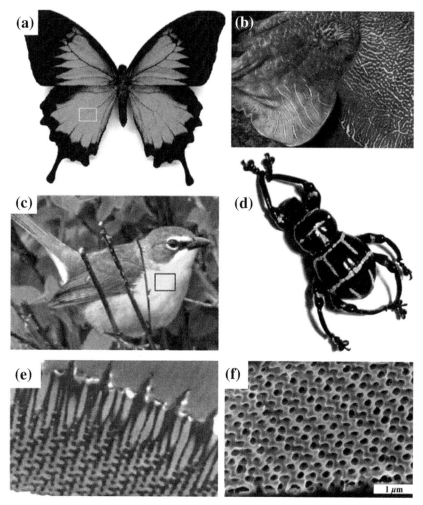

FIGURE 14.3 Examples of structural colors in biology. Photographs of (a) the blue wings of the butterfly *Papilio ulysses*, (b) the reflective skin of the cuttlefish *Sepia apama*, (c) the yellow-breasted chat *Icteria virens*, and (d) the red-striped exoskeleton pattern covering the beetle *Pachyrhynchus moniliferus*. Electron microscopy images of the periodic biopolymeric nanostructures producing structural colors in butterfly wings and beetle exoskeleton scales are shown in (e) and (f), respectively. (a) and (c) adapted from Ref. 8 and reproduced by permission of The Royal Society; (b) adapted from Ref. 57 and reproduced by permission of The Royal Society; (d) adapted from Ref. 26 and reproduced by permission of The Royal Society of Chemistry; (e) adapted from Ref. 58 and reproduced by permission of The Royal Society. (For interpretation of the references to color in this figure legend, the reader is referred to the web version of this book.)

features of replicated samples. Unfortunately, these classifications are often used very loosely, interchangeably, and sometimes incorrectly. Since this can lead to ambiguities and misinterpretation, in the following section we provide clear definitions of these terms and use them accordingly in the remainder of this chapter.

(Bio)templating. We use this term for any process translating the geometrical and surface features of a given structure (a template) into a

different material. In *solution-based* (bio)templating (the focus of this chapter), *liquid* precursors are used to coat the surface of a template or infiltrate the void space surrounding a template structure (Figure 14.4). After solidification of the precursor to form a robust material, the surrounding template is selectively removed, leaving behind a stable, self-supporting structure containing all architectural features of the original template. The final structure is called the *(bio)replica*.

Hollow vs. solid replicas. A hollow replica is a shell-like copy of the original structure and is fabricated by depositing a thin coating around the biotemplate and subsequently removing this biotemplate (see Figure 14.4b). A solid replica is a positive or negative copy of the original template, composed of a completely filled framework. Solid replicas are fabricated by completely backfilling the free space around a template with a new compound and removing the original structure (see Figures 14.4c and 14.4d).

Negative vs. positive replicas. A negative replica is an inverse copy of the original biotemplate; i.e., the original structure is inscribed into the new compound as a framework made of air (see Figure 14.4c). A positive replica is an exact (or true) copy of the original biotemplate (see Figure 14.4d). Apart from being made from different materials, the replica and original frameworks are identical. A positive replica can be fabricated simply by repeatedly forming a negative replica. A negative of the negative replica—the positive replica—is created by making a negative replica of the original template and using this structure as the new template.

14.2.2 General Infiltration and Templating Methods

An important route to new materials with unprecedented properties involves converting unique biological structures into positive and negative copies, or replicas [10–14]. The general strategy for achieving such new materials is to use the biological structure as a template and transfer its

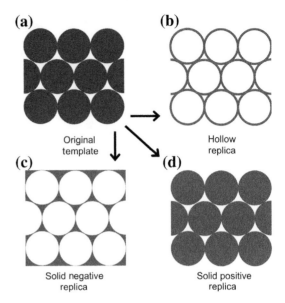

FIGURE 14.4 A given template structure (a) can yield three different types of replicas (b–d) through various infiltration and template-removal routes.

architectural features into a desirable material. Depending on its structural properties, this can be accomplished by imprinting, casting, molding, infiltration, coating, and several other techniques.

Molding and imprinting are fast and powerful ways to transfer two-dimensional surface features into polymeric and ceramic replicas [15–18]. For example, using this approach, it is possible to replicate the intricate surface structures of lotus leaves and other plants [15, 16], insect eyes [17, 18], and cicada wings [19]. Importantly, the replicated samples not only retain the surface structures of these templates, but they display desirable properties such as superhydrophobicity and antireflectivity.

Transferring more complex three-dimensional frameworks—such as those found in butterfly wings, bird feathers, and wood samples—requires infiltration and coating techniques [10–12]. For the latter, the empty space inside a given biological structure is infiltrated with a precursor species, and chemical and/or physical processes are used to convert this precursor into a solid compound surrounding the biological template. In the final step, the biological template is removed by etching or pyrolysis, leaving behind a material with the same structural features as the biological template but composed of an entirely different material.

In general, infiltration-and-coating-based biotemplating processes can be divided into low-temperature deposition/evaporation methods and solution-based methods [11]. Among the former, atomic layer deposition [20–22] and the physical vapor deposition method called *conformal evaporated film by rotation* [23–25] have proven particularly suitable for bioreplication. The common feature of these methods is the use of vapors or gaseous precursors in the infiltration step. These precursors then transform into a solid coating on the biotemplate surface by a step-wise *atom-by-atom* growth mechanism. Deposition/evaporation-based infiltration methods are thus mainly used to produce shell-like replica structures (after removal of the biotemplate). Atomic layer deposition is discussed in detail in Chapter 16 by Zhang and Knez; evaporation methods are presented in Chapter 15 by Martín–Palma and Lakhtakia.

In contrast, solution-based infiltration methods use liquid precursors and generally form solid negative copies of the original biological template. In these methods, liquid precursor solutions are infiltrated into the template void space by capillary forces. After solvent evaporation, the precursor species transform into a solid network through various chemical pathways, depending on the type of precursor and reaction conditions. The most widely applied solution-based bioreplication method is sol–gel chemistry using molecular [26–30] and colloidal nanoparticle [10, 31] precursors.

Apart from producing bioreplicas with different structural features (shell-like vs. solid frameworks), solution and deposition/evaporation-based infiltration methods also differ greatly in terms of instrumentation, processing conditions, and accessibility. For example, sol–gel chemistry methods require only a few inexpensive precursors and solvents, a chemical lab bench, and a furnace. Atomic layer deposition or conformal evaporated film-by-rotation methods, in contrast, require expensive instrumentation, gas, and vacuum lines and highly trained users. On the other hand, the precision and reproducibility of bioreplica samples obtained by evaporation/deposition-based methods are exceptional and represent a big advantage over most solution-based techniques.

For example, using atomic layer deposition, the degree of infiltration and thus the thickness of the inorganic replica shell can be controlled on an atomic-layer scale. This predictable control of shell thickness can be used to create replicas with precisely tuned optical properties. This outcome is depicted in Figure 14.5 for the photonic structure of butterfly-wing scales replicated into aluminum oxide. By carefully controlling the aluminum oxide shell thickness, for example, Wang and co-workers tuned the reflection color of the replica samples from green to yellow, pink, and purple (note that the original color of the butterfly-wing was blue) [20].

Regardless of the method used, there are a few key requirements that must be fulfilled for a system to be suitable for bioreplication: (1) simultaneous replication of large and small feature sizes; (2) preservation of framework geometry and lattice parameters; and (3) avoidance of crack formation, structural damage, and loss of long-range features.

The first requirement—simultaneous replication of features both large and small—is particularly important in bioreplication since an attractive aspect of many biological structures is their hierarchical structure, with feature sizes often spanning several orders of magnitude. In particular, in solution-based methods, the preservation of fine features is often challenging, since surface tension during solvent evaporation can easily *smooth out* nanoscale structures. Additionally, fine features can also be lost during

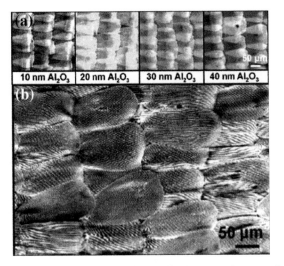

FIGURE 14.5 (a) Series of optical microscope images of wing scales of the butterfly *Morpho peleides* coated with alumina layers of increasing thickness and associated gradual color changes. (b) SEM image of alumina replicas of the wing scales. Reproduced with permission from Ref. 20. Copyright 2006 American Chemical Society.

high-temperature removal of the biotemplate (pyrolysis) or with strong etchants.

The second requirement, preservation of framework geometry and lattice parameters, is of great importance for replicating biological structures with long-range periodically ordered frameworks such as those in structural colors (Figure 14.3). As we discuss in detail in Section 14.3.2, in these materials, optical properties are a direct result of the framework geometry and the lattice parameters (the periodicity length) [9]. In fact, even small structural differences of a few percent can result in significant changes of coloration and other optical properties.

To prevent uncontrolled changes of these properties in this type of biological structure, it is imperative to keep structural deviations such as framework shrinkage or swelling to a minimum. In addition, the properties of structural colors are very sensitive to defects and loss of long-range order. Formation of cracks and

local damage of the framework during infiltration or template removal can thus greatly diminish the optical quality of the replicated structure, and, for example, can result in loss of intensity and purity of the reflected color.

To counteract these various forms of structural damage and loss of delicate features in the replicated samples, careful fine-tuning of the bioreplication process is necessary. Given the wide range of synthesis and processing parameters of the various molding, infiltration, and casting methods, unfortunately there is no one-size-fits-all solution to this process. As such, replication conditions have to be adjusted for each method individually.

To further complicate matters, biotemplates originate from a wide variety of organisms and thus have vastly different structural and mechanical properties (thermal stability, solvent resistivity, mechanical strength, etc.). Compare, for instance, the delicate wings of a butterfly with the robustness of wood. In short, no two biotemplates are exactly the same, and successful replication from a biotemplate requires fine-tuning of processing parameters, from sample preparation to templating conditions and template removal.

14.2.3 Solution-Based Bioreplication Routes

A common feature of solution-based methods is their simplicity. What they generally lack in atomic-scale precision of the replicated structure compared to evaporation and deposition methods, they compensate for by being fast, inexpensive, and broadly applicable [10–12]. Whereas atomic-scale precision and reproducibility might be of importance for electronic and semiconductor devices, for most applications in which bioreplicated surface structures and three-dimensional frameworks are of interest, the precision and reproducibility attained by solution-based bioreplication is sufficient. Such applications include the areas of optics and optoelectronics, catalysis, separation, and sorption, or as scaffolds in battery electrodes, energy absorption, and tissue engineering. In addition, solution-based methods extend the variety of accessible structures from hollow to solid replicas of the original template with a negative or positive (true replica) framework.

The process of solution-based templating has been known for thousands of years and has been applied to replicate a multitude of synthetic and natural structures with features ranging from nanometers to several meters. For example, solution-based templating can be used to replicate the nanometer-sized three-dimensional pore network of colloidal crystals into polymers and semiconductors. At much longer length scales, these same templating principles have been used for millennia to cast meter-sized statues, figures, and bells from metals and precious alloys.

These very same techniques are used in solution-based bioreplication but instead rely on biological structures such as wood, bones, insect scales, feathers, and marine animals as templates. The motivation for using biological templates lies in the enormous multitude of complex structures found in nature [6, 32–34]. Some of these structures, such as the intricate hierarchical architectures of diatoms and marine sponges [32] or the three-dimensional photonic crystal lattices of certain butterfly wings and weevil scales [6, 26], are still beyond our synthetic capabilities. Therefore, these complex materials are intriguing additions to our existing synthetic structure portfolio.

A drawback of using biotemplates is that they are not (in most cases) freestanding, unlike synthetically generated frameworks that are easily accessible. Biological structures designed to fulfill given functions within an organism are integrated into larger systems (feathers, wings, hair, skin, bones, exoskeleton, etc.). In these cases, the structural features of interest for replication are often buried and hidden, or embedded into a structure-less and sometimes impermeable cover matrix.

Such biotemplate-specific features require additional pre-infiltration steps, such as removing the component of interest from the organism by cracking, polishing, or cutting to provide access to enclosed structural frameworks. In addition, many biological structures have thin protective layers on their surface. Very often, these layers are composed of waxy, hydrophobic molecules to prevent water from wetting the surface and/or entering the inner void space of these structures. Since solution-based biotemplating often uses hydrophilic solvents (e.g., water or alcohols) and infiltration of the void space is based on capillary forces, these protective layers need to be removed through treatment with organic solvents or acids.

Once these preprocessing steps have been completed and the desired biological structure can be accessed, a given biotemplate is ready for infiltration with a precursor-containing solution. In principle, any of the techniques used to replicate synthetic templates can also be used for biological structures. These methods include colloidal nanoparticle sols, molten or supersaturated salt casting, electrochemistry, polymerization techniques of organic monomer species, and molecular sol–gel chemistry. The following section provides brief descriptions of these methods with a more detailed discussion of molecular sol–gel chemistry—the most widely used solution-based bioreplication technique.

Typical colloidal nanoparticle sols are aqueous or alcoholic solutions/suspensions of oxide nanoparticles with sizes of a few nanometers to tens of nanometers. An important parameter for this templating technique is concentration of the nanoparticle sol. On one hand, this concentration should be as high as possible to ensure a dense, solid framework after solvent evaporation: The lower the nanoparticle concentration, the higher the nanoporosity. On the other hand, high nanoparticle concentration results in enhanced particle-particle interactions, leading to an increase in the solution viscosity. Since capillary forces drive template

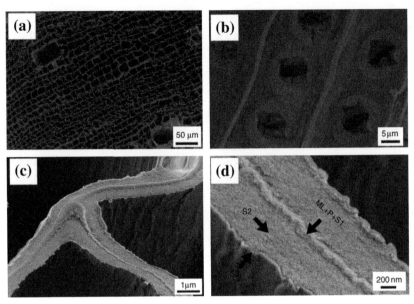

FIGURE 14.6 SEM images (a–d) of cerium/zirconium oxide replicas of spruce wood fabricated by nanoparticle sol templating. Interfaces between different wood cell-wall layers are clearly visible in the ceramic replica samples. Adapted from Ref. 33. Copyright Wiley-VCH Verlag GmbH & Co. KGaA. Reproduced with permission.

infiltration, an increased solution viscosity can significantly lower the degree of infiltration, which, in turn, results in poor replica quality. The optimum nanoparticle concentration depends on the particle type and surface chemistry, but typical values are in the micro-to-millimolar range [10, 31].

After infiltration and evaporation of the solvent, nanoparticles form an initially loosely connected framework *via* interaction of surface hydroxyl groups. This network can be further densified and solidified by thermal treatment, resulting in a stable nanoparticle-based structure held together by interparticle oxide bonds. For example, Figure 14.6 shows SEM images of ceramic replicas of wood created by infiltration of wood tissue with a sol containing cerium/zirconium oxide nanoparticle sol and calcination (i.e. heating in air) [31].

A different approach to creating inorganic replicas is to infiltrate template structures with molten or supersaturated salt solutions. For example, heated supersaturated solutions of salts, such as sodium chloride in water, are used to infiltrate polymeric templates. After solvent evaporation, solute precipitation, and thermal removal of the template, an inverse replica of the original structure composed of rock salt is obtained [35]. In a similar method, sucrose was infiltrated into diatom structures. The sucrose was then carbonized, and after dissolution of the diatom, a carbonaceous inverse replica was obtained [36].

Electrochemical deposition is an attractive method to replicate open-framework template structures into metals. Metal infiltration is achieved with a traditional electrochemical cell set-up in which the template-electrode is immersed into an electrolyte solution containing the metal salt of interest, together with a counter electrode and reference electrode. Although this technique is difficult to apply to biotemplates, because it requires the template to be deposited onto or formed on a conductive electrode, it has been successfully applied to opal templates [37, 38]. An advantage of this method is that metal deposition starts at the conducting surface and, as deposition time continues, gradually and unilaterally fills the template void space. By controlling deposition time and the current applied during the deposition process, it is thus possible to precisely control the degree of metal infiltration into the template.

An alternative to creating inorganic replicas is template infiltration with organic monomers or prepolymer solutions. In the presence of polymerization initiator compounds, the precursor species is then cross-linked by exposure to ultraviolet (UV) light or mild heating, yielding a negative polymeric replica of the original structure. Typical polymers used with this technique are elastomeric perfluoropolyether (PFPE), polydimethylsiloxane (PDMS), polymethyl methacrylate (PMMA), and epoxy resins. This technique has been extensively applied in replica molding, a soft-imprint lithographic method used to replicate two-dimensional structures such as, the surface relief of leaves, wings, and eye [15–17].

An extension of polymer templating to three-dimensional structures has been successfully demonstrated by replicating opaline structures. For this method, opal templates were infiltrated with UV-curable polyurethane or poly(acrylate—methacrylate) copolymer precursors [39]. After polymerization, the template was selectively removed by either etching (hydrofluoric acid for silica templates) or solvent extraction (toluene for polystyrene templates). Since mild processing temperatures and no or minimal amounts of solvent were required, replicas of high quality (with little framework shrinkage and damage) could be obtained. Furthermore, this method opens the door to fabricating bioreplica samples out of functional yet inexpensive organic compounds.

14.2.4 Sol–Gel Chemistry

The term *sol–gel chemistry* encompasses all processes involved in the transformation of precursor species into a solid material via a "sol" (colloidal particle suspension) to "gel" (cross-linked particles) pathway [40, 41]. Most sol–gel reactions start with molecular precursors, which first undergo a hydrolysis step followed by condensation of these molecular species into polymeric networks.

The most commonly used molecular precursors are metal alkoxides (M-OR; M is a metal atom and R is an organic group such as methyl, ethyl, isopropyl, etc.). These compounds readily react with water in a nucleophilic substitution reaction to form metal hydroxides as intermediary species in the first step of the sol–gel process; thus:

$$M - (OR)_x + x \cdot H_2O \rightarrow M - (OH)_x + x \cdot ROH. \quad (14.1)$$

The index x denotes the available alkoxide moieties for a given metal atom; for typical sol–gel compounds such as silica, alumina, titania, and zirconia, x has the following values: $x = 3$ (aluminum), $x = 4$ (silicon, titanium, zirconium). For example, for the most commonly used sol–gel compound, silica (or silicon dioxide), and for a typical starting compound such as silicon tetraethoxide (also referred to as tetraethylorthosilicate, TEOS), reaction (14.1) becomes

$$Si(OCH_2CH_3)_4 + 4H_2O \rightarrow Si(OH)_4 + 4CH_2CH_3OH, \quad (14.2)$$

with orthosilicic acid and ethanol formed as intermediary products.

These formed metal hydroxides then undergo a condensation reaction and combine either with other hydroxides moieties (oxolation, reaction 14.3) or with unhydrolized alkoxides moieties (alcoxolation, reaction 14.4). In both cases, dimeric species are formed with two silicon atoms connected by an oxo-bridge.

$$M - (OH)_x + M - (OH)_x \rightarrow (HO)_{x-1}$$
$$-M - O - M - (OH)_{x-1} + H_2O, \quad (14.3)$$

$$M - (OH)_x + M - (OR)_x \rightarrow (HO)_{x-1}$$
$$-M - O - M - (OR)_{x-1} + ROH. \quad (14.4)$$

BOX 14.1

HYDROLYSIS, CONDENSATION, AND NET REACTION FOR THE MOST COMMON SOL–GEL OXIDE COMPOUNDS.

SILICA

HYDROLYSIS:	$Si(OR)_4 + 4\,H_2O \rightarrow Si(OH)_4 + 4\,ROH$
CONDENSATION:	$Si(OH)_4 \rightarrow SiO_2 + 2\,H_2O$
NET REACTION:	$n \cdot Si(OR)_4 + 2n \cdot H_2O \rightarrow (SiO_2)_n + 4n \cdot ROH$

TITANIA

HYDROLYSIS:	$Ti(OR)_4 + 4\,H_2O \rightarrow Ti(OH)_4 + 4\,ROH$
CONDENSATION:	$Ti(OH)_4 \rightarrow TiO_2 + 2\,H_2O$
NET REACTION:	$n \cdot Ti(OR)_4 + 2n \cdot H_2O \rightarrow (TiO_2)_n + 4n \cdot ROH$

ALUMINA

HYDROLYSIS:	$Al(OR)_3 + 3\,H_2O \rightarrow Al(OH)_3 + 3\,ROH$
CONDENSATION:	$2\,Al(OH)_3 \rightarrow Al_2O_3 + 3\,H_2O$
NET REACTION:	$2n \cdot Al(OR)_3 + 3n \cdot H_2O \rightarrow (Al_2O_3)_n + 6n \cdot ROH$

ZIRCONIA

HYDROLYSIS:	$Zr(OR)_4 + 4\,H_2O \rightarrow Zr(OH)_4 + 4\,ROH$
CONDENSATION:	$Zr(OH)_4 \rightarrow ZrO_2 + 2\,H_2O$
NET REACTION:	$n \cdot Zr(OR)_4 + 2n \cdot H_2O \rightarrow (ZrO_2)_n + 4n \cdot ROH$

R = alkyl rest, e.g. CH_3 (methyl), C_2H_5 (ethyl), C_3H_7 (n-propyl and isopropyl), C_4H_9 (butyl)

These reaction pathways are identical for different sol–gel compounds as shown in Box 14.1. The rates of both hydrolysis and condensation reactions, however, depend strongly on the type of metal alkoxides used, pH of the precursor solution, temperature, presence of solvents (mainly alcohols), and reactant concentrations. Although most titanium-alkoxide precursors react readily with water (titanium tetraethoxide, for example, rapidly hydrolyzes in a strongly exothermic reaction at room temperature), the reaction of silicon alkoxides is much slower (several minutes to hours) and requires the presence of acid or base catalysts to proceed at room temperature.

Initially formed dimeric species (products in reactions 14.3 and 14.4), then undergo repeated oxolation and/or alcoxolation reactions, creating nanometer-sized colloidal particles composed of randomly cross-linked $(-O-)_{x-1}-M-O-M-(O-)_{x-1}$ species and unreacted hydroxyl residues (Figure 14.7a). During the subsequent gelation (or aging) process, these nanoparticles can form larger agglomerates by reaction of their residual hydroxyl groups (Figure 14.7b). In subsequent condensation reactions, these agglomerates then transform into a fully interconnected, three-dimensional network (Figure 14.7c). Since these agglomeration and polymerization processes occur in random fashion, the formed network generally lacks long-range order and has glass-like properties.

Control of the kinetics of gelation of the initial sol particles is of foremost importance for

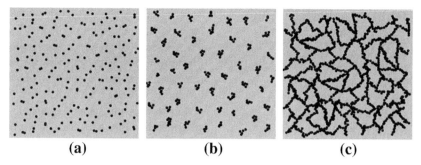

FIGURE 14.7 The three main stages of molecular sol–gel chemistry. (a) Solution stage: Partly hydrolyzed and condensed molecular precursor species forming dimers and trimers (black dots). (b) Sol stage: Dimers and trimers agglomerate into larger oligomers/nanoparticles. (c) Gel stage: Oligomers/nanoparticles further cross-link into an interconnected solid network.

processing sol–gel compounds into desired structures. To successfully transform the sol into thin films, fibers, or monoliths or to infiltrate it into template structures to create high-fidelity ceramic replicas, gelation and formation of the interconnected particle network need to be slowed. This prevents premature solidification and/or uncontrolled precipitation of the sol–gel compound. In general, gelation kinetics depends on similar parameters, as previously discussed for the hydrolysis and condensation reactions. Separating these processes is very difficult.

For templating applications, the most successful approach is to dilute the sol either during or immediately after the initial hydrolysis and condensation steps with low-boiling-point alcohols (methanol, ethanol, isopropanol). Lowering the colloidal particle concentration slows cross-linking and allows template infiltration to be separated from the gelation and solidification processes. After infiltration of the template structure, alcohol evaporates and thereby slowly concentrates colloidal particles in the sol, inducing their agglomeration and cross-linking into an extended solid network, as shown in Figure 14.7.

A disadvantage of this method is that large amounts of solvent are introduced along with the sol–gel compound into the void space of the template structure, resulting in a highly porous and crack-prone network after solvent evaporation. This interplay between gelation kinetics and sol dilution depends strongly on the sol–gel compound. Finding the appropriate balance between these competing considerations is one of the most important steps for successful replication of synthetic and natural structures.

Finally, it should be noted that for some sol–gel compounds such as titania, the hydrolysis of precursors (as well as condensation and gelation reactions) are so fast that dilution alone is not sufficient to prevent premature cross-linking and precipitation. In these systems, it is important to convert precursor molecules into intermediary species that are stable under processing conditions (coating, molding, casting, etc.) and need other stimuli (such as heat and catalysts) to undergo polymerization reactions. In the case of titania, this can be achieved by converting titanium alkoxides into stable oxy-chloro complexes under strong acidic conditions or into organic complexes by reaction with trifluoroacetic acid [42]. These entities are stable under room-temperature conditions and can be processed in an alcoholic solution. At elevated temperatures, these complexes then decompose and convert into amorphous or polycrystalline titania.

14.3 BIOREPLICA PHOTONIC CRYSTALS

In this section, we provide a specific example of solution-based bioreplication: Using silica and titania sol–gel chemistry, the three-dimensional photonic crystal structure of colored weevil scales is replicated [26–28]. This combination of structure engineering in biology—still beyond our synthetic engineering abilities—with sol–gel synthesis can result in entirely new optical materials with fascinating properties and new opportunities in energy and information technology applications [43].

14.3.1 Photonic Crystals

Upon closely examining the green color of spinach leaves or the red color of strawberries, you will find that the origin of this coloration is based on molecules. When illuminated with white light, electronic transitions in these molecules, also called *pigments*, absorb selected frequency ranges of the visible part of the electromagnetic spectrum while diffusely reflecting the rest. An excellent example of such a pigmented color is the molecule chlorophyll. It absorbs large portions of the blue and red spectral regimes, rendering the flora around us in a familiar green.

Now look closely at the green weevil *Lamprocyphus augustus* or the colorful wings of many butterflies and you will discover that the origin of this coloration stems from interesting biopolymeric structures with a high degree of periodic ordering in one, two, or three dimensions (Figures 14.3 and 14.8) [7, 26]. The mechanism behind these *structural colors* is very different. Unlike pigments, structural colors are based on diffraction and specular reflection rather than absorption and diffuse reflection [9]. Most of the light incident upon these periodically organized biopolymeric structures can travel through without being significantly affected. However,

some portions of white light—those with wavelengths having certain proportionalities to the length scale of the structure's periodicity—are directionally reflected, producing the angle-dependent sparkling colors seen in many insects, birds, and marine animals.

Let us consider the one-dimensional case of a structural color: a dielectric composite with a periodic variation of the refractive index in a single dimension, as shown in Figure 14.9 [44]. An incident wave (in the forward direction) will be reflected at each low-to-high refractive index interface. In a simplified picture, for waves with wavelengths half the periodic length of the one-dimensional lattice (Figure 14.9a), the reflecting waves are all in phase and will reinforce each other (Figure 14.9b). When they're combined with the forward-traveling wave, the result is a standing wave: i.e. waves with this wavelength do not propagate through the composite and are instead reflected (Figure 14.9c). This composite is said to have a *photonic band gap* with its width dependent on the refractive index contrast of the two dielectrics and their filling fractions.

On the other hand, for waves with wavelengths outside the band gap (Figure 14.9d), the reflected waves are out of phase and cancel each other (Figure 14.9e), resulting in a net forward traveling wave (albeit with slightly lowered intensity), as shown in Figure 14.9f. Note, however, that this effect is only valid for a given frequency range of incident light normal to the periodic structure. For off-normal directions, the periodic length changes and waves of a different range, or color, are reflected.

Such one-dimensional periodic structures have long been used as optical components such as mirrors, filters, and optical cavities, but their two and three-dimensional analogs promise to open the door to entirely new optical concepts based on photonic band structures [45]. This idea of a *photonic crystal* was independently proposed by Yablonovitch [46] and John [47] in 1987 with the goal to control radiative properties

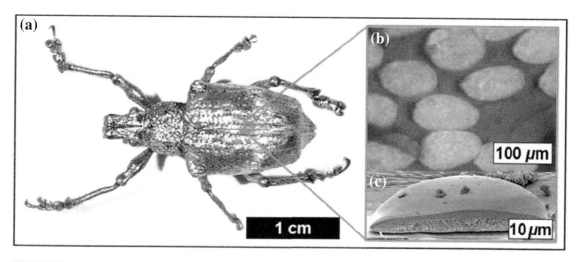

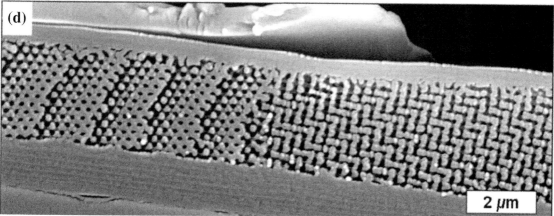

FIGURE 14.8 (a) Optical image of the Brazilian weevil *Lamprocyphus augustus* and (b) some of its exoskeleton under white-light illumination. (c, d) Cross-sectional SEM images of an entire scale and a section of the scale showing the photonic crystal structure. Reproduced with permission from Ref. 54. Copyright 2008, American Physical Society.

of materials and induce photon localization, respectively. These ideas are based on creating materials with an omnidirectional (or complete) photonic band gap—a range of frequencies for which light propagation within the photonic crystal is classically forbidden in any direction. In other words, this is a material for which the optical density of states is zero across a given frequency range while being non zero just above and below this range.

The prospects of these exciting predictions have motivated research into fabricating three-dimensional photonic crystals with band gaps at optical frequencies. The general requirements are to create a three-dimensionally periodic structure composed of dielectric compounds with different refractive indices; the larger the difference, the wider the photonic band gap. Furthermore, the periodicities of this lattice should be nearly the same in all directions (i.e. the Brillouin zones

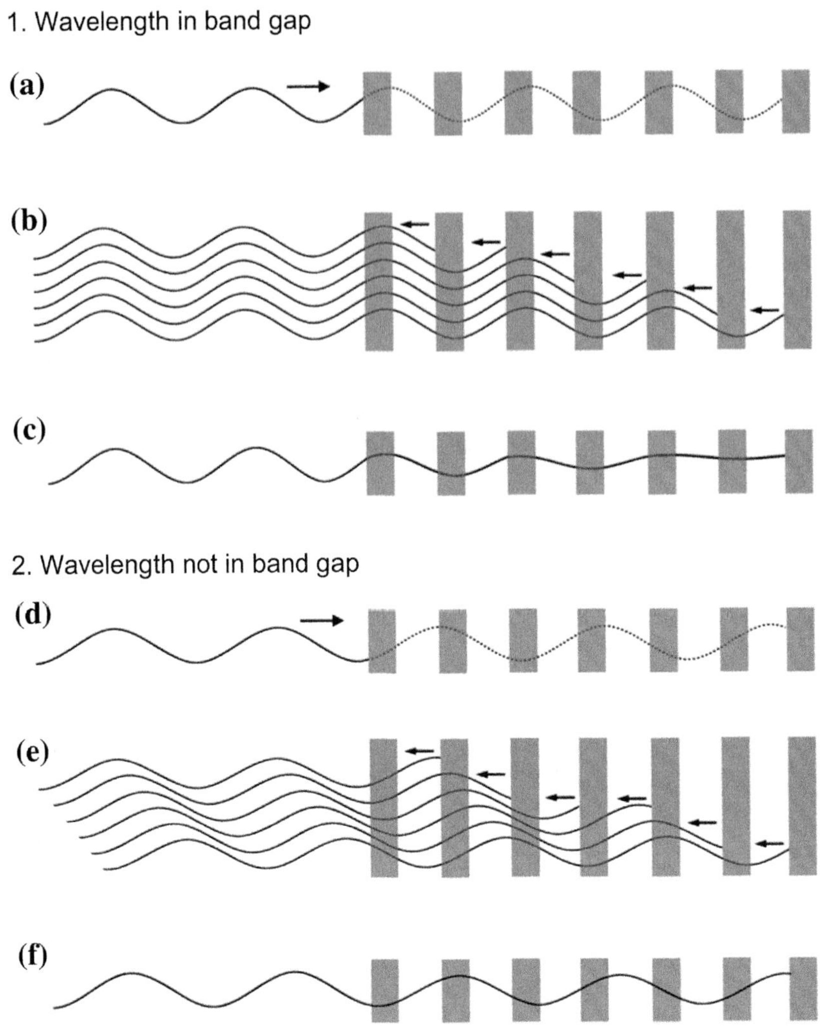

1. Wavelength in band gap

2. Wavelength not in band gap

FIGURE 14.9   Schematic depiction of light interacting with a one-dimensional photonic crystal given as a periodic array of dielectric slabs (gray rectangles) surrounded by air. The two cases shown are for light with wavelengths inside (a–c) and outside (d–f) of the photonic band gap. (a) Incident wave with wavelength inside band gap. (b) Reflected waves are all in phase and reinforce each other. (c) Reflected waves and incident wave produce a standing wave that does not propagate through the photonic crystal. (d) Incident wave with wavelength outside band gap. (b) Reflected waves are out of phase and cancel each other. (c) Forward-traveling wave propagates through the photonic crystal with only slightly reduced intensity. Adapted from Ref. 44.

should be as spherical as possible) [45]. This latter requirement makes face-centered-cubic lattices particularly interesting candidates for three-dimensional photonic crystals; indeed, modeling and photonic band structure calculations show that diamond-based lattice geometries are the "champion" structures [48]. With this guidance, photonic crystals with complete band gaps in the infrared and microwave regimes have been successfully fabricated by a number of approaches ranging from self-assembly to lithography, direct-writing techniques, and even mechanical drilling of arrays of holes into ceramic blocks [49–53].

Unfortunately, there are limitations in extending these methods to the visible range, due to inherent size limitations of patterning and writing techniques and the lack of robust assembly methods needed to yield diamond-based crystal geometries with lattice constants of a few hundred nanometers. As a consequence, a synthetic three-dimensional photonic crystal with a complete band gap in the visible spectral regime has proven elusive.

14.3.2 Surprising Weevils

In contrast to our limited abilities in engineering photonic crystals operating in the visible part of the electromagnetic spectrum, biological species have developed sophisticated structures to efficiently interact with visible light. The results of these interactions include large angular fields of view, reduced surface reflection, Bragg diffraction, and multiple scattering [6–9].

The latter two are of particular importance in the world of insects, which often rely on structural colors for defense, camouflage, and reproduction. For example, it is the periodic variation of biopolymeric compounds embedded into wings and exoskeletons that lends many butterflies and beetles their iridescent appearance, which can be used to scare off predators, hide in plain sight, or attract mates. Furthermore, photonic structures in these species were optimized to function under various illumination conditions, from glaring sunlight to the diffuse and dim lighting on forest floors.

Combined with the pure beauty of biological iridescence, these diverse applications and the ability of these materials to function in a wide range conditions have motivated biologists and physicists alike. Great efforts have been undertaken to search for the structural origins of these

coloration effects. This has led to the identification of an enormous diversity in biological photonic structures: from simple multilayer film composites to two- and three-dimensional periodic lattices with chiral, honeycomb, and cubic geometries (and combinations thereof) [6–11].

The biggest structural surprise from a photonic crystal standpoint came in 2008 with the discovery of a photonic crystal with the "champion" diamond-based lattice in the green Brazilian weevil *L. augustus* (Figure 14.8) [54]. Soon after, other weevils (*Eupholus schoenherri, Eudiagogus pulcher, Pachyrhynchus moniliferus*) were identified as creating their blue, green, and red exoskeleton colorations with diamond-based photonic crystals [26]. These structures have the same lattice geometries as *L. augustus* but different lattice constants (Figure 14.10). It should be emphasized that these biological structures are still the only examples (to date) of diamond-based photonic crystals with lattice constants comparable to wavelengths in the visible spectral regime.

Diamond-based photonic crystal structures are located within the cuticle scales of a weevil [26, 54]. These scales have a leaf-like shape (around $100 \times 50 \times 5\text{--}20 \ \mu m^3$ in size) and are attached to the exoskeleton of the beetles. Each scale has a solid outer shell and an interior diamond-based photonic crystal framework, both made out of chitin-based insect cuticle. The exact structure of this diamond-based lattice was evaluated through various structural and optical characterization tools, combined with modeling and photonic band structure calculations. The results showed a lattice composed of ABC stacked layers of hexagonally arranged air cylinders in a surrounding matrix of cuticle material. The refractive index of cuticle (around 1.5) is not high enough to form a complete photonic band gap in any of the weevil photonic crystals. However, calculations revealed that a complete band gap would open in the green part of the visible spectrum if the photonic crystal framework of the weevil *L. augustus* were made from a dielectric material with a refractive index of at least 2.1,

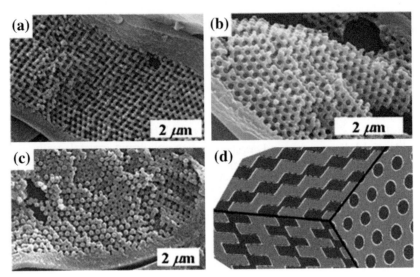

FIGURE 14.10   (a–c) SEM images of diamond-based photonic crystal structures found in exoskeleton scales of the weevils *Eupholus schoenherri*, *Pachyrhynchus moniliferus*, and *Eudiagogus pulcher*, respectively. (d) Calculated dielectric function of the diamond-based lattice showing three orthogonal planes (air: dark gray; biopolymer: light gray). Adapted from Ref. 26 and reproduced by permission of The Royal Society of Chemistry.

making this biological structure an ideal candidate for biotemplating [27].

14.3.3 Toward New Optical Materials

After the discovery of photonic crystals with the desirable diamond-based lattice geometry in weevil scales, the fabrication of the first complete photonic band gap at visible frequencies became a real possibility. Using the structure found in *L. augustus* as a basis, modeling and photonic band structure calculations revealed the geometric and dielectric properties needed for a complete band gap to form in replicated samples: (1) a positive replica of the original beetle photonic crystal lattice; (2) a framework compound (high refractive-index component) refractive index of at least 2.1 to open a band gap and at least 2.3 for a 5% wide (gap-to-midgap ratio) band gap; (3) a high refractive-index component volume fraction between 30% and 40% to optimize the band

gap width; and (4) a variation in replica lattice within 15–20% of the original structure to assure the band gap remains within visible frequencies [27].

Even though the calculations and modeling results give very clear instructions on how an optimal structure needs to be designed, translating these results into a solution-based templating process is rather challenging. Sol–gel chemistry, with its flexible processing parameters and versatility, is promising; however, it is necessary to adjust and modify the typical sol–gel infiltration and processing conditions to meet all the geometric and dielectric requirements for this replica.

The first predicted property, the creation of a positive replica of the original beetle photonic crystal structure, requires a double-templating process since a negative replica of this structure, which could be obtained by a single replication step, was predicted to lack a complete band gap. One way to meet this goal is to fabricate an

FIGURE 14.11 Schematic of the double-templating bioreplication route. The diamond-based photonic crystal structure of the weevil *L. augustus* (top) is converted into a titania positive replica *via* an intermediary silica negative replica. Gray: air; green: dielectric (silica or titania). (For interpretation of the references to color in this figure legend, the reader is referred to the web version of this book.)

intermediary, negative replica out of silica, which then serves as a sacrificial template in a second replication step (Figure 14.11). Galusha *et al.* showed the highest-fidelity negative replica structures are obtained by using a hybrid organic-inorganic silica sol for the infiltration process [26, 27]. The added polymer component lends higher flexibility to the framework, and thus reduces cracking during the drying stage. Following solidification of the silica-based framework, this biological template was removed by etching with a mixture of concentrated perchloric and nitric acids. This method was chosen over pyrolytic removal of the template to keep shrinkage of the framework to a minimum, with around 5% (acid treatment) compared to 30% (heat treatment).

The silica-based inverse replica was then used as the new (intermediary) template and

was infiltrated with a liquid titania sol–gel precursor. Titania is the compound of choice for photonic crystals operating in the visible spectrum since it combines a high refractive index of 2.1–2.8, depending on the type and degree of crystallinity, with excellent optical transparency (thereby satisfying the second requirement for a refractive index greater than 2.3 for the high-dielectric component).

Infiltration of the precursor solution was followed by heating the silica/titania composite to 500 °C to induce nanocrystallization of titania. This infiltration/heating cycle can be repeated to adjust the final filling fraction of titania (achieving the third requirement of a filling fraction between 30% and 40%) and creating a dense framework. In the final processing step, the intermediary silica-based template was selectively removed by hydrofluoric-acid etching, leaving behind a

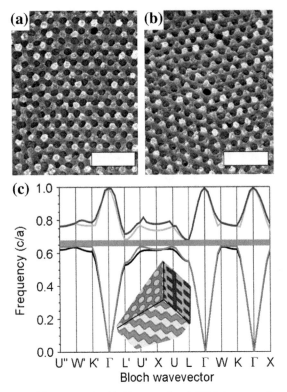

FIGURE 14.12 (a, b) Cross-sectional SEM images of a bioreplicated diamond-based photonic crystal made of titania. (c) Calculated band structure diagram for this photonic crystal lattice showing a narrow complete band gap (gray rectangle). Scale bars are 1 μm. Adapted from Ref. 27. Copyright Wiley-VCH Verlag GmbH & Co. KGaA. Reproduced with permission.

positive replica of the weevil photonic-crystal structure made of titania.

This successful replication was confirmed by structural and optical characterization techniques. For example, Figure 14.12 depicts SEM images of the positive replica of the *L. augustus* photonic crystal structure made of nanocrystalline titania with a measured refractive index of 2.3 ± 0.1. Using the structural and dielectric properties of the replica, band-structure calculations revealed the formation of a complete photonic band gap (with a gap-to-mid-gap ratio of 2–5%) in the green portion of the visible

spectrum (Figure 14.12c) [27]. However, it should be emphasized that these calculated results were obtained for a perfect photonic crystal of infinite size and free of any defects. Therefore, the width of this calculated band gap is most likely too narrow to stay open in a real material.

Regardless of whether a complete band gap is formed or not, these bioreplicated structures are by far the most efficient photonic crystals currently available for manipulating light at visible frequencies. These findings will allow experimental testing of some of the predicted new optical properties of photonic crystals with visible light. For example, one of the original motivations behind photonic crystal research was controlling dynamics of radiative processes; however, more than 20 years after it was first proposed, this effect has been rarely observed due to the lack of appropriate photonic-crystal samples.

This changed recently when the first experimental studies with titania bioreplica of diamond-based photonic crystals revealed the enormous potential of this new type of optical material. In these studies, the spontaneous emission properties of nanometer-sized light sources (nanocrystal quantum dots) within titania replicas were investigated [43]. Unprecedented modification of the dynamics of spontaneous emission at visible frequencies was observed, showing a variation of the excited-state lifetime by more than a factor of 10—approximately five times higher than previous results obtained from the best synthetic photonic crystals.

These results are extremely promising. Controlling spontaneous emission lies at the heart of many emerging applications, ranging from solar energy conversion, solid-state lighting, and lasing to quantum information processing. Bioreplication has supplied the first efficient photonic crystals toward these applications, and intriguing new optical properties will be made possible through this marriage of biological structure engineering and materials synthesis.

14.4 CONCLUDING REMARKS AND FUTURE DIRECTIONS

In general, the aim of bioreplication is to mimic biological structures in semiconductors, metals, and polymers for a wide range of technological applications beyond what nature intended. As we have discussed in this chapter, solution-based routes provide an appealing, simple pathway for generating structurally complex architectures with feature sizes covering several length scales.

With a wide range of chemistries at our disposal, scientists have begun to explore applications beyond those discussed here, such as *biohybrid* materials built upon previously formed natural or synthetic inorganic solids to host organic components. This organic counterpart can be intercalated between two-dimensional solids or embedded in three-dimensional mesoporous structures for applications in regenerative medicine, biodegradable materials for food packaging, and separation membranes [55].

In addition, it has been shown how biological species have myriad photonic structures to efficiently interact with light. The results of these interactions include large angular fields of view, reduced surface reflection, Bragg diffraction, and multiple scattering. Researchers are now generating replicas of such structures to capture and store solar energy with specific functionalities, such as light-harvesting components and catalysts that could be used for solar fuel production [56].

These and other new biomimetic and biotemplated structures and three-dimensional framework materials lend themselves to exciting new applications in optoelectronics, catalysis, separations, energy absorption, and tissue engineering, and provide fertile ground for researchers exploring technological solutions using bioreplication techniques.

Acknowledgments

We thank Matthew Jorgensen, Jeremy Galusha, and Moussa Barhoum for valuable discussions and contributions.

References

[1] M. Barberoglou, P. Tzanetakis, C. Fotakis, E. Stratakis, E. Spanakis, V. Zorba, S. Rhizopoulou and S. Anastasiadis, Laser structuring of water-repellent biomimetic surfaces, SPIE Newsroom, 19 January 2009. http://spie.org/x33323.xml?pf=true&ArticleID=x33323.

[2] A. Ressine, G. Marko-Varga, and T. Laurell, Porous silicon protein microarray technology and ultra-/superhydrophobic states for improved bioanalytical readout, *Biotech Ann Rev* **13** (2007), 149–200.

[3] W.R. Hansen and K. Autumn, Evidence for self-cleaning in gecko setae, *Proc Natl Acad Sci* **102** (2005), 385–389.

[4] E. Arzt, S. Gorb, and R. Spolanek, From micro to nano contacts in biological attachment devices, *Proc Natl Acad Sci* **100** (2003), 10603–10606.

[5] L.F. Boesel, C. Greiner, E. Arzt, and A. del Campo, Gecko-inspired surfaces: a path to strong and reversible dry adhesives, *Adv Mater* **22** (2010), 2125–2137.

[6] M. Srinivasarao, Nano-optics in the biological world: beetles, butterflies, birds, and moths, *Chem Rev* **99** (1999), 1935–1961.

[7] P. Vukusic and J.R. Sambles, Photonic structures in biology, *Nature* **424** (2003), 852–855.

[8] M.D. Shawkey, N.I. Morehouse, and P. Vukusic, A protean palette: colour materials and mixing in birds and butterflies, *J R Soc Interf* **6** (2009), S221–S231.

[9] S. Kinoshita, S. Yoshioka, and J. Miyazaki, Physics of structural colors, *Rep Prog Phys* **71** (2008), 76401–76500.

[10] O. Paris, I. Burgert, and P. Fratzl, Biomimetics and biotemplating of natural materials, *MRS Bull* **35** (2010), 219–225.

[11] M.R. Jorgensen and M.H. Bartl, Biotemplating routes to three-dimensional photonic crystals, *J Mater Chem* **21** (2011), 10583–10591.

[12] D.P. Pulsifer and A. Lakhtakia, Background and survey of bioreplication techniques, *Bioinsp Biomim* **6** (2011), 031001.

[13] G. Zuccarello, D. Scribner, R. Sands, and L.J. Buckley, Materials for bio-inspired optics, *Adv Mater* **14** (2002), 1261–1264.

[14] T.-X. Fan, S.-K. Chow, and D. Zhang, Biomorphic mineralization: from biology to materials, *Prog Mater Sci* **54** (2009), 542–659.

[15] T. Saison, C. Perez, V. Chauveau, S. Berthier, E. Sondergard, and H. Arribart, Replication of butterfly wing and natural lotus leaf structures by nanoimprint on silica sol–gel films, *Bioinsp Biomim* **3** (2008), 046004.

[16] A.J. Schulte, K. Koch, M. Spaeth, and W. Barthlott, Biomimetic replicas: transfer of complex architectures with different optical properties from plant surfaces onto technical materials, *Acta Biomater* **5** (2009), 1848–1854.

[17] D.-H. Ko, J.R. Tumbleston, K.J. Henderson, L.E. Euliss, J.M. DeSimone, R. Lopez, and E.T. Samulski, Biomimetic microlens array with antireflective "moth-eye" surface, *Soft Matter* **7** (2011), 6404–6407.

[18] D.P. Pulsifer, A. Lakhtakia, R.J. Martín-Palma, and C.G. Pantano, Mass fabrication technique for polymeric replicas of arrays of insect corneas, *Bioinsp Biomim* **5** (2010), 036001.

[19] G. Xie, G. Zhang, F. Lin, J. Zhang, Z. Liu, and S. Mu, The fabrication of subwavelength anti-reflective nanostructures using a bio-template, *Nanotechnology* **19** (2008), 095605.

[20] J. Huang, X. Wang, and Z.L. Wang, Controlled replication of butterfly wings for achieving tunable photonic properties, *Nano Lett* **6** (2006), 2325–2331.

[21] M. Knez, A. Kadri, C. Wege, U. Gösele, H. Jeske, and K. Nielsch, Atomic layer deposition on biological macromolecules: metal oxide coating of tobacco mosaic virus and ferritin, *Nano Lett* **6** (2006), 1172–1177.

[22] D.P. Gaillot, O. Deparis, V. Welch, B.K. Wagner, J.P. Vigneron, and C.J. Summers, Composite organic-inorganic butterfly scales: production of photonic structures with atomic layer deposition, *Phys Rev E* **78** (2008), 031922.

[23] A. Lakhtakia, R.J. Martín-Palma, M.A. Motyka, and C.G. Pantano, Fabrication of free-standing replicas of fragile, laminar, chitinous biotemplates, *Bioinsp Biomim* **4** (2009), 034001.

[24] R.J. Martín-Palma, C.G. Pantano, and A. Lakhtakia, Replication of fly eyes by the conformal-evaporated-film-by-rotation, *Nanotechnology* **19** (2008), 355704.

[25] R.J. Martín-Palma, C.G. Pantano, and A. Lakhtakia, Biomimetization of butterfly wings by the conformal-evaporated-film-by-rotation technique for photonics, *Appl Phys Lett* **93** (2008), 083901.

[26] J.W. Galusha, L.R. Richey, M.R. Jorgensen, J.S. Gardner, and M.H. Bartl, Study of natural photonic crystals in beetle scales and their conversion into inorganic structures via a sol–gel bio-templating route, *J Mater Chem* **20** (2010), 1277–1284.

[27] J.W. Galusha, M.R. Jorgensen, and M.H. Bartl, Diamond-structured titania photonic band gap crystals from biological templates, *Adv Mater* **22** (2010), 107–110.

[28] M.R. Jorgensen, B. Yonkee, and M.H. Bartl, Solid and hollow inorganic replicas of biological photonic crystals, *Scripta Mater* **65** (2011), 954–957.

[29] W. Zhang, D. Zhang, T. Fan, J. Gu, J. Ding, H. Wang, Q. Guo, and H. Ogawa, Novel photoanode structure templated from butterfly wing scales, *Chem Mater* **21** (2008), 33–40.

[30] S. Zhu, X. Liu, Z. Chen, C. Liu, C. Feng, J. Gu, Q. Liu, and D. Zhang, Synthesis of Cu-doped WO_3 materials with photonic structures for high performance sensors, *J Mater Chem* **20** (2010), 9126–9132.

[31] A.S. Deshpande, I. Burgert, and O. Paris, Hierarchically structured ceramics by high-precision nanoparticle casting of wood, *Small* **2** (2006), 994–998.

[32] S. Weiner, L. Addadi, and H.D. Wagner, Materials design in biology, *Mater Sci Eng C* **11** (2000), 1–8.

[33] P. Fratzl, Cellulose and collagen: from fibres to tissues, *Curr Opin Colloid Interf Sci* **8** (2003), 32–39.

[34] J. Aizenberg, J.C. Weaver, M.S. Thanawala, V.C. Sundar, D.E. Morse, and P. Fratzl, Skeleton of Euplectella sp.: structural hierarchy from the nanoscale to the macroscale, *Science* **309** (2005), 275–278.

[35] J. Wijnhoven, L. Bechger, and W.L. Vos, Fabrication and characterization of large macroporous photonic crystals in titania, *Chem Mater* **13** (2001), 4486–4499.

[36] S.M. Holmes, B.E. Graniel-Garcia, P. Foran, P. Hill, E.P.L. Roberts, B.H. Sakakini, and J.M. Newton, A novel porous carbon based on diatomaceous earth, *Chem Commun* (2006), 2784–2785.

[37] X. Yu, Y.-J. Lee, R. Furstenberg, J.O. White, and P.V. Braun, Filling fraction dependent properties of inverse opal metallic photonic crystals, *Adv Mater* **19** (2007), 1689–1692.

[38] B.H. Juárez, C. López, and C. Alonso, Formation of zinc inverted opals on indium tin oxide and silicon substrates by electrochemical deposition, *J Phys Chem B* **108** (2004), 16708–16712.

[39] B. Gates, Y. Yin, and Y. Xia, Fabrication and characterization of porous membranes with highly ordered three-dimensional periodic structures, *Chem Mater* **11** (1999), 2827–2836.

[40] D.J. Brinker and G.W. Scherrer, *Sol–gel science, the physics and chemistry of sol–gel processing*, Academic Press, San Diego, CA, USA (1990).

[41] J. Livage, M. Henry, and C. Sanchez, Sol–gel chemistry of transition metal oxides, *Prog Solid State Chem* **18** (1988), 259–341.

[42] M.H. Bartl, S.W. Boettcher, K.L. Frindell, and G.D. Stucky, 3-D molecular assembly of function in titania-based composite material systems, *Acc Chem Res* **38** (2005), 263–271.

[43] M.R. Jorgensen, J.W. Galusha, and M.H. Bartl, Strongly modified spontaneous emission rates in diamond-structured photonic crystals, *Phys Rev Lett* **107** (2011), 143902.

[44] E. Yablonovitch, Photonic crystals: semiconductors of light, *Sci Am* **285** (6) (December 2001), 47–55.

[45] J. Joannopoulos, R. Meade, and J. Winn, *Photonic crystals*, Princeton University Press, Princeton, NJ, USA (1995).

[46] E. Yablonovitch, Inhibited spontaneous emission in solid-state physics and electronics, *Phys Rev Lett* **58** (1987), 2059–2062.

[47] S. John, Strong localization of photons in certain disordered dielectric superlattices, *Phys Rev Lett* **58** (1987), 2486–2489.

[48] M. Maldovan and E.L. Thomas, Diamond-structured photonic crystals, *Nat Mater* **3** (2004), 593–600.

[49] A. Blanco, E. Chomski, S. Grabtchak, M. Ibisate, S. John, S.W. Leonard, C. Lopez, F. Meseguer, H. Miguez, J.P. Mondia, G.A. Ozin, O. Toader, and H.M. van Driel, Large-scale synthesis of a silicon photonic crystal with a complete three-dimensional bandgap near 1.5 micrometres, *Nature* **405** (2000), 437–440.

[50] Y.A. Vlasov, X.Z. Bo, J.C. Sturm, and D.J. Norris, On-chip natural assembly of silicon photonic bandgap crystals, *Nature* **414** (2001), 289–293.

[51] F. García-Santamaría, M.J. Xu, V. Lousse, S.H. Fan, P.V. Braun, and J.A. Lewis, A germanium inverse woodpile structure with a large photonic band gap, *Adv Mater* **19** (2007), 1567–1570.

[52] M.H. Qi, E. Lidorikis, P.T. Rakich, S.G. Johnson, J.D. Joannopoulos, E.P. Ippen, and H.I. Smith, A three-dimensional optical photonic crystal with designed point defects, *Nature* **429** (2004), 538–542.

[53] E. Yablonovitch, T.J. Gmitter, and K.M. Leung, Photonic band structure: the face-centered-cubic case employing nonspherical atoms, *Phys Rev Lett* **67** (1990), 2295–2298.

[54] J.W. Galusha, L.R. Richey, J.S. Gardner, J.N. Cha, and M.H. Bartl, Discovery of a diamond-based photonic crystal structure in beetle scales, *Phys Rev E* **77** (2008), 050904.

[55] E. Ruiz-Hitzky, M. Darder, P. Aranda, and K. Ariga, Advances in biomimetic and nanostructured biohybrid materials, *Adv Mater* **22** (2010), 323–336.

[56] B.D. Yuhas, A.L. Smeigh, A.P.S. Samuel, Y. Shim, S. Bag, A.P. Douvalis, M.R. Wasielewski, and M.G. Kanatzidis, Biomimetic multifunctional porous chalcogels as solar fuel catalysts, *J Am Chem Soc* **133** (2011), 7252–7255.

[57] L.M. Mäthger, E.J. Denton, N.J. Marshall, and R.T. Hanlon, Mechanisms and behavioural functions of structural coloration in cephalopods, *J R Soc Interf* **6** (2009), S149–S163.

[58] L. Poladian, S. Wickham, K. Lee, and M.C.J. Large, Iridescence from photonic crystals and its suppression in butterfly scales, *J R Soc Interf* **6** (2009), S233–S242.

ABOUT THE AUTHORS

Aditi S. Risbud is a science writer and materials scientist. She earned a doctorate in materials science and engineering from the University of California, Santa Barbara, USA, with thesis work on magnetic semiconductor materials. She also holds a certificate in science communications from the University of California, Santa Cruz. Currently, she is the senior communications and marketing officer for the University of Utah's College of Engineering. Previously, she directed communications and outreach efforts for the Molecular Foundry, a nanoscience user facility at Lawrence Berkeley National Laboratory, and started her career in science communications at global public relations firm Weber Shandwick.

Michael H. Bartl is an associate professor of chemistry and adjunct professor of physics at the University of Utah in Salt Lake City, UT, USA. He earned his doctorate in chemistry from Karl-Franzens University, Graz, Austria, and conducted postdoctoral research at the University of California, Santa Barbara, with Professors Galen Stucky and Evelyn Hu. He was the recipient of a *DuPont Young Professorship* in 2007. In 2010, he was named an *Emerging Investigator* by the *Journal of Materials Chemistry* and a *Brilliant 10* researcher by *Popular Science* magazine. His group studies functional materials for energy and information technology applications, including bioinspired photonics, nanocrystals, and thin films.

Vapor-Deposition Techniques

Raúl J. Martín-Palma[a] and Akhlesh Lakhtakia[b]

[a]Department of Materials Science and Engineering, Pennsylvania
State University, University Park, PA 16802, USA
[b]Department of Engineering Science and Mechanics, Pennsylvania
State University, University Park, PA 16802, USA

Prospectus

The term *vapor deposition* encompasses a large palette of techniques essential for both the reproduction of certain structural features of a biotemplate and the replication of a biotemplate. Physical vapor deposition, chemical vapor deposition, atomic layer deposition, and molecular beam epitaxy are succinctly described in this chapter in the context of engineered biomimicry.

Keywords

Atomic layer deposition (ALD), Chemical vapor deposition (CVD), Electron-beam evaporation, Ion-beam-assisted deposition (IBAD), Laser ablation, Molecular beam epitaxy (MBE), Oblique angle deposition (OAD), Physical vapor deposition (PVD), Pulsed laser deposition (PLD), Sputtering, Thermal evaporation

15.1 INTRODUCTION

Biological species are endowed with multiscale structures, ranging from the nano- to micro- to macroscale, which provide them with very specific functionalities. As such, for either reproducing specific geometric features of biological structures or replicating biological structures, fabrication techniques have to be suitable for the accurate replication of features at very different length scales. A variety of vapor-deposition techniques is available for both biomimetics and bioreplication.

In this chapter, the most commonly used vapor-deposition techniques are reviewed. These techniques include physical vapor deposition (thermal and electron-beam evaporation, sputtering, laser ablation, ion-beam-assisted deposition, oblique-angle deposition, and conformal-evaporated-film-by-rotation technique), chemical vapor deposition (CVD), atomic layer deposition (ALD), and molecular beam epitaxy (MBE). Given its importance for engineered biomimicry, ALD is also treated in Chapter 16 by Zhang and Knez. Although not discussed in this chapter, it is worth noting that the combined use of a focused ion beam (FIB) and a scanning electron microscope (SEM) could develop into a bioreplication technique in the near future [1].

Engineered Biomimicry
http://dx.doi.org/10.1016/B978-0-12-415995-2.00015-5

15.2 PHYSICAL VAPOR DEPOSITION

Physical vapor deposition (PVD) involves the generation of a vapor flux and its subsequent condensation in the form of a thin film on a substrate in a vacuum chamber. The term PVD encompasses several techniques, including thermal and electron-beam evaporation, sputtering, and laser ablation. The major differences between all of these PVD techniques are in the way that the vapor flux is generated from a target made of a specific material. More than one target may be used, and vapor fluxes from more than one material may be generated. At the same time, one or more gases may also be introduced to chemically modify either the vapor species or the growing thin film.

PVD techniques are used to fabricate a wide variety of thin films ranging from decorative optical coatings to high-temperature superconducting films. The thickness of the deposits can vary from a few angstroms to several millimeters, and very high deposition rates (up to 50 μm s^{-1}) can be achieved [2]. A very large number of inorganic materials (metals, alloys, compounds, and mixtures) as well as some organic materials can be deposited using PVD techniques [3].

Thus, the term PVD comprises several versatile methods for the fabrication of thin films of a wide variety of materials. PVD provides quite good structural control at the micrometer and/or nanometer length scales by carefully monitoring the processing parameters [4].

15.2.1 Thermal/Electron-Beam Evaporation

Thermal evaporation was devised by Faraday during the 1850s [5]. During this process, atoms and clusters of atoms or molecules are removed in the form of a vapor flux from a metal crucible, containing some bulk material (target) by heater the crucible, either by passing a current through it or

by a heater filament. Figure 15.1 presents a schematic representation of an evaporation system.

Alternatively, during electron-beam (e-beam) evaporation, a beam of electrons bombards the bulk material in the crucible to generate the vapor flux. The crucible and its contents are placed in a vacuum chamber, with pressure typically below 10^{-4} Torr. The vapor flux condenses on a substrate. Although the use of an electron beam to vaporize metals in vacuum is usually credited to Rühle [6], the basic process had been discovered serendipitously by von Pirani [7] slightly more than a century ago.

In a typical thermal evaporation process, the target material is heated by Joule effect to an appropriate temperature at which there is an appreciable vapor pressure. For most materials that vaporize below a temperature around 1,500 °C, evaporation can be achieved simply by putting the source material in contact with a hot surface that is resistively heated by passing a current through it. Typical resistive heating elements are carbon, molybdenum, tantalum, tungsten/wolfram, and BN/TiB$_2$ composite ceramics [8]. The heated surface may have one of many configurations—including basket, boat, crucible, and wire—for rapid heating and to realize a uniform distribution of the vapor flux. Among the major advantages of thermal evaporation, high deposition rates, relative simplicity, and low cost of the equipment must be mentioned. However, thermal evaporation is not very suitable for fabricating multicomponent thin films, since some bulk materials evaporate before others due to differences in their melting points and vapor pressures.

Electron-beam evaporation uses high-energy electron beams, typically accelerated with voltages from about 5 to 20 kV, to bombard the target material or materials that are placed in a crucible. Crucibles of copper have been widely used for many years, although crucibles of boron nitride, graphite, nickel, and tungsten are also used, depending on the target material(s) [8]. This evaporation technique can vaporize most pure

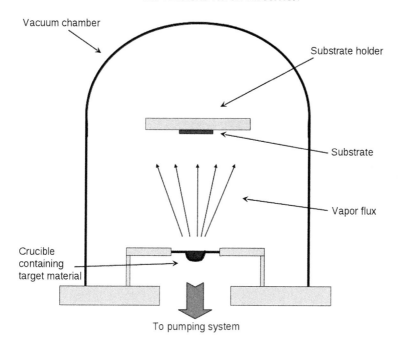

Vacuum chamber

Substrate holder

Substrate

Vapor flux

Crucible containing target material

To pumping system

FIGURE 15.1 Schematic of a typical thermal or electron-beam evaporation system. The source material is heated by an electrical current in thermal evaporation or by bombardment by an electron beam in electron-beam evaporation.

metals, including those with high melting points. E-beam evaporation is particularly suitable for the deposition of thin films of refractory materials, including most ceramics (oxides and nitrides), glasses, carbon, and refractory metals. Among all PVD techniques, e-beam evaporation provides probably the highest deposition rates. By the use of high-power e-beam sources, deposition rates as high as 50 μm s^{-1} have been achieved. Moreover, with adequate adjustment of the waist of the electron beam, uniform films of high purity can be obtained.

A biomimetic technique based on e-beam deposition was used to reproduce the blue color of the wings of butterflies of the genus *Morpho*. Multilayers composed of alternating thin films of TiO$_2$ and SiO$_2$ were evaporated onto a substrate that had been nanopatterned using e-beam lithography and etching [9, 10]. The optical characteristics of the structurally colored

Morpho wings can thus be reproduced using widely available technologies at a relatively low cost.

15.2.2 Sputtering

In the basic *sputtering* process, a cathode made of the target material is bombarded by energetic ions generated in a glow-discharge plasma situated in front of the target, as shown in Figure 15.2. The target can be an element, alloy, compound, or their mixture. The bombardment process causes the removal, i.e., sputtering, of target atoms by momentum transfer from the bombarding energetic gas ions (such as argon ions) accelerated in an electric field. The sputtered atoms form a vapor flux, which may then condense on a substrate as a thin film [11]. This process can be performed in a vacuum chamber using either low-pressure plasma

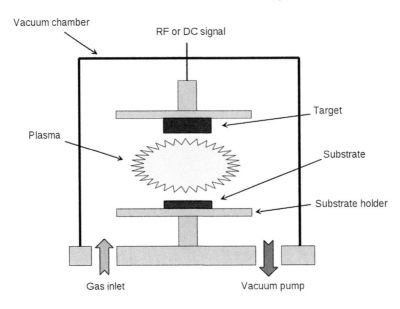

FIGURE 15.2 Schematic of a typical sputtering system in either the DC or the RF configuration.

($<5 \times 10^{-3}$ Torr) or high-pressure plasma (5×10^{-3} $- 30 \times 10^{-3}$ Torr) [8]. Secondary electrons are also emitted from the target surface as a result of ion bombardment. These electrons play an important role in maintaining the plasma.

Three widely used configurations to generate the plasma for sputtering deposition are the direct-current (DC) configuration, the radio-frequency (RF) configuration, and the magnetron-assisted configuration. The simplest of the three is the DC configuration, comprising a cathode (target), an anode (on which the substrate is placed), and a DC power source. The cathode and the anode are placed in a vacuum chamber [8]. Argon is widely used to establish a discharge. Because the plasma can be established uniformly over a large area, a solid target with a large area can be used. The surface of the target does not need to be planar, so targets with different shapes can be used to be conformal to the surface of a given substrate, resulting in improved thickness homogeneity.

The RF configuration is generally used for the deposition of electrically insulating materials

such as oxides and polymers. When a RF potential is capacitatively coupled to a target (cathode), an alternating positive/negative potential appears on its surface. In one half cycle, positively charged ions are accelerated toward the surface of the target with enough energy to cause sputtering. In the next half cycle, electrons reach the surface of the target to prevent the build-up of charge. Frequencies used for sputtering deposition are typically in the range of 0.5–30 MHz, with 13.56 MHz [8] being the most widely used. RF sputtering is used at a low pressure ($<10^{-3}$ Torr).

Finally, in the magnetron-assisted configuration, a magnetic field is imposed to increase the plasma density as well as the current density at the cathode (target), thereby effectively increasing the sputtering rate. The magnetic field is tangential to the cathode surface. The electrons ejected from the cathode are deflected to stay close to the target surface. If the magnets behind the target are arranged properly, the electrons can circulate on a closed path on the target surface. This electron-trapping effect effectively increases the collision probability between electrons and the gas molecules, thereby creating a high-density plasma. This configuration enables sputtering at low pressure with a high deposition rate.

The basic sputtering process was devised about a century and a half ago by Grove [12], who used the term *cathode disintegration*, but later researchers began to use both *spluttering* and *sputtering*. Thin films of many materials have been successfully deposited using this technique. In particular, sputtering is capable of depositing high-melting-point materials such as refractory metals and ceramics. Moreover, since the sputtered atoms usually carry more energy than the evaporated atoms, the sputter-grown films generally have higher mass density, superior adhesion to the substrate, and good crystalline structures. However, sputtering is limited by low ionization efficiencies in the plasma as well as by the heating of the substrate that often necessitates the use of cooling equipment. Significantly, the typical deposition rate of sputtering is considerably lower than that of thermal or electron-beam evaporation.

Reactive sputtering is the sputtering of elemental targets in the presence of chemically reactive gases that react with both the vapor flux ejected from the target and the target surface. It is a widely used technique for the deposition of a very wide range of thin films of compounds, including oxides, nitrides, carbides, fluorides, arsenides, and their alloys [13]. Although reactive sputtering is conceptually simple, it is in fact a complex and nonlinear process that involves many interdependent parameters.

Given its versatility, sputtering has become a process widely used for the deposition of a broad range of industrially important coatings. Examples include hard, wear-resistant coatings, low-friction coatings, corrosion-resistant coatings, decorative coatings, and coatings with specific optical or electrical properties [11].

15.2.3 Laser Ablation

In *laser ablation*, also called *pulsed laser deposition* (PLD), an intense, pulsed laser beam irradiates the target. When the laser pulse is absorbed by the target, its energy is used first for electronic excitation and then converted into thermal, chemical, and mechanical forms of energy, resulting in evaporation, ablation, plasma formation, and even exfoliation. The ejected material expands into the surrounding vacuum in the form of a plume containing many energetic species, including atoms, molecules, electrons, ions, clusters, particles, and molten globules. These diverse species finally condense onto a substrate as a thin film.

Laser ablation is often carried out in a high or ultra-high vacuum chamber. Reactive gaseous species, such as oxygen, can be introduced for the reactive deposition of oxides or other compound materials.

Generally speaking, laser ablation provides better control by simultaneous evaporation of multicomponent materials in a very short period of time. Because the ablation rate is related to the total mass ablated from the target per laser pulse [14], the development of lasers with high repetition rate and short pulse durations makes laser ablation—in combination with the condensation of an inert gas on the substrate—very attractive for the mass production of well-defined thin films with complex stoichiometry.

There are three possible growth modes in laser ablation [8]: First, the step-flow growth is often observed during deposition, either on a substrate with steps present on its surface (i.e., a highly miscut substrate) or at elevated temperatures. Upon arrival at the substrate surface, atoms diffuse to atomic step edges and form into surface islands. The growing surface is viewed as steps travelling across the surface. Second, in the layer-by-layer growth mode, islands continue to nucleate on the surface until a critical island density is reached. As more material is added, the islands continue to grow until neighboring islands begin to coalesce, resulting in a high density of pits on the surface. The addition of more atoms to the surface results in their diffusion into these pits to complete the layer. This process is repeated for each subsequent layer. Finally, the three-dimensional growth mode is similar to the layer-by-layer growth mode except that once an island is formed, an additional island will nucleate on top of the previous island. Continuing growth in one layer will not persist, leading to a roughened surface.

15.2.4 Ion-Beam-Assisted Deposition

Ion-beam-assisted deposition (IBAD) is not a deposition technique *per se*. Instead, it is a technique wherein ion implantation is combined with another PVD technique. The evaporated species produced by the chosen PVD technique are simultaneously impinged by an independently generated flux of ions [15]. Thus, while the individual atoms or molecules condense on the substrate to form a thin film, highly energetic ions (typically from 100 to 2,000 eV) are produced and directed at the growing thin film.

IBAD is particularly advantageous in that it has many independent processing parameters. The concurrent ion bombardment significantly improves adhesion and permits control over the morphology, density, internal stresses, crystallinity, and chemical composition of the thin film. Ion bombardment can also blend together coating and substrate atoms. The energy and flux of bombarding ions can be exploited to modify the size and crystallographic orientation of grains. Columnar morphology often observed in conventional, low-temperature PVD is negated by IBAD to create very dense thin films [15].

15.2.5 Oblique-Angle Deposition

Oblique-angle deposition (OAD) is a PVD method wherein the vapor flux is collimated to enable the fabrication of thin films with columnar morphology [16]. Thermal and electron-beam evaporation techniques are commonly used to generate the vapor flux. The substrate is so positioned as to receive the vapor flux at an angle χ_v greater than 0° and as high as 90° with respect to the substrate plane. The *columnar thin film* (CTF) thus formed comprises parallel, tilted nanocolumns whose assemblage is optically equivalent to a biaxial crystal in the infrared and visible regimes. The CTFs are highly dense, with the vapor flux normally incident on the substrate, but the density trails off as the vapor flux angle χ_v is reduced toward 0°.

Rocking the substrate about a tangential axis during deposition imparts the nanocolumns with a two-dimensional shape, whereas rotating the substrate about a central normal axis makes the nanocolumns acquire a three-dimensional shape. Rocking and rotation can be made to happen concurrently or sequentially. The thin films this forms are called *sculptured thin films* (STFs). The nanocolumns are made of 1–3 nm clusters, which accounts for the ease with which columnar shapes can be sculptured during deposition.

STFs are useful as polarization transformers and polarization filters, optical sensors, and vehicles for launching multiple surface-plasmon-polariton waves. Their intrinsic high porosity, in combination with optical anisotropy and possible two-dimensional electron confinement, make STFs potential candidates for electroluminescent devices, high-speed and high-efficiency electrochromic films; optically transparent conducting films sculptured from pure metals; and multistate electronic switches based on filamentary conduction.

For example, Figure 15.3 shows a cross-sectional view of a distributed Bragg reflector grown using the OAD technique. The structure comprises CTFs of two different types grown alternatingly, one with $\chi_v = 90°$ and the other with $\chi_v = 15°$. The CTFs grown with a normal vapor flux ($\chi_v = 90°$) are very dense and

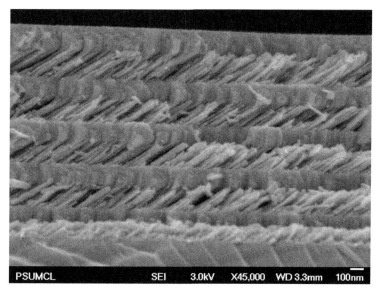

FIGURE 15.3 Cross-sectional SEM of a distributed Bragg reflector [17], consisting of alternate layers of very different CTFs, which have different porosities and different effective permittivity tensors.

comprise upright nanocolumns, whereas the CTFs grown with a highly oblique vapor flux are highly porous and comprise tilted nanocolumns. As a result, their effective permittivity tensors are very different, and the periodic stacking of the two types of CTFs leads to the exhibition of the Bragg phenomenon.

15.2.6 Conformal-Evaporated-Film-by-Rotation Technique

The *conformal-evaporated-film-by-rotation* (CEFR) technique allows fabrication of high-fidelity replicas of biotemplates with micro- and nanoscale features distributed over planar and curved surfaces [18, 19]. In the CEFR technique, the template is mounted on a substrate holder that is rotated rapidly about its central normal axis while the OAD technique is being implemented to coat the exposed surface of the template with a thin film. The vapor flux angle χ_v is fixed in the neighborhood of $5°$, as shown schematically in Figure 15.4. After the coating of thickness about

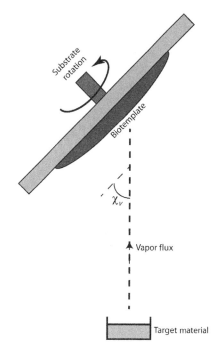

FIGURE 15.4 Schematic of the CEFR technique.

400 nm is separated from template, the coating becomes a high-fidelity replica.

To date, this technique has been applied for replicating the compound eyes of tephritid flies [18], as shown in Figure 15.5, and the wings of butterflies [19, 20], as shown in Figure 15.6, without compromising their optical characteristics that are due to nanoscale (<100 nm) structural features. The CEFR technique is particularly well suited for bioreplication because the temperatures involved during deposition are sufficiently low and the replication process occurs in a non-corrosive environment, thereby avoiding damage to the underlying biotemplate.

The CEFR technique has been modified to improve the uniform thickness of the replica by introducing a second degree of freedom to the biotemplate motion during deposition [21]. The first degree, as in the original CEFR technique, is the rotation of the biotemplate about a central normal axis. The second degree is the rocking of the biotemplate so as to continuously vary χ_v during deposition.

The modified CEFR technique is one of the two main steps of the Nano4Bio technique devised to fabricate multiple high-fidelity replicas of a single biotemplate [22]. As depicted schematically in Figure 15.7, in the first step of this technique, the modified CEFR technique is used to deposit a ~250-nm-thick conformal coating of nickel on the biotemplate. In the second step, a roughly 60-μm-thick structural layer of nickel is electroformed onto the thin layer to give it the structural integrity needed for casting or stamping. The biotemplate is then plucked off and plasma ashing is carried out to completely remove all organic material in the third step. What is left behind is a master negative made of nickel. This can be used in the fourth step as either a die for stamping or a mold for casting multiple replicas. Casting produces high fidelity at the 2-μm length scale [22], but stamping will improve the reproduction fidelity at lower length scales [1, 23]. Because the Nano4Bio technique can simultaneously produce multiple replicas of multiple biotemplates, it is suitable for industrial bioreplication.

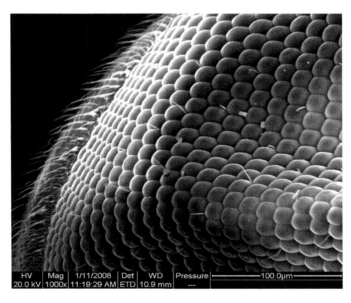

FIGURE 15.5   SEM of the eye of a *tephritid* fly (common fruit fly) coated with GeSbSe chalcogenide glass using the CEFR technique [18].

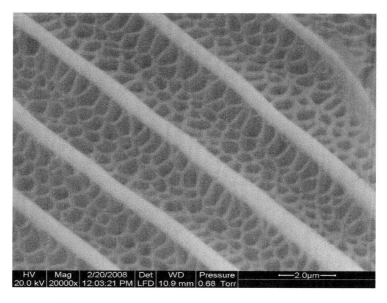

FIGURE 15.6 High-resolution SEM of a wing of the butterfly *Battus philenor* coated with GeSbSe chalcogenide glass using the CEFR technique [19]. Features at the nanoscale are evident.

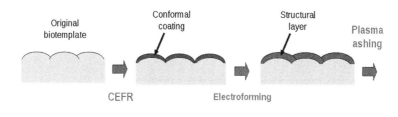

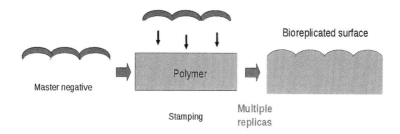

FIGURE 15.7 Schematic of the Nano4Bio technique.

15.3 CHEMICAL VAPOR DEPOSITION

Chemical vapor deposition (CVD) involves either the dissociation of a gaseous chemical and/or chemical reactions between gaseous reactants when heated, irradiated by photons, or subjected to a plasma [24]. As a product, a thin film is deposited on a surface. This technique is used to produce very pure high-performance solid materials.

Depending on the activation sources for the chemical reactions, the deposition process can be categorized into thermally activated, laser-assisted, or plasma-assisted CVD. The CVD process occurs in a vacuum chamber, with pressures ranging from the atmospheric pressure (atmospheric-pressure CVD) to below 10^{-8} Torr (ultra-high-vacuum CVD). Figure 15.8 represents a typical plasma-assisted CVD system.

The main steps that occur in a typical CVD process can be summarized as follows [14]: (1) transport of reacting gaseous species to the surface of a substrate, (2) adsorption of the species on that surface, (3) heterogeneous surface reaction catalyzed by the surface of the substrate, (4) surface diffusion of the species to growth sites, (5) nucleation and growth of the film on the substrate, and (6) desorption of gaseous reaction products and transport of reaction products away from the surface. The main CVD process parameters—such as temperature, pressure, reactant gas concentration, and total gas flow—require accurate control and monitoring. The chemical reactions include pyrolysis, oxidation, reduction, hydrolysis, or a combination of these and may be catalyzed by the substrate. The actual chemical reactions determine the operating temperature range.

CVD is a well-established technique for the deposition of metallic, ceramic, and semiconducting thin films because it offers the advantages of a relatively simple apparatus, excellent uniformity, high density, high deposition rate, and amenability to large-scale production. CVD is a more complex method of forming thin films and coatings than PVD. CVD exhibits several distinct advantages, such as the capability of producing highly pure and dense films or fine particles at reasonably high deposition rates and the capability of coating surfaces of complex shapes.

Many forms of CVD are in wide use and are frequently referenced in the literature. These processes differ in the means by which chemical reactions are initiated (e.g., activation process) and process conditions [4]. As such, atmospheric-pressure CVD, low-pressure CVD, and

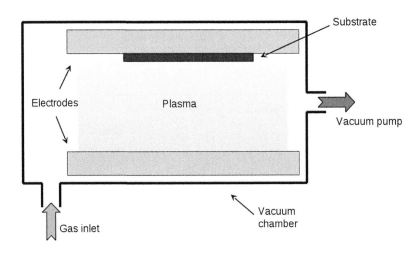

FIGURE 15.8 Schematic of a plasma-assisted chemical vapor-deposition system.

ultrahigh-vacuum CVD are named after the typical chamber pressure at which the reactions take place. Depending on the characteristics of the plasma, the following forms can be found: microwave plasma-assisted CVD, plasma-enhanced CVD, magneto-microwave plasma CVD, and remote plasma-enhanced CVD. If the characteristics of the vapor used are considered, the following two forms are commonly found: aerosol-assisted CVD and direct liquid-injection CVD. Metal-organic CVD uses metal-organic precursors, whereas in rapid thermal CVD the substrate is heated. Catalytic CVD is based on the catalytic decomposition of precursors using a resistively heated filament. This technique is also known as hot-wire CVD or hot-filament CVD. In laser-assisted CVD, a laser heats a localized spot and no other heating source is present [25].

In conventional thermally activated CVD, resistive heating of the hot-wall reactor provides sufficiently high temperatures for dissociation of the various gaseous species. This leads to the entire heating of the substrate to a high temperature before the desired reaction is achieved. It precludes the use of substrates having melting points much lower than the reaction temperature. Alternately, one could heat the reacting gases in the vicinity of the substrate by placing a hot filament of tungsten inside the chamber.

Plasma-enhanced CVD is known to exhibit a distinct advantage over thermally activated CVD owing to its lower deposition temperature. Various types of energy resources—e.g., DC, RF, microwave, and electron-cyclotron-resonance microwave (ECR-MW)—are used for plasma generation in CVD. In a DC-activated process, the reacting gases are ionized and dissociated by an electrical discharge, thereby generating a plasma consisting of electrons and ions. Microwave plasma is an attractive option because the microwave frequency (2.45 GHz) can oscillate electrons; thus, high ionization fractions are generated as electrons collide with gas atoms and molecules.

Laser-assisted CVD is associated with the deposition of chemical vapors using a laser beam generated from CO_2, Nd:YAG, or excimer lasers. Laser-assisted CVD differs from conventional CVD in that the area of growth can be limited to that of where the laser beam passes. Nevertheless, laser-assisted CVD can be used for a large variety of target materials and substrates [25].

Although CVD is a complex chemical process, it has several advantages [24]. Highly dense, very pure, uniform thin films are produced with good reproducibility and adhesion at reasonably high deposition rates. Control of crystal structure, surface morphology, and orientation of the CVD products is easily possible by controlling the CVD process parameters. The deposition rate can be adjusted readily. Low deposition rates are preferred for the growth of epitaxial thin films for microelectronic applications. However, for the deposition of thick protective coatings, a high deposition rate is preferred, and it can be greater than tens of mm per hour. High deposition rates lower the production costs.

CVD also exhibits the flexibility of using a wide range of chemical reagents such as halides, hydrides, and organometallics that enable the deposition of a large spectrum of materials, including metals, carbides, nitrides, oxides, sulfides, III–V materials, and II–VI materials. Relatively low deposition temperatures are employed in CVD, and the desired materials can be deposited *in situ* at low energies through vapor phase reactions, followed by nucleation and growth on the substrate. This enables the deposition of refractory materials at a small fraction of their melting points. For example, refractory materials such as SiC (melting point: 2,700 °C) can be deposited at 1,000 °C. Finally, CVD can be used to uniformly and conformally coat substrates with complex surfaces.

Like any deposition technique, CVD has several drawbacks as well [24]. Foremost are the chemical and safety hazards caused by the use of toxic, corrosive, flammable, and/or explosive reagent gases. However, these drawbacks have been minimized using variants of CVD, such as electrostatic spray-assisted CVD and combustion CVD, that use environmentally friendly reagents. Furthermore, it is difficult to deposit multicomponent materials with well-controlled stoichiometry using multiple reagents because different reagents have different vaporization rates.

Finally, the use of sophisticated CVD variants—such as low-pressure or ultrahigh-vacuum CVD, plasma-assisted CVD, and photo-assisted CVD—tends to increase the cost of fabrication. If production costs need to be reduced, however, simpler variants of CVD, such as aerosol-assisted CVD and flame-assisted CVD, may be employed.

Low-temperature CVD has been used for bio-replication. Controlled vapor-phase oxidation of silanes on the surface of biological structures produces exact, inorganic oxide replicas of several biological structures, including a wing of a butterfly, a wing of a housefly, and a leaf of *Colocasia esculenta* (a self-cleaning plant) [26]. Thus, CVD was used to replicate intricate and hierarchical structures on several length scales. Likewise, multifunctional zinc oxide interfaces were fabricated by the use of metal-organic CVD, with the compound eyes of butterflies serving as biotemplates [27].

Moreover, a combination of the FIB technique [28] and CVD has been used to fabricate artificial structures inspired by the scales of the *Morpho* wings [29]. The original and artificial scales show comparable optical characteristics, as discussed by Dushkina and Lakhtakia in Chapter 11. The overall reflectance spectrums of both structures for various incidence angles of light are quite similar and contain reflectance peaks at around 440-nm wavelength.

15.4 ATOMIC LAYER DEPOSITION

Atomic layer deposition (ALD) is a surface-controlled and self-limiting method for depositing thin films from gaseous precursors [30]. Although ALD can be considered a modification of CVD, ALD's distinctive feature is the self-limiting growth mechanism that imparts to it several attractive characteristics: accurate and easy control of film thickness, production of sharp interfaces, uniformity over large areas, excellent conformality with the substrate, good reproducibility, multilayer processing capability, and desirable qualities in thin films made at relatively low temperatures [31]. For nanotechnologists, the two most important characteristics of ALD are excellent conformality and the possibility of subnanometer-level control of film thickness.

ALD relies on alternate pulsing of the precursor gases and vapors onto the substrate surface—in a vacuum chamber—and subsequent chemisorption or surface reaction of the precursors. The vacuum chamber is purged with an inert gas between the precursor pulses. The ALD process is schematically depicted in Figure 16.1. With a proper adjustment of the experimental conditions, the process proceeds via saturative steps, i.e., the precursors exposed on the surface chemisorb on it (or react with the surface groups), saturatively forming a tightly bound monolayer on the surface. The subsequent purging step removes all the excess molecules from the vacuum chamber. When the next precursor is sent in to the chamber, it encounters only the surface monolayer with which it reacts, producing the desired solid product and gaseous byproducts. Under such conditions, the growth of the thin film is self-limiting, since the amount of solid deposited during one cycle is dictated by the amount of precursor molecules present in the saturatively formed surface monolayer. Therefore, the growth is stable and the thickness increase is constant in each deposition cycle. The self-limiting growth mechanism facilitates the growth of conformal thin films with accurate thickness on large areas. This technique also allows the growth of multilayer structures. However, a major limitation of ALD is its very low deposition rate.

The distinctive sequencing feature in ALD makes it an attractive method for the precise growth of crystalline compound layers, complex layered structures, superlattices, and layered alloys with precise interfaces [24]. ALD can be used to produce thin films with good conformal coverage, and it has the ability to control film thickness accurately at the subnanometer level. Such distinctive advantages have made it a potentially valuable tool for nanotechnology.

The same advantages make it useful for bioreplication. Indeed, ALD has been used to fabricate alumina replicas of the corneal layer of the compound eyes of flies [32] and butterfly wings [33]. For example, ALD was used to produce a 100-nm-thick alumina coating on a fly eye, and then the biotemplate was removed by pyrolysis. The resulting replica captured the 200-nm nipple-like features patterning the compound eye [32]. ALD has also been used to replicate the spines of the sea mouse [34], to infiltrate spider silk in order to toughen it [35], and to prepare photocatalytic replicas of the inner membranes of avian eggshells [36]. Chapter 16 provides a detailed treatment of ALD for biomimicry.

15.5 MOLECULAR BEAM EPITAXY

Molecular beam epitaxy (MBE) is a technique used to produce ultrathin films as high-quality epitaxial layers with very sharp interfaces and good control of thickness, doping, and composition [37]. Deposition usually takes place under high or ultrahigh vacuum conditions (typically below 10^{-10} Torr). Because of the high degree of control possible with MBE and the possibility of growing compound semiconductors, it is a valuable tool in the development of sophisticated electronic and optoelectronic devices.

The MBE process can be considered a refined form of evaporation. Ultrapure target materials are placed in effusion cells (also called *Knudsen cells*) and heated to their sublimation points [8]. Molecular beams thereby produced are then directed toward a single-crystal substrate, in the vicinity of which they may react chemically with each other or other gaseous species introduced into the vacuum chamber and then condense as a layer on the substrate. Figure 15.9 shows the

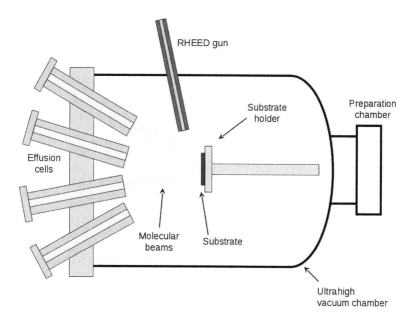

FIGURE 15.9 Schematic of a typical system for molecular beam epitaxy (MBE). Solid target materials are heated in effusion cells to produce molecular beams. The substrate is heated to the necessary temperature and, when needed, continuously rotated to improve the growth homogeneity. A reflection high-energy-electron diffraction (RHEED) gun is used for *in situ* monitoring.

schematic of a typical MBE apparatus. Each fabricated layer has a definite crystallographic relationship with the substrate. The substrate is usually heated and often rotated continuously to improve uniformity of deposition.

The molecular beams are typically obtained from thermally evaporated elemental target materials. However, metal–organic group-III compounds, gaseous group-V hydrides, organic compounds, or some combination may also be used as target materials. To obtain high-purity epitaxial layers, it is critical that the target materials be extremely pure and that the entire process be carried out in an ultrahigh-vacuum environment. Growth rates are typically on the order of a few $\mathring{A}\,s^{-1}$ and the molecular beams can be shuttered in a fraction of a second, enabling precise control of the composition, doping, microstructure, and thickness of the growing layer at the molecular level. Given the ultrahigh-vacuum environment of the system, analytical techniques such as reflection high-energy-electron diffraction (RHEED) and mass spectrometry are often used for *in situ* monitoring of the growing thin film.

15.6 CONCLUDING REMARKS

In many cases, a specific functionality of a biological specimen emerges from its particular structure at the nano-, micro-, and/or macroscale. This diversity of length scales makes it crucial to identify the most adequate bioreplication technique, depending on the characteristics of the chosen biotemplate. Vapor-deposition techniques are attractive for both biomimetics and bioreplication.

References

[1] D.P. Pulsifer and A. Lakhtakia, Background and survey of bioreplication techniques, *Bioinsp Biomim* **6** (2011), 031001.

[2] K.L. Chopra, *Thin film phenomena*, McGraw-Hill, New York, NY, USA (1969).

[3] P. Ehrhart, Film deposition methods, in *Nanoelectronics and information technology: advanced electronic materials and novel devices* (R. Waser, ed.), Wiley-VCH, Weinheim, Germany (2005).

[4] R.J. Martín-Palma and A. Lakhtakia, *Nanotechnology: a crash course*, SPIE Press, Bellingham, WA, USA (2010).

[5] M. Faraday, Experimental relations of gold (and other metals) to light, *Phil Trans R Soc Lond* **147** (1857), 145–181.

[6] R. Rühle, Verfahren zur Bedampfung im Vakuum, Deutsche Reich Patent 764927 (1939).

[7] M. von Pirani, Production of homogeneous bodies from tantalum or other metals, US Patent 848600 (26 March 1907).

[8] W. Gao, Z. Li, and N. Sammes, *An introduction to electronic materials for engineers*, 2nd ed., World Scientific, River Edge, NJ, USA (2011).

[9] A. Saito, Y. Miyamura, M. Nakajima, Y. Ishikawa, K. Sogo, Y. Kuwahara, and Y. Hirai, Reproduction of the *Morpho* blue by nanocasting lithography, *J Vac Sci Technol B* **24** (2006), 3248–3251.

[10] A. Saito, Y. Miyamura, Y. Ishikawa, J. Murase, M. Akai-Kasaya, and Y. Kuwahara, Reproduction, mass-production, and control of the *Morpho*-butterfly's blue, *Proc. SPIE* **7205** (2009), 720506.

[11] P.J. Kelly and R.D. Arnell, Magnetron sputtering: a review of recent developments and applications, *Vacuum* **56** (2000), 159–172.

[12] W.R. Grove, On the electro-chemical polarity of gases, *Phil Trans R Soc Lond* **142** (1852), 87–101.

[13] I. Safi, Recent aspects concerning DC reactive magnetron sputtering of thin films: a review, *Surf Coat Technol* **127** (2000), 203–218.

[14] S.C. Tjong and H. Chan, Nanocrystalline materials and coatings, *Mater Sci Eng R* **45** (2004), 1–88.

[15] J.R. Davis, *Handbook of materials for medical devices*, ASM International, Materials Park, OH, USA (2003).

[16] A. Lakhtakia and R. Messier, *Sculptured thin films: nanoengineered morphology and optics*, SPIE Press, Bellingham, WA, USA (2005).

[17] R.J. Martín-Palma, V. Torres-Costa, and C.G. Pantano, Distributed Bragg reflectors based on chalcogenide glasses for chemical optical sensing, *J Phys D: Appl Phys* **42** (2009), 055109.

[18] R.J. Martín-Palma, C.G. Pantano, and A. Lakhtakia, Replication of fly eyes by the conformal-evaporated-film-by-rotation technique, *Nanotechnology* **19** (2008), 355704.

[19] R.J. Martín-Palma, C.G. Pantano, and A. Lakhtakia, Biomimetization of butterfly wings by the conformal-evaporated-film-by-rotation technique for photonics, *Appl Phys Lett* **93** (2008), 083901.

[20] A. Lakhtakia, R.J. Martín-Palma, M.A. Motyka, and C.G. Pantano, Fabrication of free-standing replicas of

fragile, laminar, chitinous biotemplates, *Bioinsp Biomim* **4** (2009), 034001.

[21] D.P. Pulsifer, A. Lakhtakia, and R.J. Martín-Palma, Improved conformal coatings by oblique angle deposition for bioreplication, *Appl Phys Lett* **95** (2009), 193701.

[22] D.P. Pulsifer, A. Lakhtakia, R.J. Martín-Palma, and C.G. Pantano, Mass fabrication technique for polymeric replicas of arrays of insect corneas, *Bioinsp Biomim* **5** (2010), 036001.

[23] D.P. Pulsifer, A. Lakhtakia, M.S. Narkhede, M.J. Domingue, B.G. Post, J. Kumar, R.J. Martín-Palma, and T.C. Baker, Fabrication of polymeric visual decoys for the male emerald ash borer (*Agrilus planipennis*), *J Bionic Eng* **10** (2013), 129–138.

[24] K.L. Choy, Chemical vapour deposition of coatings, *Prog Mater Sci* **48** (2003), 57–170.

[25] S.N. Bondi, W.J. Lackey, R.W. Johnson, X. Wang, and Z.L. Wang, Laser assisted chemical vapor deposition synthesis of carbon nanotubes and their characterization, *Carbon* **44** (2006), 1393–1403.

[26] G. Cook, P.L. Timms, and C. Göltner-Spickermann, Exact replication of biological structures by chemical vapor deposition of silica, *Angew Chem Int Ed* **42** (2003), 557–559.

[27] S. Liu, Y. Yang, Y. Jin, J. Huang, B. Zhao, and Z. Ye, Multifunctional ZnO interfaces with hierarchical micro and nanostructures: bio-inspiration from the compound eyes of butterflies, *Appl Phys A* **100** (2010), 57–61.

[28] R.M. Langford, P.M. Nellen, J. Gierak, and Y. Fu, Focused ion beam micro- and nanoengineering, *MRS Bull* **32** (2007), 417–423.

[29] K. Watanabe, T. Hoshino, K. Kanda, Y. Haruyama, and S. Matsui, Brilliant blue observation from a Morpho-butterfly-scale quasi-structure, *Jpn J Appl Phys* **44** (2005), L48–L50.

[30] M. Ritala and M. Leskelä, Atomic layer epitaxy—a valuable tool for nanotechnology? *Nanotechnology* **10** (1999), 19–24.

[31] M. Leskelä and M. Ritala, Atomic layer deposition (ALD): from precursors to thin film structures, *Thin Solid Films* **409** (2002), 138–146.

[32] J. Huang, X. Wang, and Z.L. Wang, Bio-inspired fabrication of antireflection nanostructures by replicating fly eyes, *Nanotechnology* **19** (2008), 025602.

[33] J. Huang, X. Wang, and Z.L. Wang, Controlled replication of butterfly wings for achieving tunable photonic properties, *Nano Lett* **6** (2006), 2325–2332.

[34] F. Mumm, M. Kemell, M. Leskelä, and P. Sikorski, A bio-originated porous template for the fabrication of very long, inorganic nanotubes and nanowires, *Bioinsp Biomim* **5** (2010), 026005.

[35] S.-M. Lee, E. Pippel, U. Gösele, C. Dresbach, Y. Qin, C.V. Chandran, T. Bräuniger, G. Hause, and M. Knez, Greatly increased toughness of infiltrated spider silk, *Science* **324** (2009), 488–492.

[36] S.-M. Lee, G. Grass, G.-M. Kim, C. Dresbach, L. Zhang, U. Gösele, and M. Knez, Low-temperature ZnO atomic layer deposition on biotemplates: flexible photocatalytic ZnO structures from eggshell membranes, *Phys Chem Chem Phys* **11** (2009), 3608–3614.

[37] A.Y. Cho and J.R. Arthur, Molecular beam epitaxy, *Prog Solid State Chem* **10** (1975), 157–191.

ABOUT THE AUTHORS

Raúl J. Martín-Palma is Professor of Physics at the Department of Applied Physics of the Universidad Autónoma de Madrid (Spain) and adjunct professor at the Pennsylvania State University's Department of Materials Science and Engineering (University Park, PA, USA). He received his MS Degree in applied physics in 1995 and his PhD in physics in 2000, both from the Universidad Autónoma de Madrid. He has been Post-Doctoral Fellow at the New Jersey Institute of Technology (Newark, NJ, USA) and Visiting Professor at the Pennsylvania State University. He has received several awards for young scientists for his research on nanostructured materials from the Materials Research Society (USA), European Materials Research Society, and Spanish Society of Materials. He serves as an associate editor of the *Journal of Nanophotonics* and is a Fellow of SPIE.

Akhlesh Lakhtakia received degrees from the Banaras Hindu University, India (BTech and DSc), and the University of Utah, USA (MS and PhD), in electronics engineering and electrical engineering, respectively. He is the Charles Godfrey Binder (Endowed) Professor of Engineering Science and Mechanics at the Pennsylvania State University and currently serves as the editor-in-chief of the *Journal of Nanophotonics*. His current research interests include nanotechnology, bioreplication, surface multiplasmonics, complex materials, metamaterials, and sculptured thin films. He is a Fellow of SPIE, the Optical Society of America, American Physical Society, Institute of Physics (UK), and American Association for the Advancement of Science.

16

Atomic Layer Deposition for Biomimicry

Lianbing Zhang[a] and Mato Knez[a,b]

[a]CIC nanoGUNE Consolider, Tolosa Hiribidea 76,
20018 Donostia-San Sebastian, Spain
[b]Ikerbasque, Basque Foundation for Science, Alameda Urquijo 36-5,
48011 Bilbao, Spain

Prospectus

With the development of new synthetic procedures and technological processes, the interest in biomimicry has gathered rejuvenation in the past decades. One particularly interesting research method is the atomic layer deposition (ALD), which was established in various fields of technology as a vacuum-based chemical-processing technique and enabler for the deposition of extremely thin functional coatings. The benefits of this technology over similar techniques make it increasingly attractive for applications in biomimicry. In this chapter, short descriptions of the technology and its benefits and drawbacks are given. Subsequently, we summarize development in various research topics involving ALD and biomimicry.

Keywords

Anatase, Atomic layer deposition, Biocompatibility, Bioinorganic hybrid materials, Biomineralization, Biotemplate, Catalysis, Cellulose, Collagen, DNA, Dye-sensitized solar cell, Electrodes, Enzyme mimetics, Ferritin, Hydrophobicity, Nanostructure, Plasma-assisted ALD, S layer, Spider silk.

16.1 ATOMIC LAYER DEPOSITION: HISTORY AND TECHNOLOGY

In the 1960s, Kol'tsov from the Leningrad Technological Institute published a method for thin-film coating, showing the principle of the process that is now called *atomic layer deposition* (ALD) [1]. The method, at that time termed *molecular layering*, described an alternating exposure of a substrate to two reactive species in the vapor phase. The work was published in Russian and was therefore not recognized for a long time by the scientific community outside the Soviet Union.

In the 1970s, Suntola and Antson developed a similar methodology as an enabling technology to controllably produce thin-film electroluminescent displays (TFELs) [2], which were adopted quite soon thereafter. For example, the airport in Helsinki, Finland, installed an electroluminescent screen based on the thin-film coating process and kept it operational from 1983 through 1998. During this period, the term *atomic layer*

TABLE 16.1 List of abbreviations used in this chapter.

AAO	anodic aluminum oxide
ALD	atomic layer deposition
ALE	atomic layer epitaxy
CNT	carbon nanotube
CVD	chemical vapor deposition
DEZ	diethylzinc
DNA	deoxyribonucleic acid
DSSC	dye-sensitized solar cell
EDX	energy-dispersive X-ray spectroscopy
HA	hydroxyapatite
hADSC	human adipose-derived adult stem cells
MLD	molecular layer deposition
MMO	mixed metal-oxide framework
MPI	multiple pulsed vapor phase infiltration
NMR	nuclear magnetic resonance
ODTS	octadecyltrichlorosilane
SOD	superoxide dismutase
TEM	transmission electron microscopy
TFEL	thin-film electroluminescent display
TMA	trimethylaluminum
TMV	tobacco mosaic virus
XRD	X-ray diffraction

epitaxy (ALE) was used. The fact that most of the films deposited with this method do not grow epitaxially led to a change in the terminology from ALE to ALD, which is now in common use (see Table 16.1).

16.1.1 ALD Technology

The ALD process can, to a certain extent, be compared with chemical vapor deposition (CVD) [3]. In both processes, the deposition of an inorganic film occurs when two reactive chemical species (precursors) encounter each other and form a more stable compound through an exothermic reaction. Usually, the process takes place in a

vacuum chamber wherein the substrate to be coated is exposed to the two reactive precursors. Once those react, a film starts to grow. There are, however, some very significant differences between ALD and CVD, which are as follows.

The two precursors for CVD are usually injected into the vacuum chamber simultaneously. One has certainly to consider various factors such as the process temperature, the vapor pressures of the precursors, and the doses of the precursors. Properly adjusted, the CVD process will result in a formation of a film with thickness as a function of processing time. Since both precursors are usually highly reactive with each other and present in the chamber at the same time, the lifetimes of the reacting species are limited and thus the CVD process is quite rapid. The precision of the coating dimension has some limitations, however. These limitations include the step coverage of coatings on trenched structures with high aspect ratios: The bottom of pores or grooves in the substrate may not be coated with the same thickness as the upper surface of the substrate. Another aspect of the simultaneous injection of the precursors is the formation of particles in the vapor phase of the reactor and their precipitation on the substrate, which often leads to nonconformal coatings, pinholes, roughness of the coating's surface, and so on.

The principle of ALD is based on similar chemistry to that of CVD, but technologically the process shows one significant difference, which eventually results in better control [4–7]. The process is chemically split into two half-reactions (Figure 16.1). In the first stage, only one precursor is introduced to the substrate and a saturative chemisorption mechanism is used, allowing the precursor to form a layer of chemisorbed but still reactive species on the surface of the substrate. In other words, the reactive precursor attaches to every accessible surface site, provided that chemical anchor groups are present, in this way ensuring that the forthcoming step will not lead to a line-of-sight deposition. After the precursor has been

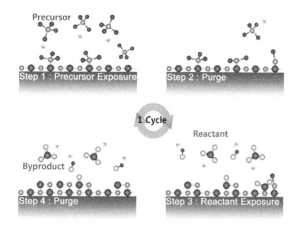

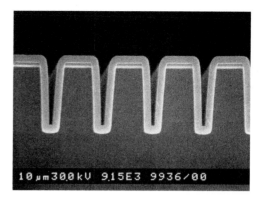

FIGURE 16.1 Schematic of an ALD process. One ALD cycle consists of four separate steps. In step 1, the substrate is exposed to the molecules of the first precursor, which adsorb ideally as a monolayer on the surface. In step 2, the excess molecules are removed from the gas by inert gas purging. In step 3, the substrate is exposed to the second precursor, which reacts with the adsorbed first precursor to form a layer of the desired material. In step 4, the excess second precursor and the reaction byproducts are removed from the gas phase by purging. This cycle is repeated (arrow) until the desired thickness of the coating is obtained. Reprinted from Ref. 6. Copyright © 2009, with permission from Elsevier.

FIGURE 16.2 Cross-sectional SEM image of an Al_2O_3 ALD film with a thickness of 300 nm on a Si wafer with a periodic trench structure, showing perfect step coverage. Reprinted from Ref. 122. Copyright © 1999, with permission from John Wiley and Sons.

16.1.2 Thermal Processing Window of the ALD Process

It is crucial for a well-operating ALD process that the two precursors are never present in the chamber at the same time. This will avoid parasitic CVD and enable conformal and reproducible growth of the film. Maintaining the self-saturating adsorption of the first precursor is another key issue for a well-performing ALD process. Each ALD reaction shows an ALD window, which describes a temperature range in which the growth is self-limiting (Figure 16.3) [9]. Within this temperature window, the first precursor will adsorb on the surface and remain there until the second precursor reacts with it.

Operating at temperatures below the ALD window might have either of two effects: (1) diminished adsorption of the first precursor due to low reactivity with the substrate, resulting in a lower growth per cycle, or (2) condensation of the first precursor due to low temperature and thus enhanced growth per cycle. Exceeding the ALD window at higher temperatures will in a similar way lead to non-self-limiting growth due to either (3) thermal decomposition of the first precursor and thus CVD-like growth, or (4) thermal desorption of the first precursor from the

chemisorbed, the excess is removed by inert gas purging.

The second precursor is subsequently injected to react with the chemisorbed first precursor, resulting in the formation of up to one monolayer of the coating. Another inert gas purge removes the reaction byproducts and the excess of the second precursor.

Repeating this procedure results in film growth with Ångström-scale precision, the increment being controlled with the number of cycles. In addition, since the lifetimes of the reacting species in the reactor are increased because the precursors are present in well-separated time slots, the ratio of the deposited film thickness at the bottom and at the top of a pore or groove is extremely high, even for structures with deep pores, trenches, or even spongy morphologies, such as aerogels or something similar (Figure 16.2) [8].

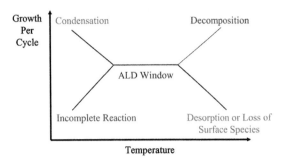

FIGURE 16.3 Schematic of the relation of the ALD growth per cycle versus temperature. The scheme shows the ALD window with possible scenarios for exceeding the temperature limits at the lower and upper ends. Reprinted from Ref. 5. Copyright © 2010, with permission from the American Chemical Society.

substrate due to the enhanced thermal budget. Therefore, the ALD process is self-limiting and reproducible only in a certain temperature range for a particular pair of precursors. However, within this range the growth is reproducible, and the self-limiting nature is based on the interface chemistry only.

16.1.3 Chemistry of ALD Processes

The chemistry involved in ALD is in most cases based on simple chemical reactions such as hydrolysis. In a few cases, the reactions are redox reactions or condensation reactions. Technologically, most of the ALD processes can be described as either thermal ALD, which is the most common case as described thus far, or the plasma-enhanced or plasma-assisted ALD, where the second precursor is pushed toward higher reactivity by generating a plasma [5].

For thermal ALD of an Al_2O_3 coating [9], a common first precursor is trimethylaluminum (TMA), which is a highly volatile and pyrophoric compound, and the second precursor is water. The pair of precursors reacts to produce Al_2O_3 and the growth can be easily controlled in a temperature window between $100\,°C$ and $300\,°C$. Each cycle adds approximately 1 Å of Al_2O_3 on top of the substrate.

Al_2O_3, however, can also be grown with plasma-assisted ALD. Here, the second precursor is not water but oxygen plasma, produced *in situ* and which, due to the oxygen radicals produced, shows enhanced reactivity toward TMA and allows a lower processing temperature. The quality of the coating may also be improved, since the radical species tend to react more vigorously with the ligands of the precursor or the reaction byproducts and remove those from the film. The plasma-assisted ALD, however, requires more complex instrumentation and is often not as easy to handle as thermal ALD. In addition, the reacting species may sometimes have a deleterious impact on the substrates, particularly when those are polymeric. Also, due to the shorter lifetime of the radical species, coatings of deep pores and trenches may not be as conformal as the corresponding thermal ALD processes, since the radicals may quickly recombine even before reaching the bottom of such pores and trenches. Nevertheless, the plasma-assisted ALD processes show good promise for coatings at lower temperatures, especially if metal coatings are required or the chemical purity of the coating is of importance for the anticipated application [10].

From the chemical point of view, two general cases may be differentiated. In the more common case, ALD will result in the growth of inorganic materials, such as metal oxides, nitrides, sulfides, or even metals [9]. This is simply dependent on the selection of the precursors and the thermal budget. In the currently less common case, organic molecules are used as precursors. The move to reactive organic molecules allows the layer-by-layer growth of polymers.

To differentiate these two growth processes, the organic ALD is called *molecular layer deposition* (MLD). The first examples of MLD showed the deposition of polyamides by alternating injection of organic acid chlorides and amines [11]. The homo-bifunctionality of the molecules used ensured that the growth was self-limiting. More recently, the MLD was combined with

ALD to grow inorganic–organic hybrid materials, which might have serious impact on the future development of materials—for example, for flexible electronics [5].

16.1.4 Limiting Factors of the ALD Process

In spite of the foregoing benefits of ALD/MLD over competing methodologies, some aspects of the process are considered less beneficial and may limit the use of ALD/MLD in combination with sensitive substrates or in mass production. Those circumstances need to be critically assessed in advance of the application of ALD/MLD.

16.1.4.1 Vacuum

Usually, the process takes place in a vacuum chamber, and this fact makes the process unwieldy or expensive for many applications. The purpose of the vacuum operation is primarily the elimination of excess precursors or byproducts to diminish parasitic CVD. The vacuum poses difficulties for many industrial applications and therefore already some prototypes exist that do not require vacuum chambers. The inert conditions are produced locally without the need to insert the substrate into a vacuum environment [12].

16.1.4.2 Temperature

The majority of the ALD processes take place at temperatures exceeding 100 °C. For most polymers and biomaterials, this is too high a temperature. However, the processing temperature strongly depends on the precursor pair and their volatility, reactivity, and thermal stability. Generally, ALD/MLD precursors need to be very volatile at the desired processing temperature, or lower, but they must not decompose at the processing temperature, since otherwise the deposition will be affected with parasitic CVD. They should also be highly reactive with the substrate and with each other at the processing temperature.

Often the first precursors are metal organics. A good example is TMA, which reacts with water in ambient conditions. The drawback of this particular process for operation at ambient temperatures is not related to the TMA but rather to the second precursor: water. The excess water during each cycle condenses on the substrate and leads to the simultaneous presence of both precursors in the next cycle. Extended purging times overcome this problem, but they increase the process duration and are therefore uneconomical.

The reactivity of some precursors at low temperatures is not satisfactory. Chemists are very actively and continuously synthesizing novel compounds with the goal of pushing down the processing temperatures for many materials. As a result, a growing number of materials are being processed significantly below 100 °C. Table 16.2 summarizes the currently available processes that can be performed at temperatures of 100 °C or less.

16.1.4.3 Process Duration

ALD/MLD is considered a slow process. One cycle will lead to a coating thickness in the Å range and the cycle itself, depending on the temperature, the nature of the precursor, and the throughput of the pump, might not be quicker than one second. CVD outperforms ALD/MLD by a huge margin, in this respect.

However, ALD is not a line-of-sight deposition method (as described earlier), which enables an upscaling of the process in an efficient manner. Substrates can easily be stacked inside a processing chamber and the chamber can be constructed to have a large volume. This in turn can easily speed up the coating process on a mass scale and in this way outperform nearly any competing coating technology.

For example, one step during the computer processor manufacturing by Intel® relies on ALD; thus an industrial-scale application is shown to be feasible. Roll-to-roll processing of flexible substrates is another example of an industrial-scale application of ALD and the

TABLE 16.2 Low-temperature (RT–100 °C) ALD processes.

Material	Precursor 1	Precursor 2	Temperature	References
Al_2O_3	TMA	H_2O	33 °C	13
Al_2O_3	TMA	O_3	RT	14
Al_2O_3	TMA	O_2-plasma	RT	15
B_2O_3	BBr_3	H_2O	RT	16
CdS	$Cd(CH_3)_2$	H_2S	RT	17
HfO_2	$Hf[N(Me_2)]_4$	H_2O	90 °C	18
Pd	$Pd(hfac)_2$	H_2	80 °C	19
Pd	$Pd(hfac)_2$	H_2-plasma	80 °C	20
Pt	$MeCpPtMe_3$	O_2-plasma + H_2	100 °C	21
PtO_2	$MeCpPtMe_3$	O_2-plasma	100 °C	21
SiO_2	$Si(NCO)_4$	H_2O	RT	22
SiO_2	$SiCl_4$	H_2O	RT[a]	23
SnO_2	TDMASn	H_2O_2	50 °C	24
Ta_2O_5	$TaCl_5,$	H_2O	80 °C	25
Ta_2O_5	$Ta[N(CH_3)_2]_5$	O_2-plasma	100 °C	26
Ta	$TaCl_5$	H-plasma	RT	27
Ti	$TiCl_4$	H-plasma	RT	28
TiO_2	$Ti[OCH(CH_3)]_4$	H_2O	35 °C	29
TiO_2	$TiCl_4$	H_2O	100 °C	30
V_2O_5	$VO(OC_3H_9)_3$	O_2	90 °C	31
ZnO	$Zn(CH_2CH_3)_2$	H_2O	60 °C	32
ZnO	$Zn(CH_2CH_3)_2$	H_2O_2	RT	33
ZrO_2	$Zr(N(CH_3)_2)_4)_2$	H_2O	80 °C	34

[a]Wth pyridine as catalyst.

corresponding machinery is being independently developed by several vendors.

16.1.4.4 Cost of the Precursors

The remaining drawbacks relate to the costs of the precursors, which depend on the target material and often are very high. The unavoidable waste of the precursors even enhances the cost factor: The precursors are injected in an excess dose to saturate the substrate surface. The excess is usually purged and lost. Optimized processing parameters can reduce the waste to a great extent, but there are no efficient commercial solutions available that allow a complete recovery of the excess precursors used. Both cost and waste are serious issues for many industrial applications.

16.2 APPLICATION OF ALD TO BIOMATERIALS

As described in Section 16.1, ALD is a technology formulated to produce devices such as displays. With that origin, it is axiomatic that the research and development on and with ALD mainly affect technological fields. The use of substrates with limited thermal stability, primarily polymers and biomaterials, was not considered to be possible until the processing temperatures of some ALD processes were pushed down to ranges that can be tolerated by such sensitive substrates.

The initiating work for this novel application field of ALD was the publication from the group of George [13], describing the coating of a polymer bottle with Al_2O_3 at temperatures as low as 33 °C. From the thermal point of view, the process showed great promise for application to biological substrates and biomaterials. Of course, several other factors play a role, such as the required vacuum, which could easily lead to destruction of a biological substrate by dehydration, but also the unknown reactivity of the precursor with the biomaterial, which in the worst case would be destructive to proteins.

The first experiments applying ALD to biomaterials were based on trial-and-error approaches but performed surprisingly well. The remainder of this section summarizes the work of the past years, in which ALD was applied to structural and/or functional biomimicry, biocompatibility, and biomineralization.

16.2.1 Structural Mimicry

The use of ALD for mimicking the structural properties of biological substrates is the most common route during the past several years. A very common case of a biological nanostructure is DNA, which resembles a one-dimensional fiber. DNA is considered a promising template for the synthesis of metallic nanowires [35], but technological applications with DNA-based nanowires are still lacking. Due to the great stability of DNA, a coating of the molecule by ALD appeared feasible. DNA has been used as a template for metal-oxide deposition, either for curiosity [36] or as means of functionalization of a substrate [18].

The latter approach may be used to produce a carbon nanotube (CNT)–based transistor. The function of the DNA in this case is not related to its intrinsic physical properties but merely to the chemical functionality the molecule offers. The CNT itself cannot be easily coated by ALD, since it lacks functional anchor groups for chemisorption of the precursor. If deposited, the films usually do not adhere very well to the CNTs and show enhanced surface roughness due to the initial island growth on defect sites [37]. To enable a coating of the CNT, Lu et al. wrapped the CNT with DNA and the adhesion was provided via stacking of π-electrons [18]. The chemical functionalities of the DNA subsequently acted as anchor groups for the ALD process. Those groups enabled a uniform coating of the CNT with the high dielectric–constant material HfO_2, which, without the wrapped DNA molecules, is not easily possible (Figure 16.4). However, approaches using DNA as a template for ALD coating do not truly relate to structural mimicry, because the resulting material does not reflect the original DNA structure.

Some peptides in certain experimental conditions form nanostructure networks through self-assembly. With diphenylalanine, even millimeter-long fibers can be obtained by electrospinning [38]. Organogel formation will result in ribbons with lengths of hundreds of micrometers, widths of some hundreds of nanometers, and thicknesses of some tens of nanometers. Removal of the solvent (e.g., chloroform) will yield xerogels.

Kim and coworkers made use of such ribbons, which they obtained after gelation of diphenylalanine, as templates for a coating with TiO_2 [39–41]. The TiO_2 coating was deposited at 140 °C and the ribbons were subsequently calcinated at temperatures exceeding 300 °C. Thermal degradation of the peptides and recrystallization of the resulting hollow TiO_2 coat as anatase occurred in a way similar to that shown earlier with electrospun polymer fibers as templates [42]. The choice of TiO_2 for this approach was based on two important aspects: (1) TiO_2 is one of those materials that can be easily handled by ALD and also deposited at low temperatures, and (2)

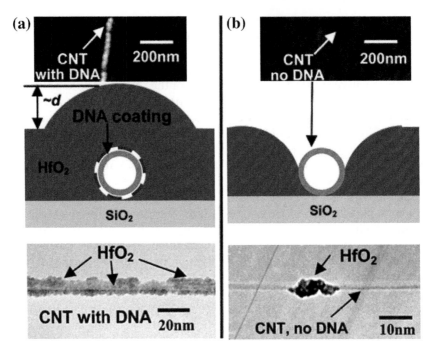

FIGURE 16.4 ALD of HfO$_2$ coatings on single-walled carbon nanotubes (SWNTs) (a) with and (b) without DNA functionalization. (Top panel) AFM images of ≈5-nm–thick HfO$_2$ coatings on SWNTs suspended on SiO$_2$. (Middle panel) Cross-sectional schematics of the coating profiles. (Bottom panel) TEM images of nominally 5-nm–thick ALD–HfO$_2$ coating on suspended SWNTs. Reprinted from Ref. 18. Copyright © 2006, with permission from the American Chemical Society.

depending on the crystalline phase, TiO$_2$ exhibits numerous interesting physical and chemical properties.

Kim's group investigated the wetting properties [39], electrochemical properties [40], and application potential for dye-sensitized solar cells [41] of the replicated TiO$_2$ ribbons in a series of works. The wetting behavior of the anatase phase TiO$_2$ was switched with ultraviolet light. Though the switchable wetting effect was known [43], the novelty of this approach was primarily the very interesting replica of the structure with a large surface area.

The follow-up papers from the same group showed some more interesting aspects of the produced structures from the technological point of view. The replicated nanoribbons were proposed as potential electrodes for Li-ion batteries [40].

The highly porous structure with only 15-nm–thick TiO$_2$ layers permits efficient diffusion of an electrolyte and thus the Li$^+$ ions into the TiO$_2$ matrix. Compared to TiO$_2$ nanopowders, the nanoribbons show great enhancement of the specific capacity, reaching a fivefold increase.

The last proposed application by the same authors with the same structures relates to dye-sensitized solar cells (DSSCs) [41]. In the simplest case, a DSSC consists of crystalline TiO$_2$, an organic dye, and an electrolyte sandwiched between two electrodes [44]. The nanoribbons were used as the anode, and cis-bis(isothiocyanato) bis(2,2'-bipyridi-4,4'-dicarboxylato)ruthenium (II) bis-tetrabutyl-ammonium (Ru535-bisTBA) as the dye. Depending on the calcination temperature, structures consisting of pure anatase or mixed anatase-rutile phases were obtained.

Again, the ribbon structures performed better than TiO$_2$ powders. The best power conversion efficiency was obtained with the mixed phase and amounted to 3.8%. In comparison, the powder-based approach showed an efficiency of 3.6%. The large overall surface of the ribbons, together with a large crystallite size, the reduced grain boundaries, and densely packed crystallites, are assumed to be the sources of the improvement.

ALD was also implemented on collagen-fiber networks. Although it is not very pure, such a collagen-fiber network can easily be obtained after peeling an avian egg. The soft tissue located between the exterior calcite shell and the egg white protects the embryo from bacterial invasion while allowing for effective gas exchange. The main constituent of this membrane is collagen. Such a membrane was coated by ALD with TiO$_2$ or ZnO at temperatures ranging from 70 °C to 300 °C. The main objective was to figure out whether or not the ALD process will affect the structural morphology [45]. The higher deposition temperatures lead to denaturation, but it was found that the initial deposition of a protective layer at lower temperatures results in stabilization of the structure for subsequent deposition at higher temperatures.

Investigations of the crystallinity of the ZnO coating confirmed that crystalline features of a wurtzite type occur at a processing temperature as low as 70 °C. This is an important observation because, although polycrystalline or even nanocrystalline, the ZnO coating may readily exhibit photocatalytic effects. In contrast, TiO$_2$ coatings only show reasonable indications of anatase at processing temperatures above 160 °C.

The photocatalytic efficiency of the coating was deduced from the bactericidal effect the membrane exhibits upon illumination with ultraviolet light. The membranes were built into cells containing *E. coli* bacteria. Samples of the bacteria-containing solution were taken at diverse instants of time and cultured to count the populations. As expected from the crystallinity data, the membranes coated with ZnO at 100 °C already have a good photocatalytic efficiency, whereas TiO$_2$-coated membranes required processing temperatures exceeding 160 °C for a similar effect. Thus, ZnO is apparently a reasonable alternative for TiO$_2$, particularly if the substrate is thermally sensitive.

The mechanical properties of the ALD-treated collagen-based membranes changed. Tensile tests showed a simultaneous increase in strength and ductility of the membranes, thus increasing the toughness threefold [46]. Such a behavior is remarkable and unusual in physics, as discussed in Section 16.3.2.

A further class of fibers used as substrates for ALD processing consists of cellulose fibers. Initial work on cellulose fibers derived from paper was performed by coating with TiO$_2$ or bilayers of Ir/Al$_2$O$_3$ or Ir/TiO$_2$, respectively [47, 48]. Cellulose shows much higher resistance to thermal treatment than most protein-based materials, thereby enabling higher processing temperatures. The metal oxides were deposited at 150 °C or 250 °C and iridium at 250 °C, below the decomposition temperature of the cellulose. The TiO$_2$ coatings consisted of crystalline anatase and were photocatalytically active. Additional Ir coating even improved the photocatalytic activity.

A protective effect, induced by the metal-oxide coating, was observed. The byproduct during the Ir deposition process is atomic oxygen, which is expected to decompose the cellulose. But the metal oxide prevents the decomposition, possibly due to the chemical reactivity of the metal oxide with the atomic oxygen.

The group of Persons performed a more directed approach toward functionalization of cellulose-based fibers. Their target was cotton fibers, which, after coating with thin inorganic films by ALD should result in fabrics with enhanced wear resistance. They initially deposited a 50-nm-thick coating of Al$_2$O$_3$ at 100 °C [49].

Subsequent work focused on the wetting behavior of the fabrics as a function of the thickness of the ALD coating [50]. An interesting observation was that a single ALD cycle of TMA/water abruptly switched the wetting behavior of cotton from wetting to nonwetting. Processing with further cycles caused a reverse switch.

The effect can be explained as follows: Native cotton fibers contain nanoscale fibrils on the surface. Because of the initial nucleation of the inorganic film upon processing, the first ALD cycles might increase the surface roughening. Further cycling will merge nucleation clusters and eventually result in a continuous film, which will become smoother with an increasing cycle number. The initial surface roughening might be the reason for the rather hydrophobic surface, and the smoothing during the continuation of the process might be the reason for the surface becoming hydrophilic again.

The S layer is a two-dimensional biological template frequently used in nanoscience or materials science [51]. It is a molecular sheet with regularly arranged pores that self-assembles from proteins of cell envelopes of certain bacteria, if assembly conditions are properly chosen. As in many other approaches besides ALD, the goal here was also to use the S layer as a mask for the synthesis of nanodots. The difficulty, however, is that ALD is a non-line-of-sight deposition method but relies on the surface functionality of the substrate. Moreover, the S layer consists of proteins with a plethora of functional groups, and thus it was expected that the coating would not be selective to the pores. To avoid deposition on the protein sheet, the functional sites were passivated with octadecyltrichlorosilane (ODTS) and the subsequent deposition of HfO_2 occurred exclusively on the underlying Si wafer through the tiny pores [52]. After the process, the S layer was removed by thermal treatment at 600 °C in air. The resulting nanodots on the Si wafer surface had diameters of about 9 nm and a very regular distribution.

Many biological materials are nanostructures with precise sizes, structures, and content. *Ferritin* is among biological nanoparticles of great interest to materials scientists. Ferritin is a globular protein assembled of 24 subunits, with a diameter of 12 nm, and it contains a cavity with a diameter of approximately 7 nm. In nature, ferritin acts as a storage container and transport vehicle for iron, which is stored within the ferritin cavity as ferrihydrite. The iron-containing core can be removed from the protein capsule through small, 3–4 Å channels, which are located at the boundaries of the protein subunits. The hollow ferritin remaining is named *apoferritin*.

ALD deposition of Al_2O_3 or TiO_2 on spread layers of ferritin resulted in freestanding metal-oxide films with embedded ferritin [29, 36]. Given the fact that numerous materials were wet-chemically synthesized within the apoferritin cavity, ALD promises a good approach to thin films with embedded luminescent, magnetic, or plasmonic nanoparticles. The importance of this work lies in the fact that the ferritin proteins appear to be robust enough to withstand the process conditions, which are considered rather harsh for biomaterials. A follow-up work showed that indeed apoferritin is not destroyed during the process [53].

The channels perforating the protein shell deposit TiO_2 within the hollow cavity. Control of the deposition toward either the outer surface of the apoferritin or the inside of the cavity was made possible by varying the pretreatment conditions of the apoferritin (Figure 16.5). The amount of water bound to the surface of the above-mentioned 3–4 Å channels was of crucial importance for the precursors to enter the cavity. If present, the water hydrolyzes the precursor already at the entrance of the channel, and further diffusion of precursor molecules is hindered due to clogging. Dehydrating the channels leads to an increased lifetime of the precursors and enhanced possibility to enter the cavity.

The described clogging can easily happen if the channel diameters are sufficiently tiny.

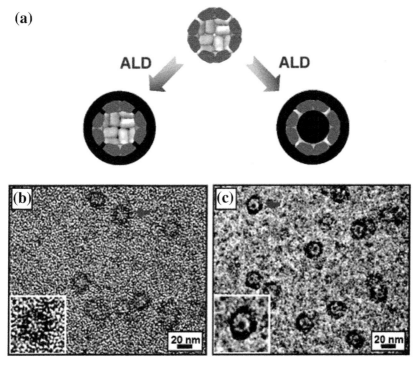

FIGURE 16.5 (a) Schematic depiction of two different TiO_2 nanostructures obtained by the ALD process. Typical bright-field TEM images of (b) hollow-shell and (c) core-shell nanoparticles templated by apoferritin. Reprinted from Ref. 53. Copyright © 2009, with permission from the American Chemical Society.

Fewer difficulties appear if the channels are of larger diameter, as is the case with the tobacco mosaic virus (TMV). This plant virus was actually the first protein-based structure to which an ALD process was applied [29]. The structure of TMV is tubular with a length of 300 nm and an outer diameter of 18 nm. Building blocks consist of approximately 2,130 identical protein subunits, which are helically stacked and form a hollow channel with a diameter around 4 nm along the axis of the virus. The TMV is the first known virus [54] and the initiator of the whole research field of modern virology. This particular virus tolerates temperatures up to 80 °C, is not harmful to mammals, and is easy to handle in laboratories. For those reasons, even nonbiologists developed great interest in this virus,

either for the development of electron microscopy in 1939 [55] or for nanostructure fabrication [56–62], or more recently for various technological applications [63–65].

The scientific question for treating TMV with an ALD process was whether or not the virus would resist the reactive precursor treatment at elevated temperatures in a vacuum. To minimize the impact, ALD processes were chosen that do not require high temperatures—that is, deposition of Al_2O_3 and TiO_2 at 35 °C [29, 36]. The morphological features of the TMV were preserved after the process, which could be deduced because the outer diameter of the TMV remained 18 nm. Having been suspended on a substrate, the coating affected the surfaces of the viruses as well as the substrates they were suspended on, thus resulting in a continuous film of the metal oxides with embedded TMV, in analogy to the films produced with ferritin.

The important observation after treating the TMV was that the coating also occurred within the hollow central channel of the virus. Transmission electron microscopy (TEM) observations clearly showed an enhanced contrast in the channel, which was not only proof of the presence of the metals but also that the structure of the virus did not collapse.

The TMV is prone to break into smaller subunits if mechanically treated. Such subunits were also observed in the TEM after the ALD process. Small subunits, dependent on their aspect ratio, tend to lay upright, which confirmed that the central channel of 4 nm in diameter was coated, leaving behind a much narrower channel in the center (Figure 16.6). This can be explained as follows: The precursors used for the ALD process can access any available surface. In the case of the TMV, the channel of 4 nm in diameter is accessible to the precursors that attach and form the metal oxide during the process. Once the deposit in the central channel starts growing, the channel opening shrinks. When it becomes too narrow for the precursors to enter, the deposit will close the opening and encapsulate the channel. The size of the remaining channel should be a function of the size of the molecular precursors.

The use of TMV as a template for ALD processing enables additional functionalization. This can be done by combining ALD with approaches based on wet-chemical methods, such as redox reactions or electroless deposition [57–62]. Also, adsorption of TMV on various

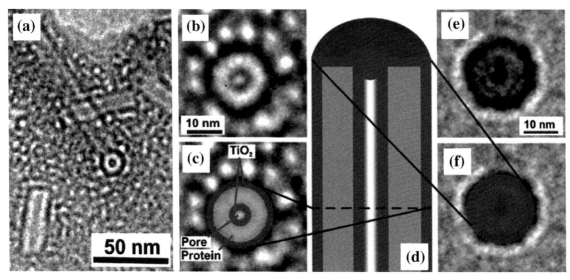

FIGURE 16.6   (a) TEM (200 kV) image of TMV treated with TiO$_2$ by ALD. A disk from a broken TMV (circular particle) embedded in an amorphous TiO$_2$ film presents an axial view. The TiO$_2$ covering the interior channel appears hollow with a channel diameter of 1–1.5 nm and a wall thickness of 1 nm. The covered inner channel of the virus appears brighter along the axis, indicating a hollow TiO$_2$ nanotube. (b) Magnification of a further TiO$_2$-covered disk showing a hollow area inside the TiO$_2$-coated interior channel of the virus. (c) Optically enhanced (b). The bright circle represents the viral protein sheath. Darker circles show the TiO$_2$ coating of the viral surface (outer surface and channel surface). The surrounding gray area is the embedding amorphous TiO$_2$ film. (d) Sketch of a cross-section of a TiO$_2$-covered TMV. In the top part of the virus, no channel is visible in the center; this part represents the assumed clogged area of the inner viral channel. (e) Magnification of a further TiO$_2$-covered disk showing a clogged interior channel of the virus. (f) Optically enhanced (e). Reprinted from Ref. 29. Copyright © 2006, with permission from the American Chemical Society.

substrates can be achieved, even in a patterned way [66, 67]. Gerasopoulos *et al.* combined those strategies to produce TMV patterns, which were metallized and subsequently coated by ALD [68]. The metallization enhanced the stability of the viruses for the subsequent deposition of TiO_2 or Al_2O_3 so that the processing temperature could be increased to 220 °C for Al_2O_3 and 150 °C for TiO_2. The resulting patterns are attractive for applications in catalysis, dye-sensitized solar cells (DSSC), etc.

We stated earlier that the ALD deposition is not directed (non-line-of-sight) but that all available surfaces are coated because of the chemical saturation mechanism. Nevertheless, there are limitations for the uniformity of coatings, which are mainly related to the diffusivity of the precursor molecules. Long, narrow channels require a seriously longer diffusion time for the precursor to traverse. The diffusion time increases as the diameter of the channels decreases with each ALD cycle. As excess precursor is applied to the substrate and the diffusion of the excess precursor out of the channels needs to be considered, the purging time is affected as well; it can easily increase the whole processing time by several orders of magnitude.

A very recent approach to apply ALD to biomaterials with extreme aspect ratios made use of bristles of a sea mouse, a species in the genus *Polychaeta* [69]. The bristles exhibit even, centimeter-long, parallel, hollow channels with diameters of around 200 nm. The structures served as templates for Al_2O_3 deposition with the goal to synthesize high-aspect-ratio nanotubes. The wall thickness of the resulting nanotubes was around 20 nm, but the total length of the nanotubes could not be determined, since it was difficult to completely release the nanotubes from the template. It remains for future work to enhance the efficiency of ALD for coating such extremely high-aspect-ratio structures.

Compared to all the aforementioned biomaterials, biominerals are much easier to process by ALD because they consist of inorganic materials.

Among the most prominent and beautiful structures are the exoskeletons of diatoms [70]. Many exoskeletons show perfect arrangements of pores and protrusions. A coating by ALD is easily possible for such templates, even at higher temperatures and with aggressive precursors [71]. After ALD processing, the pore size of the frustule valves shrinks from 40 nm to about 5 nm, with applications of such pore arrangements in molecular separation. A more demanding approach involved an initial modification of the biosilica with germanium and the formation of a photonic crystal slab [72]. The results appear promising for future investigation and optimization, particularly with respect to emission color and brightness.

16.2.2 Functional Mimicry

16.2.2.1 Optical Properties

Many natural materials show fascinating optical effects, whether for attracting insects (e.g., to flowers) or warning off enemies. In the case of flowers, one can observe a plethora of colors during the spring blooming period. The physics behind the coloration is more complex than usually considered by the observer. The human eye can recognize coloration in the visible spectral regime but is not aware of optical absorption or reflection in other spectral regimes. Some of those attract certain insects.

The optical properties of a red rose petal come from two coloration effects: the chemicals or pigments and the structural colors. Whereas the chemical coloration influences the optical appearance of the petal in the visible spectral regime, the structural colors are primarily based on the nano- and microstructure of the petal surface and are seen in the ultraviolet regime. Structural colors are discussed in detail in Chapter 11 by Dushkina and Lakhtakia. By twofold inversion of the rose petal with a Ni mold and a polymer, the structure of the rose petal was copied from a polymer, but without the pigmentation [73].

Amazingly, the polymer replica attracted insects, even without the pigment. ALD was applied to the rose petal to fine-tune the reflectance in the lower-ultraviolet regime. By coating with thin films of alumina (Al_2O_3), a shift of the reflectance peak as a function of the coating thickness was observed. The shift may be related to two effects: the modification of the structure's effective refractive index by adding alumina on top, and changing structural parameters by adding an inorganic film on top. Presumably, both of these effects have an influence on the shift of the reflectance peak.

A more exciting form of coloration is found in natural photonic crystals. Excellent examples are furnished by many butterfly wings with over-lays of micro- and nanostructures. In the work of Huang et al. [74], the wings of a blue-colored butterfly (*Peleides morpho*) were coated with Al_2O_3 by ALD. This technique for such approaches is attractive because of the non-line-of-sight deposition and the extreme conformity of the coating. The deposited films had thicknesses of 10–40 nm in steps of 10 nm. With increasing thickness of the coating, the optical appearance of the butterfly wing shifted from the original blue coloration toward pink (Figure 16.7). The optical behavior is influenced by the addition of Al_2O_3, which has a different refractive index than the chitin that the wing consists of. A removal of the chitin by thermal treatment resulted in an Al_2O_3 replica, which, again due to the change of the refractive index, showed a blue shift of the main reflectance peak from 550 nm.

The fascination with the colors of butterfly wings as well as the possibility of replicating and/or tuning those colors by ALD resulted in several investigations by other research groups. Gaillot et al. investigated the wings of another species of butterfly, *Papilio blumei*, which shows green iridescence with blue-colored tails [75]. The work was focused on the accessibility of the air pockets in the photonic structures to the metal–organic precursors. TiO_2 was the material of choice for the ALD process because of its

refractive index being higher than that of Al_2O_3. For the coating, two differing cases—(1) deposition on the top and bottom surfaces of the wing only, and (2) coating of the surfaces of the embedded air pockets—were implemented. For distinguishing those two cases, the shift of the main reflectance peak at 524 nm was observed in detail. When the structure is coated externally only, this peak shifts toward larger wavelengths by about 1 nm per nm TiO_2 coating thickness. When the air pockets are also coated, the peak shifts by 12 nm per nm TiO_2 coating thickness. The control of the coating, however, still seems difficult. The accessibility of the air pockets for the precursor is perhaps related to structural defects or cracks in the chitin, or it proceeds from the edges of the structure through inter-connected channels. Thus, coating of only the top and bottom surfaces can be guaranteed.

Further tuning of the ALD deposition on butterfly wings led to additional functionalities. In a recent work, Liu et al. showed that besides the optical features, the antiwetting properties of the butterfly wings could also be replicated [76]. The hydrophobicity is an important effect for the survival of a butterfly, since any humidity on the wing will make flying impossible. The nanoscale roughness of the structure appears to be more detrimental for hydrophilicity than the material the structure is composed of and therefore can be easily reproduced by ALD. More recently, even photovoltaic applications have been proposed. Through replication of the structure of the butterfly wing (here *Hypochrysops polycletus*), antireflective structures were produced in a solar cell stack [77].

Antireflective properties are also found in the eye of a household fly. The structure of the eye is complicated and consists of small lenslets (*ommatidia*), which contain further nanoscale protuberances. This combination results in anti-reflective properties above a wavelength of 400 nm with a reflectance peak at a wavelength of 330 nm. Similar to the butterfly wings, the fly eye was coated with Al_2O_3 at 100 °C [78].

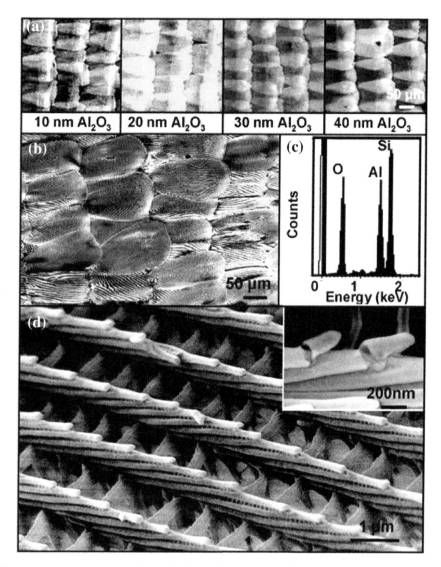

FIGURE 16.7 Images of the alumina replicas of the butterfly wing scales. (a) An optical microscope image of the alumina-coated butterfly wing scales, of which the color changed from original blue to pink. (b) A low-magnification SEM image of the alumina replicas of the butterfly wing scales on silicon substrate after the butterfly template was completely removed. (c) The energy-dispersive X-ray (EDX) spectrum of the alumina replica shown in (b). (d) A higher-magnification SEM image of an alumina-replicated scale, where the replica exhibits exactly the same fine structures. (e) An SEM image of two broken rib tips on an alumina replica. Reprinted from Ref. 74. Copyright © 2006, with permission from the American Chemical Society.

By annealing at 500 °C in air, the template was removed and the resulting alumina replica showed antireflective properties. Those were in agreement with the original structure, with a slight shift of the reflectance peak from 330 nm to 375 nm. From the point of view of ALD processing, this strategy could be adapted for technological applications, but the small size of the eyes limits the practical applicability, since competing strategies are currently more cost effective [79].

16.2.2.2 Wetting Behavior

One of the most important properties of structured surfaces in nature is the wetting behavior. The very delicate structure of butterfly wings shows water-repelling properties, which is important for the survival of the insect. A remarkable experiment showed the use of amorphous Al_2O_3, which is intrinsically hydrophilic, for replication of hydrophobic structures—namely, butterfly wings and water-strider legs. The wetting properties of the coated wings were subsequently compared to coated water-strider legs [80]. In spite of bulk Al_2O_3 being hydrophilic, the Al_2O_3 replicas were hydrophobic.

The hydrophobic nature of structured surfaces is characterized by two model states, the Cassie state [81] and the Wenzel state [82], which differ in their contact-angle hysteresis [83]. For antiwetting behavior, the Cassie state is favored. Both the butterfly wing and the water-strider leg show a Cassie-like behavior prior to ALD coating. The thin film of alumina on top of the structures induces some changes. Whereas the coated water-strider leg persists in the Cassie state, the coated butterfly wing changes to the Wenzel state. The reason for this differential behavior can be found in the chemical and structural composition of those two differing materials. In the case of the butterfly wing, the surface is natively coated with wax, which strongly contributes to the hydrophobic behavior. In contrast, the wetting behavior of the water-strider leg appears to be dominated by the structural properties. Namely, the aspect ratio of the structures in the water-strider legs is much higher than in the case of the butterfly wings, enabling the trapping of air, which strongly favors the superhydrophobic behavior.

For those experiments, ALD plays an important role because the distortion of the structural properties of the substrate should be reduced to a minimum to discriminate the impact of the surface chemistry and the structure on the water-repelling effect. In the cases of both the butterfly wings and the water-strider legs, the thin Al_2O_3 film seriously modified the surface chemistry, whereas the structure was only barely changed. The outcome shows that, with such experiments, one can quite easily determine the contributions of chemistry and structure to wetting phenomena of a surface.

16.2.3 Biocompatibility

Biocompatible coatings play an increasingly important role for biology and medicine, as discussed in detail in Chapters 7 and 8. Artificial substrates are commonly used for growth of cells and tissue or for implants. In many cases, those artificial materials can satisfy the required preconditions in terms of mechanical stability, weight, etc., but are not biocompatible. On the other hand, biocompatible materials are often too expensive, too heavy, or too brittle. A good example is furnished by titanium implants, which show very good biocompatibility and stability but are very expensive.

Thin-film coatings provide one possible solution for this problem. An implant or a substrate may be designed from a material that is not biocompatible but satisfies all other requirements. A thin film could be applied on top for biocompatibility, which could seriously decrease the cost of the production, provided that the thin film is biocompatible, stable, free of pinholes, and firmly attached to the substrate.

The latter two requirements are intrinsically provided by ALD. The thin-film coating will be conformal and pinhole free because of the self-limiting growth mechanism described in Section 16.1. The firm attachment of the coating to the surface of the substrate is provided by the chemical anchoring of the precursors to the functional groups of the substrate. Two remaining questions occur: Are ALD-deposited coatings biocompatible? Are those coatings mechanically and chemically stable? Stability can be obtained by the proper choice of thickness and material. Thus, the biocompatibility of ALD coatings seems to be the most important question for biological and medical applications of ALD, and it is astonishing that to date investigations have been sparse. Most of the published work concentrates either on the biocompatibility of materials or on the biocompatibility of the coated structures.

The most common material for ALD deposition is Al_2O_3. Finch et al. [84] investigated the proliferation of coronary artery smooth muscle cells on an Al_2O_3-coated glass with 60-nm–coating thickness and compared it to the proliferation on uncoated glass and a silane-terminated surface. The Al_2O_3 surface is terminated by hydroxyl groups and shows hydrophilic behavior similar to the glass surface. Consequently, the cell proliferation on the Al_2O_3 surface and the glass surface were similar, whereas the hydrophobic control surface (i.e., the silane-terminated glass) showed opposite behavior. This rather simple test indicated that Al_2O_3 films deposited by ALD are not less biocompatible than ordinary glass surfaces.

Putkonen et al. took a more direct approach to ALD-deposited biocompatible coatings [85]. The idea behind their work was to develop an ALD process for hydroxyapatite (HA). Being a ternary compound with many variation possibilities in stoichiometry, HA poses serious problems for ALD processing, since alternating pulses of two or three precursors cannot be applied to synthesize this material.

The approach of Putkonen et al. is more complex in comparison to a normal ALD process and is chemically very sophisticated. Their four-precursor ALD process consists of the precursor pairs $Ca(thd)_2/O_3$ and $(CH_3O)_3PO/H_2O$ and is performed at 300 °C. The first precursor pair produced a film of calcium carbonate. In the second stage, the carbonate groups were exchanged with phosphate groups by applying the second precursor pair. The carbon content after this process was still high, and thermal annealing in dry or moist N_2 was necessary to further reduce it. Eventually, HA films were obtained once the annealing temperature was above 500 °C. The films were tested for biocompatibility by attaching preosteoblast MC 3T3-E1 cells (Figure 16.8). Good biocompatibility was observed, especially with the annealed films; this was derived from the fact that the cells

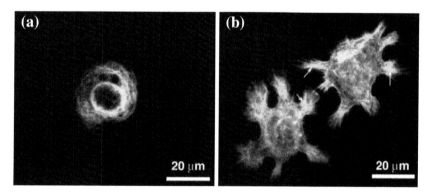

FIGURE 16.8 MC 3T3-E1 cells were grown for three hours on (a) as-deposited and (b) annealed hydroxyapatite films formed on Si(100) by ALD. On the annealed, crystallized film, the cells show clear lamellipodia and filopodia structures. Cytoskeletal actin was stained by Alexa 488–labeled phalloidin (green) and vinculin with a monoclonal antibody followed by Alexa 546–labeled anti-mouse antibody (red). Reprinted from Ref. 85. Copyright © 2009, with permission from Elsevier. (For interpretation of the references to color in this figure legend, the reader is referred to the web version of this book.)

started spreading and forming lamellipodia, which are actin projections on some motile cells involved in the process of cell migration. In spite of being a complicated process, the deposition of HA by ALD is a promising first step toward truly biocompatible thin films. Further optimization and simplification of the process will be required to make this approach economical.

Further investigations on the biocompatibility of ALD-deposited coatings involve structured substrates in addition to the chemical properties of the film. The wetting properties of surfaces have a strong influence on the biocompatibility. The wetting properties can be modified through the chemical functionalities on the surface but also strongly depend on the structural features of the surface [81, 82].

Porous anodic alumina (AAO) is a material with nanoscale pores on the surface and is used commonly by biomedical research groups worldwide. Besides the wet-chemical modification of AAO, porous alumina also has been coated with Pt, TiO_2, or ZnO by ALD. The coated porous alumina was tested for the proliferation of neonatal human epidermal keratinocytes [86–88]. Improvement in cell proliferation was observed when AAO was coated with TiO_2 and ZnO but not with platinum.

Hyde *et al.* investigated the functionalization of cotton fabrics by ALD for the sake of improved biocompatibility [89]. The cotton fabrics were coated with thin films of TiN at process temperatures around 150°C. Because cotton provides hydroxyl groups, the TiN film gets covalently bound to the cotton fibers. The TiN coating showed some oxidation, presumably on the surface, after storage in air for 10–15 days. Therefore, the films are better described as titanium oxynitride. The biocompatibility of the coated fabrics was investigated by the adherence of human adipose-derived adult stem cells (hADSC). The coating thickness and the wetting properties of the surface were taken into consideration. The biocompatibility was good for all coating thicknesses and the cell adhesion was maximized on the most hydrophobic surfaces with only about 2 nm thickness of TiN. It is not clear, however, to which extent the surface of the coating consisted of the nitride or the oxy-nitride.

TiO_2 is known to be biocompatible; thus, the designation of the nitride for biocompatibility has to be considered with care in this particular case. Nevertheless, with respect to the unlimited natural resources for cotton and the easy handling of the material, this approach seems to be very promising and, upon further optimization, could indeed lead to the inexpensive production of biocompatible substrates.

16.3 FUTURE PERSPECTIVES OF ALD IN BIOMIMETICS

16.3.1 Enzyme Mimetics

Thus far, our discussion has been confined to various approaches toward structural or functional mimetics involving thin films produced by ALD. One important aspect of the ALD technology is the processing of catalytically active materials. The example of the photocatalytically active ZnO and TiO_2 coatings applied to collagen was already mentioned. The resulting bioinorganic composite was able to generate H_2O_2 under ultraviolet irradiation and exhibited an antibactericidal effect [45].

Similarly, photocatalytically active structures were produced from pods (legumes) as templates [90]. The used legume contained microscaled tubular trichomes, which were coated with Al_2O_3 and ZnO. After calcination, a ZnAl-mixed metal-oxide framework (MMO) formed, showing crystalline phases of ZnO and $ZnAl_2O_4$. The latter ternary compound falls under the family of spinels, a group of minerals with cubic crystal structure and a chemical formulation of AB_2O_4 with A and B being metal ions and O being oxygen. The formation of $ZnAl_2O_4$ during calcination is expected because Al_2O_3 and ZnO have a great tendency for a nonequilibrated interfacial diffusion: The Zn ions diffuse much more quickly into Al_2O_3 than the Al ions into ZnO, which is compensated by a diffusion of voids from the Al_2O_3 into ZnO, eventually resulting in a void formation, the Kirkendall voids [91]. The resulting large surface area per unit mass improves the photocatalytic activity, which was confirmed with investigations of the degradation of two dyes, poly{1–4[4-(3-carboxy-4-hydroxyphenylazo) benzenesulfonamido]-1,2-ethanediyl sodium salt (PAZO) and sulforhodamine B, upon ultraviolet irradiation.

Besides the metal-oxide compounds, ALD offers the possibility to synthesize noble-metal nanoparticles. Although ALD is designed to produce thin films, there is a possibility to grow nanoparticles with good control to a certain level because, during the nucleation and initial growth, the deposited material forms islands on the substrate. Dependent on the substrate and the deposited material, the compact and pinhole-free film may be formed within few cycles, as expected from the ALD process, but also the film formation can be hindered and the initial clusters or islands grow to form particles. Examples of the latter case were shown with the deposition of Pt nanoparticles on carbon aerogels and strontium titanate nanocubes [92, 93]. In a similar fashion, catalytic Pd and bimetallic PtRu nanoparticles with a narrow size distribution were also synthesized (Figure 16.9) [94, 95].

Nanosized structures from those metals are highly desired for their catalytic properties [96, 97]. A significant application of such catalysts is in biomedicine, where there is now a focus on nanoparticles as enzyme mimetics [98]. For example, Pt nanoparticles show catalytic activities that resemble those of enzymes superoxide dismutase (SOD) and catalase, which help cells survive under oxidative stress [99, 100]. Nanoalloys of AgM (M = Au, Pd, and Pt) mimic the enzymes peroxidase and oxidase, which

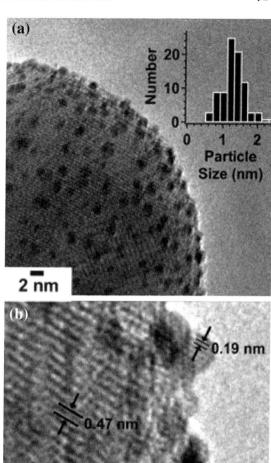

FIGURE 16.9 (a) TEM images of ALD-grown metal nanoparticles decorating an alumina sphere. The histogram gives the nanoparticle size distribution measured from TEM where the mean particle diameter is 1.2 nm with a distribution width of 0.3 nm. (b) High-resolution TEM image showing lattice fringes for the Al_2O_3 and the PtRu nanoparticles. Reprinted from Ref. 95. Copyright © 2010, with permission from the American Chemical Society.

have potential in immunoassays [101]. The same activity has been observed with CeO_2 and magnetic Fe_3O_4 nanoparticles [102, 103]. Deposited by ALD, all of these materials are potential enzyme-analog catalysts [104, 105].

However, investigations into the catalytic activities of metal nanoparticles synthesized by ALD are mainly focused on technological applications, such as for fuel cells [94, 95, 106]. The reason lies in the fact that most research groups making use of ALD have a background in various nonbiomedical areas, such as electronics. In addition, the majority of the recent ALD processes were developed and optimized to satisfy the needs of the electronic industry, which until now is the only large-scale industrial application field of ALD.

Further development of the precursors and processes is required for biomedical applications. The best candidates are Ag and Au, which can in principle be deposited by ALD, but either the processes are still very demanding or the required precursors are not commercially available. ALD as enabler technology for biomedical applications (e.g., enzyme mimetics) is an undiscovered field with great potential. Compared with conventional nanoparticle synthetic methods (e.g., wet chemistry), ALD provides complementary approaches, including precise control of both size and processing of complicated structures. Future growth is envisaged.

16.3.2 Biomineralization

Bio-organic molecules or materials show defined but complicated chemical compositions or arrangements. The adaptation of biomaterials to specific needs leads to insertion of metals into protein structures [107, 108]. Those insertions are in most cases related to the enhancement of mechanical properties. Proteins containing certain metal ions or compounds can become hard, stiff, tough, and so on. Bryan and Gibbs, who found that Zn incorporated into the protein matrix of the jaws of the marine polychaete worm *Nereis* contributes to its hardness, showed a good example of metal incorporation to change mechanical properties [109, 110].

Interactions of the ALD precursors with organic functional groups present on or in the maternal substrates may occur in many different ways. Such interactions are often the origin of certain improved or unexpected properties. This is to a certain extent in good agreement with many naturally occurring processes of biomineralization.

The pioneering work of Lee *et al.* showed an infiltration of metals into a protein by means of diffusion of ALD precursors into a spider silk [111]. The spider silk was chosen because the mechanical properties of natural spider silk are remarkable. In toughness it outperforms most humanmade materials, such as carbon fibers, poly-aramides, nylon, etc.

The materials of choice for the ALD process are Al2O3, TiO2, and ZnO because of their processability at low temperatures. Exposure of the spider silk to the first precursor is seriously extended from subseconds to minutes to enable the precursor to diffuse into the protein matrix and attach to functional groups such as alcohols, amines, etc. Metals became homogeneously distributed within the silk. Instead of expected hardening and stiffening, the silks turned more ductile and stronger, resulting in a tenfold increase in toughness (Figure 16.10).

A series of characterization experiments, including nuclear magnetic resonance (NMR), X-ray diffraction (XRD), transmission electron microsopy–energy-dispersive X-ray spectroscopy (TEM-EDX), Raman spectroscopy, etc., showed that the metal originating from the ALD precursor is indeed also found in the bulk of the spider silk. This proves infiltration of the protein from the gas phase. The proposed mechanism is a diffusion of the precursor into the bulk protein, interruption of hydrogen bonds between some protein chains, and insertion of metals into those bonds. After the metal infiltration was performed by ALD, the β sheets appeared to be

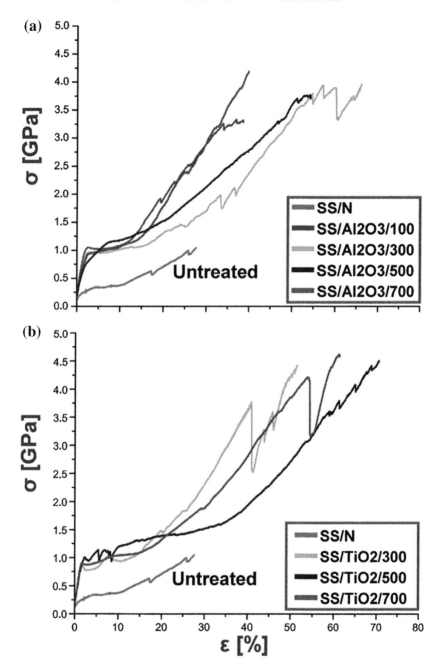

FIGURE 16.10 Tensile test curves of silk fiber samples treated by TMA/H_2O and TIP/H_2O precursor pairs and comparison to untreated samples and samples treated with various numbers of cycles. (a, b) Stress (σ) and strain (ε) curves of silk fibers treated with TMA/H_2O and TIP/H_2O pulse pairs with various numbers of cycles ranging from 100 to 700. Reprinted from Ref. 111. Copyright © 2009, with permission from the American Association for the Advancement of Science.

seriously affected, i.e., their sizes were reduced (Figure 16.11). The size, number, and distribution of the β sheets seem to determine the mechanical properties of the silk [112], showing good agreement between theory and experiment in this particular case.

Similarly, increased mechanical toughness has been observed when dried collagen is processed by ALD diethylzinc or titanium(tetraisopropoxide) [46]. More in-depth investigation has shown that the effect results from the incorporation of metal into the bulk of the collagen during the ALD (Figure 16.12). Evidently, the lack of solvents (due to the vacuum-based processing) permits a diffusion of the reactive species into the collagen structure and induces small changes with significant effects.

The overall mechanical toughness was increased threefold, simultaneously increasing the stress and strain of the measured substrates. The origin of the property change has not been resolved in detail yet. However, investigations until now have shown that, besides the presence of the metal inside the protein matrix, some chemical and physical changes occur.

Based on investigations of the infiltrated collagen by XRD, Raman spectroscopy, infrared spectroscopy, etc., a model has been proposed that considers chemical bonding of the metal to the protein helix and the corresponding change in the crystallinity of the material [46]. The structural and chemical properties of collagen differ from those of spider silk. Spider silk shows a mixture of amorphous protein and β sheets; collagen

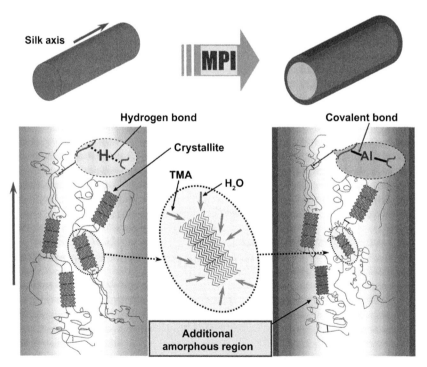

FIGURE 16.11 Schematic description of proposed molecular changes in the silk modified by MPI. Together with the water pulses, Al^{3+} weakens the hydrogen bonds and inserts into the broken bonding sites, thereby resulting in the formation of metal-coordinated or even covalent bonds with the Al. Reprinted from Ref. 111. Copyright © 2009, with permission from the American Association for the Advancement of Science.

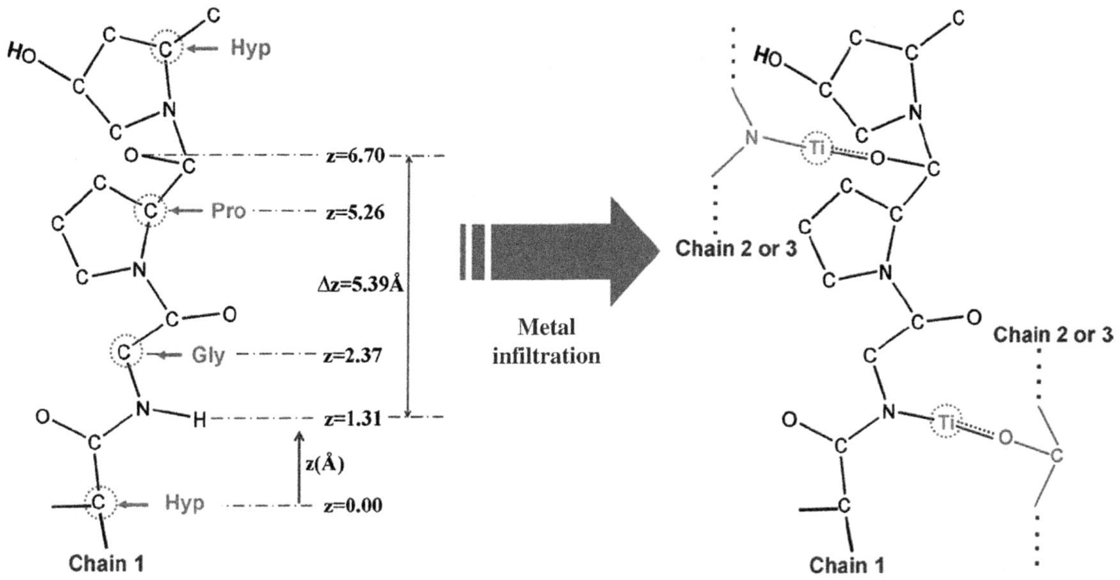

FIGURE 16.12 Speculative conformational change of collagen chains after metal infiltration. The repeating sequence is Gly-Pro-Hyp. The z coordinates are listed (left side). The right side shows a schematic description conjectured from WAXS and Raman spectra. Reprinted from Ref. 46. Copyright © 2010, with permission from the American Chemical Society.

mainly consists of protein helices. The hydrogen bonding between the functional groups of the collagen seems to be affected by intercalation of the metals, which eventually leads to new reflections observed by XRD as well as a new bonding situation that was observed by Raman spectroscopy.

Again, indications could be found that chemical interactions of the metal-organic precursor with the protein occur. One may argue that this process will inevitably lead to a coating of the fiber in parallel and that the ceramic (titania or alumina) coating may have strong impact on the mechanical properties of the resulting product. For this reason, the same process was also performed in a regular coating mode (exposure time equal to pulsing time) with very minor influence on the mechanical properties of the fiber [113]. The process with extended exposure times for infiltration of the proteins with metals was named *multiple-pulsed vapor phase infiltration* (MPI) to distinguish between the original coating mode and the infiltration mode.

Processes of improving or modifying mechanical properties by controlled and organized incorporation of inorganic minerals into an organic matrix exist in biology. Such processes are collectively known as *biomineralization*. A very common example is the formation of bone. The different mechanical properties of various types of bone (woven bone, cortical bone, etc.) are derived from the varying amount of mineral content (e.g., HA) within a matrix of collagen and other types of proteins [114]. Because the ALD processing of spider silk and collagen leads to similar outcomes, this process mimics biomineralization. Another example of biomineralization mimicry was discussed in Section 16.2.1: the ALD of TiO_2 within the cavity of apoferritin. The process is in a sense comparable to the mineralization of apoferritin to ferritin [53].

Those initial approaches toward the mimicry of biomineralization by ALD were rather curiosity-driven. The infiltration is a very interesting side effect of the ALD process to be considered if soft

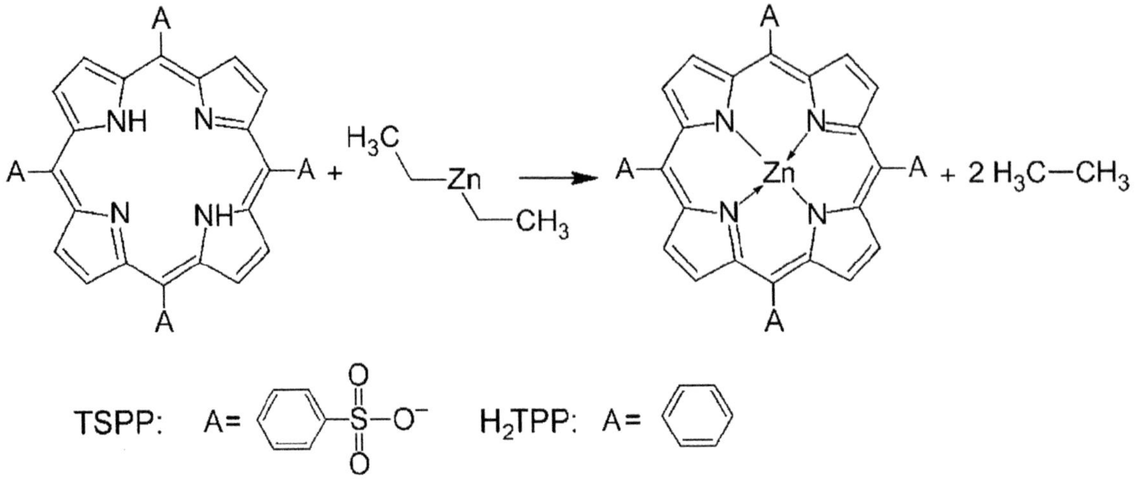

FIGURE 16.13 Chemical reaction between diethylzinc and porphyrins during ALD. Reprinted from Ref. 117. Copyright © 2009, with permission from Wiley-VCH.

materials are to be processed. Indications for infiltration by ALD had been observed earlier [115] and considered to be a drawback of the process for coating of polymers. For a more controlled and precise design of materials, detailed investigations and understanding of the chemistry and physics behind the observed phenomena are indispensible.

Site-specific reactions of the ALD precursors with bio-organic functionalities are gaining importance [116], but research has been scanty. For example, after processing porphyrin with diethylzinc (DEZ), a site-specific reaction of the precursor with the pyrrolic-NH groups was observed (Figure 16.13) [117]. This result indicated that primary and secondary amines are potential reaction sites for the Zn precursor. Further studies have shown that the infiltration, particularly with DEZ, can also be beneficially applied to produce nanostructures from block copolymers [118] or indicated that C=C and ester bonds are in some cases reactive with TMA [119, 120]. Even C–F bonds appear reactive with ALD precursors, at least with DEZ [121]. However, one has to consider that identical functional groups have differing chemical reactivity in differing chemical environments. Therefore, the mentioned few experiments are only a small indicator for possible chemical modifications.

16.4 SUMMARY

The application of atomic layer deposition to biomaterials or for mimicking biological processes is an emerging research direction. It is apparent from the published work that ALD comes into focus whenever the other coating technologies reach their limits. A very obvious example is the delicate structure of butterfly wings. ALD is the method of choice to uniformly coat all visible and invisible surfaces as long as those are accessible by the vaporized precursors. Further examples of the application of the ALD process include mineralization processes within protein-based structures. The lifetime of the metal-organic molecules is increased, allowing the molecules to diffuse into soft matter and react with functional groups within the bulk.

The various aspects of the ALD process, structural replication, enhancement or introduction of functionalities, biocompatible coatings, catalysis, and mineralization are all in a very early stage of development. The tendency toward multidisciplinary research will, sooner rather than later, lead to new and fascinating scientific insights as well as exciting new applications.

References

[1] S.I. Kol'tsov, Production and investigation of reaction products of titanium tetrachloride with silica gel, *J. Appl. Chem. USSR* **42** (1969), 1023.

[2] T. Suntola and J. Antson, *Method for producing compound thin films*, US Patent 4,058,430 (issued 25 November 1977).

[3] N. Pinna and M. Knez, *Atomic layer deposition of nanostructured materials*, Wiley, Weinheim, Germany (2012).

[4] M. Knez, K. Nielsch, and L. Niinisto, Synthesis and surface engineering of complex nanostructures by atomic layer deposition, *Adv Mater* **19** (2007), 3425–3438.

[5] S.M. George, Atomic layer deposition: an overview, *Chem Rev* **110** (2010), 111–131.

[6] H. Kim, H.-B.-R. Lee, and W.-J. Maeng, Applications of atomic layer deposition to nanofabrication and emerging nanodevices, *Thin Solid Films* **517** (2009), 2563–2580.

[7] M. Leskelä and M. Ritala, Atomic layer deposition chemistry: recent developments and future challenges, *Angew Chem Int Ed* **42** (2003), 5548–5554.

[8] J.T. Korhonen, P. Hiekkataipale, J. Malm, M. Karppinen, O. Ikkala, and R.H.A. Ras, Inorganic hollow nanotube aerogels by atomic layer deposition onto native nanocellulose templates, *ACS Nano* **5** (2011), 1967–1974.

[9] R.L. Puurunen, Surface chemistry of atomic layer deposition: a case study for the trimethylaluminum/water process, *J Appl Phys* **97** (2005), 121301–121352.

[10] H.B. Profijt, S.E. Potts, M.C.M. van de Sanden, and W.M.M. Kessels, Plasma-assisted atomic layer deposition: basics, opportunities, and challenges, *J Vac Sci Technol A* **29** (2011), 050801.

[11] T. Yoshimura, S. Tatsuura, and W. Sotoyama, Polymer-films formed with monolayer growth steps by molecular layer deposition, *Appl Phys Lett* **59** (1991), 482–484.

[12] D.H. Levy, D. Freeman, S.F. Nelson, P.J. Cowdery-Corvan, and L.M. Irving, Stable ZnO thin film transistors by fast open air atomic layer deposition, *Appl Phys Lett* **92** (2008), 192101.

[13] M.D. Groner, F.H. Fabreguette, J.W. Elam, and S.M. George, Low-temperature Al$_2$O$_3$ atomic layer deposition, *Chem Mater* **16** (2004), 639–645.

[14] S.K. Kim, S.W. Lee, C.S. Hwang, Y.-S. Min, J.Y. Won, and J. Jeong, Low temperature (<100 °C) deposition of aluminum oxide thin films by ALD with O$_3$ as oxidant, *J Electrochem Soc* **153** (2006), F69–F76.

[15] E. Langereis, M. Creatore, S.B.S. Heil, M.C.M. van den Sanden, and W.M.M. Kessels, Plasma-assisted atomic layer deposition of Al$_2$O$_3$ moisture permeation barriers on polymers, *Appl Phys Lett* **89** (2006), 081915.

[16] M. Putkonen and L. Niinisto, Atomic layer deposition of B$_2$O$_3$ thin films at room temperature, *Thin Solid Films* **514** (2006), 145–149.

[17] Y. Luo, D. Slater, M. Han, J. Moryl, and R.M.J. Osgood, Low-temperature, chemically driven atomic-layer epitaxy: in situ monitored growth of CdS/ZnSe (1 0 0), *Appl Phys Lett* **71** (1997), 3799–3801.

[18] Y. Lu, S. Bangsarutip, X. Wang, L. Zhang, Y. Nishi, and H. Dai, DNA functionalization of carbon nanotubes for ultrathin atomic layer deposition of high k dielectrics for nanotubes transistors with 60 mV/decade switching, *J Am Chem Soc* **128** (2006), 3518–3519.

[19] G.A.T. Eyck, S. Pimanpang, H. Bakhru, T.-M. Lu, and G.-C. Wang, Atomic layer deposition of Pd on an oxidized metal substrate, *Chem Vap Deposit* **12** (2006), 290–294.

[20] G.A.T. Eyck, J.J. Senkevich, F. Tang, D. Liu, S. Pimanpang, T. Karaback, G.-C. Wang, T.-M. Lu, C. Jezewski, and W.A. Lanford, Plasma-assisted atomic layer deposition of palladium, *Chem Vap Deposit* **11** (2005), 60–66.

[21] H.C.M. Knoops, A.J.M. Mackus, M.E. Donders, M.C.M. van den Sanden, P.H.L. Notten, and W.M.M. Kessels, Remote plasma ALD of platinum and platinum oxide films, *Electrochem Solid-State Lett* **12** (2009), G34–G36.

[22] W. Gasser, Y. Uchida, and M. Matsumura, Quasi-monolayer deposition of silicon dioxide, *Thin Solid Films* **250** (1994), 213–218.

[23] J.W. Klaus, O. Sneh, and S.M. George, Growth of SiO$_2$ at room temperature with the use of catalyzed sequential half-reactions, *Science* **278** (1997), 1934–1936.

[24] J.W. Elam, D.A. Baker, A.J. Hryn, A.B.F. Martinson, M.J. Pellin, and J.T. Hupp, Atomic layer deposition of tin oxide films using tetrakis(dimethylamino) tin, *J Vac Sci Technol A* **26** (2008), 244–252.

[25] K. Kukli, J. Aarik, A. Aidla, O. Kohan, T. Uustare, and V. Sammelselg, Properties of tantalum oxide thin films grown by atomic layer deposition, *Thin Solid Films* **260** (1995), 135–142.

[26] S.B.S. Heil, F. Roozeboom, M.C.M. van den Sanden, and W.M.M. Kessels, Plasma-assisted atomic layer deposition of Ta$_2$O$_5$ from alkylamide precursor and remote O$_2$ plasma, *J Vac Sci Technol A* **26** (2008), 472–480.

[27] S.M. Rossnagel, A. Sherman, and F. Turner, Plasma-enhanced atomic layer deposition of Ta and Ti for interconnect diffusion barriers, *J Vac Sci Technol A* **18** (2000), 2016–2020.

[28] H. Kim and S.M. Rossnagel, Growth kinetics and initial stage growth during plasma-enhanced Ti atomic layer deposition, *J Vac Sci Technol A* **20** (2002), 802–808.

[29] M. Knez, A. Kadri, C. Wege, U. Gösele, H. Jeske, and K. Nielsch, Atomic layer deposition on biological macromolecules: Metal oxide coating of tobacco mosaic virus and ferritin, *Nano Lett* **6** (2006), 1172–1177.

[30] J. Aarik, A. Aidla, T. Uustare, and V. Sammelselg, Morphology and structure of TiO₂ thin films grown by atomic layer deposition, *J Cryst Growth* **148** (1995), 268–275.

[31] J. Keranen, C. Guimon, E. Iiskola, A. Auroux, and L. Niinisto, Surface-controlled gas-phase deposition and characterization of highly dispersed vanadia on silica, *J Phys Chem B* **107** (2003), 10773–10784.

[32] E. Guziewicz, I.A. Kowalik, M. Godlewski, K. Kopalko, V. Osinniy, A. Wójcik, S. Yatsunenko, E. Łusakowska, W. Paszkowicz, and M. Guziewicz, Extremely low temperature growth of ZnO by atomic layer deposition, *J Appl Phys* **103** (2008), 033515.

[33] D.M. King, X. Liang, P. Li, and A.W. Weimer, Low-temperature atomic layer deposition of ZnO films on particles in a fluidized bed reactor, *Thin Solid Films* **516** (2008), 8517–8523.

[34] J. Meyer, P. Görrn, F. Bertram, S. Hamwi, T. Winkler, H.H. Johannes, T. Weimann, P. Hinze, T. Riedl, and W. Kowalsky, Al₂O₃/ZrO₂ nanolaminates as ultrahigh gas-diffusion barriers—a strategy for reliable encapsulation of organic electronics, *Adv Mater* **21** (2009), 1845–1849.

[35] J. Richter, R. Seidel, R. Kirsch, M. Mertig, W. Pompe, J. Plaschke, and H.K. Schackert, Nanoscale palladium metallization of DNA, *Adv Mater* **12** (2000), 507–510.

[36] M. Knez, K. Nielsch, A.J. Patil, S. Mann, and U. Gösele, Atomic layer deposition on biological macromolecules, *ECS Trans* **3** (2007), 219–225.

[37] D.B. Farmer and R.G. Gordon, Atomic layer deposition on suspended single-walled carbon nanotubes via gas-phase noncovalent functionalization, *Nano Lett* **6** (2006), 699–703.

[38] G. Singh, A.M. Bittner, S. Loscher, N. Malinowski, and K. Kern, Electrospinning of diphenylalanine nanotubes, *Adv Mater* **20** (2008), 2332–2336.

[39] T.H. Han, J.K. Oh, J.S. Park, S.-H. Kwon, S.-W. Kim, and S.O. Kim, Highly entangled hollow TiO₂ nanoribbons templating diphenylalanine assembly, *J Mater Chem* **19** (2009), 3512–3516.

[40] S.-W. Kim, T.H. Han, J. Kim, H. Gwon, H.-S. Moon, S.-W. Kang, S.O. Kim, and K. Kang, Fabrication and electrochemical characterization of TiO₂ three-dimensional

nanonetwork based on peptide assembly, *ACS Nano* **3** (2009), 1085–1090.

[41] T.H. Han, H.-S. Moon, J.O. Hwang, S.I. Seok, S.H. Im, and S.O. Kim, Peptide-templating dye-sensitized solar cells, *Nanotechnology* **21** (2010), 185601.

[42] G.-M. Kim, S.-M. Lee, G.H. Michler, H. Roggendorf, U. Gösele, and M. Knez, Nanostructured pure anatase titania tubes replicated from electrospun polymer fiber templates by atomic layer deposition, *Chem Mater* **20** (2008), 3085–3091.

[43] R. Wang, K. Hashimoto, A. Fujishima, M. Chikuni, E. Kojima, A. Kitamura, M. Shimohigoshi, and T. Watanabe, Light-induced amphiphilic surfaces, *Nature* **388** (1997), 431–432.

[44] U. Bach, D. Lupo, P. Comte, J.E. Moser, F. Weissortel, J. Salbeck, H. Spreitzer, and M. Grätzel, Solid-state dye-sensitized mesoporous TiO₂ solar cells with high photon-to-electron conversion efficiencies, *Nature* **395** (1998), 583–585.

[45] S.-M. Lee, G. Grass, G.-M. Kim, C. Dresbach, L. Zhang, U. Gösele, and M. Knez, Low-temperature ZnO atomic layer deposition on biotemplates: flexible photocatalytic ZnO structures from eggshell membranes, *Phys Chem Chem Phys* **11** (2009), 3608–3614.

[46] S.-M. Lee, E. Pippel, O. Moutannabir, I. Gunkel, T. Thurn-Albrecht, and M. Knez, Improved mechanical stability of dried collagen membrane after metal infiltration, *ACS Appl Mater Interf* **2** (2010), 2436–2441.

[47] M. Kemell, V. Pore, M. Ritala, M. Leskelä, and M. Linden, Atomic layer deposition in nanometer-level replication of cellulosic substances and preparation of photocatalytic TiO₂/cellulose composites, *J Am Chem Soc* **127** (2005), 14178–14179.

[48] M. Kemell, V. Pore, M. Ritala, and M. Leskelä, Ir/oxide/cellulose composites for catalytic purposes prepared by atomic layer deposition, *Chem Vap Deposit* **12** (2006), 419–422.

[49] D.K. Hyde, K.J. Park, S.M. Stewart, J.P. Hinestroza, and G.N. Parsons, Atomic layer deposition of conformal inorganic nanoscale coatings on three-dimensional natural fiber systems: effect of surface topology on film growth characteristics, *Langmuir* **23** (2007), 9844–9849.

[50] G.K. Hyde, G. Scarel, J.C. Spagnola, Q. Peng, K. Lee, B. Gong, K.G. Roberts, K.M. Roth, C.-A. Hanson, C.K. Devine, S.M. Stewart, D. Hojo, J.-S. Na, J.S. Jur, and G.N. Parsons, Atomic layer deposition and abrupt wetting transitions on nonwoven polypropylene and woven cotton fabrics, *Langmuir* **26** (2010), 2550–2558.

[51] R. de la Rica and H. Matsui, Applications of peptide and protein-based materials in bionanotechnology, *Chem Soc Rev* **39** (2010), 3499–3509.

[52] J. Liu, Y. Mao, E. Lan, D.R. Banatao, G.J. Forse, J. Lu, H.-O. Blom, T.O. Yeates, B. Dunn, and J.P. Chang, Generation of oxide nanopatterns by combining self-assem-

bly of S-layer proteins and area-selective atomic layer deposition, *J Am Chem Soc* **130** (2008), 16908–16913.

[53] H. Kim, E. Pippel, U. Gösele, and M. Knez, Titania nanostructures fabricated by atomic layer deposition using spherical protein cages, *Langmuir* **25** (2009), 13284–13289.

[54] M.J. Beijerinck, Ueber ein contagium vivum fuidum als Ursache der Fleckenkrankheit der Tabaksblätter, *Verh Kon Akad Wetesch* **65** (1898), 3–21.

[55] G.A. Kausche and H. Ruska, The visualisation of adsorption of metal colloids on protein bodies: the reaction between colloidal gold—tobacco mosaic virus, *Kolloid Z* **89** (1939), 21–26.

[56] W. Shenton, T. Douglas, M. Young, G. Stubbs, and S. Mann, Inorganic–organic nanotube composites from template mineralization of tobacco mosaic virus, *Adv Mater* **11** (1999), 253–256.

[57] C.E. Fowler, W. Shenton, G. Stubbs, and S. Mann, Tobacco mosaic virus liquid crystals as templates for the interior design of silica mesophases and nanoparticles, *Adv Mater* **13** (2001), 1266–1269.

[58] E. Dujardin, C. Peet, G. Stubbs, J.N. Culver, and S. Mann, Organization of metallic nanoparticles using tobacco mosaic virus templates, *Nano Lett* **3** (2003), 413–417.

[59] M. Knez, M. Sumser, A.M. Bittner, C. Wege, H. Jeske, S. Kooi, M. Burghard, and K. Kern, Electrochemical modification of individual nano-objects, *J Electroanal Chem* **522** (2002), 70–74.

[60] M. Knez, A.M. Bittner, F. Boes, C. Wege, H. Jeske, E. Maiss, and K. Kern, Biotemplate synthesis of 3-nm nickel and cobalt nanowires, *Nano Lett* **3** (2003), 1079–1082.

[61] M. Knez, M. Sumser, A.M. Bittner, C. Wege, H. Jeske, T.P. Martin, and K. Kern, Spatially selective nucleation of metal clusters on the tobacco mosaic virus, *Adv Funct Mater* **14** (2004), 116–124.

[62] S. Balci, A.M. Bittner, K. Hahn, C. Scheu, M. Knez, A. Kadri, C. Wege, H. Jeske, and K. Kern, Copper nanowires within the central channel of tobacco mosaic virus particles, *Electrochim Acta* **51** (2006), 6251–6257.

[63] K. Gerasopoulos, M. McCarthy, E. Royston, J.N. Culver, and R. Ghodssi, Nanostructured nickel electrodes using the Tobacco mosaic virus for microbattery applications, *J Micromech Microeng* **18** (2008), 104003.

[64] E. Royston, A. Ghosh, P. Kofinas, M.T. Harris, and J.N. Culver, Self-assembly of virus-structured high surface area nanomaterials and their application as battery electrodes, *Langmuir* **24** (2008), 906–912.

[65] R.J. Tseng, C.L. Tsai, L.P. Ma, and J.Y. Ouyang, Digital memory device based on tobacco mosaic virus conjugated with nanoparticles, *Nat Nanotechnol* **1** (2006), 72–77.

[66] M. Knez, M.P. Sumser, A.M. Bittner, C. Wege, H. Jeske, D.M.P. Hoffmann, K. Kuhnke, and K. Kern, Binding the tobacco mosaic virus to inorganic surfaces, *Langmuir* **20** (2004), 441–447.

[67] S. Balci, D.M. Leinberger, M. Knez, A.M. Bittner, F. Boes, A. Kadri, C. Wege, H. Jeske, and K. Kern, Printing and aligning mesoscale patterns of tobacco mosaic virus on surfaces, *Adv Mater* **20** (2008), 2195–2200.

[68] K. Gerasopoulos, M. McCarthy, P. Banerjee, X. Fan, J.N. Culver, and R. Ghodssi, Biofabrication methods for the patterned assembly and synthesis of viral nano templates, *Nanotechnology* **21** (2010), 055304.

[69] F. Mumm, M. Kemell, M. Leskelä, and P. Sikorski, A bio-originated porous template for the fabrication of very long, inorganic nanotubes and nanowires, *Bioinsp Biomim* **5** (2010), 026005.

[70] N. Kröger and N. Poulsen, Diatoms—from cell wall biogenesis to nanotechnology, *Annu Rev Genetics* **42** (2008), 83–107.

[71] D. Losic, G. Triani, P.J. Evans, A. Atanacio, J.G. Mitchell, and N.H. Voelcker, Controlled pore structure modification of diatoms by atomic layer deposition, *J Mater Chem* **16** (2006), 4029–4034.

[72] G.L. Rorrer, C. Jeffryes, C.-H. Chang, D.-H. Lee, T. Gutu, J. Jiao, and R. Solanki, Biological fabrication of nanostructured silicon-germanium photonic crystals possessing unique photoluminescent and electroluminescent properties, *Proc SPIE* **6645** (2007), 66450A.

[73] S.-M. Lee, J. Üpping, A. Bielawny, and M. Knez, The structure based color of natural petals discriminated by polymer replication, *ACS Appl Mater Interf* **3** (2011), 30–34.

[74] J. Huang, X. Wang, and Z.L. Wang, Controlled replication of butterfly wings for achieving tunable photonic properties, *Nano Lett* **6** (2006), 2325–2331.

[75] D.P. Gaillot, O. Deparis, V. Welch, B.K. Wagner, J.P. Vigneron, and C.J. Summers, Composite organic-inorganic butterfly scales: production of photonic structures with atomic layer deposition, *Phys Rev E* **78** (2008), 031922.

[76] F. Liu, Y.P. Liu, L. Huang, X.H. Hu, B.Q. Dong, W.Z. Shi, Y.Q. Xie, and X.A. Ye, Replication of homologous optical and hydrophobic features by templating wings of butterflies *Morpho menelaus*, *Opt Commun* **284** (2011), 2376–2381.

[77] X. Tang, L.A. Francis, P. Simonis, M. Haslinger, R. Delamare, O. Deschaume, D. Flandre, P. Defrance, A.M. Jonas, J.P. Vigneron, and J.P. Raskin, Room temperature atomic layer deposition of Al_2O_3 and replication of butterfly wings for photovoltaic application, *J Vac Sci Technol A* **30** (2012), 01A146.

[78] J. Huang, X. Wang, and Z.L. Wang, Bio-inspired fabrication of anti-reflection nanostructures by replicating fly eyes, *Nanotechnology* **19** (2008), 025602.

[79] R. Brunner, A. Deparnay, M. Helgert, M. Burkhardt, T. Lohmuller, and J.P. Spatz, Product piracy from nature:

biomimetic microstructures and interfaces for high-performance optics, *Proc SPIE* **7057** (2008), 705705.

[80] Y. Ding, S. Xu, Y. Zhang, A.C. Wang, M.H. Wang, Y. Xiu, C.P. Wong, and Z.L. Wang, Modifying the anti-wetting property of butterfly wings and water-strider legs by atomic layer deposition coating: Surface materials versus geometry, *Nanotechnology* **19** (2008), 355708.

[81] A.B.D. Cassie and S. Baxter, Wettability of porous surfaces, *Trans Faraday Soc* **40** (1944), 546–551.

[82] R.N. Wenzel, Resistance of solid surfaces to wetting by water, *Ind Eng Chem* **28** (1936), 988–994.

[83] D. Quere, Non-sticking drops, *Rep Prog Phys* **68** (2005), 2495–2532.

[84] D.S. Finch, T. Oreskovic, K. Ramadurai, C.F. Herrmann, S.M. George, and R.L. Mahajan, Biocompatibility of atomic layer-deposited alumina thin films, *J Biomed Mater Res A* **87A** (2008), 100–106.

[85] M. Putkonen, T. Sajavaara, P. Rahkila, L. Xu, S. Cheng, L. Niinistö, and H.J. Whitlow, Atomic layer deposition and characterization of biocompatible hydroxyapatite thin films, *Thin Solid Films* **517** (2009), 5819–5824.

[86] R.J. Narayan, N.A. Monteiro-Riviere, R.L. Brigmon, M.J. Pellin, and J.W. Flam, Atomic layer deposition of TiO$_2$ thin films on nanoporous alumina templates: medical applications, *JOM* **61** (2009), 12–16.

[87] R.J. Narayan, S.P. Adiga, M.J. Pellin, L.A. Curtiss, S. Stafslien, B. Chisholm, N.A. Monteiro-Riviere, R.L. Brigmon, and J.W. Elam, Atomic layer deposition of nanoporous biomaterials, *Mater Today* **13** (3) (March 2010), 60–64.

[88] R.J. Narayan, S.P. Adiga, M.J. Pellin, L.A. Curtiss, A.J. Hryn, S. Stafslien, B. Chisholm, C.-C. Shih, C.-M. Shih, S.-J. Lin, Y.-Y. Su, C. Jin, J. Zhang, N.A. Monteiro-Riviere, and J.W. Elam, Atomic layer deposition-based functionalization of materials for medical and environmental health applications, *Phil Trans R Soc Lond A* **368** (2010), 2033–2064.

[89] G.K. Hyde, S.D. McCullen, S. Jeon, S.M. Stewart, H. Jeon, E.G. Loboa, and G.N. Parsons, Atomic layer deposition and biocompatibility of titanium nitride nano-coatings on cellulose fiber substrates, *Biomed Mater* **4** (2009), 025001.

[90] Y. Zhao, M. Wei, Z.L. Wang, and X. Duan, Biotemplated hierarchical nanostructure of layered double hydroxides with improved photocatalysis performance, *ACS Nano* **3** (2009), 4009–4016.

[91] H.J. Fan, M. Knez, R. Scholz, K. Nielsch, E. Pippel, D. Hesse, M. Zacharias, and U. Gösele, Monocrystalline spinel nanotube fabrication based on the Kirkendall effect, *Nat Mater* **5** (2006), 627–631.

[92] J.S. King, A. Wittstock, J. Biener, S.O. Kucheyev, Y.M. Wang, T.F. Baumann, S.K. Gin, A.V. Hamza, M. Bauemer, and S.F. Bent, Ultralow loading Pt nanocatalysts prepared by atomic layer deposition on carbon aerogels, *Nano Lett* **8** (2008), 2405–2409.

[93] S.T. Christensen, J.W. Elam, F.A. Rabuffetti, Q. Ma, S.J. Weigand, B. Lee, S. Seifert, P.C. Stair, K.R. Poeppelmeier, M.C. Hersam, and M.J. Bedzyk, Controlled growth of platinum nanoparticles on strontium titanate nanocubes by atomic layer deposition, *Small* **5** (2009), 750–757.

[94] H. Feng, J.W. Elam, J.A. Libera, W. Setthapun, and P.C. Stair, Palladium catalysts synthesized by atomic layer deposition for methanol decomposition, *Chem Mater* **22** (2010), 3133–3142.

[95] S.T. Christensen, H. Feng, J.L. Libera, N. Guo, J.T. Miller, P.C. Stair, and J.W. Elam, Supported RuPt bimetallic nanoparticle catalysts prepared by atomic layer deposition, *Nano Lett* **10** (2010), 3047–3051.

[96] S. Ikeda, S. Ishino, T. Harada, N. Okamoto, T. Sakata, H. Mori, S. Kuwabata, T. Torimoto, and M. Matsumura, Ligand-free platinum nanoparticles encapsulated in a hollow porous carbon shell as a highly active heterogeneous hydrogenation catalyst, *Angew Chem Int Ed* **45** (2006), 7063–7066.

[97] M. Moreno-Manas and R. Pleixats, Formation of carbon-carbon bonds under catalysis by transition-metal nanoparticles, *Acc Chem Res* **36** (2003), 638–643.

[98] X.N. Hu, J.B. Liu, S. Hou, T. Wen, W.Q. Liu, K. Zhang, W.W. He, Y.L. Ji, H.X. Ren, Q. Wang, and X.C. Wu, Research progress of nanoparticles as enzyme mimetics, *Sci China* **54** (2011), 1749–1756.

[99] L. Zhang, L. Laug, W. Muenchgesang, E. Pippel, M. Brandsch, and M. Knez, Reducing stress on cells with apoferritin-encapsulated platinum nanoparticles, *Nano Lett* **10** (2010), 219–223.

[100] L. Zhang, W. Fisher, E. Pippel, G. Hause, M. Brandsch, and M. Knez, Receptor-mediated cellular uptake of nanoparticles: a switchable delivery system, *Small* **7** (2011), 1538–1541.

[101] W.W. He, X.C. Wu, and J.B. Liu, Design of AgM bimetallic alloy nanostructures (M = Au, Pd, Pt) with tunable morphology and peroxidase-like activity, *Chem Mater* **22** (2010), 2988–2994.

[102] A. Asati, S. Santra, and C. Kaittanis, Oxidase-like activity of polymer-coated cerium oxide nanoparticles, *Angew Chem Int Ed* **48** (2009), 2308–2312.

[103] L. Gao, J. Zhuang, and L. Nie, Intrinsic peroxidase-like activity of ferromagnetic nanoparticles, *Nat Nanotechnol* **2** (2007), 577–583.

[104] J. Paivasaari, M. Putkonen, and L. Niinisto, Cerium dioxide buffer layers at low temperature by atomic layer deposition, *J Mater Chem* **12** (2002), 1828–1832.

[105] J. Bachmann, J. Jing, M. Knez, S. Barth, H. Shen, S. Mathur, U. Gösele, and K. Nielsch, Ordered iron oxide nanotube arrays of controlled geometry and tunable magnetism by atomic layer deposition, *J Am Chem Soc* **129** (2007), 9554–9555.

[106] X. Jiang, T.M. Gür, F.B. Prinz, and S.F. Bent, Atomic layer deposition (ALD) co-deposited PtRu binary and

Pt skin catalysts for concentrated methanol oxidation, *Chem Mater* **22** (2010), 3024–3032.

[107] H.A. Lowenstam, Minerals formed by organisms, *Science* **211** (1981), 1126–1131.

[108] S. Porter, The rise of predators, *Geology* **39** (2011), 607–608.

[109] P.E. Gibbs and G.W. Bryan, Copper—the major metal component of *Glycerid polychaete* jaws, *J Mar Biol Assoc UK* **60** (1980), 205–214.

[110] G.W. Bryan and P.E. Gibbs, Zinc—the major inorganic component of *Nereid polychaete* jaws, *J Mar Biol Assoc UK* **59** (1979), 969–973.

[111] S.-M. Lee, E. Pippel, U. Gösele, C. Dresbach, Y. Qin, C.V. Chandran, T. Bräuniger, G. Hause, and M. Knez, Greatly increased toughness of infiltrated spider silk, *Science* **324** (2009), 488–492.

[112] Y. Termonia, Molecular modeling of spider silk elasticity, *Macromolecules* **27** (1994), 7378–7381.

[113] S.-M. Lee, E. Pippel, and M. Knez, Metal infiltration into biomaterials by ALD and CVD: a comparative study, *Chem Phys Chem* **12** (2011), 791–798.

[114] S. Mann, *Biomineralization principles and concepts in bioinorganic materials chemistry*, Oxford University Press, New York, NY, USA (2005).

[115] C.A. Wilson, R.K. Grubbs, and S.M. George, Nucleation and growth during Al_2O_3 atomic layer deposition on polymers, *Chem Mater* **17** (2005), 5625–5634.

[116] G.N. Parsons, S.M. George, and M. Knez, Progress and future directions for atomic layer deposition and ALD-based chemistry, *MRS Bull* **36** (2011), 865–871.

[117] L. Zhang, A.J. Patil, L. Li, A. Schierhorn, S. Mann, U. Gösele, and M. Knez, Chemical infiltration during atomic layer deposition: metalation of porphyrins as model substrates, *Angew Chem Int Ed* **48** (2009), 4982–4985.

[118] Y. Wang, Y. Qin, A. Berger, E. Yau, C. He, L. Zhang, U. Gösele, M. Knez, and M. Steinhart, Nanoscopic morphologies in block copolymer nanorods as templates for atomic-layer deposition of semiconductors, *Adv Mater* **21** (2009), 2763–2766.

[119] C.Y. Chang, F.Y. Tsai, S.J. Jhou, and M.J. Chen, Enhanced OLED performance upon photolithographic patterning by using an atomic-layer-deposited buffer layer, *Org Electron* **9** (2008), 667–672.

[120] B. Gong, Q. Peng, J.S. Jur, C.K. Devine, K. Lee, and G.N. Parsons, Sequential vapor infiltration of metal oxides into sacrificial polyester fibers: Shape replication and controlled porosity of microporous/mesoporous oxide monoliths, *Chem Mater* **23** (2011), 3476–3485.

[121] S.-M. Lee, V. Ischenko, E. Pippel, A. Masic, O. Moutanabbir, P. Fratzlm, and M. Knez, An alternative route towards metal-polymer hybrid materials prepared by vapor-phase processing, *Adv Func Mater* **21** (2011), 3047–3055.

[122] M. Ritala, M. Leskela, J.P. Dekker, C. Mutsaers, P.J. Soininen, and J. Skarp, Perfectly conformal TiN and Al_2O_3 films deposited by atomic layer deposition, *Chem Vap Deposit* **5** (1999), 7–9.

ABOUT THE AUTHORS

Lianbing Zhang studied biochemistry at Martin-Luther-University Halle, Germany. Afterwards, he carried out his PhD studies at the Max-Planck-Institute of Microstructure Physics in Halle. His studies were related to the synthesis of hybrid-materials with atomic layer deposition (ALD) and further synthetic strategies. He finished his doctoral thesis on the biomedical application potential of ferritin-templated nanoparticles in 2011. In 2010 he received the Young Scientist Award from the European Materials Research Society. In the same year, he was awarded the Chinese Government Award for Outstanding Self-Financed Students Abroad.

In February 2012, he joined the research institute CIC nanoGUNE in San Sebastian (Spain) as post-doctoral research scholar. His research is focused on the biomimicry of nanomaterials and their application potential in biomedicine.

Mato Knez studied chemistry at the University of Ulm. After receiving his diploma in organic chemistry, he moved to the Max-Planck-Institute of Solid State Research in Stuttgart to work on his doctoral thesis on the use of plant viruses as biological templates for nanostructuring, which was completed in 2003. Thereafter, he spent two years as a post-doctoral research scholar at the Max-Planck-Institute of Microstructure Physics in Halle, investigating the application potential of thin film coating technologies for nanomaterials and materials science in general. In 2006, he was awarded a grant from the German ministry of education and research (BMBF) in the framework of the program Nanofutur to form and guide a research group. In 2009, he was a visiting professorship to Brescia, Italy. In January 2012, he joined the CIC NanoGUNE research institute in San Sebastian (Spain) as group leader and Ikerbasque research professor. In the same year, he was awarded the Gaede prize of the German Vacuum Society for his work on nanomaterials synthesized and functionalized by atomic layer deposition.

His major research interests are vapor phase processes for thin film coatings, particularly ALD, and bio-inorganic or organic-inorganic materials and their physical properties.

Evolutionary Computation and Genetic Programming

Wolfgang Banzhaf

Department of Computer Science, Memorial University of Newfoundland,
St. John's, NL A1B 3X5, Canada

Prospectus

This chapter focuses on evolutionary computation, in particular genetic programming, as examples of drawing inspiration from biological systems. We set the choice of evolution as a source for inspiration in context and discuss the history of evolutionary computation and its variants before looking more closely at genetic programming. After a discussion of methods and the state of the art, we review application areas of genetic programming and its strength in providing human-competitive solutions.

Keywords

Algorithms, Artificial intelligence, Automatic programming, Bioinspired computing, Breeding, Crossover, Differential evolution, Evolutionary computation, Evolutionary programming, Evolution strategies, Generation, Genetic algorithms, Genetic programming, Human-competitive, Machine learning, Mutation, Natural selection, Population, Reproduction, Search space

17.1 BIOINSPIRED COMPUTING

One of the more prominent examples of bioinspiration is the application of this philosophy to the development of new ways to organize computation. Life can be paraphrased by describing it as "information processing in a body." Hence biology, the science of the living, has long been concerned with these two key aspects of life: structure and dynamics. The structure of the body, from a single-celled organism such as a bacterium to extended, highly complex, multicellular, intelligent beings such as mammals,[1] is a study object of evolutionary, developmental, and molecular biology, among other subjects. Other branches of biology are concerned with the dynamics of behavior of organisms in relation to an inanimate environment as well as in relation to other organisms that often provide

---

[1] The spatially largest organism is a fungus, *Armillaria solipides*, with an extension of 8.9 km^2, whereas the largest genome of a vertebrate is that of a fish, *Protopterus aetiopicus*, with a size of 130 Giga bases (as compared to the 3.2 Giga bases of *Homo sapiens*). A base is one of four nucleotides in the alphabet of DNA.

Engineered Biomimicry
http://dx.doi.org/10.1016/B978-0-12-415995-2.00017-9

an important and dynamically changing part of their environment. Ultimately, behavior requires intense processing of information, both for survival and for the benefit of an organism. Behavior of individuals is studied in a branch of biology called *ethology*, the behavior of species in their interaction with the environment is studied in ecology, and the dynamics of species over time is the subject of population and evolutionary biology. Molecular biology considers the regulation of behavior on the molecular level. From the lowest level of molecules to the highest level of evolution of species, this dynamic is about reception and processing of information and the appropriately executed actions following from the results of such *computation*.

Given this context, it is no wonder that computer scientists and engineers have embraced the paradigms of biology and tried to extract ideas from the living world to apply them in man-made computing environments such as computers and robots. Robots are actually the application area of bioinspiration closest to actual living organisms, since they can be said to possess a body, a structure that has to act in the real world. Less obvious, yet very active, is the area of bioinspired computing, where researchers try to extract more or less abstract principles and procedures from living organisms and realize them in a computational (algorithmic, software) setting.

There is full agreement in the sciences now that the generation of successive sequences of species in what has been called the *tree of life* is a product of evolution, governed by the principles of Darwin's theory of natural selection [1]. Evolution and its models are the source of bioinspiration that we shall discuss in this chapter in more detail. In a way, this is the most fundamental part of biology because it is the driving mechanism for the diversity of life on our planet. However, to set this in context, we want to at least mention in the remainder of this section several other examples of bioinspired computing, not necessarily in the temporal order of their development.

All aspects of adaptation of organisms to their environment, including the appearance of intelligent behavior, have been used as (bio-) inspiration. The field of *neural networks* [2], for example, has taken inspiration from the structure and function of nervous tissue in higher-order animals, including the brains of humans. The field of *fuzzy logic* [3] took inspiration from human cognitive processes with the ability to think in noncrisp terms. The field of *artificial immune systems* [4] has taken inspiration from the elaborate adaptive systems within higher-order beings that enable them to defend themselves against intruders. The field of *ant colony optimization* [5] has taken inspiration from the distributed nature of ant colonies and their apparently purposeful molding of the environment. The field of *swarm intelligence* [6] has taken inspiration from different sorts of animals organizing their behavior in swarms, flocks, or schools in order to achieve macro-effects (on the level of the entire swarm) from micro-causes (behavior of individuals). The field of *artificial life* [7] has taken inspiration from the very beginnings and basics of life to find ways to produce behavior akin to living behavior for the benefit of, e.g., computer games or for the simulation of alternatives to living matter, in order to better understand life on Earth.

May this list suffice for the moment. It is not exhaustive, and year after year new ideas are being proposed for computation derived from biological systems. Probabilistic reasoning, machine learning, emergence of novelty, complex adaptive systems, social behavior, intelligence, sustainability, and survival are all terms that can be related to and studied in models of bioinspired computing. Essential to these models is the idea that a distributed system of interacting entities can bring about effects that are not possible for single entities or entities isolated from each other to produce.

In this chapter let us focus, however, on one particular paradigm within the area of bio-inspired computing, an area that is very intimately connected to all signs of life: evolution. After a general discussion of algorithms derived from evolution (*evolutionary algorithms* or *evolutionary computing*), we consider in more detail the most modern branch of this area, *genetic programming*.

17.2 HISTORY AND VARIANTS OF EVOLUTIONARY COMPUTING

Evolutionary algorithms or evolutionary computing is an area of computer science that applies heuristic search principles inspired by natural evolution to a variety of different domains, notably to parameter optimization or other types of problem solving traditionally considered in artificial intelligence.

Early ideas in this field developed at a time when computers were barely commercially sold. Alan Turing was one of the first authors to correctly identify the power of evolution for the purpose of solving problems and exhibiting intelligent behavior. In his 1950 essay entitled "Computing Machinery and Intelligence," [8] Turing considered the question of whether machines could think. He was concerned that digital computers, despite their power and universality, would only be capable of executing programs deterministically. He felt that this would not be sufficient to produce intelligent behavior. "Intelligent behavior presumably consists in a departure from the completely disciplined behavior involved in computation, but a rather slight one, which does not give rise to random behavior" [8], p. 457. Computation here refers to the only known form of digital computation at the time: deterministic computation. Turing pointed out that a digital computer with a random element would be an interesting variant of such a machine, especially "when we are searching for a solution of some problem." It was clear to Turing that machines at some point

would need to be able to learn in a fashion similar to children—a very clear indication of him taking inspiration from biology. "Now the learning process may be regarded as a search [...]. Since there is probably a large number of satisfactory solutions, the random method seems to be better than the systematic. It should be noticed that it is used in the analogous process of evolution" [8, p. 459].

So, already in 1950 several ideas were voiced that would lead the way to evolutionary algorithms. The notion of a soft kind of randomness, which would later become mutation and crossover, the notion of intelligent behavior as the goal of these algorithms, the notion of a search process to achieve learning and problem solving, and the notion of kinship to evolution in Nature were all entertained in this early article by Alan Turing.

In 1962, a budding computer scientist[2] from the University of Michigan published a paper entitled "Outline for a Logical Theory of Adaptive Systems" in which he proposed most of what later became known as *genetic algorithms* [9]. Holland wrote: "The study of adaptation involves the study of both the adaptive system and its environment," thus foreshadowing the necessity to define a fitness function as a stand-in for the environment. He then proposed to look at the adaptive system as a "population of programs" and emphasized the advantage of looking at adaptation from the viewpoint of a population: "There is in fact a gain in generality if the generation procedure operates in parallel fashion, producing sets or populations of programs at each moment rather than individuals" [9, p. 298]. Here Holland correctly identified the strength that populations of solutions bring to a problem when applied and tested in parallel. "The generated population of programs will act upon a population of problems (the environment) in an attempt to produce solutions. For adaptation to take place the adaptive system must at least be able to compare generation procedures as to their efficiency in producing solutions." What Holland

[2] Computer science as a discipline did not even exist then. It is only in hindsight that we call it that; officially, it was called *communication science* at the time.

called *generation procedures* were later termed—in the context of supervisory programs, that is, programs that allow a system to adapt to different environmental conditions—*mutation*. After introducing differential selection as a key driving force for adaptation, "Adaptation, then, is based upon differential selection of supervisory programs. That is, the more 'successful' a supervisory program, in terms of the ability of its problem-solving programs to produce solutions, the more predominant it is to become (in numbers) in a population of supervisory programs. [...] Operation of the selection principle depends upon continued generation of new varieties of supervisory programs. There exist several interesting possibilities for producing this variation" [9, pp. 300–301]. "The procedure described [...] requires that the supervisory program duplicate, with some probability of variation or mutation [...]" [9, p. 309].

Thus, as early as 1962, a sketch of evolutionary algorithms was in place that could only become more pronounced over the years [10]. The randomness of Turing's paper, later reflected in Friedberg's work [11], gave way to a variation-selection loop with accumulation of beneficial variations in a population and the regular information exchange between individuals.

Other paradigmatic developments in evolutionary algorithms at the time include *evolutionary programming* [12] and *evolutionary strategies* [13]. For a more thorough review of early work in evolutionary computing, the reader is pointed to Ref. 14, discussing a selection of papers from the fossil record of evolutionary computing.

All evolutionary algorithms follow the Darwinian principle of differential natural selection. This principle states that the following preconditions must be fulfilled for evolution to occur via (natural) selection:

1. There are entities called individuals that form a population. These entities can reproduce or can be reproduced.
2. There is heredity in reproduction; that is to say that individuals of this population produce similar offspring.

3. In the course of reproduction there is variety that affects the likelihood of survival and, therefore, the reproducibility of individuals. This variety is produced by stochastic effects (such as random mutation and recombination) as well as by systematic effects (mating of like with like, etc.).
4. There are finite resources that cause the individuals to compete. Due to overreproduction of individuals, not all can survive the struggle for existence. Differential natural selection is a result of this competition exerting a continuous pressure toward adapted or improved individuals relative to the demands of their environment.

In evolutionary algorithms, there is a population of individual solutions. Usually this population is initialized as a random population, i.e., from random elements determined to be potentially useful in this environment. Next is a determination of fitness in the process of evaluation. This could take any form of a measurement or calculation to determine the relative strength of an individual solution. The outcome of this measurement or calculation is then used in the selection step to determine which individual solutions are to survive the competition for resources and

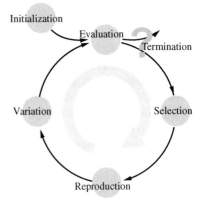

FIGURE 17.1 The general process of an evolutionary algorithm is a cycle of evaluation, selection, reproduction, and variation that accumulates beneficial changes. Variation operators could be mutation, duplication, or crossover/recombination.

which are to be replaced by copies or variants of the surviving individual solutions. The last step is to apply variation to these individual solutions so that the population is complete again and ready for the next round of evaluation. Figure 17.1 summarizes the general process.

Let us now turn our attention to genetic programming.

17.3 GENETIC PROGRAMMING: HISTORY, PRINCIPLES, AND METHODS

John Holland became known for the *genetic algorithm* [10], but he already had spoken of programs in his seminal 1962 paper. It was, however, the long-held mainstream view in computer science that subjecting computer programs to the random forces of mutation and recombination would not yield viable programs. The notion was that computer code is too brittle to be improved by randomness recruited in evolution.

The term *genetic programming (GP)* [15] describes a research area within the field of evolutionary computation that deals with the evolution of computer code. Its algorithms aim either to approximate solutions to problems in machine learning or to induce precise solutions in the form of grammatically correct (language) structures for the automatic programming of computers. Again, the same general process depicted in Figure 17.1 is applied, but this time to structures that determine the behavior of a computer.

It took a long time to realize that it is not impossible to evolve computer code. This exciting development needed a number of different (smaller) intermediate steps, starting from the original genetic algorithms using fixed-length bit strings to represent numbers in optimization problems. The first realization was that rule systems, instead of numbers representing problem

solutions, could be subjected to evolution. In another seminal paper, Holland and Reitman introduced the *classifier system* [16] that allowed *if-then rules* under evolution. Classifier systems have since taken on a life of their own [17], but the key contribution to the future field of GP was that *if-then* rules are a hallmark of programming languages and execution of more complex computer code. However, a classifier system cannot be regarded as a system for evolutionary program induction because there are no program individuals under evolution.[3]

Not long after Holland and Reitman, Smith introduced a variable-length representation, allowing it to concatenate rules into rule-based *programs* that can solve a task defined by a fitness function [18]. In 1981, Forsyth [19] published a logical rule system with parameters that allows us to classify examples into different classes. Here a training set of data samples can be used to train the classifier. The programs would logically and numerically evaluate data samples to conclude on the class of the example. Forsyth summarized in wonderful prose: "I see three justifications for this kind of exercise. Firstly, it is interesting in its own right; secondly, the rules behave in an interesting fashion; and thirdly, it seems to work. In the first place it is fun to try a little abstract gardening, growing an orchard of binary trees. And it might be fruitful in another sense. After all, we are only here by courtesy of the principle of natural selection, AI workers included, and since it is so powerful in producing natural intelligence it behooves us to consider it as a method for cultivating the artificial variety" [19, pp. 163–164]. One cannot escape the impression that the author of these lines could not at first believe that his method was so effective at classifying samples.

In the second half of the 1980s, the number of early GP systems proliferated. Cramer introduced in 1985 two evolutionary programming systems based on different simple languages he

---

[3] Note that Holland's first paper [9] already discussed programs!

designed [20]. Hicklin [21], and Fujiki and Dickinson [22] wrote precursor systems for particular applications using the standard programming language LISP before Koza in 1989 finally documented a method that both used a universal language and was applied to many different problems [23]. GP came into its own with the publication of John Koza's book in 1992 [15]. It is his achievement to have recognized the power and generality of this method and to document, with numerous examples, how the approach can be used in different application areas. In the introduction to his book he wrote: "In particular, I describe a single, unified, domain-independent approach to the problem of program induction—namely genetic programming."

Now that we have reviewed the gradual development of ideas, it is time to discuss the principles of GP. GP works with a population of computer programs that are executed or interpreted in order to judge their behavior. Usually, fitness measurements sample the behavioral space of a program to determine the degree to which the outcome of the behavior of a program individual is what it is intended for. For instance, the deviation between the quantitative output of a program and its target value (defined through an error function) could be used to judge the behavior of the program. This is a straightforward procedure if the function of the target program can be clearly defined. Results may also be defined as side effects of a program, such as consequences of the physical behavior of a robot controlled by a genetically developed program. Sometimes an explicit fitness measure is missing altogether—for instance, in a game situation—and the results of the game (winning or losing) are taken to be sufficient scoring for the program's strategy. Again, very much following the original intuition of Holland, various programs are applied to the same problem and their performances relative to each other are used to determine which programs are to be conserved for future generations and which are to be discarded.

The outcomes of fitness evaluation are used to select programs. There are several different methods for selection, both deterministic and stochastic. Selection determines (a) which programs are allowed to survive (overproduction selection) and (b) which programs are allowed to reproduce (mating selection). Once a set of programs has been selected for further reproduction, the following operators are applied:

- reproduction,
- mutation,
- crossover.

Reproduction simply copies an individual to an offspring population (the next generation), *mutation* varies the structure of an individual under control of a random-number generator, and *crossover* or *recombination* mixes the structures of two (or more) programs to generate one or more new programs for the offspring population. Additional variation operators are applied in different applications. Most of these contain knowledge in the form of heuristic search recipes that are adapted to the problem domain.

The material under evolution is, as we said, computer code. However, the representation of this code and the implementation of variation operators is important in order to avoid the brittleness of the code.[4] The two most popular representations for computer programs under evolution nowadays are expression trees of functional programming languages and (linear) sequences of instructions from imperative programming languages [24]. Figure 17.2 shows how these two representations can be subjected to a crossover or recombination operation. There are many other representations for GP, notably

---

[4] The *brittleness* of a computer code refers to the fact that the code can be easily broken, even with the slightest (possibly random) variation, with the result that it does not work at all.

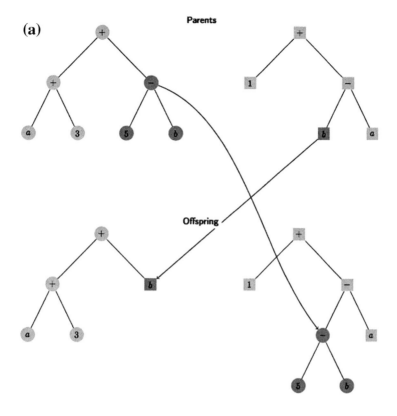

Linear GP

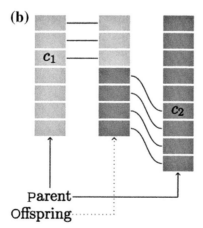

FIGURE 17.2 Tree-based and sequence-based representation of programs and their respective recombination. (a) Exchanging subtrees. (b) Exchanging blocks of code.

those that make use of developmental or generative processes that grow programs; see, for instance, Refs. 25 and 26.

Once programs have been generated, they are interpreted or compiled to produce behavior, which is subsequently measured in terms of its fitness. In this way, fitness advantages of individual programs are exploited in a population to lead to better solutions. In GP, as well as in other evolutionary algorithms, differential selection can be realized in many different ways.

The simplest approach is a generational selection scheme called *fitness-proportional* or *roulette-wheel selection*. The fitness of all individuals in the population is summed up. The fitness f_i of each individual i is then normalized by the sum of all fitness values found, and this normalized value determines the probability p_i of the individual i of being selected for reproduction/mutation/crossover. Thus, $p_i = f_i/\sum_j f_j$. Based on the stochasticity of random events, fitness proportional selection also allows weak individuals to succeed in reproduction some of the time.

Another selection method is called *tournament selection*. A subset of individuals of the population is drawn randomly, and those individuals are compared to each other in terms of fitness. The individuals with higher fitness are allowed to replace (directly as in reproduction, or with a variation as in mutation and crossover) the individuals with lower fitness. Normally, tournament selection is done with a minimum number of $k = 4$ individuals, which carries the lowest selection pressure. Larger tournaments, however, have been used in the literature, up to the extreme of holding tournaments among the whole population, at which point the selection scheme is called *truncation selection*.

Ranking selection is another selection scheme introduced to address some weaknesses of fitness proportional selection. Each individual is assigned a rank in the population, and selection proceeds by selecting either by a linearly or an exponentially associated probability based on the rank of an individual. A detailed comparison of different selection schemes is given in Ref. 27.

The entire process of selection can be seen in close analogy to the way humans breed animals. To obtain animals with the desired characteristics, the breeder has to select those individuals from the population that carry the targeted traits to a higher degree than others.[5]

All selection schemes rely on a sufficiently accurate determination of fitness to work properly. Therefore, one of the most important ingredients in GP is the definition of a fitness measure to determine the appropriateness of a program individual's behavior. Sometimes the fitness measure has to be iteratively improved in order for the evolved solutions to actually perform the function they were intended for.

Fitness calculation is actually one of the areas where GP and other evolutionary algorithms differ. GP has to judge the behavior of a program, a structure that determines the behavior of a computer under different inputs, i.e., an active entity. This executed program has to produce outputs that adhere as closely as possible to a prescribed behavior. Thus, while a GA (or another optimization technique) is used to optimize a particular instance of an optimization problem, GP has to find behavior in a larger search space, providing the correct behavior under a multitude of input/output pairs. The situation can be depicted by considering the difference between (1) optimizing a function, i.e., finding the minimum or maximum of a function, as in finding the center ($x = y = 0$) as the function maximum in Figure 17.3 in the GA task and (2) constructing the function whose samples are provided by the data points in Figure 17.4 in a typical GP task.

---

[5] Darwin's original theory of evolution was inspired by humans breeding animals, especially pigeons and cattle.

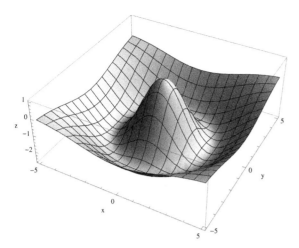

FIGURE 17.3   Finding the highest point in a fitness landscape of a function $z = z(x, y)$: A typical GA task.

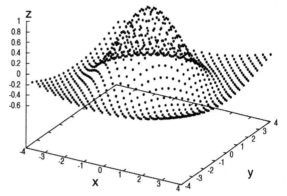

FIGURE 17.4   Finding the function $z = z(x, y)$ fitting the data points: a typical GP task.

Fitness measurement is thus a more extended affair in GP, and cannot usually be completed with one sample measurement in the behavioral space. Instead, multiple instances of input/output pairs are presented to the program population, and the results for all pairs are usually averaged to arrive at a fitness value for programs that subsequently forms the basis for selection.

A few more words on the difference between GA tasks and GP tasks are in order. One might argue that the GP task is nothing more than an optimization task, looking to minimize the error of the function approximation for Figure 17.4. If one were to have a fixed-sized genome consisting of, say, coefficients of a polynomial in x and y, assuming an expression of the type

$$z = \sum_{i,j=0}^{n} a_{ij} x^i y^j \qquad (17.1)$$

with a_{ij} being the genetic variables (alleles) that are subject to a GA, this can be legitimately considered a GA. Note, however, that the functional dependencies need to be determined beforehand, both in terms of the dimensionality as well

as in terms of the order (in this case, dependency on x and y with terms up to order n). Some of these terms might still be close to $a_{ij} = 0$, which would indicate either no or only a minute contribution of the term to the overall outcome z.

Conversely, though, one could argue that searching for the maximum in Figure 17.3 could be construed as a GP task. This could be done by assuming a growing set of fitness cases through probing the landscape for the maximum, based on a model of the landscape. By then trying to find a symbolic expression for the function, based on the existing points probed, calculating its first- and second-order derivatives, and solving for appropriate conditions to find the maximum, a new prediction could be made. This new point would be visited, but with an initially inaccurate model for the function it can be assumed that it would be off somewhat from the real maximum of the function. Using this kind of modeling approach (where GP is in charge of developing and refining the model), it can be easily imagined that such a process could yield a faster approach to the maximum than just sampling the space.

The difference between a GA approach and a GP approach to optimizing fitness of respective solutions is thus best tied to the representation used for the task. Is it a length-changing representation (as in the foregoing example, the expression for a function), or is it a fixed-length

TABLE 17.1 Comparison of GA versus GP.

	GA	GP
Representation	Fixed length	Variable length
Individual	Passive	Active
Genome	Parameters	Program/algorithm
Input	None	Input values
Fitness	One value	Many fitness cases
Typical application	Function optimization	Function approximation
Typical size of search space	10^{100}	$10^{100,000}$

representation, as in Eq. (17.1)? Does it use multiple fitness cases or is there only one fitness measurement? The former is a GP approach, the latter a GA approach. Naturally, a GP approach will search a larger space of possibilities because the combinatorics of its structures is much larger, with typical GP search spaces $10^{1,000}$ times larger than typical GA search spaces. Table 17.1 summarizes the issues.

17.4 ADVANCES AND STATE OF THE ART

In his seminal work of 1992, Koza established the field of GP by arguing convincingly that manipulation of structures of symbolic expressions is possible with evolutionary algorithms and that the resulting technique would have a wide variety of applications. In subsequent years, the field experienced both broadening and deepening [24]. Many different representations for GP were studied, among them other generic data structures such as sequences of instructions or directed graphs, as well as more exotic data structures such as memory stacks or neural networks. Today, different approaches are considered GP, from the evolution of expression trees to the evolution of electronic circuits or even

architectural designs (*structures*, for short). The overarching principle is to subject all these kinds of structures with variable complexity to forces of evolution by applying mutation, crossover, and fitness-based selection. The results must not necessarily be programs, but they could be descriptions for designs (e.g., structures of bridges) or other manipulatable elements.

An ever-present difficulty with GP is that the evolution of structures of variable complexity (e.g., program code) often leads to individuals with a large number of elements, often with considerable redundancy. Notably it was found that variable complexity often leads to inefficient code that requires a lot of memory space. Several researchers subsequently observed that the evolutionary forces seem to exert a pressure toward more complex solutions, parts of which could be removed after evolution without doing any harm to the behavior of the evolved solution. By drawing an analogy from biological evolution of genomes, this phenomenon was originally called *code bloat*, *intron growth*, or *growth of ineffective code* [28]. It was found that code growth is not the only unintended result of evolutionary processes, but it has been the most examined emergent phenomenon to date [29].[6] At least three different influences are at work promoting the growth of complexity during evolution. The most

---

[6] Another emergent phenomenon in GP is the emergence of repetitive code [30].

important influence has to do with the protection effect of redundant code if subjected to the action of crossover or mutation. Redundant code is more resistant to crossover and mutation and allows its carrier solution to survive better, compared to other individuals that do not possess this redundancy [31]. Removal bias in crossover operations [32] that describes the fact that code can grow to infinity from any size but only be reduced to zero from a particular size is another explanation. Finally, a genetic drift toward larger solutions [33] has been named as an important influence.

Over the last decade, the relation between robustness of organisms and their evolvability has been under intense study in biology (see, for example, Ref. 34). A seeming paradox between these two features frequently found in Nature has been resolved. It was found that neutral evolution is a key aspect of robustness. *Neutral evolution* refers to the capability of evolution to change genotypes without changes to phenotypes of individuals.[7] This capability has been known for decades, and had previously been discovered to be an important process in evolution [35]. The understanding of neutrality led to new mathematical models like the idea of neutral networks [36]. These ideas are beginning to exert influence in the EC community [37], and it turns out that, also in GP, solutions that are more robust are preferred through the evolutionary process, another emergent phenomenon [38, 39].

Although the GA theory has been well established, theoretical progress in GP has been more difficult to achieve since GP works with variable complexity and multiple fitness cases for fitness scoring. Many researchers are working to produce results for GP by gleaning from GA research. The schema theory of GA [40, 41] has been a primary target of knowledge transfer.[8] In the meantime, several different schema theorems have been formulated for GP, and theory has progressed substantially [42].

As researchers analyzed search spaces of programs, it was realized that their size is many orders of magnitude larger than search spaces of combinatorial optimization problems. A typical size for a program search space might be $10^{100,000}$, as opposed to a typical search space for a combinatorial optimization problem being of the order of 10^{100}. Although this finding might be interpreted as discouraging for search mechanisms, it was also realized that the solution density in program spaces is, above a certain threshold of complexity, constant with changing complexity [43]. In other words, there are proportionally many more valid solutions in program spaces than in the spaces of combinatorial optimization problems.

GP has made great strides over the last two decades, but many issues are still open and require continued investigation. The theory of GP [42]:

- has succeeded in finding appropriate schema theorems that allow to understand how the search space and the population representation interact,
- has started to analyze Markov chain models of the search process dynamics, and
- has found ways to characterize search spaces (difficulty, complexity) and relate them to the performance of GP systems.

In coming years, GP theory is expected to make progress on the treatment of dynamical

---

[7] The genotype of an individual is its genetic make-up, potentially subjected to mutation and crossover, whereas the phenotype of an individual is the resulting program (behavior).

[8] A *schema* in a GA is a subpattern of the genotype that takes the form of a template and can be used to identify several genotypes. For instance, in a binary string genotype with 4 bits: 1 * 0 *, all genotypes with a 1 in position 1 and a 0 in position 3 belong to this schema.

problems, proofs of convergence of the search algorithms, and classifying problem spaces.

On the practical side, GP research will target [44]:

- the identification of appropriate representations for GP in particular problems,
- the design of open-ended evolutionary systems with GP,
- the problem of generalization in GP,
- the establishment of benchmarks for measuring and comparing GP performance, and
- modularity and scalability in GP.

There is also more room for adding bioinspiration to GP. For instance, the relation between evolution and development has been studied for decades in biology [45]. It was found that the time-dependent process of gene expression and gene regulation through both internal and external cues is the mechanism by which both processes can be unified [46]. Some progress has also been made in GP to couple evolution and development. The developmental approach in GP takes the form of a recipe that, upon its execution, generates a structure that is subjected to fitness tests. Thus, it is not the GP program itself that is tested but the result of its execution.

Similar to the coupling between evolution and development, the coupling between development and learning was considered an important link for understanding the mechanisms of development and learning processes. Cognitive neuroscience has presented evidence for this coupling by finding that there are critical periods in development in which certain learning tasks are facilitated (and only sometimes possible). If the critical period is missed, learning success in a task is substantially reduced [48].

The coupling between development (or evolution) and learning has only recently been explored in GP. The problem is to clearly separate adaptations or fitness gains resulting from development or evolution versus those from learning. Its reason is the less stringent separation of time scales among evolution, development, and learning in GP systems. Whereas biological evolution can happen over many thousands or millions of years, development over the lifetime of an organism, and learning over phases of that lifetime, all three mechanisms are on similar time scales in GP, usually tied into single runs of a GP system. In addition, in most GP systems there is no notion of species (and their evolution). Rather, the entire population is essentially mixed and therefore belongs to one single species. Finally, the goal under the influence of evolution is behavior, the same entity usually associated with learning. First attempts to examine learning as a separate task for which evolution/development have to provide the means have been made [49], yet this area requires much more investigation.

17.5 APPLICATIONS

The textbook of Banzhaf et al. from 1998 lists 173 GP applications from A to Z that already existed at the time [24]. Fifteen years have passed, and the field has continued to develop rapidly. The main application areas of GP are (from narrow to wide) [24]:

- computer science,
- science,
- engineering,
- business and finance, and
- art and entertainment.

Koza has contributed many interesting applications to some of these areas, demonstrating the breadth of the method. However, a more detailed look is warranted.

In computer science, much effort has gone into the development of algorithms using GP. By being able to manipulate symbolic structures, GP is one of the few heuristic search methods for algorithms. Sorting algorithms, caching algorithms, compression algorithms [51], random-number generators, and algorithms for automatic parallelization of code [52], to name a few, have been studied. The spectrum of applications in computer science spans from the generation of proofs for predicate calculus to the evolution of machine code for accelerating function evaluation. The general tendency is to try to automate the design process for algorithms of different kinds. Recently the process of debugging code, i.e., the correction of errors, has been added to the list of applications [53]. Computer science itself has many applications, and it is natural that those areas also benefit indirectly by improving methods in computer science. For instance, in the area of computer vision, GP has been used, among others, for

- object detection (for example, Refs. 54 and 55),
- filter evolution (for example, Refs. 56 and 57),
- edge detection (for example, Ref. 58),
- interest point detection (for example, Ref. 59), and
- texture segmentation (for example, Ref. 60).

In addition, the area of software engineering is a field very fruitful for applications of GP [61]. Query optimization for database applications is a widespread application of evolutionary computation techniques (see their use in PostgreSQL and H2, Refs. 62 and 63).

Typical applications for GP in science are those to modeling and pattern recognition. Modeling certain processes in physics and chemistry with the unconventional help of evolutionary creativity supports research and understanding of the systems under study [64, 65]. For instance, parameters of models in soil science can be readily estimated by GP [66]. Predictions based on models generated with GP have widespread applications. An example from climate science is [67], where seawater level is forecast by a GP modeling technique using past time series. Many modeling applications for EC methods in general exist in

astronomy and astrophysics; see, for instance, Refs. 68 and 69.

Modeling is, however, but one of the applications of GP in science. Pattern recognition is another key application used in molecular biology and other branches of biology and medicine as well as in science in general [70, 71]. Here, GP has delivered results that are competitive if not better than human-generated results [72, 73], a special area of applications we return to in the next section. Classification and data mining are other applications in which GP is in prominent use [74, 75].

In engineering, GP and other evolutionary algorithms are used as standalone tools [76] or sometimes in competition or cooperation with other heuristic methods such as neural networks or fuzzy systems. The general goal is, again, to model processes such as material properties [77] or production plants or to classify results of production. In recent years, design in engineering has regained some prominence [78]. Control of manmade apparatus is another area in which GP has been used successfully, with process control and robot control (e.g., Ref. 79) the primary applications.

In business and finance, GP has been used to predict financial data, notably bancruptcy of companies [80, 81]. The entire area of computational finance is ripe with applications for GP (and other evolutionary techniques); see Refs. 82 and 83. For an early bibliography of business applications of GP and GAs, the reader is referred to Ref. 84. Since, generally speaking, modeling and prediction are core applications in economic contexts, GP is an important nonlinear modeling technique to consider; see, e.g., Refs. 85 and 86.

In art and entertainment, GP is used to evolve realistic animation scenes and appealing visual graphics (see [87] for an early example). Computer games are another active area of research and application for GP (see, for instance, [88]). Board games have been studied with GP-developed strategies, too [89, 90]. GP also has been used in visual art and music [91].

In music, for example, GP was used to extract structural information from musical composition in order to model the process so that automatic composition of music pieces becomes possible [92].

Many of these problems require a huge amount of computational power on the part of the GP systems. Parallel evolution has hence been a key engineering aspect of developments in GP. As a paradigm, GP is very well suited for a natural way of parallelization. With the advent of inexpensive parallel hardware in recent years, in particular through graphics processing units [93–95], a considerable proliferation of results is expected from GP systems [96].

17.6 HUMAN-COMPETITIVE RESULTS OF GENETIC PROGRAMMING

In the last decade, a substantial number of results have been published in various fields that claim to have produced human-competitive results by the application of GP as a problem-solving method [97].

These claims are based on a comparison between the currently best-known human solutions to a problem and their respective counterparts produced by GP. Applications are from areas such as quantum computing algorithms, analog electrical circuit design and other mechanical and electrical designs, game-playing applications, finite algebras and other mathematical systems, bioinformatics and other scientific pattern recognition problems, reverse engineering of systems, and empirical model discovery.

The claims of human-competitiveness are based on criteria that Koza *et al.* proposed in 2003:

1. The result was patented as an invention in the past, is an improvement over a patented invention, or would qualify today as a patentable new invention.

2. The result is equal to or better than a result that was accepted as a new scientific result at the time when it was published in a peer-reviewed scientific journal.

3. The result is equal to or better than a result that was placed into a database or archive of results maintained by an internationally recognized panel of scientific experts.

4. The result is publishable in its own right as a new scientific result, independent of the fact that the result was mechanically created.

5. The result is equal to or better than the most recent human-created solution to a long-standing problem for which there has been a succession of increasingly better human-created solutions.

6. The result is equal to or better than a result that was considered an achievement in its field at the time it was first discovered.

7. The result solves a problem of indisputable difficulty in its field.

8. The result holds its own or wins a regulated competition involving human contestants (in the form of either live human players or human-written computer programs).

Some of the similarities of these successes have been summarized by Koza [73] as follows:

- Usually, a large amount of computational power has to be invested in order to gain human-competitive results from GP runs.
- Most times, a dedicated representation for the solution, known to be efficient by the specialist, has been applied to allow the full power of expression of solutions to be born on the problem.
- The GP system has been equipped with dedicated growth or development operators such that the adaptation of complexity of a description can be achieved smoothly.

Due to the ability of the human mind to quickly grasp the recipes of a problem's solution that an artificial system has applied, the question remains open whether solutions found by a GP system will remain qualitatively better than solutions discovered by humans over the long term. Perhaps the best area to consider for this kind of attempt is mathematics. First results have been achieved that seem to indicate that, under very special circumstances, certain mathematical problems can be solved more efficiently using GP [98].

17.7 CONCLUSIONS

Implementation of GP will continue to benefit in coming years from new approaches that include results from developmental biology and epigenetics.

Application of GP will continue to broaden. Many applications focus on engineering applications. In this role, GP may contribute considerably to creative solutions to long-held problems in the real world.

Since GP was first used around 1990, raw computational power has increased by roughly a factor of 40,000 following Moore's law of doubling of transistor density every 18 months. As Koza points out, although initially only toy problems were amenable to solution through GP, subsequent increases in computational power and methodological progress of GP have allowed new solutions to previously patented inventions as well as, more recently, completely new inventions that are by themselves patentable. A milestone in this regard was reached in 2005 when the first patent was issued for an invention produced by a GP system [99].

The use of bioinspiration, notably through lessons from our understanding of natural evolution, has led to some very substantial progress in the implementation of artificial systems that show human-level problem-solving abilities. Achieving *artificial intelligence* in computing machines is still far in the future; however, in restricted steps in that direction have been taken. It is the firm conviction of the author of this chapter that a major component of any future system that could truly lay claim to the property of intelligence will be bioinspiration.

Acknowledgments

I express my sincere gratitude to my students, collaborators, and colleagues with whom it is such a pleasure to work on various projects in evolutionary computation. I also acknowledge funding agencies that financed many projects over the course of nearly two decades. Specifically, I mention the German Science Foundation (DFG), the Technical University of Dortmund, and the state of Northrhine-Westphalia for funding from 1993 to 2003 and NSERC (Canada), Memorial University of Newfoundland, and the government of Newfoundland for funding from 2003 to present.

References

[1] C. Darwin, *On the origin of species*, John Murray, London, UK (1859).

[2] C.M. Bishop, *Neural networks for pattern recognition*, Clarendon Press, Oxford, UK (1995).

[3] G.J. Klir and B. Yuan, *Fuzzy sets and fuzzy logic*, Prentice-Hall, Englewood Cliffs, NJ, USA (1995).

[4] L.N. De Castro and J. Timmis, *Artificial immune systems: a new computational intelligence approach*, Springer-Verlag, Heidelberg, Germany (2002).

[5] M. Dorigo and C. Blum, Ant colony optimization theory: a survey, *Theor Comput Sci* **344** (2005), 243–278.

[6] E. Bonabeau, M. Dorigo, and G. Theraulaz, *Swarm intelligence: from natural to artificial systems*, Oxford University Press, New York, NY, USA (1999).

[7] C. Adami, *Introduction to artificial life*, Telos, Springer-Verlag, New York, NY, USA (1998).

[8] A.M. Turing, Computing machinery and intelligence, *Mind* **59** (1950), 433–460.

[9] J.H. Holland, Outline for a logical theory of adaptive systems, *J ACM* **9** (1962), 297–314.

[10] J. Holland, *Adaptation in natural and artificial systems*, University of Michigan Press, Ann Arbor, MI, USA (1975).

[11] R.M. Friedberg, A learning machine: Part I, *IBM J Res Dev* **2** (1958), 2–13.

[12] L. Fogel, A. Owens, and M. Walsh, *Artificial intelligence through simulated evolution*, Wiley, New York, NY, USA (1966).

[13] I. Rechenberg, *Evolution strategy*, Holzmann Froboog, Stuttgart, Germany (1974).

[14] D.B. Fogel (ed.), *Evolutionary computation: the fossil record*, IEEE Press, New York, NY, USA (1998).

[15] J. Koza, *Genetic programming*, MIT Press, Cambridge, MA, USA (1992).

[16] J.H. Holland and J. Reitman, Cognitive systems based on adaptive algorithms, in *Pattern directed inference systems* (D.A. Waterman and F. Hayes-Roth, eds.), Academic Press, New York, NY, USA (1978), 313–329.

[17] R.J. Urbanowicz and J.H. Moore, Learning classifier systems: a complete introduction, review, and roadmap, *J Artif Evol Appl* **2009** (2009), 736398.

[18] S.F. Smith, *A learning system based on genetic adaptative algorithm*, PhD thesis, University of Pittsburgh (1980).

[19] R. Forsyth, BEAGLE: a Darwinian approach to pattern recognition, *Kybernetes* **10** (1981), 159–166.

[20] N. Cramer, A representation for the adaptive generation of simple sequential programs, *Proceedings of first international conference on genetic algorithms*, Lawrence Erlbaum, Pittsburgh, PA, USA (1985), 183–187.

[21] J.F. Hicklin, *Application of the genetic algorithm to automatic program generation*, Master's thesis, University of Idaho (1986).

[22] C. Fujiki and J. Dickinson, Using the genetic algorithm to generate lisp source code to solve the prisoner's dilemma, *Proceedings of second international conference on genetic algorithms*, Lawrence Erlbaum, Pittsburgh, PA, USA (1987), 236–240.

[23] J. Koza, Hierarchical genetic algorithms operating on populations of computer programs, *Proceedings of 11th international joint conference on artificial intelligence*, Morgan Kaufmann, San Mateo, CA, USA (1989), 768–774.

[24] W. Banzhaf, P. Nordin, R. Keller, and F. Francone, *Genetic programming – an introduction*, Morgan Kaufmann, San Francisco, CA, USA (1998).

[25] C. Ryan, J.J. Collins, and M. Neill, Grammatical evolution: evolving programs for an arbitrary language, *Proceedings of EuroGP 1998* (W. Banzhaf, R. Poli, M. Schoenauer, and T. Fogarty, eds.), Springer-Verlag, Heidelberg, Germany (1998), 83–96.

[26] J.F. Miller, An empirical study of the efficiency of learning boolean functions using a cartesian genetic programming approach, *Proceedings of the genetic and evolutionary computation conference (GECCO-1999)* (W. Banzhaf, ed.), Morgan Kaufmann, San Francisco, CA, USA (1999), 1135–1142.

[27] T. Blickle and L. Thiele, A comparison of selection schemes used in evolutionary algorithms, *Evol Comput* **4** (1996), 361–394.

[28] P. Angeline, Genetic programming and emergent intelligence, in *Advances in genetic programming* (K.E. Kinnear, ed.), MIT Press, Cambridge, MA, USA (1994), 75–98.

[29] P. Nordin, W. Banzhaf, and F.D. Francone, Introns in nature and in simulated structure evolution, *Proceedings of biocomputing and emergent computation (BCEC97), Skovde, Sweden* (D. Lundh, B. Olsson, and A. Narayanan, eds.), World Scientific, Singapore (1–2 September, 1997), 22–35.

[30] W. Langdon and W. Banzhaf, Repeated patterns in genetic programming, *Nat Comput* **7** (2008), 589–613.

[31] P. Nordin and W. Banzhaf, Complexity compression and evolution, in *Genetic algorithms: proceedings of the sixth international conference (ICGA95)* (L. Eshelman, ed.), Morgan Kaufmann, San Francisco, CA, USA (1995), 310–317.

[32] T. Soule and J.A. Foster, Removal bias: a new cause of code growth in tree-based evolutionary programming, *IEEE international conference on evolutionary computation*, IEEE Press, New York, NY, USA (1998), 781–786.

[33] W. Langdon, *Fitness causes bloat*, Technical report, CSRP-97-22, University of Birmingham, UK (1997).

[34] A. Wagner, *Robustness and evolvability in living systems*, Princeton University Press, Princeton, NJ, USA (2005).

[35] M. Kimura, *The neutral theory of molecular evolution*, Cambridge University Press, Cambridge, United Kingdom (1983).

[36] C. Forst, C. Reidys, and J. Weber, Evolutionary dynamics and optimization, in *Advances in artificial life* (F. Moran, A. Moreno, J.J. Merelo, and P. Chacon, eds.), Springer-Verlag, Berlin, Germany (1995), 128–147.

[37] T. Hu and W. Banzhaf, Evolvability and speed of evolutionary algorithms in light of recent developments in biology, *J Artif Evol Appl* (2010), 568375. http:// www.hindawi.com/archive/2010/568375/(accessed 14 March 2013).

[38] T. Hu, J. Payne, J. Moore, and W. Banzhaf, Robustness, evolvability, and accessibility in linear genetic programming, *Proceedings of EuroGP 2011* (S. Silva and J.A. Foster, eds.), Springer, Berlin, Germany (2011), 13–24.

[39] T. Hu, J.L. Payne, W. Banzhaf, and J.H. Moore, Evolutionary dynamics on multiple scales, *Genet Program Evol Mach* **13** (2012), 305–337.

[40] D. Goldberg, *Genetic algorithms in search, optimization and machine learning*, Addison Wesley, Reading, MA, USA (1989).

[41] M. Vose, *The Simple genetic algorithm: foundations and theory*, MIT Press, Cambridge, MA, USA (1999).

[42] R. Poli, L. Vanneschi, W.B. Langdon, and N. Freitag, Theoretical results in genetic programming: the next ten years, *Genet Program Evol Mach* **11** (2010), 285–320.

[43] W. Langdon and R. Poli, Boolean functions fitness spaces, *Proceedings of EuroGP'99* (R. Poli, P. Nordin, W. Langdon, and T. Fogarty, eds.), Springer-Verlag, Berlin, Germany (1999), 1–14.

[44] M. O'Neill, L. Vanneschi, S. Gustafson, and W. Banzhaf, Open issues in genetic programming, *Genet Program Evol Mach* **11** (2010), 339–363.

[45] M.J. West-Eberhard, *Developmental plasticity and evolution*, Oxford University Press, New York, NY, USA (2003).

[46] S. Ben-Tabou de Leon and E.H. Davidson, Gene regulation: gene control network in development, *Annu Rev Biophys Biomol Struct* **36** (2007), 191–212.

[47] L. Spector and K. Stoffel, Ontogenetic programming, *Proceedings of genetic programming 1996* (J. Koza, D.E. Goldberg, D.B. Fogel, and R. Riolo, eds.), MIT Press, Cambridge, MA, USA (1996), 394–399.

[48] N.W. Daw, N.E. Berman, and M. Ariel, Interaction of critical periods in the visual cortex of kittens, *Science* **199** (1978), 565–567.

[49] S. Harding, J. Miller, and W. Banzhaf, Evolution, development and learning using self-modifying cartesian genetic programming, *Proceedings of the 11th International conference on genetic and evolutionary computation* (F. Rothlauf, ed.), ACM Press, New York, NY, USA (2009), 699–706.

[50] J. Koza, F.H. Bennett, D. Andre, and M.A. Keane, *Genetic programming III: Darwinian invention and problem solving*, Morgan Kaufmann, San Francisco, CA, USA (1999).

[51] A. Kattan and R. Poli, Evolution of human-competitive lossless compression algorithms with gp-zip2, *Genet Program Evol Mach* **12** (2011), 1–30.

[52] C. Ryan, *Automatic re-engineering of software using genetic programming*, Kluwer Academic, Boston, MA, USA (2000).

[53] S. Forrest, T.V. Nguyen, W. Weimer, and C. Le Goues, A genetic programming approach to automated software repair, *Proceedings of the 11th international conference on genetic and evolutionary computation* (F. Rothlauf, ed.), ACM Press, New York, NY, USA (2009), 947–954.

[54] B. Bhanu and Y. Lin, Object detection in multi-modal images using genetic programming, *Appl Soft Comput* **4** (2004), 175–201.

[55] M. Zhang, U. Bhowan, and N. Bunna, Genetic programming for object detection: a two-phase approach with an improved fitness function, *Electronic Lett Comput Vision Image Anal* **6** (2007), 27–43.

[56] R. Poli, Genetic programming for feature detection and image segmentation, in *Evolutionary computation* (T. Fogarty, ed.), Springer-Verlag, Berlin, Germany (1996), 110–125.

[57] S. Harding and W. Banzhaf, Genetic programming on gpus for image processing, *Int J High-Perform Syst Archit* **1** (2008), 231–240.

[58] W. Fu, M. Johnston, and M. Zhang, Genetic programming for edge detection: a global approach, *Proceedings of CEC 2011*, IEEE Press, New York, NY, USA (2011), 254–261.

[59] G. Olague and L. Trujillo, Evolutionary-computer-assisted design of image operators that detect interest points using genetic programming, *Image Vision Comput* **29** (2011), 484–498.

[60] A. Song and V. Ciesielski, Texture segmentation by genetic programming, *Evol Comput* **16** (2008), 461–481.

[61] M. Harman and A. Mansouri, Search-based software engineering: introduction to the special issue, *IEEE Trans Softw Eng* **36** (2010), 737–741.

[62] PostgreSQL, Manual, version 9.1. http://www.postgresql.org/docs/9.1/static/geqo-pg-intro.html (accessed 25 June 2012).

[63] H2, H2 database – features. http://www.h2database.com/html/features.html (accessed 25 June 2012).

[64] R. Stadelhofer, W. Banzhaf, and D. Suter, Evolving black-box quantum algorithms using genetic programming, *Artif Intell Eng Design and Manuf* **22** (2008), 285–297.

[65] F. Archetti, I. Giordani, and L. Vanneschi, Genetic programming for QSAR investigation of docking energy, *Appl Soft Comput* **10** (2010), 170–182.

[66] N. Naderi, P. Roshani, M.Z. Samani, and M.A. Tutunchian, Application of genetic programming for estimation of soil compaction parameters, *Appl Mech Mater* **147** (2012), 70–74.

[67] M. Ali Ghorbani, R. Khatibi, A. Aytek, O. Makarynskyy, and J. Shiri, Sea water level forecasting using genetic programming and comparing the performance with artificial neural networks, *Comput Geosci* **36** (2010), 620–627.

[68] P. Charbonneau, Genetic algorithms in astronomy and astrophysics, *Astrophys J Suppl Ser* **101** (1995), 309–334.

[69] J. Li, X. Yao, C. Frayn, H. Khosroshahi, and S. Raychaudhury, An evolutionary approach to modeling radial brightness distributions in elliptical galaxies, in *Parallel problem solving from Nature: PPSN VIII* (X. Yao, E. Burke, J.A. Lozano, J. Smith, J.J. Merelo-Guervs, J.A. Bullinaria, J. Rowe, P. Tino, A. Kabn, and H.-P. Schwefel, eds.), Springer-Verlag, Berlin, Germany (2004), 591–601.

[70] W.P. Worzel, J. Yu, A.A. Almal, and A.M. Chinnaiyan, Applications of genetic programming in cancer research, *Int J Biochem Cell Biol* **41** (2009), 405–413.

[71] S.M. Winkler, M. Affenzeller, and S. Wagner, Using enhanced genetic programming techniques for evolving classifiers in the context of medical diagnosis, *Genet Program Evol Mach* **10** (2009), 111–140.

[72] S. Silva and L. Vanneschi, State-of-the-art genetic programming for predicting human oral bioavailability of drugs, *Adv Bioinformat* **74** (2010), 165–173.

[73] J. Koza, Human-competitive results produced by genetic programming, *Genet Program Evol Mach* **11** (2010), 251–284.

[74] M. Brameier and W. Banzhaf, A comparison of linear genetic programming and neural networks in medical data mining, *IEEE Trans Evol Comput* **5** (2001), 17–26.

[75] P.G. Espejo, S. Ventura, and F. Herrera, A survey on the application of genetic programming to classification, *IEEE Trans Syst Man Cybern, Part C: Appl Rev* **40** (2010), 121–144.

[76] D. Dasgupta and Z. Michalewicz (eds.), *Evolutionary algorithms in engineering applications*, Springer-Verlag, Heidelberg, Germany (1997).

[77] L. Gusel and M. Brezocnik, Application of genetic programming for modelling of material characteristics, *Expert Syst Appl* **38** (2011), 15014–15019.

[78] J. Lohn, G.S. Hornby, and D.S. Linden, An evolved antenna for deployment on NASA's space technology 5 mission, in *Genetic programming theory and practice II* (U.M. O'Reilly, R.L. Riolo, G. Yu, and W. Worzel, eds.), Kluwer Academic, Boston, MA, USA (2004), 301–315.

[79] J.U. Dolinsky, I.D. Jenksinson, and G.J. Cloquhoun, Application of genetic programming to the calibration of industrial robots, *Comput Ind* **58** (2007), 255–264.

[80] H. Iba and T. Sasaki, Using genetic programming to predict financial data, *Proceedings of the 1999 congress on evolutionary computation (CEC-1999)*, IEEE Press, New York, NY, USA (1999), 244–251.

[81] T.E. McKee and T. Lensberg, Genetic programming and rough sets: a hybrid approach to bankruptcy classification, *Eur J Oper Res* **138** (2002), 436–451.

[82] S.H. Chen, *Genetic algorithms and genetic programming in computational finance*, Kluwer Academic, Boston, MA, USA (2002).

[83] J. Wang, Trading and hedging in S&P 500 spot and futures markets using genetic programming, *J Futures Markets* **20** (2000), 911–942.

[84] B.K. Wong and T.A. Bodnovich, A bibliography of genetic algorithm business application research: 1988–June 1996, *Expert Syst* **15** (1998), 75–82.

[85] M. Álvarez-Díaz and G. Caballero Miguez, The quality of institutions: a genetic programming approach, *Econ Model* **25** (2008), 161–169.

[86] M. Álvarez-Díaz, J. Mateu-Sbert, and J. Rosselló-Nadal, Forecasting tourist arrivals to balearic islands using genetic programming, *Int J Comput Econ Econ* **1** (2009), 64–75.

[87] L. Gritz and J. Hahn, Genetic programming for articulated figure motion, *J Visual Comput Animat* **6** (1995), 129–142.

[88] K.T. Sun, Y.C. Lin, C.Y. Wu, and Y.M. Huang, An application of the genetic programming technique to strategy development, *Expert Syst Appl* **36** (2009), 5157–5161.

[89] Y. Azaria and M. Sipper, GP-gammon: Genetically programming backgammon players, *Genet Program Evol Mach* **6** (2005), 283–300.

[90] M. Sipper and M. Giacobini, Introduction to special section on evolutionary computation in games, *Genet Program Evol Mach* **9** (2008), 279–280.

[91] C.G. Johnson and J.J.R. Cardalda, Genetic algorithms in visual art and music, *Leonardo* **35** (2002), 175–184.

[92] B. Johanson and R. Poli, GP-Music: an interactive genetic programming system for music generation with automated fitness raters, *Genetic programming: proceedings of 3rd annual conference*, Morgan Kaufman, San Francisco, CA, USA (1998), 181.

[93] W. Banzhaf, S. Harding, W.B. Langdon, and G. Wilson, Accelerating genetic programming through graphics processing units, *Genetic programming theory and practice VI* (R. Riolo, T. Soule, and B. Worzel, eds.), Springer-Verlag, New York, NY, USA (2009), 1–23.

[94] D. Robilliard, V. Marion, and C. Fonlupt, High performance genetic programming on gpu. *Proceedings of the 2009 workshop on Bio inspired algorithms for distributed systems*, ACM Press, New York, NY, USA (2009), 85–94.

[95] W. Langdon, A many threaded CUDA interpreter for genetic programming, *Genet Program Evol Mach* **11** (2010), 146–158.

[96] W.B. Langdon and A.P. Harrison, GP on SPMD parallel graphics hardware for mega bioinformatics data mining, *Soft Comput* **12** (2008), 1169–1183.

[97] J. Koza, M. Keane, M.J. Streeter, W. Mydlowec, J. Yu, G. Lanza, and D. Fletcher, *Genetic programming IV: routine human-competitive machine intelligence*, Kluwer Academic, Norvell, MA, USA (2003).

[98] L. Spector, D.M. Clark, I. Lindsay, B. Barr, and J. Klein, Genetic programming for finite algebras, *Proceedings of the genetic and evolutionary computation conference (GECCO)* (M. Keijzer et al., eds.), ACM Press, New York, NY, USA (2008), 1291–1298.

[99] M. Keane, J. Koza, and M. Streeter. *Improved general-purpose controllers*, US Patent 6,847,851 (25 January 2005).

ABOUT THE AUTHOR

Wolfgang Banzhaf is currently University Research Professor at Memorial University of Newfoundland. From 2003 to 2009, he served as Head of the Department of Computer Science at Memorial. From 1993 to 2003, he was an Associate Professor for Applied Computer Science at Technical University of Dortmund, Germany. He also worked in industry, as a researcher with the Mitsubishi Electric Corporation in Japan and the US. He holds a PhD in Physics from the University of Karlruhe in Germany. His research interests are in the field of bio-inspired computing, notably evolutionary computation and complex adaptive systems.

Index

Printed and bound by CPI Group (UK) Ltd, Croydon, CR0 4YY

08/05/2025

01864886-0003